APPLICATIONS OF SOLID PHASE
MICROEXTRACTION

RSC Chromatography Monographs

Series Editor: Roger M. Smith, *University of Technology, Loughborough, UK*

Advisory Panel: J.C. Berridge, *Sandwich, UK*; G.B. Cox, *Indiana, USA*; I.S. Lurie, *Virginia, USA*; P.J. Schoenmaker, *Eindhoven, The Netherlands*; C.F. Simpson, *London, UK*; G.G. Wallace, *Wollongong, Australia*

This series is designed for the individual practising chromatographer, providing guidance and advice on a wide range of chromatographic techniques with the emphasis on important practical aspects of the subject.

Supercritical Fluid Chromatography
edited by Roger M. Smith, *University of Technology, Loughborough, UK*

Packed Column SFC
by T.A. Berger, *Berger Instruments, Newark, Delaware, USA*

Chromatographic Integration Methods, Second Edition
by Norman Dyson, *Dyson Instruments Ltd., UK*

Separation of Fullerenes by Liquid Chromatography
edited by Kiyokatsu Jinno, *Toyohashi University of Technology, Japan*

HPLC: A Practical Guide
by Toshihiko Hanai, *Health Research Foundation, Kyoto, Japan*

Applications of Solid Phase Microextraction
edited by Janusz Pawliszyn, *University of Waterloo, Ontario, Canada*

How to obtain future titles on publication

A standing order plan is available for this series. A standing order will bring delivery of each new volume immediately upon publication. For further information, please write to:

The Royal Society of Chemistry, Turpin Distribution Services Ltd., Blackhorse Road, Letchworth, Hertfordshire SG6 1HN, UK.
Telephone: +44 (0) 1462 672555; Fax: +44 (0) 1462 480947

RSC
CHROMATOGRAPHY
MONOGRAPHS

Applications of Solid Phase Microextraction

Edited by
Janusz Pawliszyn
Department of Chemistry, University of Waterloo
Waterloo, Ontario, Canada

ROYAL SOCIETY OF CHEMISTRY

ISBN 0-85404-525-2

A catalogue record for this book is available from the British Library.

Published by The Royal Society of Chemistry,
Thomas Graham House, Science Park, Milton Road, Cambridge CB4 0WF, UK

For further information see our web site at www.rsc.org

Typeset by Paston PrePress Ltd, Beccles, Suffolk
Printed by MPG Books Ltd, Bodmin, Cornwall, UK

Foreword

The development and implementation of new sample preparation technologies are very slow compared to other parts of the analytical process. For example, today we have very powerful modern instrumentation available, such as GC/MS and LC/MS, to perform separation and quantitation of complex extraction mixtures. At the same time many analytical methods use very traditional extraction techniques, such as liquid–liquid or Soxhlet extractions. The reasons for this situation are many and likely include the low level of research activity in academia in this area, largely caused by the complexity of natural matrices, which discourages more basic research. This has created the situation where the time required for the sample preparation step determines overall analysis time. Traditionally, scientists defined the objective of the sample preparation step quite narrowly, as isolation of target components from the sample matrix in the laboratory. However, it becomes evident that in the near future more demand will be placed on technologies of sample preparation to be able to meet such objectives as moving analysis to the field or facilitating continuous process and on-site monitoring. Also, more emphasis will be placed on getting more information about the sample, such as speciation or more complete characterization of the natural distribution of analytes in the system.

Solid Phase Microextraction (SPME) has been introduced as a modern alternative to current sample preparation technology, and is able to address some of the requirements put forward by analytical researches. It eliminates the use of organic solvents, and substantially shortens the time of analysis and allows easy automation of the sample preparation step. It is also suitable for on-site analysis and process monitoring. The configuration and operation of the SPME device is very simple. One who knows how to use a syringe is able to operate a SPME device. However, this feature creates a false impression that the extraction is a simple, almost trivial process. This misunderstanding frequently results in disappointments. It should be emphasized that the fundamental processes involved in solid phase microextraction are very similar to more traditional techniques and therefore challenges to develop successful methods are analogous. The nature of target analytes and complexity of a sample matrix determine the level of difficulties in accomplishing a successful extraction. The simplicity and convenience of the extraction devices primarily impacts the costs of practical implementation of the developed methods.

The objective of this book is to introduce the reader to a range of successful SPME applications developed by a number of international research groups.

SPME has been interfaced to several types of analytical instrumentation. The technique has been applied to extraction of a wide range of analytes from air, water and solid matrices. The areas covered in the book include environmental, food, forensic, clinical and pharmaceutical applications. Emphasis of the book is placed on quantitative analysis. In addition, examples focusing on characterization of analyte distribution in natural systems, speciation and investigation of reaction intermediates are also discussed. The reader is encouraged to focus on the details of analytical procedures, which are required to ensure that the technique performs successfully.

Contents

Environmental Applications

Food, Flavour, Fragrance and Pheromone Applications

Pharmaceutical, Clinical and Forensic Applications

Reaction Monitoring

Related Techniques

Contributors

Jiu Ai, *United States Tobacco Manufacturing Company Inc., 800 Harrison Street, Nashville, TN 37203, USA*

Josep M. Bayona, *Department of Environmental Chemistry, CID-CSIC, Jordi Girona 18, 08034 Barcelona, Spain*

Emilio Benfenati, *Istituto di Ricerche Farmacologiche 'Mario Negri', Via Eritrea 62, 20157 Milano, Italy*

Anna A. Boyd-Boland, *AGAL, 1 Suakin Street, Pymble, 2073 Australia*

Terry J. Braggins, *MIRINZ Food Technology & Research, PO Box 617, Hamilton, New Zealand*

Jennifer S. Brodbelt, *Department of Chemistry and Biochemistry, University of Texas at Austin, Austin, TX 78712, USA*

Jochen Bürck, *Institute für Instrumentelle Analytik, Forschungszentrum Karlsruhe, Postfach 3640, D-76021 Karlsruhe, Germany*

Kok Kay Chee, *Department of Chemistry, National University of Singapore, Republic of Singapore 119260*

Sau L. Chong, *Department of Chemistry, University of South Florida, 4202 E. Fowler Avenue, Tampa, FL 33620, USA*

W. M. Coleman III, *RJ Reynolds Tobacco Company, Winston-Salem, NC 27102, USA*

Steve Crook, *Zeneca Agrochemicals, Jealott's Hill Research Station, Bracknell, Berkshire RG12 6EY, UK*

Hiro Daimon, *Department of Chemistry, University of Waterloo, Waterloo, ON N2L 3G1, Canada*

Richard Dams, *Laboratory of Analytical Chemistry, Ghent University, Proeftuinstraat 86, B-9000 Gent, Belgium*

John R. Dean, *Department of Chemical and Life Sciences, University of Northumbria at Newcastle, Newcastle upon Tyne NE1 8ST, UK*

Demetrio De la Calle García, *Institute of Inorganic and Analytical Chemistry, Friedrich-Schiller-Universität, Lessingstrasse 8, 07747 Jena, Germany*

Tom De Smaele, *Laboratory of Analytical Chemistry, Ghent University, Proeftuinstraat 86, B-9000 Gent, Belgium*

Ralf Eisert, *Department of Chemistry, University of Waterloo, Waterloo, ON N2L 3G1, Canada*

Anett Georgi, *Department of Remediation Research, Centre for Environmental Research, Permoserstrasse 15, 04318 Leipzig, Germany*

Tadeusz Górecki, *Department of Chemistry, University of Waterloo, Waterloo, ON N2L 3G1, Canada*

Casey C. Grimm, *USDA-ARS, Southern Regional Research Center, 1100 Robert E. Lee Boulevard, New Orleans, LA 70124, USA*

Christoph Grote, *Fraunhofer Institute of Toxicology and Aerosol Research, Nikolai-Fuchs-Strasse 1, D-30625 Hannover, Germany*

Maurizio Guidotti, *PMP Rieti, Via Salaria per L'Aquila 8, 02100 Rieti, Italy*

Brad J. Hall, *Department of Chemistry and Biochemistry, University of Texas at Austin, Austin, TX 78712, USA*

P. Hancock, *Department of Chemical and Life Sciences, University of Northumbria at Newcastle, Newcastle upon Tyne NE1 8ST, UK*

Jalal A. Hawari, *NRC, Biotechnology Research Institute, 6100 Royalmount Avenue, Montreal, Quebec H4P 2R2, Canada*

Makiko Hayashida, *Department of Legal Medicine, Nippon Medical School, Tokyo 113–0022, Japan*

Joop L.M. Hermens, *Research Institute of Toxicology, Utrecht University, PO Box 80176, NL-3508 TD Utrecht, The Netherlands*

Kiyokatsu Jinno, *School of Materials Science, Toyohashi University of Technology, Toyohashi 441–8122, Japan*

Sys Stybe Johansen, *Institute of Forensic Chemistry, University of Copenhagen, PO Box 2713, DK-2100 Copenhagen O, Denmark*

Tsuyoshi Kaneko, *Chiba Prefectural Police Headquarters, Forensic Science Laboratory, 71–1, Chuoko 1-chome, Chuo-ku, Chiba-shi, Chiba 260–0024, Japan*

Tohru Kojima, *Department of Legal Medicine, Hiroshima University School of Medicine, 1–2–3 Kasumi, Minami-ku, Hiroshima 734–8551, Japan*

Frank-Dieter Kopinke, *Department of Remediation Research, Centre for Environmental Research, Permoserstrasse 15, 04318 Leipzig, Germany*

Mette Krogh, *Institute of Pharmacy, University of Oslo, PO Box 1068, Blindern, N-0316 Oslo, Norway*

Takeshi Kumazawa, *Department of Legal Medicine, Showa University School of Medicine, 1–5–8 Hatanodai, Shinagawa-ku, Tokyo 142–8555, Japan*

Hian Kee Lee, *Department of Chemistry, National University of Singapore, Republic of Singapore 119260*

Xiao-Pen Lee, *Department of Legal Medicine, Showa University School of Medicine, 1–5–8 Hatanodai, Shinagawa-ku, Tokyo 142–8555, Japan*

Karin C.H.M. Legierse, *Research Institute of Toxicology, Utrecht University, PO Box 80176, NL-3508 TD Utrecht, The Netherlands*

Karsten Levsen, *Fraunhofer Institute of Toxicology and Aerosol Research, Nikolai-Fuchs-Strasse 1, D-30625 Hannover, Germany*

Heather Lord, *Department of Chemistry, University of Waterloo, Waterloo, ON N2L 3G1, Canada*

Imelda McCann, *School of Chemistry, The Queen's University of Belfast, David Keir Building, Belfast BT9 5AG, UK*

Abdul Malik, *Department of Chemistry, University of South Florida, 4202 E. Fowler Avenue, Tampa, FL 33620, USA*

Venkatachalam Mani, *Supelco Inc., Supelco Park, Bellefonte, PA 16823, USA*

Perry A. Martos, *Department of Chemistry, University of Waterloo, Waterloo, ON N2L 3G1, Canada*

Adam J. Matich, *Horticuture and Food Research Institute of New Zealand Ltd., Palmerston North Research Centre, Tennent Drive, Private Bag 11 030, Palmerston North, New Zealand*

Monika Möder, *Department of Analytical Chemistry, Centre for Environmental Research Leipzig-Halle, Permoserstrasse 15, D-04318 Leipzig, Germany*

Luc Moens, *Laboratory of Analytical Chemistry, Ghent University, Proeftuinstraat 86, B-9000 Gent, Belgium*

Gloriano Moneti, *Centro Interdipartimentale di Spettometria di Massa, Università di Firenze, Viale G. Pieraccini 6, 50139 Firenze, Italy*

M.C. Montel, *Station de Recherches sur la Viande, INRA de Theix, Ministry of Agriculture, 63122 St. Genès-Champanelle, France*

Laura Müller, *Department of Chemistry, University of Waterloo, Waterloo, ON N2L 3G1, Canada*

Akira Namera, *Department of Legal Medicine, Hiroshima University School of Medicine, 1–2–3 Kasumi, Minami-ku, Hiroshima 734–8551, Japan*

Lay-Keow Ng, *Revenue Canada, 79 Bentley Avenue, Ottawa, ON K1A 0L2, Canada*

Torben Nilsson, *European Commission Joint Research Centre, Environment Institute, TP 290, I-21020, Ispra, Italy*

B. Denis Page, *Health Canada, Sir F. Banting Research Centre, 2203D, Ottawa, ON K1A 0L2, Canada*

Albrecht Paschke, *Department of Chemical Ecotoxicology, Centre for Environmental Research, Permoserstrasse 15, 04318 Leipzig, Germany*

Janusz Pawliszyn, *Department of Chemistry, University of Waterloo, Waterloo, ON N2L 3G1, Canada*

Stig Pedersen-Bjergaard, *Institute of Pharmacy, University of Oslo, PO Box 1068, Blindern, N-0316 Oslo, Norway*

L. Perani, *Istituto di Ricerche Farmacologiche 'Mario Negri', Via Eritrea 62, 20157 Milano, Italy*

G. Pieraccini, *Centro Interdipartimentale di Spettometria di Massa, Università di Firenze, Viale G. Pieraccini 6, 50139 Firenze, Italy*

P. Pierucci, *Istituto di Ricerche Farmacologiche 'Mario Negri', Via Eritrea 62, 20157 Milano, Italy*

Peter Popp, *Department of Analytical Chemistry, Centre for Environmental Research Leipzig-Halle, Permoserstrasse 15, D-04318 Leipzig, Germany*

Jürgen Pörschmann, *Department of Remediation Research, Centre for Environmental Research, Permoserstrasse 15, 04318 Leipzig, Germany*

Knut E. Rasmussen, *Institute of Pharmacy, University of Oslo, PO Box 1068, Blindern, N-0316 Oslo, Norway*

Manfred Reichenbächer, *Institute of Inorganic and Analytical Chemistry, Friedrich-Schiller-Universität, Lessingstrasse 8, 07747 Jena, Germany*

Pat Sandra, *RIC BVBA, Kennedypark 20, B-8500 Kortrijk, Belgium*

Keizo Sato, *Department of Legal Medicine, Showa University School of Medicine, 1–5–8 Hatanodai, Shinagawa-ku, Tokyo 142–8555, Japan*

Hirokazu Sawada, *School of Materials Science, Toyohashi University of Technology, Toyohashi 441–8122, Japan*

Shu Li, *Department of Chemistry, University of Pittsburgh, Pittsburgh, PA 15260, USA*

M. Sledge, *Dipartimento di Biologia Animale e Genetica, Università di Firenze, Via Romana 17, 50125 Firenze, Italy*

Nicholas H. Snow, *Department of Chemistry, Seton Hall University, South Orange, NJ 07079, USA*

Danese C. Stahl, *Department of Chemistry, PO Box 9024, University of North Dakota, Grand Forks, ND 58202, USA*

Osamu Suzuki, *Department of Legal Medicine, Hamamatsu University School of Medicine, 3600 Handa-cho, Hamamatsu 431–3192, Japan*

Régine Talon, *Station de Recherches sur la Viande, INRA de Theix, Ministry of Agriculture, 63122 St.Genès-Champanelle, France*

Masahiro Taniguchi, *School of Materials Science, Toyohashi University of Technology, Toyohashi 441–8122, Japan*

David C. Tilotta, *Department of Chemistry, PO Box 9024, University of North Dakota, Grand Forks, ND 58202, USA*

S. Turillazzi, *Dipartimento di Biologia Animale e Genetica, Università di Firenze, Via Romana 17, 50125 Firenze, Italy*

Eñaut Urrestarazu Ramos, *Research Institute of Toxicology, Utrecht University, PO Box 80176, NL-3508 TD Utrecht, The Netherlands*

Wouter H.J. Vaes, *Research Institute of Toxicology, Utrecht University, PO Box 80176, NL-3508 TD Utrecht, The Netherlands*

Frank R. Visser, *New Zealand Dairy Research Institute, Private Bag 11 029, Palmerston North, New Zealand*

Matteo Vitali, *Environmental Hygiene, Istituto di Igene, Università di Roma 'La Sapienza', P.le A. Moro, 5–00185 Roma, Italy*

Stephen G. Weber, *Department of Chemistry, University of Pittsburgh, Pittsburgh, PA 15260, USA*

Chen-Wen Whang, *Department of Chemistry, Tunghai University, Taichung, Taiwan 407*

Ming Keong Wong, *Department of Chemistry, National University of Singapore, Republic of Singapore 119260*

Ke-Wu Yang, *Research Centre for Eco-Environmental Sciences, Chinese Academy of Sciences, PO Box 2871, Beijing 100085, China*

Mikio Yashiki, *Department of Legal Medicine, Hiroshima University School of Medicine, 1–2–3 Kasumi, Minami-ku, Hiroshima 734–8551, Japan*

Glossary

a	Fibre coating inner radius
A	Area of needle opening
b	Fiber coating outer radius
BTEX	Abbreviation for benzene, toluene, ethylbenzene, and three xylene isomers: m-xylene, o-xylene, and p-xylene
C_f^∞	Equilibrium concentration of analyte in fiber coating
C_g^∞	Equilibrium concentration of analyte in gas
C_h^∞	Equilibrium concentration of analyte in gaseous headspace above sample
C_s^∞	Equilibrium concentration of analyte in sample
C_w^∞	Equilibrium concentration of analyte in water
C_0	Initial concentration of analyte in sample
CE	Capillary electrophoresis
CGC	Capillary gas chromatography, or gas chromatograph
d	Vial inner radius
D_f	Diffusion coefficient of analyte in fiber coating
D_g	Diffusion coefficient of analyte in gas
D_s	Diffusion coefficient of analytes in sample matrix
EPA	Environmental Protection Agency of the United States of America
FID	Flame ionization detector, commonly used in a gas chromatograph
GC	Gas chromatography, or gas chromatograph
HPLC	High performance liquid chromatography
i.d.	Inside diameter
ICPMS	Inductively coupled plasma mass spectrometry
ITMS	Ion trap mass spectrometer
k	Pseudo first order rate constant of chemical reaction
k	Rate constant of chemical reaction
k_p	Partition ratio ($k_p = K_{fs}V_f/V_v$)
K_{fg}	Fiber/gas distribution constant ($K_{fg} = C_f^\infty/C_g^\infty$)
K_{fh}	Fiber/headspace distribution constant ($K_{fh} = C_f^\infty/C_h^\infty$)
K_{fs}	Fiber/sample matrix distribution constant ($K_{fs} = C_f^\infty/C_s^\infty$)
K_{fw}	Fiber/water distribution constant ($K_{fw} = C_f^\infty/C_w^\infty$)
K_{hs}	Headspace/sample distribution constant ($K_{hs} = C_h^\infty/C_s^\infty$)
L	Fibre coating length

LOD	Limit of detection
MESI	Membrane extraction with a sorbent interface
MIP-AED	Microwave induced plasma atomic emission detection
MS	Mass spectrometry or mass spectrometer
n	Amount of analyte extracted onto the coating
o.d.	Outside diameter
PA	Poly(acrylate)
PAH	Polynuclear aromatic hydrocarbon
PDMS	Poly(dimethylsiloxane)
PDAM	Pyrenyldiazomethane
PTV	Programmable temperature vaporizer (GC injector)
RSD	Relative standard deviation
SD	Standard deviation
SFC	Supercritical fluid chromatography
SFE	Supercritical fluid extraction
S/N	Signal to noise ratio
SPE	Solid phase extraction
SPI	Septum-equipped temperature programmable injector, used in the Varian GC
SPME	Solid phase microextraction
SS	Stainless steel
t_e	Equilibration time
$t_{95\%}$	Time required to extract 95% of analyte amount at equilibrium conditions
u	Velocity of sample matrix in respect to extracting phase
V_f	Volume of fiber coating
V_h	Volume of gaseous headspace above sample
V_s	Sample volume
V_v	Void volume of the tubing containing extracting phase
VOC	Volatile organic compounds
Z	Distance between needle opening and position of coating
δ	Boundary layer thickness

Calibration and Quantitation by SPME

CHAPTER 1

Quantitative Aspects of SPME

JANUSZ PAWLISZYN

1 Introduction

The objective of this chapter is to introduce the basic concepts facilitating accurate and precise quantitation using SPME technology. The information presented below is a summary of the comprehensive discussion of the topic covered in the recently published book.[1]

Solid Phase Microextraction (SPME) was introduced as a solvent-free sample preparation technique. The basic principle of this approach is to use a small amount of the extracting phase, usually less than 1 μL. Sample volume can be very large, when the investigated system is sampled directly, for example air in a room or lake water. The extracting phase can be either high molecular weight polymeric liquid, similar in nature to stationary phases in chromatography, or it can be a solid sorbent, typically of a high porosity to increase the surface area available for adsorption.

To date, the most practical geometric configuration of SPME utilizes a small fused silica fibre, usually coated with a polymeric phase. The fibre is mounted for protection in a syringe-like device. The analytes are absorbed, or adsorbed, by the fibre phase (depending on the nature of the coating) until an equilibrium is reached in the system. The amount of an analyte extracted by the coating at equilibrium is determined by the magnitude of the partition coefficient (distribution ratio) of the analyte between the sample matrix and the coating material.

In SPME, analytes typically are not extracted quantitatively from the matrix. However, equilibrium methods are more selective because they take full advantage of the differences in extracting-phase/matrix distribution constants to separate target analytes from interferences. Exhaustive extraction can be achieved in SPME when the distribution constants are large enough. This can be accomplished for most compounds by the application of internally cooled fibre.[2] In exhaustive extraction, selectivity is sacrificed to obtain a quantitative transfer of target analytes into the extracting phase. One advantage of this approach is that, in principle, it does not require calibration, since all the analytes of interest are transferred to the extracting phase. On the other hand, the equilibrium approach usually requires calibration for complex samples. This is usually accomplished by using surrogates, or standard addition technique, to quantify the analytes and to compensate for matrix-to-matrix variations and their effect on distribution constants.

Since equilibrium rather than exhaustive extraction occurs in the microextraction methods, SPME is ideal for field monitoring. It is unnecessary to measure the volume of the extracted sample, and therefore the SPME device can be exposed directly to the investigated system for quantitation of target analytes. In addition, extracted analytes are introduced to analytical instrument simply by placing the fibre in the desorbtion unit (Figures 1b and 1c). This convenient, solvent free process facilitate sharp injection bands and rapid separations.[3] These features of SPME result in integration of the first steps in analytical process: sampling, sample preparation and introduction of extracted mixture to an analytical instrument.

The equilibrium nature of the technique also facilitates speciation in natural systems since the presence of a minute fibre, which removes small amounts of target analytes, is not likely to disturb the system. Because of the small size, coated fibres can be used to extract analytes from very small samples. For example, SPME has been used to probe for substances emitted by a single flower bulb during its lifespan.

Figure 1a illustrates the commercial SPME device, manufactured by Supelco, Inc. (Bellefonte, PA). The fibre, glued into a piece of stainless steel tubing, is mounted in a special holder. The holder is equipped with an adjustable depth gauge, which makes it possible to control repeatably how far the needle of the device is allowed to penetrate the sample container (if any) or the injector. This is important, as the fibre can be easily broken when it hits an obstacle. The movement of the plunger is limited by a small screw moving in the z-shaped slot of the device. For protection during storage or septum piercing, the fibre is withdrawn into the needle of the device, with the screw in the uppermost position. During extraction or desorption, the fibre is exposed by depressing the plunger, which can be locked in the lowered (middle) position by turning it clockwise (the position depicted in Figure 1a). The plunger is moved to its lowermost position only for replacement of the fibre assembly. Each type of fibre has a hub of a different colour. The hub-viewing window enables a quick check of the type of fibre mounted in the device.

If the sample is placed in a vial, the septum of the vial is first pierced with the needle (with the fibre in the retracted position), and the plunger is lowered, which exposes the fibre to the sample. The analytes are allowed to partition into the coating for a predetermined time, and the fibre is then retracted back into the needle. The device is next transferred to the analytical instrument of choice. When gas chromatography (GC) is used for analyte separation and quantitation, the fibre is inserted into a hot injector, where thermal desorption of the trapped analytes takes place (Figure 1c). The process can be automated by using an appropriately modified syringe autosampler. For HPLC applications, a simple interface mounted in a place of the injection loop can be used to re-extract analytes into the desorption solvent (Figure 1b).

The SPME device is capable for both spot and time-averaged sampling. As

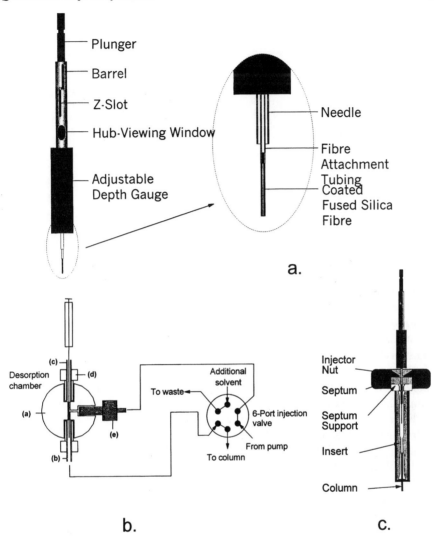

Figure 1 (a) *Design of the commercial SPME device.* (b) *SPMEMPLC interface: (a) stainless steel (SS) 1/16" tee, (b) 1/16" SS tubing, (c) 1/16" polyetheretherketone (PEEK) tubing (0.02" i.d.), (d) two-piece finger-tight PEEK union, (e) PEEK tubing (0.005" i.d.) with a one-piece PEEK union.* (c) *SPME/GC interface*

described above, for spot sampling, the fibre is exposed to a sample matrix until the partitioning equilibrium is reached between the sample matrix and the coating material. In a time-average approach, on the other hand, the fibre remains in the needle during the exposure of the SPME device to the sample. The coating works as a trap for analytes that diffuse into the needle, resulting in an accumulated mass of analyte proportional to an integral of concentration over time (see equation 10).

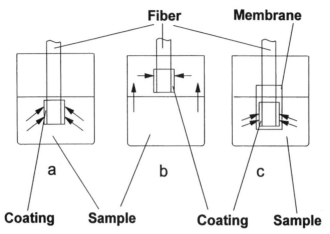

Figure 2 *Modes of SPME operation: direct extraction* (a), *headspace extraction* (b) *and membrane-protected SPME* (c)

SPME sampling can be performed in three basic modes: direct extraction, headspace extraction, and extraction with membrane protection. Figure 2 illustrates the differences between these modes. In direct extraction mode (Figure 2a), the coated fibre is inserted into the sample and the analytes are transported directly from the sample matrix to the extracting phase. To facilitate rapid extraction, some level of agitation is required to transport the analytes from the bulk of the sample to the vicinity of the fibre. For gaseous samples, the natural flow of air (*e.g.* convection) is frequently sufficient to facilitate rapid equilibration, but for aqueous matrices, more efficient agitation techniques, such as fast sample flow, rapid fibre or vial movement, stirring or sonication are required to reduce the effect of a 'depletion zone' produced close to the fibre as a result of slow diffusional transport of analyte through the stationary layer of liquid matrix surrounding the fibre.

In the headspace mode (Figure 2b), the analytes are extracted from the gas phase equilibrated with the sample. The primary reason for this modification is to protect the fibre from adverse effects caused by non-volatile, high molecular weight substances present in the sample matrix (*e.g.* humic acids or proteins). The headspace mode also allows matrix modifications, including pH adjustment, without affecting the fibre. In a system consisting of a liquid sample and its headspace, the amount of an analyte extracted by the fibre coating does not depend on the location of the fibre in the liquid phase or in the gas phase, therefore the sensitivity of headspace sampling is the same as the sensitivity of direct sampling as long as the volumes of the two phases are the same in both sampling modes. Even when no headspace is used in direct extraction, a significant sensitivity difference between direct and headspace sampling can occur only for very volatile analytes. However, the choice of sampling mode has a very significant impact on the extraction kinetics. When the fibre is in the headspace, the analytes are removed from the headspace first, followed by indirect extraction from the matrix. Therefore, volatile analytes are extracted faster than semivolatiles. Temperature has a significant effect on the kinetics of the process, since it determines the vapour pressure of analytes. In general, the equilibration times for volatile compounds are shorter for headspace SPME extraction than for direct extraction under similar agitation conditions, because of the following three reasons: a substantial portion of the analytes is present in the headspace prior to the beginning of the extraction process, there is typically large interface between sample matrix and headspace, and the diffusion coefficients in the gas phase are typically higher by four orders of magnitude than in liquids. As the concentration of semivolatile compounds in the gaseous phase at room temperature is small, headspace extraction rates for those compounds are substantially lower. They can be improved by using very efficient agitation or by increasing the extraction temperature.

In the third mode (SPME extraction with membrane protection, Figure 2c), the fibre is separated from the sample with a selective membrane, which lets the analytes through while blocking the intereferences. The main purpose for the use of the membrane barrier is to protect the fibre against adverse effects caused by high molecular weight compounds when very dirty samples are analysed. While extraction from headspace serves the same purpose, membrane protection enables the analysis of less volatile compounds. The extraction process is substantially slower than direct extraction because the analytes need to diffuse through the membrane before they can reach the coating. The use of thin membranes and an increase in the extraction temperature result in shorter extraction times.

2 Theoretical Aspects of Solid Phase Microextraction Optimization and Calibration

Thermodynamics

Solid phase microextraction is a multiphase equilibration process. Frequently, the extraction system is complex, as in a sample consisting of an aqueous phase with suspended solid particles having various adsorption interactions with analytes, plus a gaseous headspace. In some cases specific factors have to be considered, such as analyte losses by biodegradation or adsorption on the walls of the sampling vessel. In the discussion below we will only consider three phases: the fibre coating, the gas phase or headspace, and a homogeneous matrix such as pure water or air. During extraction, analytes migrate between all three phases until equilibrium is reached.

The mass of an analyte extracted by the polymeric coating is related to the overall equilibrium of the analyte in the three-phase system. Since the total mass of an analyte should remain constant during the extraction, we have:

$$C_0 V_s = C_f^\infty V_f + C_h^\infty V_h + C_s^\infty V_s \qquad (1)$$

where C_0 is the initial concentration of the analyte in the matrix; C_f^∞, C_h^∞, and C_s^∞ are the equilibrium concentrations of the analyte in the coating, the headspace, and the matrix, respectively; V_f, V_h, and V_s, are the volumes of the coating, the headspace, and the matrix, respectively. If we define the coating/gas distribution constant as $K_{fh} = C_f^\infty / C_h^\infty$, and the gas/sample matrix distribution constant as $K_{hs} = C_h^\infty / C_s^\infty$, the mass of the analyte absorbed by the coating, $n = C_f^\infty V_f$, can be expressed as:

$$n = \frac{K_{fh} K_{hs} V_f C_0 V_s}{K_{fh} K_{hs} V_f + K_{hs} V_h + V_s} \tag{2}$$

Also

$$K_{fs} = K_{fh} K_{hs} = K_{fg} K_{gs} \tag{3}$$

since the fibre/headspace distribution constant, K_{fh} can be approximated by the fibre/gas distribution constant K_{fg}, and the headspace/sample distribution constant, K_{hs}, by the gas/sample distribution constant, K_{gs}, if the effect of moisture in the gaseous headspace can be neglected. Thus, equation 2 can be rewritten as:

$$n = \frac{K_{fs} V_f C_0 V_s}{K_{fs} V_f + K_{hs} V_h + V_s} \tag{4}$$

The equation states, as expected from the equilibrium conditions, that the amount of analyte extracted is independent of the location of the fibre in the system. It may be placed in the headspace, or directly in the sample, as long as the volumes of the fibre coating, headspace, and sample are kept constant. There are three terms in the denominator of equation 4 which give a measure of the analyte capacity of each of the three phases: fibre ($K_{fs} V_f$), headspace ($K_{hs} V_h$), and the sample itself (V_s). If we assume that the vial containing the sample is completely filled (no headspace), the term $K_{hs} V_h$ in the denominator, which is related to the capacity ($C_h^\infty V_h$) of the headspace, can be eliminated, resulting in:

$$n = \frac{K_{fs} V_f C_0 V_s}{K_{fs} V_f + V_s} \tag{5}$$

Equation 5 describes the mass absorbed by the polymeric coating after equilibrium has been reached in the system. In most of determinations, K_{fs}, is relatively small compared to the phase ratio of sample matrix to coating volume ($V_f \ll V_s$). In that situation, the capacity of the sample is much larger compared to the capacity of the fibre, resulting in a very simple relationship:

$$n = K_{fs} V_f C_0 \tag{6}$$

The above equation emphasizes the field sampling capability of the SPME technique. It is not necessary to sample a well defined volume of the matrix since the amount of analyte extracted is independent of V_s, as long as $K_{fs} V_f \ll V_s$. So, an SPME device can be placed directly in contact with the investigated system to allow quantitation.

Strictly speaking, the above discussion (equations 1–6) is limited to partitioning equilibrium involving liquid polymeric phases such as poly(dimethylsiloxane). The method of analysis for solid sorbent coatings is analogous for low analyte concentration, since the total surface area available for adsorption is proportional to the coating volume if we assume constant porosity of the sorbent. For high analyte concentration the saturation of the surface can occur resulting in nonlinear isotherms as discussed in Chapter 7. Similarly, high concentration of the competitive interference compound can displace the target analyte from the surface of the sorbent. The simplest way to consider these high concentration effects is to replace the volume of the fibre coating, V_f in the above equations as a measure of the total fibre surface area by a fraction of the original coating volume corresponding to a free surface area available for adsorption.

Prediction of Distribution Constants

In many cases, the distribution constants present in equations 2–6 which determine the sensitivity of SPME extraction can be estimated from physicochemical data and chromatographic parameters. This approach eliminates the need for a separate calibration step. For example, distribution constants between a fibre coating and gaseous matrix (*e.g.* air) can be estimated using isothermal GC retention times on a column with a stationary phase identical to the fibre coating material.[4] This is possible because the partitioning process in gas chromatography is analogous to the partitioning process in solid phase microextraction, and there is a well-defined relationship between the distribution constant and the retention time. The nature of the gaseous phase does not affect the distribution constant, unless the components of the gas, such as moisture, swell the polymer, thus changing its properties. A most useful method for determining the coating-to-gas distribution constants uses the linear temperature programmed retention index (LTPRI) system, which indexes compounds' retention times relative to the retention times of *n*-alkanes. This system is applicable to retention times for temperature-programmed gas-liquid chromatography. The logarithm of the coating-to-air distribution constants of *n*-alkanes can be expressed as a linear function of their LTPRI values. For PDMS, this relationship is $\log K_{fg} = 0.00415 \times \text{LTPRI} - 0.188$.[5] Thus, the LTPRI system permits interpolation of the K_{fg} values from the plot of $\log K_{fg}$ versus retention index. The LTPRI values for many compounds are available in the literature, hence this method allows estimation of K_{fg} values without experimentation. If the LTPRI value for a compound is not available from published sources, it can be determined from a GC run. Note that the GC column used to determine LTPRI should be coated with the same material as the fibre coating.

Estimation of the coating/water distribution constant can be performed using equation 5. The appropriate coating/gas distribution constant can be found by applying the techniques discussed above, and the gas/water distribution constant (Henry's constant) can be obtained from physicochemical tables, or can be estimated by the structural unit contribution method.[6]

Some correlations can be used to anticipate trends in SPME coating/water distribution constants for analytes. For example, a number of investigators have reported the correlation between octanol/water distribution constant K_{ow} and K_{fw}. This is expected, since K_{ow} is a very general measure of the affinity of compounds to the organic phase. It should be remembered, however, that the trends are valid only for compounds within homologous series, such as aliphatic hydrocarbons, aromatic hydrocarbons or phenols; they should not be used to make comparisons between different classes of compounds, because of different analyte activity coefficients in the polymer.

Effect of Extraction Parameters

Thermodynamics theory predicts the effects of modifying certain extraction conditions on partitioning, and indicates parameters to control for reproducibility. The theory can be used to optimize the extraction conditions with a minimum number of experiments, and to correct for variations in extraction conditions without the need to repeat calibration tests under the new conditions. For example, SPME analysis of outdoor air may be done at ambient temperatures that can vary significantly. Relationships that predicts the effect of temperature on the amount of analyte extracted allow calibration without the need for extensive experimentation.[7] Extraction conditions that affect K_{fs} include temperature, salting, pH, and organic solvent content in water. A brief discussion about the use of extraction parameters in the SPME method of optimization can be found the next section.

Kinetics

Kinetic theory is very useful in optimization of the extraction conditions by identifying 'bottlenecks' of solid phase microextraction. It indicates strategies to increase extraction speed. In the discussion below we will limit our consideration to direct extraction (Figure 3).

Perfect Agitation

Let us first consider the case where the liquid or gaseous sample is perfectly agitated. In other words, the sample phase moves very rapidly with respect to the fibre, so that all the analytes present in the sample have access to the fibre coating. In this case, the equilibration time, t_e, defined as the time required to extract 95% of the equilibrium amount (Figure 4) of an analyte from the sample, corresponds to:

$$t_e = t_{95\%} = \frac{2(b - a)^2}{D_f} \tag{7}$$

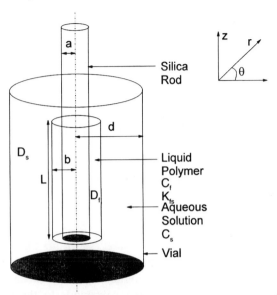

Figure 3 *Graphic representation of the SPME/sample system configuration, with dimensions and parameters labelled as follows: a, fibre coating inner radius; b, fibre coating outer radius; L, fibre coating length; d, vial inner radius; C_f, analyte concentration in the fibre coating; D_f, analyte diffusion coefficient in the fibre coating; C_s, analyte concentration in the sample; D_s, analyte diffusion coefficient in the sample; K_{fs}, analyte distribution coefficient between fibre coating and sample; $K_{fs} = C_f/C_s$*

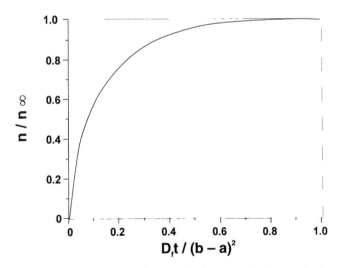

Figure 4 *Mass absorbed versus time from perfectly agitated solution of infinite volume*

Using this equation, one can estimate the shortest equilibration time possible for the practical system by substituting appropriate data for the diffusion coefficient of an analyte in the coating (D_f) and the fibre coating thickness ($b - a$). For example, the equilibration time for the extraction of benzene from a perfectly stirred aqueous solution with a 100 μm PDMS film is expected to be about 20 seconds. Equilibration times close to those predicted for perfectly agitated samples have been obtained experimentally for extraction of analytes from air samples (because of high diffusion coefficients in gas) or when very high sonication power was used to facilitate mass transfer in aqueous samples. However, in practice, there is always a layer of unstirred water around the fibre. A higher stirring rate will result in a thinner water layer around the fibre.

Practical Agitation

Independent of the agitation level, fluid contacting a fibre's surface is always stationary, and, as the distance from the fibre surface increases, the fluid movement gradually increases until it corresponds to the bulk flow in the sample. To model mass transports, the gradation in fluid motion and convection of molecules in the space surrounding the fibre surface can be simplified by a zone of a defined thickness in which no convection occurs, and perfect agitation in the bulk of the fluid. This static layer zone is called the *Prandtl* boundary layer (see Figure 5).[8] Its thickness is determined by the agitation conditions and the viscosity of the fluid.

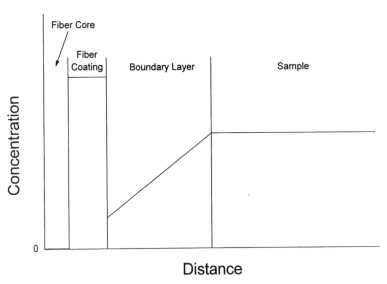

Figure 5   *Boundary layer model configuration showing the different regions considered and the assumed concentration versus radius profile for the case when the boundary layer determines the extraction rate*

The equilibration time can be estimated for practical cases from the equation below:

$$t_e = t_{95\%} = 3\frac{\delta K_{fs}(b-a)^2}{D_s} \tag{8}$$

where $(b-a)$ is the fibre coating's thickness, D_s is the analyte's diffusion coefficient in the sample fluid, K_{fs} is the analyte's distribution constant between the fibre and the sample. This equation can be used to predict equilibration times when the extraction rate is controlled by the diffusion in the boundary layer. In other words, the extraction time calculated by using equation 8 must be longer than the corresponding time predicted by equation 7.

Time-weighted Average Sampling Using the SPME Device

In addition to the analyte concentration measurement at a well defined place in space and time, obtained by using the approaches discussed above, an integrating sampling is possible with the SPME device. This is particularly important in field measurements when changes of analyte concentration over time, and place-to-place variations, must often be taken into account.

When the extracting phase is not exposed directly to the sample, but is contained in the protective tubing (needle) without any flow of the sample through it the extraction occurs by the process of diffusion through the static gas phase present in the needle. The integrating system can consist of an externally coated fibre withdrawn into the needle as shown in Figure 6. This simple geometric arrangement represents a very convenient method which is able to generate a response proportional to the integral of the analyte concentration over time and/or space (when the needle is moved through the space). The only mechanism of analyte transport to the extracting phase is diffusion through the gaseous phase contained in the tubing. During this process, a linear concentration profile is established in the tubing between the small needle opening, characterized by surface area A, and the distance Z between the needle opening and the position of the extracting phase. The amount of analyte extracted, dn, during time interval, dt, can be calculated by considering the first Fick's law of diffusion:[1]

$$dn = -AD_g\frac{dc}{dx}dt = -AD_g\frac{\Delta C(t)}{Z}dt \tag{9}$$

where $\Delta C(t)/Z$ is a value of the gradient established in the needle between needle

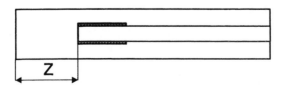

Figure 6 *Time-weighted average sampling with the coated fibre retracted in the needle*

opening and the position of the extracting phase, Z; $\Delta C(t) = C_z - C(t)$ where $C(t)$ is a time-dependent concentration of analyte in the sample in the vicinity of the needle opening, and C_z concentration of the analyte in the gas phase in the vicinity of the coating. If C_z is close to zero for a high coating/gas distribution constant capacity, then: $\Delta C(t) = -C(t)$. The concentration of analyte at the coating position in the needle, C_z, will increase with integration time, but it will be kept low compared to the sample concentration because of the presence of the sorbing coating. Therefore the accumulated amount over time can be calculated as:

$$n = D_g \frac{A}{Z} \int C(t)\mathrm{d}t \qquad (10)$$

As expected, the extracted amount of analyte is proportional to the integral of a sample concentration over time, the diffusion coefficient of analytes in gaseous phase, D_g, area of the needle opening, A, and inversely proportional to the distance of the coating position in respect of the needle opening, Z. It should be emphasized that this application of the SPME device is not an equilibrium measurement. Equation 10 is valid only in a situation where the amount of analyte extracted onto the sorbent is a small fraction (below RSD of the measurement, typically 5%) of the equilibrium amount in respect to the lowest concentration in the sample. To extend integration times, the coating can be placed deeper into the needle (larger Z), the opening of the needle can be reduced by placing an additional orifice (smaller A), or a high capacity sorbent can be used. The first two solutions will result in a low measurement sensitivity. An increase of sorbent capacity presents a more attractive opportunity. It can be achieved by either increasing the volume of the coating, or its affinfity towards the analyte. An increase of the coating volume will require an increase in the size of the device. The optimum approach to increased integration time is to use sorbents characterized by large coating/gas distribution constants, or derivatization reagent in the sorbent.

3 Practical Aspects of SPME Calibration and Optimization

A properly optimized method ensures good accuracy and precision together with detection limits. Below, I briefly summarize the most important steps which should be considered when developing SPME methods.

Selection of Fibre Coating

The chemical nature of the target analyte determines the type of coating used. A simple general rule, 'like dissolves like', applies very well for the liquid coatings. Selection of the coating is based primarily on the polarity and volatility character-istics of the analyte. Poly(dimethylsiloxane) (PDMS) is the most useful coating and should be considered first. It is very rugged and able to withstand high injector temperatures, up to about 300 °C. PDMS is a non-polar liquid phase; thus it extracts non-polar analytes very well with a wide linear dynamic range. However, it can also be applied successfully to more polar compounds, particularly after optimizing extraction conditions. An additional advantage of this phase is the possibility of estimating the distribution constants for organic compounds from retention parameters on PDMS-coated GC columns.

Both the coating thickness and the distribution constant determine the sensitivity of the method and the extraction time. Thick coatings offer increased sensitivity, but require much longer equilibration times. As a general rule, to speed up the sampling process, the thinnest coating offering the sensitivity required should be used.

Selection of the Derivatizing Reagent

Derivatization performed before and/or during extraction can enhance sensitivity and selectivity of both extraction and detection, as well as enable SPME determination of analytes normally not amenable to analysis by this method. Post extraction methods can only improve chromatographic behaviour and detection. Incorporation of the derivatization step complicates the SPME procedure, therefore should only be considered when necessary. Selective reactions producing specific analogues result in less interference during quantitation. This approach can be used for analyte determination in complex matrices. Additionally sensitivity enhancement can be achieved when the derivatizing reagent contains moieties that enhance detection.

Selection of the Extraction Mode

Extraction mode selection is based on the sample matrix composition, analyte volatility, and its affinity to the matrix. For very dirty samples, the headspace or fibre protection mode should be selected. For clean matrices, both direct and headspace sampling can be used. The latter is applicable for analytes of medium to high volatility. Headspace extraction is always preferential for volatile analytes because the equilibration times are shorter in this mode compared to direct extraction. Extraction conditions for many compounds, including polar and ionic ones, can be improved by matrix modifications. Application of headspace SPME can be extended to semi-volatile compounds and analytes strongly bound to the matrix, by increasing the extraction temperature. Fibre protection should be used only for very dirty samples in cases where neither of the first two modes can be applied.

Selection of the Agitation Technique

Equilibration times in gaseous samples are short and frequently limited only by the rate of diffusion of the analytes in the coating. A similar situation occurs when analytes characterized by large air/water distribution constants are determined in water by the headspace technique. When the aqueous and gaseous phases are at equilibrium prior to the beginning of the sampling process, most of the analytes are in the headspace. As a result, the extraction times are short even when no agitation is used. However, for aqueous samples, agitation is required in most cases to facilitate mass transport between the bulk of the aqueous sample and the fibre.

Magnetic stirring is most commonly used in manual SPME experiments. Care must be taken when using this technique to ensure that the rotational speed of the stirring bar is constant and that the base plate does not change its temperature during stirring. This usually implies the use of high quality digital stirrers. Alternately, with cheaper stirrers, the base plate should be thermally insulated from the vial containing the sample to eliminate variations in sample temperature during extraction. Magnetic stirring is efficient when fast rotational speeds are applied.

Needle vibration technique uses an external motor and a cam to generate a vibrating motion of the fibre and the vial. This technique has been implemented by Varian in the SPME autosampler. This technique provides good agitation, resulting in equilibration times similar to those obtained for magnetic stirring; however, good performance is limited to small vials and direct extraction mode. This technique can be conveniently applied to process a large number of samples since the sample vials do not require any manipulations, such as the introduction of stirring bars.

Selection of Separation and/or Detection Technique

So far, most SPME applications have been developed for gas chromatography, but other separation techniques, including HPLC, capillary electrophoresis (CE) and supercritical fluid chromatography (SFC) can also be used in conjunction with this technique. The complexity of the extraction mixture determines the proper quantitation device. Regular chromatographic and CE detectors can normally be used for all but the most complex samples, for which mass spectrometry should be applied. As selective coatings become available, the direct coupling to MS/MS and ICPMS becomes practical.

Optimization of Desorption Conditions

Standard gas chromatographic injectors, such as popular split-splitless types, are equipped with large volume inserts to accommodate the vapours of the solvent introduced during liquid injections. As a result, the linear flow rates of the carrier gas in those injectors are very low in splitless mode, and the transfer of the volatilized analytes onto the front of the GC column is very slow. No solvent is introduced during SPME injection, therefore the large insert volume is unnecessary. Opening the split line during SPME injection is not practical, since it results in reduced sensitivity. Efficient desorption and rapid transfer of the analytes from the injector to the column require high linear flow rates of the carrier gas around the coating. This can be accomplished by reducing the internal diameter of the injector insert, matching it as closely as possible to the outside diameter of the coated fibre. Narrow bore inserts for SPME are commercially available from Supelco for a range of GC instruments.[9]

Parameters which control the desorption process in the HPLC interface are analogous to those in GC applications. In addition to temperature and flow rate, the composition of the mobile phase also affects the process. In many cases, it is possible to use the mobile phase without any modifications. In some instances, addition of an appropriate solvent to the interface will assist desorption. The linear flow rate of the mobile phase should be maximized by choosing a small i.d. tubing for the desorption chamber. Temperature also plays an important role in accelerating the desorption.

Optimization of Sample Volume

The volume of the sample should be selected based on the estimated distribution constant K_{fs}. The distribution constant can be estimated by using literature values for the target analyte or a related compound with the coating selected. K_{fs} can be calculated or determined experimentally by equilibrating the sample with the fibre and determining the amount of analyte extracted by the coating. Care must be taken to avoid analyte losses *via* adsorption, evaporation, microbial degradation *etc.*, when very long extraction times are required to reach the equilibrium.

The sensitivity of the SPME method is proportional to the number of moles of the analyte n extracted from the sample and, for direct extraction method, is given by equation 5. As the sample volume V_s increases, so does the amount of analyte extracted until the volume of the sample becomes significantly larger than the product of the distribution constant and volume of the coating (fibre capacity $K_{fs} \ll V_s$).

Determination of the Extraction Time

An optimal approach to SPME analysis is to allow the analyte to reach equilibrium between the sample and the fibre coating. The equilibration time is defined as the time after which the amount of analyte extracted remains constant and corresponds within the limits of experimental error to the amount extracted after infinite time. Care should be taken when determining the equilibration times, since, in some cases, a substantial reduction of the slope of the curve might be wrongly taken as the point at which equilibrium is reached. Such a phenomenon often occurs in headspace SPME determinations of aqueous samples, where a rapid rise of the equilibration curve corresponding to extraction from the gaseous phase only is followed by a very slow increase related to analyte transfer from water through the headspace to the fibre. Figure 7 illustrates an example of such a curve for phenanthrene. The amount extracted increases rapidly in the first five minutes, but the equilibrium is not reach until much later. Determination of the amount extracted at equilibrium allows the calculation of the distribution constants.

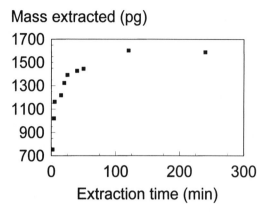

Figure 7 *Extraction time profile corresponding to extraction of [²H₁₀]phenanthrene from water at 50 °C under magnetic stirring. Fibre coating: 7 µm PDMS*

When equilibration times are excessively long, shorter extraction times can be used. However, in such cases the extraction time and mass transfer conditions have to be strictly controlled to assure good precision. At equilibrium, small variations in the extraction time do not affect the amount of the analyte extracted by the fibre. On the other hand, at the steep part of the curve, even small variations in the extraction time may result in significant variations of the amount extracted. The relative error is the larger, the shorter is the extraction time. Autosamplers can measure the time very precisely, and the precision of analyte determination can be very good, even when equilibrium is not reached in the system. However, this requires the mass transfer conditions and the temperature to remain constant during all experiments.

Calculation of the Distribution Constant

The target analyte's distribution constant defines the sensitivity of the method. It is not necessary to calculate the fibre coating/sample matrix distribution constant, K_{fs}, when the calibration is based on isotopically labelled standards or standard addition, or when identical matrix and headspace volumes are used for the standard and for the sample with external calibration. However, it is always advisable to determine K_{fs} since this value gives more information about the experiment and aids optimization. K_{fs} can be used for calculation of the headspace volume, sample volume, and coating thickness required to reach the desired sensitivity.

The distribution constant for direct extraction mode can be calculated from the following dependence obtained from equation 5:

$$K_{fs} = \frac{nV_s}{V_f(C_0 V_s - n)} \tag{11}$$

Optimization of Extraction Conditions

An extraction temperature increase causes an increase in the extraction rate, but simultaneously a decrease in the distribution constant. In general, if the extraction rate is of major concern, the highest temperature which still provides satisfactory sensitivity should be used.

Adjustment of the pH of the sample can improve the sensitivity of the method for basic and acidic analytes. This is related to the fact that, unless ion exchange coatings are used, SPME can extract only neutral (non-ionic) species from water. By properly adjusting the pH, weak acids and bases can be converted to their neutral forms, in which they can be extracted by the SPME fibre. To make sure that at least 99% of the acidic compound is in the neutral form, the pH should be at least two units lower than the pK_a of the analyte. For basic analytes, the pH must be larger than pK_b by two units.

Determination of the Linear Dynamic Range of the Method

Modification of the extraction conditions affects both the sensitivity and the equilibration time. It is advisable, therefore, to check the previously determined extraction time before proceeding to the determination of the linear dynamic range. This step is required if substantial changes of the sensitivity occur during the optimization process.

SPME coatings include polymeric liquids such as PDMS, which by definition have a very broad linear range. For solid sorbents, such as Carbowax/DVB or PDMS/DVB, the linear range is narrower because of a limited number of sorption sites on the surface, but it still can span over several orders of magnitude for typical analytes in pure matrices. In some rare cases when the analyte has extremely high affinity towards the surface, saturation can occur at low analyte concentrations. In such cases, the linear range can be expanded by shortening the extraction time.

Selection of the Calibration Method

Standard calibration procedures can be used with SPME. A fibre blank should be checked first to ensure that neither the fibre nor the instrument cause interferences with the determination. The fibre should be conditioned prior to the first use by desorption in a GC injector, or a specially designed conditioning device. This process ensures that the fibre coating itself does not introduce interferences. Fibre conditioning may have to be repeated after analysis of samples containing significant amounts of high-molecular weight compounds, since such compounds may require longer desorption times than the analytes of interest.

When the matrix is simple (*e.g.* air or groundwater), the distribution constants are very similar to those for pure matrix. It has been shown, for example, that typical moisture levels in ambient air, as well as the presence of salt and/or alcohol in water in concentrations lower than 1%, usually do not change the K values beyond the 5% RSD typical for SPME determinations. In many such instances, calibration might not be necessary since the appropriate distribution constants which define the external calibration curve are available in the literature, or can be calculated from chromatographic retention parameters. External calibration can also be used successfully when the matrix is more complex, but well defined (*e.g.* process streams of relatively constant composition). Of course, calibration standards have to be prepared in such cases in the matrix of the same composition rather than in the pure medium (*e.g.* water).

A special calibration procedure, such as isotopic dilution or standard addition, should be used for more complex samples. In these methods, it is assumed that the target analytes behave similarly to spikes during the extraction. This is usually a valid assumption when analysing homogeneous samples. However, it might not be true when heterogeneous samples are analysed, unless the native analytes are completely released from the matrix under the conditions applied. Moreover, whenever any of these method is used, an inherent assumption is made that the response is linear in the concentration range between the original analyte concentration and the spiked concentration. While this is usually true for fibres extracting the analytes by absorption (PDMS, PA) and detectors with wide linear range (FID), problems may arise when porous polymer fibres (PDMS/DVB, Carbowax/DVB) are used, or when the detector applied has a narrow linear dynamic range. It is important, therefore, to check the linearity of the response using standard solutions, before applying standard addition or isotopic dilution for calibration. To improve the accuracy and precision, multipoint standard addition should be used whenever practical.

Precision of the Method

The most important factors affecting precision in SPME are listed below:

- agitation conditions
- sampling time (if non-equilibrium conditions are used)
- temperature
- sample volume
- headspace volume
- vial shape
- condition of the fibre coating (cracks, adsorption of high MW species)
- geometry of the fibre (thickness and length of the coating)
- sample matrix components (salt, organic material, humidity *etc.*)
- time between extraction and analysis
- analyte losses (adsorption on the walls, permeation through Teflon, absorption by septa)
- geometry of the injector
- fibre positioning during injection
- condition of the injector (pieces of septa)

- stability of the detector response
- moisture in the needle

To ensure good reproducibility of the SPME measurement, the experimental parameters listed above should be kept constant.

Automation of the Method

SPME is a very powerful investigative tool, but it can also be a technique of choice in many applications for processing a large number of samples. To accomplish this task would require automation of the methods developed. As automated SPME devices with more advanced features and capabilities become available, automation of the methods developed becomes easier. The currently available SPME autosampler from Varian enables direct sampling with agitation of the sample by fibre vibration, and static headspace sampling. In some cases, custom modifications to the commercially available systems can facilitate operation of the method closer to optimum conditions

References

1. J. Pawliszyn, *Solid Phase Microextraction. Theory and Practice*, Wiley-VCH, New York, 1997.
2. Z. Zhang and J. Pawliszyn, *Anal. Chem.*, 1995, **67**, 34.
3. T. Gorecki and J. Pawliszyn, *Anal. Chem.*, 1995, **67**, 3265.
4. Z. Zhang and J. Pawliszyn, *J. Phys. Chem.*, 1996, **100**, 17648.
5. P. Martos, A. Saraullo and J. Pawliszyn, *Anal. Chem.*, 1997, **69**, 1992.
6. R. Schwarzenbach, P. Gschwend and D. Imboden in *Environmental Organic Chemistry*, John Wiley & Sons, New York, 1993, pp. 109–123.
7. P. Martos and J. Pawliszyn, *Anal. Chem.*, 1997, **69**, 206.
8. A.D. Young, *Boundary Layers*, BSP Professional Books, Oxford, 1989.
9. R. Shirey, *HRC*, 1995, **18**, 495.

CHAPTER 2

Quantitation by SPME before Reaching a Partition Equilibrium

JIU AI

1 Quantitative Aspects of SPME at Equilibrium

The ideal way to carry out a quantitative analysis with a sampling technique is to completely transfer an analyte from the sample matrix to the analytical apparatus. As is discussed in Chapter 1, quantitative transfer is highly improbable for solid phase microextraction (SPME). In general, what we are seeking is a linear proportional relationship between the amount of the extracted analyte, n, and its initial concentration in the sample matrix, C_0. That is $n \propto C_0$. With this proportional relationship, the extracted quantity correctly reflects the concentration of the analyte in the sample matrix. For direct SPME extraction in sample matrices, when partition equilibrium is reached, the $n \propto C_0$ relationship is in effect as described by equation 5 in Chapter 1. For headspace SPME sampling at partition equilibrium, n is related to C_0 as expressed in equation 2 of Chapter 1. There is also a $n \propto C_0$ relation in this equation. The quantitative relationship between the extracted analyte and its initial concentration in the sample matrix in both direct and headspace SPME processes is the foundation for quantitative SPME when partition equilibrium is attained.

It takes certain sampling time for an analyte to reach partition equilibrium between the sample matrix and the SPME fiber. The equilibrium time ranges from a few minutes to a couple of hours depending on the sampling conditions and the nature of the analyte. It is unpractical to wait until partition equilibrium is reached if the equilibrium time is too long. Since equations 2 and 5 in Chapter 1 are derived from the equilibrium states, the real time process of the analyte transfer from the sample matrix to the SPME polymer is not considered in deriving them. The extraction process has to be explored with another approach, that is investigating the dynamics of SPME. The initial dynamic study of SPME by Pawliszyn and co-workers[1-3] involved complicated mathematical treatment in solving diffusion equations. Here a simple model is given which can describe the dynamic process of SPME and predict the experimental results.

2 Dynamic Process of Direct SPME in the Sample Matrix

Direct SPME in the sample matrix involves analyte transfer in two phases, from the sample matrix to the SPME polymer film. In most cases, mass transfer from the sample matrix to the surface of the SPME polymer film is a process of mass diffusion. The extraction of the analyte by the polymer film can either be an absorption process, in which analyte molecules are absorbed by the polymer film, or an adsorption process where analyte molecules are adsorbed on the surface of the polymer film, depending on the nature of the SPME polymer film. If the polymer film is a liquid phase, analyte molecules are absorbed into the film and the extraction process is the analyte partitioning between two phases, the sample matrix and the polymer film, respectively. In this case, the mass transfer process is the analyte diffusion from the sample matrix bulk to the SPME polymer film surface and then from the surface to the inner layers of the polymer film. According to Fick's first law of diffusion, the mass flow rate can be expressed as follows for a continuous flow at the sample matrix/SPME polymer interface:

$$F = -D_s\left(\frac{\partial C_s}{\partial x}\right)|_{x=\delta_1} = -D_f\left(\frac{\partial C_f}{\partial x}\right)|_{x=\delta_1} \tag{1}$$

Here F is the mass flow rate of the analyte from the sample matrix bulk to the SPME polymer surface that should be equal to the flow rate from the polymer surface to its inner layers for a balanced mass transfer. D_s is the diffusion coefficient of the analyte in the sample matrix phase. D_f is the diffusion coefficient of the analyte in the polymer phase. C_s and C_f are concentrations of the analyte in sample matrix and polymer film, respectively. δ_1 is the position of the interface between the sample matrix and the SPME polymer film. F must be proportional to the SPME extraction rate of the analyte and can be expressed in the following way:

$$F = \frac{1}{A_f}\frac{dn}{dt} \tag{2}$$

Equation 1 becomes

$$\frac{1}{A_f}\frac{\partial n}{\partial t} = -D_s\left(\frac{\partial C_s}{\partial x}\right)|_{x=\delta_1} = -D_f\left(\frac{\partial C_f}{\partial x}\right)|_{x=\delta_1} \tag{3}$$

A_f is the surface area of the SPME polymer film.

In order to solve equation 3, two assumptions are made for the common practice of direct SPME. The first assumption is that the diffusion layer in the sample matrix phase is kept constant. When the sample matrix phase (aqueous solution in most cases) is effectively agitated, a steady-state mass diffusion, in which the thickness of diffusion layer is a constant, can be attained. Therefore, this assumption is reasonable for an effectively agitated sample matrix. In the polymer film phase, the diffusion layer is the film thickness and a steady-state diffusion is in effect in this phase. The second assumption is that the concentration gradient in the SPME polymer film is linear. This so-called linear diffusion layer assumption is commonly used in dealing with diffusion problems.[4] Then, the average concentration of the analyte in the polymer film is approximately $(C_f + C_f')/2$. Based on these two assumptions, the partial differential equation 3 can be simplified to an ordinary differential equation:

$$\frac{1}{A_f}\frac{dn}{dt} = m_s(C_s - C_s') = m_f(C_f - C_f') \tag{4}$$

And we have:

$$n \approx V_f \frac{C_f + C_f'}{2} \tag{5}$$

Here C_s is the analyte concentration in the bulk of the sample matrix. C_s' is the surface concentration of the analyte in the sample matrix. C_f is the analyte concentration in the polymer film at the surface with the sample matrix. C_f' is the analyte concentration in the polymer film in contact with the silica fiber. The variables m_s and m_f are the mass transfer coefficients of the analyte in sample matrix phase and polymer film phase, respectively. This parameter is the ratio of diffusion coefficient over the thickness of the diffusion layer. Figure 1 shows a schematic diagram of the sample matrix/polymer film interface.

Mass diffusion is considered as the rate-determining step that slows down the partition process. Partition equilibrium is supposed to be quickly reached at the boundary between the polymer film and the sample matrix. Therefore we have:

$$K_{fs} = \frac{C_f}{C_s'} \tag{6}$$

In the bulk of the sample matrix, we have:

$$C_s = C_0 - \frac{n}{V_s} \tag{7}$$

Combining equations 4, 5, 6 and 7, we have:

$$\frac{1}{A_f}\frac{dn}{dt} = m_f(C_f - C_f') = m_f\left(\frac{2m_s K_{fs}}{m_s + 2m_f K_{fs}}C_0 - \frac{2m_s K_{fs}V_f + 2m_s V_s}{m_s V_s V_f + 2m_f K_{fs}V_s V_f}n\right) \tag{8}$$

Equation 8 can be easily solved with the initial condition: $n|_{t=0} = 0$

$$n = \left[1 - \exp\left(-A_f\frac{2m_s m_f K_{fs}V_f + 2m_s m_f V_s}{m_s V_s V_f + 2m_f K_{fs}V_s V_f}t\right)\right]\frac{K_{fs}V_f V_s}{K_{fs}V_f + V_s}C_0 \tag{9}$$

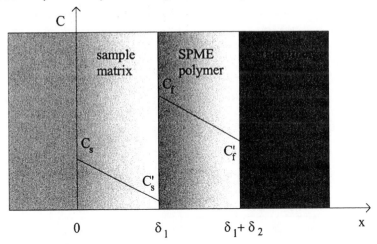

Figure 1 *Interface of sample matrix/SPME polymer film. A steady-state diffusion is assumed when the sample matrix (aqueous phase) is effectively agitated. Linear diffusion layers are assumed in both the sample matrix phase and the polymer film phase*

Equation 9 expresses a relation between the amount of extracted analyte, n, and its initial concentration in the sample matrix, C_0, as a function of sampling time, t. The $n \propto C_0$ relationship exists if the sampling time, t, is held constant for each sampling. One interesting result of equation 9 is that it becomes equation 5 in Chapter 1 when sampling time goes to infinity although equation 9 is derived from a pure dynamic process and equation 5 in Chapter 1 is obtained from an equilibrium state. Therefore, equation 9 can be written as follows:

$$n = n^\infty[1 - \exp(-a_s t)] \tag{10}$$

Here n^∞ is the extracted amount at equilibrium as described in equation 5 in Chapter 1 and:

$$a_s = 2A_f \frac{m_s m_f K_{fs} V_f + m_s m_f V_s}{m_s V_s V_f + 2m_f K_{fs} V_s V_f} \tag{11}$$

The parameter a_s defined by equation 11 is a measure of how fast partition equilibrium can be reached in the SPME process. It is determined by mass transfer coefficients, equilibrium constant and physical dimensions of the sample matrix and the SPME polymer film. The above mathematical treatment has been described in more detail elsewhere.[5]

Figure 2a shows extraction time profiles of three flavor chemicals in an aqueous solution. The points are experimental data and lines are from equation 10. The experimental data fit the theoretical model (equation 10) for up to 160 min of extraction. Parameter a_s defined by equation 11 and $t_{95\%}$, which is

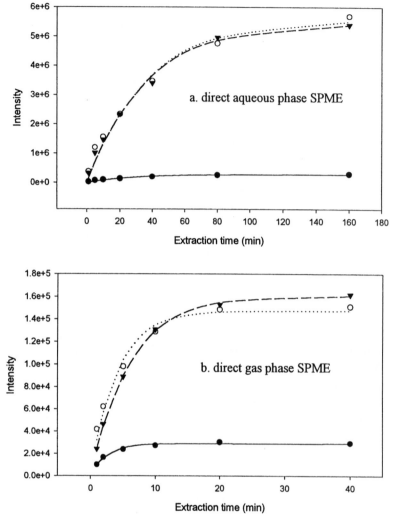

Figure 2 *Extraction time profiles of direct SPME at room temperature with 85 µm polyacrylate film; ● 2-phenylethanol (0.62 µg mL⁻¹), ○ 2,4-dimethylphenol (0.47 µg mL⁻¹) and ▼ eugenol (0.68 µg mL⁻¹). (a) aqueous phase extraction with magnetic bar stirring; (b) gas phase extraction*

the time to extract 95% of analyte amount compared at equilibrium conditions, are listed in Table 1 for these profiles. In fact, most of the SPME extraction time profiles reported in literature have the shape that can be described by equation 10. The plot in Figure 2a is just an example of direct SPME in an aqueous solution.

Figure 2b shows the extraction time profiles of direct SPME in a gas phase. Experiments were carried by using a device which could isolate the condensed phase and the gas phase.[5] During the SPME sampling, the condensed phase was

Table 1 *Parameters a_s and $t_{95\%}$ of direct SPME*

Compound	Concentration ($\mu g\ mL^{-1}$)	a_s (min^{-1})		$t_{95\%}$ (min)	
		Aqueous	Gas	Aqueous	Gas
2-Phenylethanol	0.620	0.0365	0.379	82	7.9
2,4-Dimethylphenol	0.466	0.0270	0.241	111	12.4
Eugenol	0.679	0.0280	0.161	107	18.6

not involved and the partition of the analytes was between the gas phase and the SPME polymer film. The experimental data also fit the theoretical model (equation 10). Parameter a_s, and $t_{95\%}$ are listed in Table 1, too. Since mass diffusion in the gas phase is much faster than in the aqueous phase, we have $m_s \gg m_f$ and the parameter a_s defined by equation 11 becomes:

$$a_s = 2A_f m_f \frac{K_{fs} V_f + V_s}{V_s V_f} \qquad (12)$$

This is not dependent upon the diffusion in the sample matrix phase (gas phase). Compared to aqueous/polymer interface partition, the parameter a_s at the gas/polymer interface is about an order of magnitude higher. The $t_{95\%}$ is also about an order of magnitude shorter for the gas/polymer partition compared with the aqueous/polymer partition.

One important aspect of the above model is that the $n \propto C_0$ relationship holds before partition equilibrium is reached once the sampling time is held constant. This linear proportional relationship between n and C_0 is the key requirement for quantitative analysis of SPME. Figure 3 is the experimental data plots of n vs. C_0 in aqueous solutions. The sampling time for each aqueous solution is 10 min, which is far shorter than the time needed to reach partition equilibrium as is shown in Table I ($t_{95\%} \gg 10$ min for those three chemicals). There is an excellent linear relationship between n and C_0 for all the three chemicals. The linear correlation coefficients (r) for all the three lines are better than 0.999 with C_0 ranging from 0.2 $\mu g\ mL^{-1}$ to about 6 $\mu g\ mL^{-1}$. The linear relationship between n and C_0 before equilibrium has been reported in literature[6] before and equation 9 provides a theoretical explanation to the experimental results.

3 Dynamic Process of Headspace SPME

During a headspace SPME extraction, there are three phases involved, the condensed phase, its headspace and the SPME polymer. Theoretical treatment of the diffusion process from the condensed phase to the SPME polymer film through the headspace is very complicated and the analytical solution to such a diffusion problem as headspace SPME is also very complex.[3]

In this section, a different approach is taken to tackle the problem of the mass

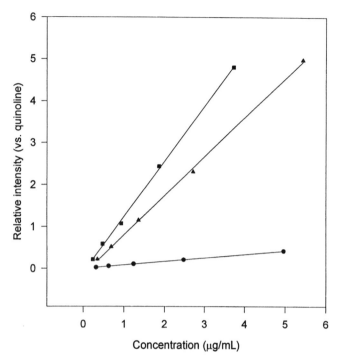

Figure 3   *Linear regressions of extracted amount* vs. *initial concentration in aqueous solutions agitated with magnetic stirring bar at room temperature.* ● *2-phenylethanol (r = 0.9998);* ■ *2,4-dimethylphenol (r = 0.9995);* ▲ *eugenol (r = 0.9996)*

transfer process in headspace SPME.[7] There should be two interfaces, the condensed/headspace interface and the headspace/polymer interface respectively, involved in the mass transfer during headspace SPME. The focus of the approach is on the mass transfer at these two interfaces. At the bulk of the headspace, it is assumed that there is no concentration gradient. This assumption is reasonable because diffusion coefficients are usually more than a hundred thousand times higher in a gas phase than in a condensed phase. If a steady-state mass transfer is attained, the mass flow rates at the above mentioned two interfaces should be equal to each other.

At the headspace/polymer interface as is shown in Figure 4, Fick's first law of diffusion describes the mass flow rate. Similar to equations 1 and 4, the extraction rate of the analyte is proportional to the mass flow rate at the interface. We have:

$$\frac{1}{A_f}\frac{dn}{dt} = m_f(C_f - C_f')$$
(13)

At the condensed/headspace interface, the net analyte transfer from the condensed phase to its headspace is related to the headspace concentration

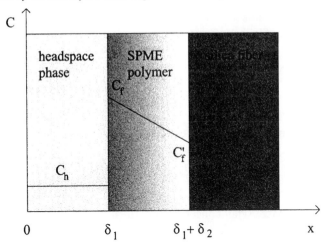

Figure 4 *Interface of headspace/SPME polymer film. The concentration gradient in headspace phase is negligible. A linear diffusion layer is assumed in the polymer film phase*

variations. Before SPME sampling, equilibrium exists between the condensed phase and its headspace for the analyte. There is no net mass transfer at its interface. During a headspace SPME sampling, analytes in the headspace are extracted into the SPME polymer film. The original equilibrium of the analyte between the condensed phase and its headspace before SPME extraction is disturbed. Once the analyte concentration in the headspace is depleted due to the SPME absorption, the analyte in the condensed phase will evaporate to the headspace. Therefore, the driving force of net analyte evaporation is its headspace concentration deviation from the equilibrium value. Let C_h^0 be the headspace concentration of the analyte at equilibrium and C_h be the analyte concentration in the headspace during the SPME sampling, the driving force of net analyte transfer at the condensed/headspace interface is $C_h^0 - C_h$. Thus the evaporation rate of the analyte at this interface can be assumed to be proportional to $C_h^0 - C_h$:

$$r = k(C_h^0 - C_h) \tag{14}$$

Here r is the evaporation rate of the analyte and k is its evaporation rate constant. k is dependent on a lot of factors such as temperature, surface area and agitation conditions of the condensed phase.

When a steady-state SPME extraction is attained, the mass flow rate at the condensed/headspace interface equals that at the headspace/polymer interface. Therefore, the evaporation rate should be proportional to the extraction rate. Combining equations 13 and 14, we have:

$$\frac{dn}{dt} = A_f m_f (C_f - C_f') = k V_h (C_h^0 - C_h) \tag{15}$$

Let K_{hs} be the equilibrium partition constant of the analyte between the condensed phase and its headspace. We have:

$$K_{hs} = \frac{C_h^0}{C_s} \quad \text{or} \quad C_h^0 = K_{hs} C_s \tag{16}$$

At the headspace/polymer boundary, equilibrium is assumed to attain quickly between the polymer surface and the headspace. We have:

$$C_f = K_{fh} C_h \tag{17}$$

The analyte in the headspace and in the SPME polymer film are originally from the condensed phase. The concentration of the analyte remaining in the condensed phase can be expressed as:

$$C_s = C_0 - \frac{n + C_h V_h}{V_s} \tag{18}$$

Using equations 5 and 15–18 to express $C_h^0 - C_h$ in terms of C_0 and n, and substituting the result into equation 15, we have:

$$\frac{dn}{dt} = \frac{2 A_f m_f k K_{fh} K_{hs} V_h V_s}{2 A_f m_f K_{fh} V_s + k K_{hs} V_h^2 + K V_h V_s} C_0 - \left(\frac{2 A_f m_f k V_h}{V_f} \right)$$
$$\times \left(\frac{K_{fh} K_{hs} V_f + K_{hs} V_h + V_s}{2 A_f m_f K_{fh} V_s + k K_{hs} V_h^2 + k V_h V_s} \right) n \tag{19}$$

Equation 19 can be solved with the initial condition: $n|_{t=0} = 0$

$$n = \left\{ 1 - \exp\left[-\left(\frac{2 A_f m_f k V_h}{V_f} \right) \left(\frac{K_{fh} K_{hs} V_f + K_{hs} V_h + V_s}{2 A_f m_f K_{fh} V_s + k K_{hs} V_h^2 + k V_h V_s} \right) t \right] \right\}$$
$$\times \frac{K_{fh} K_{hs} V_f V_s}{K_{fh} K_{hs} V_f + K_{hs} V_h + V_s} C_0 \tag{20}$$

Equation 20 is the analytical solution to the dynamic process of headspace SPME in the situation when a steady-state mass transfer between the two interfaces is attained. This equation correlates extracted amount, n, to the initial concentration, C_0, of an analyte in the condensed phase. The linear proportional relationship between n and C_0 exists when sampling time is held constant. The $n \propto C_0$ relationship in equation 20 meets the basic requirement of quantitative analysis for headspace SPME before reaching partition equilibrium. As sampling time t goes to infinity, the exponential term vanishes in equation 20 and it becomes equation 2 in Chapter 1. Once again, equation 20 derived from a pure dynamic process generates equation 2 in Chapter 1, an equation obtained from an equilibrium state.

Let

$$a_h = \left(\frac{2A_f m_f k V_h}{V_f}\right)\left(\frac{K_{fh} K_{hs} V_f + K_{hs} V_h + V_s}{2A_f m_f K_{fh} V_s + k K_{hs} V_h^2 + k V_h V_s}\right) \qquad (21)$$

Equation 20 becomes:

$$n = n^\infty[1 - \exp(-a_h t)] \qquad (22)$$

This is an equation similar to equation 10 with a more complicated parameter a_h (equation 21) and n^∞ (equation 2 in Chapter 1). Parameter a_h is dependent upon the mass transfer coefficient, the evaporation rate constant, partition constants and the physical dimension of the headspace SPME system. The larger the parameter a_h, the faster the partition equilibrium can be reached.

In the above treatment, mass transfer at both condense/gas and gas/polymer interfaces was taken into consideration as rate-determining steps. In reality, mass transfer in one of the interfaces may play the major role and become the bottle neck in the extraction process.

Diffusion in the SPME Polymer Film as the Rate-determining Step

When analyte diffusion in the SPME polymer film is very slow compared to the analyte evaporation from the condensed phase to its headspace, we have $k \gg m_f$. The parameter a_h defined by equation 21 can be simplified as:

$$a_h = 2A_f m_f \frac{K_{fh} K_{hs} V_f + K_{hs} V_h + V_s}{K_{hs} V_f V_h + V_f V_s} \qquad (23)$$

It is independent from the evaporation rate constant k. This applies to the headspace SPME of very volatile chemicals. Organic chemicals become more volatile at elevated temperatures and this equation may describe headspace SPME when the condensed phase is heated above ambient temperature.

Evaporation from the Condensed Phase as the Rate-determining Step

If the evaporation of the analyte from the condensed phase to its headspace becomes the rate-determining step, the mass transfer at the gas/polymer interface is considered a relatively fast process. In this case, we have $k \ll m_f$. The parameter a_h defined by equation 21 is simplified as:

$$a_h = k V_h \frac{K_{fh} K_{hs} V_f + K_{hs} V_h + V_s}{K_{fh} V_f V_s} \qquad (24)$$

It is not related to the analyte diffusion in the SPME polymer film. This describes a situation when the analyte has a slow evaporation rate.

At room temperature, less volatile chemicals have relatively low volatility. Their evaporation rates are very low. Their concentrations in the gas phase are far too small compared with their concentrations in the condensed phase. That means $C_h \ll C_s$ and $C_0 \approx C_s$. If the sampling time t is not long, then we have $C_h^0 - C_h \approx K_{hs}C_0$ and equation 15 becomes

$$\frac{dn}{dt} \approx kK_{hs}V_hC_0 \tag{25}$$

Solve this equation with the initial condition: $n|_{t=0} = 0$

$$n = kK_{hs}V_hC_0t \tag{26}$$

The extracted analyte n is not only directly proportional to C_0, but also directly proportional to the SPME sampling time t.

Equation 26 can also be derived directly from equation 22. If k is very small, and V_h and t are not very large, we have $a_h t \gg 1$. Equation 22 becomes:

$$n = n^\infty[1 - \exp(-a_h t)] \approx n^\infty a_h t \tag{27}$$

Substituting a_h with equation 24 and n^∞ with equation 2 in Chapter 1, we have equation 26.

The solid line in Figure 5 shows the extraction time profile of headspace SPME of 2,4-dimethylphenol in an aqueous solution in a 5 mL vial at room

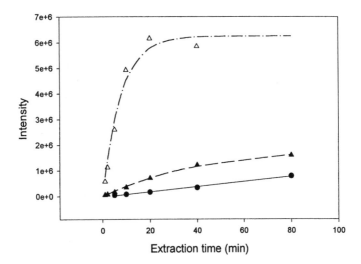

Figure 5 *Extraction time profiles of 2,4-dimethylphenol in an aqueous solution (0.78 µg mL^{-1}). ● Headspace SPME at room temperature with 4 mL V_h and 1 mL V_s. Solid line is the linear regression (r = 0.997). ▲ Headspace SPME at room temperature with 40 mL, V_h and 15 mL V_s. Dashed line is the fit with equation 22. △ Headspace SPME at 80 °C with 4 mL V_h and 1 mL V_s. Dash-dotted line is the fit with equation 22*

temperature. The extracted analyte n is linearly proportional to the sampling time t up to 80 min of extraction. The linear correlation coefficient (r) is better than 0.99. This is the situation described by equation 26. At room temperature, the evaporation of the chemical from the aqueous phase to its headspace is a very slow process.

If a larger container is used, both the condensed/headspace interface area and V_h are larger. A larger evaporation surface area leads to a larger k. Then, the extraction time profile should not follow equation 26 but equation 22. The dashed line in Figure 5 shows the headspace SPME time profile of the same aqueous solution in a 50 mL Erlenmeyer flask. The aqueous solution was agitated with a magnetic stirring bar to enhance analyte evaporation. The extraction time profile of 2,4-dimethylphenol is not a straight line any more but shows the shape that can be described by equation 22. Parameter a_h is 0.023 min^{-1} and $t_{95\%}$ is 130 min. As is predicted, when k and V_h increase, equation 22 instead of equation 26 is applicable to describe the headspace SPME process.

The dash-dotted line in Figure 5 shows the extraction time profile of headspace SPME over the same aqueous solution at 80 °C. Parameter a_h is 0.129 min^{-1} and $t_{95\%}$ is 23 min. Compared to room temperature headspace SPME, parameter a_h is much larger. Higher temperature accelerates the evaporation rate of the analyte from the aqueous phase. The time needed to reach the partition equilibrium is much shorter. This observation confirms that the evaporation of 2,4-dimethylphenol from the aqueous phase to the headspace at room temperature is the rate-determining step during the headspace SPME.

For all the circumstances discussed in this section, there is always a direct proportional relationship between n and C_0 no matter at which interfaces the mass transfer is the rate-determining step. The $n \propto C_0$ relationship meets the key requirement applying headspace SPME for quantitative analysis. Figure 6 shows the plots of n vs. C_0 for headspace SPME over aqueous solutions at 80 °C. The sampling time was held at 5 min, far shorter than the time required to reach partition equilibrium ($t_{95\%}$ for 2,4-dimethylphenol was 23 min). Straight lines are obtained as predicted. Linear correlation coefficients (r) are better than 0.999 for all the three chemicals. Headspace SPME quantification is feasible before reaching the partition equilibrium once the sampling time t is held constant.

4   Headspace SPME with a Non-steady-state Mass Transfer

In the previous section dealing with the dynamic process of headspace SPME, it was assumed that there is a balanced mass transfer between the two interfaces. The mass flow rate at the condense/headspace interface is equal to that at the headspace/polymer interface. Analyte concentration in the headspace, C_h, remains at a constant level. Once the SPME polymer film is exposed to the

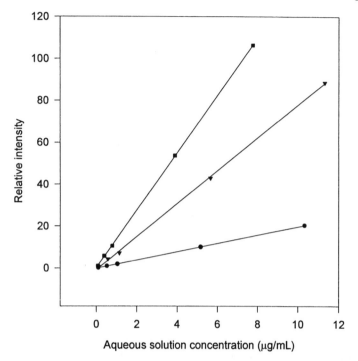

Figure 6 *Headspace SPME over aqueous solutions at 80 °C. Linear regression of extracted amount n vs. the initial concentration in aqueous solution C_0. The sampling time was 5 minutes.* ● *2-phenylethanol (r = 1.0000).* ■ *2,4-dimethyl-phenol (r = 1.0000).* ▲ *eugenol (r = 0.9998). r is the linear correlation coefficient*

headspace over a condensed phase, the disturbance and the following relaxation process may not follow the balanced mass transfer at the two interfaces. At the initial stage of the headspace SPME, such a mass flow balance is unlikely to attain in short period of time. There is a variation of headspace concentration of the analyte. The change of the headspace concentration as a function of time can be expressed as follows:

$$\frac{dC_h}{dt} = r - \frac{A_f}{V_h} m_f(C_f - C'_f) \tag{28}$$

Here r is the analyte evaporation rate as described in equation 14. It is the mass transfer rate at the condensed/headspace interface. It has a positive sign in equation 28 since the analyte evaporation increases its concentration in the headspace. The second term of the above equation is the mass transfer rate at the headspace/polymer interface as described by equation 13. It is the rate of analyte absorbed by the SPME polymer film. It has a negative sign because this process depletes the analyte concentration in the headspace. At the initial stage of SPME sampling, we have $C_s \approx C_0$ and $C_f' \approx 0$. Then equation 28 can be re-written as:

$$\frac{dC_h}{dt} = k(K_{hs}C_0 - C_h)\frac{A_f}{V_h}m_f C_f \tag{29}$$

Here C_f is the surface concentration of the analyte in contact with the headspace. There is a partition equilibrium at the headspace/polymer film boundary, $C_f = K_{fh}C_h$. Therefore we have:

$$\frac{dC_h}{dt} = kK_{hs}C_0 - \left(k + \frac{A_f m_f K_{fh}}{V_h}\right)C_h \tag{30}$$

Solve this equation with the initial condition: $C_h|_{t=0} = K_{hs}C_0$

$$C_h = a + b\exp(-ct) \tag{31}$$

with

$$a = \frac{kK_{hs}V_h}{kV_h + A_f m_f K_{fh}}C_0 \tag{32}$$

$$b = \frac{A_f m_f K_{fh}K_{hs}}{kV_h + A_f m_f K_{fh}}C_0 \tag{33}$$

and

$$c = k + \frac{A_f m_f K_{fh}}{V_h} \tag{34}$$

The headspace concentration varies according to equation 31 when headspace SPME extraction starts.

In order to obtain an expression related to the extracted amount n, equation 31 is used to express the analyte concentration on the SPME polymer surface: $C_f = K_{fh}C_h$. Use this expression along with equation 5 to express $C_f - C_f'$ in terms of n and t and substitute the result into equation 13, we have:

$$\frac{dn}{dt} + c_1 n = a_1 + b_1 \exp(-ct) \tag{35}$$

This is a first-order linear nonhomogeneous equation with:

$$a_1 = 2A_f m_f K_{fh}a \tag{36}$$

$$b_1 = 2A_f m_f K_{fh}b \tag{37}$$

$$c_1 = \frac{2A_f m_f}{V_f} \tag{38}$$

Solve this equation with initial condition: $n|_{t=0} = 0$

$$n = \frac{b_1}{c - c_1}[1 - \exp(-ct)] + \left(\frac{a_1}{c_1} - \frac{b_1}{c - c_1}\right)[1 - \exp(-c_1 t)] \tag{39}$$

If $\alpha = \dfrac{b_1}{c - c_1}$ and $\beta = \dfrac{a_1}{c_1} - \dfrac{b_1}{c - c_1}$, equation 39 can be re-written as:

$$n = \alpha[1 - \exp(-ct)] + \beta[1 - \exp(-c_1 t)] \tag{40}$$

The amount of the analyte extracted by the SPME polymer film can be expressed as the sum of two exponential terms. This equation is derived with the assumption that there is no steady-state flow of the analyte from the condensed phase to the SPME polymer film through the headspace. It is the real case at initial stages of headspace SPME. Parameters c and c_1 are two parameters to determine the relaxation period when the SPME headspace extraction starts. Parameter c, defined by equation 34, is a measure of how fast a steady-state mass flow from the condensed phase to the SPME polymer film can be attained. It is related to both the evaporation of the analyte from the condensed phase to its headspace and the diffusion of the analyte in the SPME polymer film. Parameter c_1, defined by equation 38, is only related to the diffusion of the analyte in the SPME polymer film.

As we can see from equations 32, 33, 36 and 37, both parameters a_1 and b_1 are proportional to C_0. Therefore we have $\alpha \propto C_0$ and $\beta \propto C_0$. Thus we have $n \propto C_0$ in equation 40 when sampling time is held constant. This $n \propto C_0$ relation makes headspace SPME feasible for quantitative analysis even before a steady-state mass flow of the analyte from the condensed phase to the SPME polymer film is established.

Matich *et al.*[8] applied SPME for quantitative headspace sampling of apple volatiles. They found that the experimental data fit an expression with double exponential terms (equation 40) better than the model with a single exponential term. In this section, theoretical treatment of the non-steady-state flow dynamics of headspace SPME results in equation 40, which has two exponential terms and provides a better fit to the experimental data of Matich *et al.*

5   Conclusion

Application of SPME for quantitative analysis is proved to be feasible no matter whether the partition equilibrium is retained or not. When partition equilibrium is reached, the extracted amount of an analyte is proportional to its initial concentration in the sample matrix phase as described in equations 2 and 5 in Chapter 1 for headspace SPME and direct aqueous phase SPME, respectively. Such linear proportional relationships between n and C_0 hold before partition equilibrium as is proved in this chapter (equations 9, 20 and 40). Compared with the equilibrium situation, the amount of extracted analyte is dependent upon the sampling time in non-equilibrium situations and the time dependence can be expressed with a simple exponential term. Once the sampling time is held constant, the extracted amount is proportional to the initial concentration of the analyte in the sample matrix. The quantitative relationship between n and C_0 in non-equilibrium situations broadens the applications of SPME and shortens the experimental time if the sensitivity of the analysis is not the major concern.

References

1. D. Louch, S. Motlagh and J. Pawliszyn, *Anal. Chem.*, 1992, **64**, 1187.
2. Z. Zhang and J. Pawliszyn, *Anal. Chem.*, 1993, **65**, 1843.
3. J. Pawliszyn, *Solid Phase Microextraction, Theory and Practice*, Wiley-VCH, New York, 1997.
4. J. Crank, *The Mathematics of Diffusion*, Clarendon Press, Oxford, 1975.
5. J. Ai, *Anal. Chem.*, 1997, **69**, 1230.
6. C.L. Arthur and J. Pawliszyn, *Anal. Chem.*, 1990, **62**, 2145.
7. J. Ai, *Anal. Chem.*, 1997, **69**, 3260.
8. A.J. Matich, D.D. Rowan and N.H. Banks, *Anal. Chem.*, 1996, **68**, 4114.

Coatings and Interfaces

CHAPTER 3

SPME Coupled to Capillary Electrophoresis

CHEN-WEN WHANG

1 Introduction

Over the past decades, new advances in microcolumn separation techniques made by analytical chemists have generated significant enhancement in resolving capability and speed of analysis, in comparison with the conventional chromatographic techniques using wide bore columns. On the other hand, these high-resolution techniques are very susceptible to interferences present in complex sample matrices, particularly in the analysis of biological and environmental samples. Since most sophisticated instruments cannot handle the matrix directly, insufficient sample cleanup will cause rapid deterioration of analytical systems, thereby precluding reliable results. Generally, analysis time, sample size and quality of the analytical results are much more influenced by sample preparation rather than by the separation step followed, and therefore sample preparation becomes an indispensable part of the analytical method in nearly every case. A sample preparation step not only isolates the components of interest from a sample matrix, it also brings the analytes to a suitable concentration level. While preconcentration may be the most important factor for clean, dilute samples, interferences removal (or selectivity) is more important in the analysis of complex samples. The development of sample preparation methods suitable for a class of target analytes, type of matrix and analytical technique used is a significant challenge for the analytical chemist.

In recent years, solid phase microextraction (SPME)[1] has been developed and successfully applied to analyse a variety of analytes in different matrices, including air, water, soil, and biological samples. SPME can be regarded as a combined sampling and sample concentration technique, requiring very simple and inexpensive devices. The basic concept of SPME resides in absorbing the target analytes from the sample matrix onto a thin coat of polymer attached to a fused silica fiber. For protection the fiber is mounted in a syringe-like device. During the extraction stage, the fiber is exposed to the sample matrix and the analytes partition between the sample and the coating until an equilibrium is reached. The amount of analyte extracted in SPME is dependent on the partition coefficient of an analyte between the coating and the sample matrix.[2] After extraction, the fiber is withdrawn into the protective needle of the syringe which enables a fast, simple transfer of extracted analytes to analytical instrumentation. For gas chromatographic (GC) analysis, the fiber is introduced into the GC injector port where the volatile analytes can be thermally desorbed from the coating. For high performance liquid chromatographic (HPLC) analysis, the semi- or non-volatile analytes are released into an HPLC column by solvent desorption *via* an external interface constructed from a 'T' joint.[3]

Capillary electrophoresis (CE) is one of the recently developed microseparation techniques which has been widely accepted as a fast, powerful and efficient analytical separation technique. CE is characterized by the use of narrow bore, small volume capillaries, typically 10–100 μm i.d. and 20–100 cm in length. Sample volumes of 1–100 nL are injected directly into the separation capillary, which dramatically reduces system dead volumes because injector valves or column connectors are not required. Application of a high voltage (5–30 kV) across the CE capillary, which contains a suitable electrolyte or buffer solution, causes analytes to migrate and effects analyte separation. However, the small capillary dimensions encountered and the minuscule sample zone generated in CE often lead to a relatively poor concentration limit of detection (CLOD). Several preconcentration strategies, such as using a solid-phase concentrator[4] or membrane preconcentration cartridge[5] on-line connected with the CE capillary, have been developed. These devices usually consist of an adsorptive phase at the inlet of the CE capillary and serve to enrich trace levels of analytes, as well as allow on-line sample cleanup prior to CE analysis. However, while this approach enables the analysis of much larger sample volumes (up to 100 μL) than conventional CE, the devices sometimes suffer from clogging problem with complex samples and injection band broadening caused by the excess column connections and solvent washing procedures.

Because SPME is an ideal sampling and sample preparation technique for microcolumn separation, both off- and on-line hyphenation between SPME and CE have been attempted. The extraction stage in SPME–CE is the same as that in SPME–GC or SPME–HPLC, but the desorption stage is different. In off-line SPME–CE method, no interfacing device is required and the analytes are desorbed from the SPME fiber into a small sample vial using a minimum amount of an appropriate desorption solvent, followed by electrokinetic or hydrodynamic injection into a CE capillary. On the other hand, an interface is generally required in on-line coupled SPME–CE, and the analytes-bearing fiber is directly connected to the injection end of a CE capillary *via* the interface.

2   Off-line Connection of SPME and CE

Li and Weber[6] reported an off-line combination of SPME and CE for the determination of barbiturates. A plasticized poly(vinyl chloride) (PVC)-coated steel rod (1.1 mm diameter) is used to extract the analytes from an aqueous sample. A Teflon tube, with an inside diameter just larger than the PVC-coated extraction rod, is terminated at one end by a syringe. A few microliters of back-extraction aqueous buffer is placed into the open end of the Teflon tube. After exposure to sample, the PVC-coated extraction rod is inserted into the back-extraction solution-containing tube and the analytes are back-extracted into the aqueous buffer inside the tube. The extraction rod is taken out after a specific time. The back-extraction solution droplet is then collected into an injection vial, followed by conventional CE analysis.

The use of a commercial poly(acrylate) (PA)-coated silica fiber for collection of toxic drugs (*e.g.* nitrazepam, flunitrazepam and triazolam) from a urine sample has been reported by Jinno *et al.*[7] After extraction, the fiber with the concentrated drugs is transferred to a capillary tube containing 20 μL of the acetonitrile desorption solution and maintained for 30 min to desorb the extracted drugs. The desorption solution is then electrokinetically injected into a capillary for separation by micellar electrokinetic chromatography (MEKC).

Since there is no special interface required for off-line connection of SPME and CE, extra-column band broadening will not be a matter of concern, and the high efficiency of CE can be preserved intact. However, off-line operation is generally more laborious and difficult to automate. In addition, the analyte preconcentration factor is low because only a minimal fraction of the desorbed analytes is actually injected into the CE capillary for separation and detection.

3   On-line Connection of SPME and CE

Despite the fact that liquid mobile phase is used in both CE and HPLC, it is difficult to couple SPME on-line to CE as is done with SPME–HPLC, considering that the injection volume in CE is of the order of nanoliters compared with microliters for HPLC. Efficiency loss due to extra-column band broadening can be very significant with an improperly designed interface. Therefore, realization of a zero-dead volume connection will be the prerequisite for a successful hyphenation between SPME and CE. Recently, Nguyen and Luong[8] reported the first on-line connection of SPME and CE. Polycyclic aromatic hydrocarbons (PAHs) in aqueous sample are first extracted onto a 150 μm o.d. $\times$ 1 cm length poly(dimethylsiloxane) (PDMS)-coated glass fiber, which is then attached to the injection end of a 50 μm i.d. $\times$ 350 μm o.d. separation capillary *via* an adapter. The absorbed PAHs are released into CE buffer stream by injecting a short plug of methanol, followed by cyclodextrin-modified CE separation. However, because the analytes are desorbed outside the injection end of the capillary during the solvent desorption stage, a zero-dead volume connection is not achieved. Significant variations on migration time and peak area for several PAHs are sometimes observed.

One way to eliminate the problem of extra-column dead volume is to position the analyte-bearing fiber inside the injection end of the separation capillary. With the analytes desorbed directly in the capillary, a zero-dead volume connection can be accomplished. Figure 1 shows the detailed construction of a

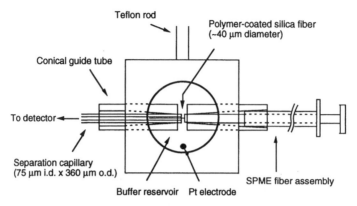

Figure 1 *Schematic diagram of the SPME–CE interface*

direct SPME–CE interface based on the above principle.[9] With the aid of two perfectly aligned inner conical guide tubes, direct insertion of a thin SPME fiber into the injection end of the CE capillary can be achieved easily and reliably. However, commercial SPME fibers for CE analysis are not available at present. In addition, the diameters of SPME fibers for GC or HPLC analysis (100–300 μm) are too large to fit in with the commonly used CE capillaries (10–100 μm i.d.). Therefore, thin fiber with an appropriate diameter has to be custom-made by the user. Fabrication of a thin SPME fiber can be easily performed by stripping the original polymer buffer or cladding from a short piece of optical fiber containing a silica core, followed by etching the exposed silica fiber down to a specific diameter using aqueous HF solution. The thin silica fiber is then dipped in a concentrated organic solvent solution of the coating material to be deposited. After removal of the fiber from the solution, the solvent is evaporated by drying and a thin layer of deposited material is formed. Figure 2a shows the electropherogram of 10 priority pollutant phenols obtained using on-line coupled SPME–CE analysis. A 40 μm diameter PA-coated silica fiber is used to extract phenols as neutral species from an acidic (pH 2) and salt-saturated aqueous sample for a specific time. Before transferring the analyte-bearing fiber to the interface, a short solution plug of 0.2 M NaOH is injected into the inlet end of a buffer-filled CE capillary and the capillary is positioned in the interface through a guide tubing. The fiber is then transferred to the interface and is carefully inserted into the NaOH-filled end of capillary through the opposite guide tubing (see Figure 1). In the presence of a strong base, phenols rapidly ionize as phenolates and desorb from the fiber into the solution. After filling the reservoir with running buffer, high voltage is applied to effect electrophoresis. Sharp and symmetrical peaks for all analytes are observed in Figure 2a, which clearly demonstrates the zero-dead volume characteristic of the interface as well as the fast desorption kinetics of phenols from the fiber coating into the liquid phase. With an ultraviolet (UV) absorbance detector, the CLOD for pentachlorophenol obtained by on-line

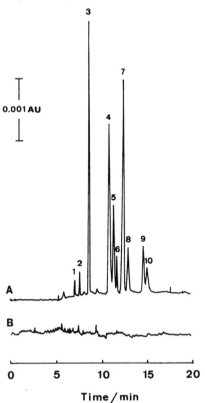

Figure 2 *Separation of 10 priority pollutant phenols by SPME–CE using* (a) *a PA-coated silica fiber and* (b) *a bare silica fiber: (1) 2,4-dimethylphenol, (2) phenol, (3) 4-chloro-3-methylphenol, (4) pentachlorophenol, (5) 2,4,6-trichlorophenol, (6) 2-methyl-4,6-dinitrophenol, (7) 2,4-dichlorophenol, (8) 2-chlorophenol, (9) 4-nitrophenol, (10) 2-nitrophenol*

coupled SPME–CE is 2 ppb, which is about two orders of magnitude lower than that obtained using the conventional CE–UV detection without sample pre-concentration. SPME is therefore effective in concentrating certain analytes present in a dilute sample for CE analysis.

It is also interesting to note that without a PA-coating, the bare fused silica fiber cannot extract sufficient amount of phenols from an aqueous medium for detection, despite its polar characteristic (Figure 2b). Obviously, it is crucial in SPME to use the appropriate coating for a given application.

Recently, CE in narrow bore (2–75 μm i.d.) capillaries has become an important technique for bioseparation. In addition, CE coupled with highly sensitive laser-induced fluorescence (LIF)[10,11] or an electrochemical (EC)[12,13] detector is ideally suited for single-cell analysis. Among the various analytes commonly investigated in single-cell analysis, catecholamines are an important group of neurotransmission compounds. However, extraction of catechol-

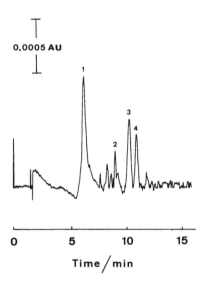

Figure 3 *Electropherogram of serotonin and catecholamines obtained by SPME–CE using a Nafion-coated silica fiber: (1) serotonin, (2) dopamine, (3) epinephrine, (4) norepinephrine*

amines from aqueous media is difficult owing to their high solubility in water. On-line cell injection by either electroosmosis[12] or hydrodynamic flow[11] is generally employed for the determination of catecholamines in a neuronal cell. On the other hand, SPME with an ultrathin fiber of micro- or sub-micrometer diameter may be an alternative way of sample preparation/injection method for single-cell analysis. Preliminary results show that catecholamines may be collected by a cation-exchanger-coated fiber from a neutral solution. Figure 3 illustrates the separation of serotonin and three catecholamines (*viz.* dopamine, epinephrine and norepinephrine) by on-line coupled SPME–CE using a Nafion®-coated silica fiber.[14] Owing to the small partition coefficients between the Nafion® coating and the aqueous matrix, the recovery of catecholamines is relatively poor. A more specific coating material or a different extraction strategy is probably needed to improve the recovery.

4   On-line Connection of SPME and MEKC

Apart from SPME–CE applications, the interface shown in Figure 1 should also find use in on-line connection between SPME and other microcolumn separation techniques, such as MEKC or capillary electrokinetic chromatography (CEC). Figure 4 shows the electropherogram of phenols obtained by on-line coupled SPME–MEKC. Six phenols are first extracted as neutral species from an acidic (pH 2), salt-saturated aqueous solution using a PA-coated silica fiber. The absorbed phenols are then desorbed into the capillary by a small amount of methanol, followed by MEKC separation using a phosphate/borate buffer

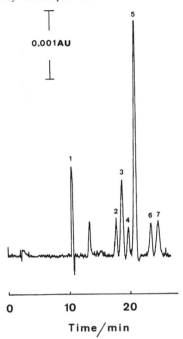

Figure 4 *Separation of phenols by SPME–MEKC: (1) methanol, (2) 2-chlorophenol, (3) 2,4-dichlorophenol, (4) 2,4,6-trichlorophenol, (5) pentachlorophenol, (6) phenol, (7) 4-chloro-3-methylphenol*

containing sodium dodecyl sulfate (SDS). Various phenols are separated according to their differential partitioning between a mobile phase and a pseudostationary phase formed by the SDS micelle. Because the separation mechanism in MEKC is different from that in CE, the migration orders of various phenols in Figure 4 (obtained by SPME–MEKC) and Figure 2 (obtained by SPME–CE) are totally different.

Acknowledgements

The author is indebted to Professor J. Pawliszyn of the University of Waterloo for his assistance during this work. The author also acknowledges a fellowship from the National Science Council of the Republic of China. The experimental results presented herein were obtained at the Department of Chemistry, University of Waterloo.

References

1. J. Pawliszyn, *Solid Phase Microextraction. Theory and Practice*, Wiley-VCH, New York, 1997.
2. C.L. Arthur and J. Pawliszyn, *Anal. Chem.*, 1990, **62**, 2145.

3. J. Chen and J. Pawliszyn, *Anal. Chem.*, 1995, **67**, 2530.
4. I. Morita and J. Sawada, *J. Chromatogr.*, 1993, **641**, 375.
5. A.J. Tomlinson, L.M. Benson, S. Jameson and S. Naylor, *Electrophoresis*, 1996, **17**, 1801.
6. S. Li and S.G. Weber, *Anal. Chem.*, 1997, **69**, 1217.
7. K. Jinno, Y. Han, H. Sawada and M. Taniguchi, *Chromatographia*, 1997, **16**, 309.
8. A.-L. Nguyen and J.H.T. Luong, *Anal. Chem.*, 1997, **69**, 1726.
9. C.W. Whang and J. Pawliszyn, *Anal. Commun.*, 1998, **35**, 353.
10. B.L. Hogan and E.S. Yeung, *Anal. Chem.*, 1992, **64**, 2841.
11. H.T. Chang and E.S. Yeung, *Anal. Chem.*, 1995, **67**, 1079.
12. T.M. Olefirowicz and A.G. Ewing, *Anal Chem.*, 1990, **62**, 1872.
13. T.M. Olefirowicz and A.G. Ewing, *Chimia*, 1991, **45**, 106.
14. C.W. Whang and J. Pawliszyn, unpublished results.

CHAPTER 4

Selectivity in SPME

SHU LI AND STEPHEN G. WEBER

1 Introduction

Selectivity has two meanings depending on the ability of a system to discriminate. When the ability to discriminate is high, as in a high-resolution separation or other high-resolution technique such as atomic spectroscopy, then selectivity describes the probability that a particular signal represents the desired analyte and not an interference. In techniques or operations in which the resolution is poorer, the expectation is that the separation of analytes and interferences is never complete. In extraction, which falls into the latter category, selectivity can be described as the ratio of the relative concentration of analyte and interferant in the two phases.[1] It is a relative enrichment factor. One objective of the design of new extraction methods is to increase the selectivity of the extraction for the group of analytes desired. Other objectives are to improve speed, to manage smaller samples, to improve the mechanics to allow automation, and to decrease the volume of waste solvents. SPME is a good candidate for development efforts because it meets the latter set of objectives, while allowing for some degree of control over selectivity.

There are both thermodynamic and kinetic aspects to selectivity. The thermodynamic driving force for molecular distribution between phases is the decrease of free energy that it experiences in the favored phase. The kinetic processes are foremost mass transfer. It is more common to control selectivity of extraction through control of the thermodynamic aspects. For an extraction involving a gas phase as one of the phases, the chemical interaction of the solute with the other phase dominates the distribution process at one temperature. As there is a large volume entropy change in going from a condensed phase to a gas phase, temperature is also important. Higher temperatures favor the gas phase. For an extraction involving two condensed phases, the chemical interactions of the solute with each condensed phase are important. Temperature probably has an influence on the partition coefficient as well. A considerable degree of insight can be gained into such processes by considering the cycle: solute$_{phase\,1}$ → solute$_{gas}$ → solute$_{phase\,2}$, as has recently been done for reversed phase HPLC.[2] Actually, this process has practical significance in SPME as well, as described below.

49

In many applications of SPME, the sample is a gas phase (head space analysis). In such a situation, the reasoning about selectivity and extractability is similar to that taken in GLC. A large number of studies using probe solutes has given a consistent picture of stationary phases in GLC.[3] The free energy of distribution between the gas phase and a liquid phase can be described as being a sum of products.

$$\log(K) = c + \sum_n a_{n1} a_{n2} \tag{1}$$

Here, K is an equilibrium constant for the distribution process and c is a constant. The terms a_{ni} ($i = 1$ or 2) reflect complementary properties of solvent (1) and solute (2). Examples are solute size and solvent enthalpy of vaporization per volume at the solvents boiling point (Hildebrand solubility parameter squared), or solute acidity and solvent basicity. For example, Li and Carr[4] correlated solute retention [log(k′)] on eight stationary phases of increasing 'polarity'. One of the solute properties was the logarithm of the partition coefficient for the solute between the gas phase and hexadecane, L^{16}. Another was its dipolarity, π^*_2. For the more nonpolar phases, the term $l \cdot \log(L^{16})$ (*i.e.* $a_{11} a_{12}$) was more important; l ranged from 0.769 to 0.606 ($T = $ 40–45 °C) as the phases became more 'polar'. At the same time, the importance of solute dipolarity increased, as expressed in the term $s \cdot \pi^*_2$ (*i.e.* $a_{21} a_{22}$). Values of s ranged from 0.4 to 2.5 at the same temperatures. This expresses the general tendency of solvents: more polar solvents tend to interact more strongly both with themselves and with solutes in comparison to more nonpolar solutes. The result is that formation of a cavity for a solute becomes harder, but the interaction free energy with a polar solute becomes more negative, as the solvent becomes more polar. For example, the solubility of nitrogen, argon, and ethane are lowered by roughly factors of 3, 4 and 6 in N-methylacetamide[5] compared with cyclohexane.[6] This reflects the polar, self-associating nature of the former solvent. As the solute size increases from nitrogen to ethane, the unrequited free energy cost of cavity formation increases.

Other means of understanding solubility are through structural and functional group contributions. For example, the functional group contributions to $\log(L^{16})$ have been determined.[7] It shows that selectivity among hydrocarbons is poor: regression coefficients for $-CH_2-$, $>CH-$, $>C<$, $-HC=$, $>C=$, $-C=$ are respectively 0.50, 0.49, 0.46, 0.47, 0.57, 0.60. Thus, the free energy for the solvation of all of these types of carbon is about the same in hexadecane. Selectivity among terminal carbons is better: CH_3-, $H_2C=$, $HC=$ are 0.33, 0.20, 0.08. Selectivity among halogen substituents is marked: F^-, Cl^-, Br^-, I^-: -0.11, 0.89, 1.29, 1.77. The values can be used to determine selectivity between homologs. Thus, while the data given show that selectivity of the partitioning for hexane and 2-hexene would be negligible, the partition coefficient for heptane is $10^{0.5}$ greater than that for hexane, *i.e.* the free energy is about -3 kJ mol^{-1} for each added CH_2 group. So, while there is little selectivity among hydrocarbons with the same number of carbons and similar branching, there is selectivity for compounds with more carbon. Many examples of this sort of correlation exist in the literature.

Liquid–liquid partitioning is best thought of as consisting of two stages with the intermediate, imaginary vapor phase. Often the partitioning of interest in SPME is from water to the SPME device. It is thus of interest to understand vapor phase/aqueous solution partitioning.[2,8] For small solutes (hydrogen to *n*-butane) the free energy of partitioning to water actually decreases with increasing solute size. In the series methane ... butane there is a slight increase in free energy of partitioning with an increase in the number of CH_2 groups such that the free energy increment is $+0.75$ kJ mol^{-1}.[9] The combination of the strong tendency of CH_2 to dissolve from the gas phase into hexadecane and the weak tendency for CH_2 not to dissolve in water leads to the overall strong tendency, nearly 4 kJ mol^{-1}, for CH_2 to partition from water into the gas phase.

Polar functional groups like water. In a short but information-packed review, Wolfenden describes several partitioning processes that are most relevant to biological molecules, that also apply to this discussion as well.[9] Gas phase to water partition coefficients for a number of small molecule models of functional groups common in biomolecules are given. They range remarkably from 0.1 for ethane and ethylene to $10^{8.5}$ for methylguanine.

McAuliffe[10] and Hansch[11] reported on the solubilities of organic compounds in water. McAuliffe,[10] and later Tanford,[12] related water solubility of nonpolar species to partitioning of solutes between a nonpolar liquid, *n*-octanol, and water. Wolfenden[9] has shown that the range of $\log_{10}$(partition coefficient) is lower in water/nonpolar solvent partitioning than water/solute vapor partitioning: the former is about 3 while the latter is about 5.5. Thus polar solutes prefer water over nonpolar liquids, and nonpolar liquids over gas phase.

The largest influence on partitioning to water is charge. As the Born equation will easily show, the enthalpy of dissolving a charged species in water (coming from the gas phase), or for that matter any solvent, is large and negative. Experimental data for the free energy of hydration of ions are indicative of this. The free energies (kJ mol^{-1}) for the hydration of K^+ and Ca^{2+} are -338 and -1593 respectively.[13] Just as the net dipole moment of a molecule is rarely a predictor of solute solubility, the net charge on a multiply charged molecule is not the controlling factor in solubility. Zwitterions, such as betaines, sulfobetaines, and glycine, are highly hydrophilic but neutral.

2 The Selectivity of SPME

The selectivity of SPME can be considered from the process of solid phase microextraction itself. The first step is sample extraction from matrices to a solid or liquid membrane, where the analyte–membrane and analyte–solvent interaction can be tailored to suit to selectivity; the second step is the desorption of sample from membrane to the instrument, where the selectivity can be increased by carefully choosing desorption conditions.

The selection of the extracting phase is the most important step governing the selectivity. The original SPME[14] used thermally activated polyimide film and uncoated fused silica as the extracting phases. The polyimide had to be treated at 350 °C prior to the extraction. Both can be applied to the determination of volatile chlorinated organic compounds in water. Benzene, toluene, ethylbenzene, and xylenes are also able to adsorb on the bare silica fiber. Several kinds of coatings have become commercially available, namely poly(dimethylsiloxane) (PDMS), poly(acrylate) (PA), poly(dimethylsiloxane)/divinylbenzene (DVB), Carboxen/PDMS, Carbowax/template resin fibers. Among those, PDMS and PA coating are the most well-studied and characterized coatings. For a specific application, the coating is chosen based on the polarity of the analyte.[15] PDMS is less polar than PA; thus it is widely used for the extraction of non-polar compounds, such as substituted benzenes[16,17] and polyaromatic hydrocarbons.[18] For polar compounds like ketones and alcohols, polar coatings like PA and Carbowax work better. The selectivity difference of polar and nonpolar coatings has been studied through analyzing organophosphorus insecticides[19,20] and Maillard reaction products.[21] Among the organophosphorus compounds tested, triazophos and methylparathion have a larger affinity for PA than for PDMS, while diazinon and prothiofos have comparable or even higher affinity to PDMS. Volatile aldehydes, pyrazines, pyridines and thiazoles were extracted by PDMS and Carbowax/divinylbenzene (CWDVB). Although these fibers show different selectivity to different groups of compounds, both fibers extract solutes with more alkyl substitutions better for all four groups of compounds. These results coincide with results for some pesticides, including *N*-substitued amines, *N*-heterocyclic compounds and organophosphorus compounds. They are extracted with PA.[22] A linear correlation of extraction efficiency (relative peak response after SPME to the standard chromatogram) is obtained to the octanol–water partition coefficient for some triazine pesticides.

The sample matrix can affect the SPME process in many ways and sometimes in such a drastic way that a several hundred percent error will be generated by ignoring it. SPME is particularly sensitive to matrix effects, because it is a microextraction: the extraction of solute does not necessarily shift equilibria towards more free, unbound solute as would occur in an 'ordinary' extraction. Different matrix problems exist from different sample origins, for example, salts from sea water,[19] alcohols from wine or industry products, proteins from biological samples. In order to reduce negative matrix effects, sometimes a new component may have to be added to the sample. For example, salts are sometimes added in order to reduce protein–drug interactions. However, the difficulty is that the matrix is usually not reproducible even for samples from the same origins.

The effects of salt added to the sample on SPME have been tested on many occasions. The variation in ionic strength or salt concentration may change the extraction efficiency. For some analytes, unsalted solution is not significantly different from the same solution but with sodium chloride added up to a concentration of 10% at the 95% confidence level.[14] Further studies showed that adding NaCl increased the extraction of some compounds, such as phenols,[23] fatty acids[24] and pyrazines.[25] Adding NaCl to saturation will increase the extraction of triazine and aniline pesticides but decrease the extraction of organophosphorus pesticides by PA. The largest effects are observed for the most polar compounds.[22] However, the extraction of organophosphorus compounds from sodium chloride solutions of concentration ranging from 0 to 40% in 10% intervals shows that the trend is not obvious: the initial decrease in extraction by adding salt may be followed by an increase at some concentration.[20] Adding sodium perchlorate (0.2 g ml^{-1}) increased the extraction of most of the Maillard reaction products. However, different increases are observed for PDMS and CWDVB fibers, and significant decreases in the extraction of thiazole, pyridine and acetylpyrazine were observed when using PDMS for extraction.[21]

The effect of dissolved organic carbon generally decreases the extraction effeciency.[19] The presence of 15 mg l^{-1} SDS increases the recoveries of fenthion (110%) and ethion (120%).[19] The presence of methanol even below 0.5% results in a significant reduction in the recovery of amphetamines ($>60\%$).[26] The reduction caused by methanol was also observed for the extraction of pesticides such as triazine.[22] The influence of ethanol on the extraction of pesticides by PDMS was studied.[27] Again, a reduction of the extraction efficiency was observed. The reduction was much more important in the concentration range 0–5% than for 5–15% ethanol. The polarity[28] of the solution does not completely explain the observations.

Since SPME consumes little sample and it is more easily operated than SPE, there will be more and more biological samples prepared by this method. However, there are difficulties to be faced. Sample matrices from biological samples are very complicated. Macromolecules such as proteins are likely to adsorb on to the surface of the microextraction device, thus changing the surface properties of the coating. Some proteins that are known to bind drugs can shift the extraction equilibrium greatly. SPME has been used to extract amphetamines from urine. The analyte is volatile so that head space extraction is used to increase the selectivity. External calibration with spiked urine gave poor precision even after controlling pH and ionic strength.[26] Antidepressant drugs in human plasma have been analyzed by SPME–GC. The influence of plasma protein has been discussed in detail.[29] The recovery of drugs such as imipramine or amitriptyline varied from 15% in water to 0.3–0.8% in plasma. The low recovery was ascribed to the protein–drug binding. Tuning of the selectivity by adding salt or organics to the sample is probably worthwhile. The effect, however, is usually mild if the penalty of sacrificing the preconcentration effect is to be avoided. Nonconventional coatings with a large variety of selectivities could alleviate the problem.

Silica bonded phases have been used for a long time in HPLC separations in solid phase extraction. Recently they have also been used in SPME. The silica bonded phase was glued to the surface of fiber in multiple layers and applied to the extraction of PAHs.[30,31] Phenyl bonded phases have a larger affinity for the substituted benzene and naphthalene than other C_8, monomeric C_{18} and polymeric C_{18} phases. Furthermore, the bonded phases of different alkyl chain lengths have different selectivities to PAHs of different size.

Various organic solvents have different affinities to different molecules, and in principle these differences could be exploited to improve the range of selectivities available in SPME. The difficult part is how to coat rods with solvents and have a reasonably large phase volume. Poly(vinyl chloride)-plasticized membranes have been used as a supports for ionophores in ion selective electrodes for a long time. They are easily made by casting a THF solution of PVC and plasticizer. Plasticizers are organic compounds which usually have higher molecular weights than normal organic solvents. It has been shown that the solvent properties of such a membrane are dependent on the plasticizers.[32] A large variety of plasticizers is available. Studies have shown that their properties vary from strong hydrogen bonding to weak hydrogen bonding, from polar to nonpolar, as shown from using the Kamlet–Taft π^* method to correlate partitioning and other free energies.[32] A suitable plasticizer is easily chosen based on the molecular structure and property of the analyte. For example, for the extraction of barbiturates, which are strong HB donors, a strong HB acceptor plasticizer can be chosen to increase the selectivity.[33] The advantage with plasticized PVC is that a large variety of possible coatings is available. The disadvantage is they are usually not as rugged as the other membranes because they are not covalently crosslinked and care must be taken both during preparation and extraction.

With a limited number of available phases, it will not be possible to selectively extract every class of analyte. By making the target analytes distinct from other solutes, selectivity can be improved. One way of doing this is by derivatization. Acids and phenols can be derivatized in gaseous or aqueous matrices by adding derivatizing reagent to the matrix.[23, 24] Amines can also be derivatized.[35] For air analysis, less polar amides are produced by reacting amines with 4-nitrophenyl trifluoroacetate. For aqueous samples, pentafluorobenzylaldehyde can be used to convert primary amines into less polar forms. A new method is to adsorb the derivatizing reagent in the SPME fiber before using it to extract acids in the gaseous or aqueous phase.[36] In this method, both extraction and reaction happen simultaneously on the SPME fiber, and SPME offers a support for both the derivatizing reagent and reaction products. In a similar way, long chain fatty acids can be adsorbed to the fiber first and then derivatized by the volatile derivatizing reagent in the gas phase.[24] Methylating ion pair reagents (tetramethylammonium hydrogen sulfate or tetramethylammonium hydroxide) and C_{10}–C_{22} acids can be extracted by PA fibers. The methyl esters of these acids are formed in the gas phase at the elevated temperature of the injection port and subjected to the subsequent separation.[24] Thus derivatization can be achieved for several analytes and in almost every stage of SPME analysis for acids. By increasing the hydrophobicity, polar analytes like acids can be extracted better and the subsequent separation or detection is improved.

Molecular selectivity is needed when large amounts of hydrophobic interferences exist in the sample matrix. Developments in SPME using artificial receptors,[37–39] immunosorbents[40] and molecularly imprinted polymers[41] are certain to expand. A preliminary study[42] showed that when an artificial receptor to barbiturates is dissolved in a plasticized PVC membrane, selectivity and yield are greatly improved. Only molecules having the structure with which the receptor is designed to interact show a large improvement in extraction yield in the presence of the receptor. Since the extraction of analytes can be enhanced by adding receptor, the supporting medium where the receptor is doped can be chosen based on the interference in the matrix. A good supporting medium in this case would extract few matrix components and be compatible with the receptor. Techniques involving molecular recognition will be more and more important in extraction methods, including SPME.

We have already mentioned for all the solid phase extraction methods, extraction is only one of the analytical steps. Back extraction steps are needed to desorb the analyte from the solid phase. However, it is possible to desorb the analyte selectively. For the thermal desorption of volatile samples, temperature-programmed desorption is one of the choices.[43] Carefully chosen solvents are expected to help increase the selectivity of solvent desorption, too. Yet in most cases, since the most difficult part is to adsorb the analytes in relatively large quantity, this second step is not the subject of much scrutiny.

Acknowledgement

We are grateful to the National Science Foundation for support of this work.

References

1. J.C. Giddings, *Unified Separation Science*, Wiley Interscience, New York, 1991.
2. P. Carr, J. Li, A.J. Dallas, D.I. Eikens and L.C. Tan, *J. Chromatogr.*, 1993, **656** 113.
3. E. Forgás and T. Cserháti, *Molecular Basis of Chromatographic Separation*, CRC Press, Boca Raton, Ch. 2, 1997.
4. J. Li and P.W. Carr, *J. Chromatogr. A*, 1994, **659**, 367–80.
5. R.H. Wood and D.E. Delaney, *J. Phys. Chem.*, 1968, **72**, 4651.
6. J.H. Dymond, *J. Phys. Chem.*, 1967, **71**, 1829.
7. P. Havelec and J.G.K. Sevcik, *J. Chromatogr. A*, 1994, **677**, 319–329.
8. Recently reviewed: W. Blokzijl and J.B.F.N. Engberts, *Angew. Chem., Intl. Ed. Engl.*, 1993, **32**, 1545.
9. R. Wolfenden, *Science*, 1983, **222**, 1087.
10. C. McAuliffe, *J. Phys. Chem.*, 1966, **70**, 1267.
11. C. Hansch, J.E. Quinlan and G.L. Lawrence, *J. Org. Chem.*, 1968, **33**, 347.
12. C. Tanford, *The Hydrophobic Effect*, Wiley, New York, 1980.
13. H.L. Freidman and C.V. Krishnan in *Water, A Comprehensive Treatise*, ed. F. Franks, Vol. 3, Ch. 1, Plenum, London, 1973.
14. C.L. Arthur and J. Pawliszyn, *Anal. Chem.*, 1990, **62**, 2145.
15. Z. Zhang, M.J. Yang and J. Pawliszyn. *Anal. Chem.*, 1994, **66**, 844A.
16. W. Potter and J. Pawliszyn, *J. Chromatogr.*, 1992, **603**, 247.
17. B. MacGillivray, J. Pawliszyn, P. Fowlie and C.J. Sagara, *J. Chromatogr. Sci.*, 1994, **32**, 317.
18. D. Potter and J. Pawliszyn, *Environ. Sci. Technol.*, 1994, **28**, 298.
19. I. Valor, J.C. Moltó, D. Apraiz and G.J. Font, *J. Chromatogr.*, 1997, **767**, 195.
20. S. Magdic, A. Boyd-Boland, K. Jinno and J. Pawliszyn, *J. Chromatogr.*, 1996, **736**, 219.

21. W.M. Coleman, III, *J. Chromatogr. Sci.*, 1997, **35**, 245.
22. R. Eisert and K. Levsen, *J. Am. Soc. Mass Spectrom.*, 1995, **6**, 1119.
23. K.D. Buchholz and J. Pawliszyn, *Anal. Chem.*, 1994, **66**, 160.
24. L. Pan, M. Adams and J. Pawliszyn, *Anal. Chem.*, 1995, **67**, 4396.
25. W.M. Coleman, III, *J. Chromatogr. Sci.*, 1996, **34**, 213.
26. H.L. Lord and J. Pawliszyn, *Anal. Chem.*, 1997, **69**, 3899.
27. L. Urruty and M. Montury, *J. Agric. Food Chem.*, 1996, **44**, 3871.
28. L.R. Snyder, *J. Chromatogr.*, 1974, **92**, 223.
29. S. Ulrich and J. Martens, *J. Chromatogr.*, 1997, **696**, 217.
30. Y. Liu, Y. Shen and M. Lee, *Anal. Chem.*, 1997, **69**, 190.
31. Y. Liu, M.L. Lee, K.J. Hageman, Y. Yang and S.B. Hawthorne, *Anal. Chem.*, 1997, **69**, 5001.
32. J.N. Valenta and S.G. Weber, *J. Chromatogr.*, 1996, **722**, 47.
33. S. Li and S.G. Weber, *Anal. Chem.*, 1997, **69**, 1217.
34. T.J. Clark and J.E. Bunch, *J. Chromatogr. Sci.*, 1997, **35**, 209.
35. L. Pan, J.M. Chong and J. Pawliszyn, *J. Chromatogr.*, 1997, **773**, 249.
36. L. Pan, M. Adams and J. Pawliszyn, *Anal. Chem.*, 1995, **67**, 4396.
37. P. Bühlemann, M. Badertscher and W. Simon, *Tetrahedron*, 1993, **49**, 595.
38. B. Linton and A.D. Hamilton, *Chem. Rev.*, 1997, **97**, 1669.
39. J.J. Rebek, *J. Mol. Recognit.*, 1992, **5**, 83.
40. I. Ferrer, M.C. Hennion and D. Barceló, *Anal. Chem.*, 1997, **69**, 4508.
41. B. Sellergren, *Trends Anal. Chem.*, 1997, **16**, 310.
42. S. Li, Y. Chung, L. Sun and S.G. Weber, *Anal. Chem.*, in press.
43. P.A. Frazey, R.M. Barkley and R.E. Sievers, *Anal. Chem.*, 1998, **70**, 638.

CHAPTER 5

Properties of Commercial SPME Coatings

VENKATACHALAM MANI

1 Introduction

Solid Phase Microextraction (SPME) is a unique sample preparation technique which uses a one cm length of fused silica fiber coated with a polymeric phase.[1,2] The polymeric phase extracts volatile, semivolatile, or nonvolatile analytes from different matrices and transfers them to the injection port of a separating instrument (GC, HPLC, GC–MS, CE *etc.*). The selectivity of extraction is mostly due to the polymeric coating on the fiber. The initial research on SPME was carried out at the University of Waterloo, by Dr. Pawliszyn's group, with fused silica fiber coated with polydimethylsiloxane (PDMS) or polyacrylate (PA). Produced in large quantities for fiber optics technology, the fused silica fiber is chemically inert and very stable, even when it is heated to high temperature (up to 250 °C). By comparison with gas chromatography, in which the polymeric phase is coated on the inside of capillary tubes, in SPME the polymeric phase is coated on the outside of a solid rod. Like gas chromatography, in which different types of coatings are used for different selectivity, different thermal stability and different polarity, different coatings on the SPME fiber provide the same advantages. Since a few decades ago only a few coated phases were available for GC, a number of different phases are available at present. In the same way, SPME started with a single polymeric coating, PDMS, but has begun to grow, with approximately ten phases available at present. It is anticipated that many additional phases will be developed in the near future. A discussion of different polymeric coatings and their applications should lead to new polymeric coatings for various applications.

2 What is an SPME Fiber?

SPME is a simple, inexpensive technique. As shown in Figure 1, a 1 cm length of fused silica fiber coated with a polymeric phase is connected to a length of stainless steel tubing that ensures the mechanical strength of the fiber assembly

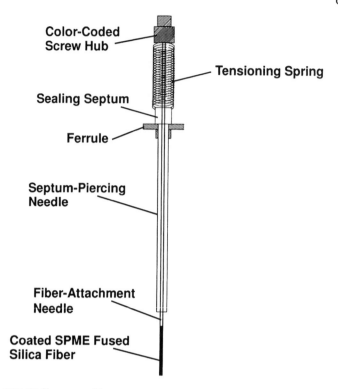

Figure 1   *SPME fiber assembly*

for repeated sampling. The fiber and stainless steel tubing are contained in a specially designed syringe. Sample preparation by SPME consists of two processes, sample extraction and sample desorption. In the extraction process, the coated fiber initially is inside the syringe needle. After the needle pierces the septum of the sample vial, the fiber is exposed to either the headspace or the liquid matrix. Analytes partition between the coating and the sample. After an equilibrium is reached, the fiber is withdrawn into the needle, and the needle is withdrawn from the septum. In the desorption process, the needle is inserted into the injection port of the analytical instrument, the fiber is exposed, and the target analytes are desorbed to the instrument for analysis. In comparison with solid phase extraction (SPE), SPME has all of the same advantages, while eliminating the disadvantages of plugging and solvent consumption. A cleanup step in SPE, requiring selective solvents, is eliminated by using a selective polymeric coating on the SPME fiber. Thus SPME eliminates the use and disposal of toxic solvents.

3   Fiber Production

The commercially available SPME fibers are manufactured at present by two processes. Some fibers are drawn by means of a fiber optics tower and thereby thousands of meters of fiber are produced. In the tower, a fused silica rod is melted and drawn into a thin rod of 110 ± 5 μm thickness. The drawn fiber cools to atmospheric temperature and is passed through an applicator of known orifice containing the coating solution. The polymeric phase is coated on the fiber as the fiber is pulled through the solution. The polymer then is cured thermally or by exposure to UV light, and is coiled into a spool. The diameter of the fused silica core and the thickness of the coating are computer controlled, and thereby the fibers are made reproducibly. Only a two centimeter length of fiber is used to make each fiber assembly. The three currently available PDMS fibers and the polyacrylate fiber are manufactured through this process.

Other fibers, containing heterogeneous phases (PDMS/divinylbenzene, Carboxen/PDMS, Carbowax/divinylbenzene and Carbowax/templated resin fibers) are produced by hand coating in multiple steps. The fused silica fiber core is made in the tower, to a diameter of 110 μm, and coated fibers are measured with a Micro Vu Spectra 14 Optical Comparator to assure uniform coating thickness. Coating thickness is determined by subtracting the radius of the fused silica core from the radius of the coated fiber:

$$\text{coating thickness} = \frac{\text{diameter of coated fiber} - \text{diameter of fused silica fiber core}}{2}$$

The fibers are conditioned at specific temperatures in an inert atmosphere and tested with standard test mixes to confirm reproducibility. Individual fibers are cut into 2 cm lengths, the phase is stripped from 1 cm, and the stripped portion of the fiber is fixed to the fiber assembly with epoxy super glue or by crimping. Thus, the fiber that is exposed for analysis is 1 cm in length. It is also possible to make fiber assemblies with longer fibers. For some theoretical calculations, the phase volume of a fiber may be required. Phase volume is calculated using the formula:

$$\text{phase volume} = \text{total volume of coated fiber} - \text{volume of fused silica core}$$

The volume of the coated fiber or the fused silica core is calculated by using the standard formula for the volume of a cylinder:

$$V = \pi r^2 h$$

where

V = volume of fiber
r = radius of uncoated fiber
h = length of fiber
V_f = volume of coating
R = radius of coated fiber
$V_f = \pi h (R^2 - r^2)$

Phase volumes for various fibers are listed in Table 1.

Table 1 *Phase volumes of commercially available SPME fibers*

Fiber coating	Fused silica diameter (mm)	Fused silica radius (mm)	Core volume (mm³)	Total diameter (mm)	Total volume (mm³)	Phase volume (mm³)
7 μm PDMS	0.110	0.055	0.095	0.124	0.121	0.026
30 μm PDMS	0.110	0.055	0.095	0.170	0.227	0.132
100 μm PDMS	0.110	0.055	0.095	0.300	0.707	0.612
85 μm Polyacrylate	0.110	0.055	0.095	0.280	0.616	0.521
65 μm PDMS/DVB	0.110	0.055	0.095	0.240	0.452	0.357
75 μm Carboxen/PDMS	0.110	0.055	0.095	0.260	0.531	0.436
65 μm Carbowax/DVB	0.110	0.055	0.095	0.240	0.452	0.357
60 μm PDMS/DVB*	0.160	0.080	0.201	0.280	0.616	0.415
50 μm Carbowax/TPR*	0.160	0.080	0.201	0.260	0.531	0.330

*Fiber designed for SPME/HPLC applications.
Note: fiber thickness varies by ± 10 μm.

4 Theoretical Aspects of SPME

The principle behind SPME is the partitioning of analytes between the sample matrix and the polymeric phase on the fiber. The amount of an analyte extracted by the coated surface is described by Nernst's partition law. For a liquid polymeric coating, such as PDMS, the amount of analyte absorbed by the coating at equilibrium is directly related to its concentration in the sample, per the equation:

$$n = \frac{K_{fs} V_f C_0 V_s}{K_{fs} V_f + V_s}$$

where

n = mass analyte adsorbed by coating
V_f = volume of coating
V_s = volume of sample
K = partition coefficient of analyte between coating and sample.

5 Different Types of Coatings

Why do we need different types of fiber coatings? The answer to this question lies in the following question: Why are there capillary columns with different types of coatings in GLC? There is no universal GC coating that will separate all analytes. To achieve better separations, different types of coatings, such as PDMS, PDMS containing 5%, 20%, 35%, 50%, and 65% phenyl, trifluoro-propyl, or cyanopropyl groups, Carbowax, chiral coatings, *etc.* are used. The same situation is present in SPME. There is no single fiber coating that will extract all analytes to the same extent. Polar fibers are effective for extracting polar analytes and nonpolar fibers are effective for extracting nonpolar analytes from different matrices. At present, the types of coatings present on the commercially available fibers can be classified as nonpolar, semipolar, and polar coatings.

Nonpolar Coatings
    100 μm polydimethylsiloxane (PDMS)
    30 μm polydimethylsiloxane
    7 μm polydimethylsiloxane
Semipolar Coatings
    65 μm polydimethylsiloxane/divinylbenzene
Polar Coatings
    85 μm polyacrylate
    55 μm Carbowax/divinylbenzene
    50 μm Carbowax/TPR100

Fibers with different polarity provide the following advantages:

- Extraction selectivity: increase recovery of specific analytes with matched-polarity fiber
- Reduce possibility of extracting interferences
- Extract polar compounds from an organic matrix

The coatings commercially available at present from Supelco may be classified into two types: (a) Homogeneous pure polymer coatings and (b) Porous particles imbedded in a polymeric phase

A polar polymeric coating with ion exchange properties, the Nafion fiber, is under development.

(a) *Homogeneous Polymer Coatings.* At present two homogeneous polymer coatings are available: polydimethylsiloxane and polyacrylate. The polydimethylsiloxane coating is made in three film thicknesses, 7 μm, 30 μm, and 100 μm, and in two forms, nonbonded and bonded. The nonbonded PDMS fiber is cured thermally or by UV treatment and is available in two thicknesses: 100 μm, which is more popular, and 30 μm, which is selective for semivolatiles. The bonded version of the PDMS coating contains a crosslinking functionality in the polymeric backbone which is catalytically crosslinked in the fiber. The main difference between the bonded and nonbonded PDMS fibers is in the organic solvent compatibility of the coating. Bonding gives higher thermal stability to the coating: up to 320 °C, compared to 270 °C for nonbonded fibers. The stability of the fiber coating is determined by the ability of the phase to crosslink with itself and bond to the core material. The introduction of vinyl groups gives the phase greater ability to crosslink with itself, to form more stable fibers.

The other homogeneous polymer coating is polyacrylate, available in 85 μm coating thickness. This fiber coating is partially crosslinked and highly polar, making it suited for extracting polar analytes from aqueous solutions.[3]

Table 2 *Commercial SPME Coatings*

Homogeneous polymer	PDMS	nonbonded	100 μm, 30 μm
		bonded	7 μm
	Polyacrylate	crosslinked	85 μm
Porous particle/polymer	PDMS/DVB	partially crosslinked	65 μm
	PDMS/Carboxen	partially crosslinked	75 μm
	Carbowax/DVB	partially crosslinked	65 μm
	Carbowax/TPR	partially crosslinked	50 μm

The polymeric-coated fiber that is under development, the Nafion fiber, is polar with ion-exchange properties. When it becomes commercially available, in the near future, it will be the most polar SPME fiber prepared to date. It will be used to extract very polar analytes.

(b) *Porous Particles Imbedded in a Polymeric Phase.* The fibers available in this category are coated with various porous particles imbedded in partially cross-linked polymeric phases. Generally, these fibers have lower mechanical stability than homogeneous polymer phases, but have high selectivity. At present four fibers are commercially available (Table 2), at film thicknesses of 50–65 μm:

(i)   Polydimethylsiloxane/divinylbenzene (PDMS/DVB)
(ii)   Polydimethylsiloxane/Carboxen[TM] (PDMS/Carboxen)
(iii) Carbowax/divinylbenzene (CW/DVB)
(iv) Carbowax/templated resin (CW/TPR)

From initial experiments with fibers coated with different porous polymer particles blended in different polymer matrices, the following conclusions were derived:

1. Increasing the porosity of the fiber coating increases the total capacity of the fiber.
2. Increasing the porosity of the polymer particle in the fiber coating retains analytes more tightly.
3. Increasing the pore size of the polymer particle in the fiber coating increases the analyte selectivity of the fiber.

The divinylbenzene polymer particles used in the study were macroporous (100 Å). The synthetic carbons used were microporous and mesoporous (5–50 Å). Information about the commercially available SPME fibers is summarized in Table 3, including polarity of phase, type of bonding, maximum operating temperature, and color of hub for phase identification. The last two fibers in the table were developed specifically for HPLC applications. These fibers are made by coating the phase onto specially pre-coated fused silica fibers. The thickness of the fused silica and precoating is 160 μm. The coating partially bonds to the precoating. Because the precoating is pliable, the fiber can be

Table 3 *Commercially available SPME fibers*

Fiber coating	Polarity	Coating stability	Maximum temperature	Hub color
7 μm PDMS	nonpolar	bonded	340 °C	green
30 μm PDMS	nonpolar	nonbonded	280 °C	yellow
100 μm PDMS	nonpolar	nonbonded	280 °C	red
85 μm Polyacrylate	polar	crosslinked	320 °C	white
65 μm PDMS/DVB	bipolar	crosslinked	270 °C	blue
75 μm Carboxen/PDMS	bipolar	crosslinked	340 °C	black
65 μm Carbowax/DVB	polar	crosslinked	260 °C	orange
60 μm PDMS/DVB	polar	crosslinked	270 °C	purple*
50 μm Carbowax/TPR	polar	crosslinked	240 °C	brown*

*Also notched, to indicate HPLC use.

crimped into the metal sheath, rather than glued to the stainless steel rod, as is done with GC fibers. This eliminates the potential for dissolution of glue in the solvents used in HPLC, and potential appearance of extraneous peaks in the chromatogram. It also gives more stability and flexibility to the fiber, resulting in less breakage. The precoating has very little effect on the extraction of analytes.

Polydimethylsiloxane/Divinylbenzene Copolymer (PDMS/DVB)

This fiber, a blend of porous divinylbenzene polymer particles with liquid polydimethylsiloxane polymer, was introduced commercially in 1996. The surface area of DVB is approximately 750 m^2 g^{-1}. DVB is mainly mesoporous, with some macropores and micropores. DVB micropores are fairly large, relative to micropores in Carboxen particles, with an average diameter of 17 Å. The polymer has a high degree of porosity, 1.5 mL g^{-1}. The pores in the polymer particles physically retain analytes, producing a strong retention of analytes that fit tightly into the pores. This makes SPME fibers containing porous materials well suited for trace level analyses. DVB mesopores are ideal for trapping C_6–C_{15} analytes.

Because DVB is a solid polymer particle, it must be blended in a liquid phase like PDMS or Carbowax. DVB blended with PDMS retains smaller analytes better than PDMS alone. The combination has better affinity for polar analytes. Several improvements have been made to this fiber to produce good selectivity in extraction of amines and alcohols. Recently this fiber was made more durable by employing an improved coating technology—adding a cross-linking agent to the coating solution to crosslink the polymer.

Carbowax/Divinylbenzene Copolymer (CW/DVB)

Carbowax has been used in gas chromatography as a moderately polar phase. Blending DVB with Carbowax increases the polarity of the fiber, relative to

Table 4 *Comparison of SPME fibers for extraction of nonpurgeables (25 ppm in 27% NaCl/water, pH 7)*

	SPME Phase/counts			
Component	*100 PDMS*	*Acrylate*	*PS/DVB*	*CW/DVB*
Methylamine	12000	11000	35000	12000
Methanol	23000	27000	21000	28000
Dimethylamine	19000	23000	72000	17000
Ethanol	40000	34000	27000	39000
Acetonitrile	43000	91000	40000	65000
lsopropanol	62000	69000	49000	66000
Propanol	32000	171000	26000	170000
Diethylamine	17000	36000	101000	11000
2-Methoxyethanol	3000	1000	2000	500
Ethylene glycol	0	300	4000	< 100
Triethylamine	427000	27000	327000	54000
DMSO	2000	2000	2000	1000

PDMS/DVB. This fiber was made to study the polarity of the liquid polymer blend on the extraction efficiency of polar analytes. To extract polar analytes, the polarity of the fiber coating must be increased in order to shift the equilibrium in favor of the fiber. The physical properties of the fiber coating also may help to shift the equilibrium toward the fiber. Carbowax tends to swell and is soluble in water. This will strip the phase from the fiber. In order to overcome this problem, and to achieve high polarity of the blending polymer, a special polyethylene glycol polymer containing highly crosslinking functionalities was prepared. This polymer swells to a lesser extent in water. As Carbowax is sensitive to oxygen at elevated temperatures, this coating will be oxidized in presence of air and oxygen. The fiber will darken and the coating will become powdery. A few precautions will improve the life of the fiber. Injection port temperatures in the range of 180–240 °C and catalytic purifiers that keep the carrier gas free of oxygen will improve the lifetime of the fiber. The extraction efficiency of this fiber with respect to alcohols and small polar analytes is depicted in Table 4; analytical conditions are shown in Table 5.

Polydimethylsiloxane/Carboxen (PDMS/Carboxen 1006)

This fiber was introduced onto the market in 1997.[4] The porous Carboxen particle is blended in the liquid polymer, similar to the PDMS/DVB fiber. Several other types of porous synthetic materials have been coated on SPME fibers. Characteristics of these materials are shown in Table 6.

The size of the pores determines which analytes are retained by a porous particle on an SPME fiber. As a rule of thumb, the pore diameter should be twice the size of the analyte molecule that is to be extracted.[5] Carboxen is a line

Table 5 *Conditions for analysis of PMI analytes (nonpurgeables)*

Sample	analytes in (water + 25% NaCl, pH 11) in 4 mL vial (nominal 4.6 mL)
	immersion: 3.9 mL sample + 0.1 mL 50% Na_3PO_4/1 g NaCl
	headspace: 2.4 mL sample + 0.06 mL 50% Na_3PO_4/0.6 g NaCl
SPME Fiber	65 μm polydimethylsiloxane/divinylbenzene
Extraction	immersion, 15 min (rapid stirring) or
	headspace, 12 min, 55 °C
Desorption	270 °C, 5 min
Column	SPB^{TM}-1 SULFUR, 30 m 0.32 mm i.d., 4.0 μm film
Oven	50 °C (2 min) to 180 °C at 10 °C min^{-1}
Carrier	helium, 50 cm s^{-1}
Detection	FID or GC–MS (selected ion mode)
Injection	splitless (3 min), 270 °C (0.75 mm i.d. liner)

Table 6 *Physical characteristics of divinylbenzene, Carboxen 1006 and templated resin (TPR-100) particles*

Porous material	Surface area $(m^2 g^{-1})$	Porosity $(mL\ g^{-1})$			
		Macro	Meso	Micro	Total
Divinylbenzene	750	0.58	0.85	0.11	1.54
Carboxen 1006	715	0.23	0.26	0.29	0.78
TPR-100	341	–	0.80	–	0.80

Macropore: > 500 Å; mesopore: 20–500 Å; micropore: 2–20 Å.

of porous synthetic carbon materials having distinctive pore designs. Carboxen 1006 has an even distribution of micro, meso and macro pores. In contrast, pores in TPR-100 resin are primarily mesopores. Carboxen 1006 has a total pore volume of 0.78 mL g^{-1}. The mean micropore diameter is slightly larger than 10 Å, and the pores have a distribution of 2–20 Å—ideal for extracting small molecules. Thus a PDMS/Carboxen fiber will be ideal for SPME analyses of molecules in the C_2–C_{12} range.[6] Molecules larger than C_{12} are strongly retained on the surface of the particle, and are difficult to desorb. Because Carboxen 1006 is synthetically produced, the pore volume and particle size can be carefully controlled. As the size of Carboxen 1006 particles is approximately 2 μm (1–5 μm), multiple layers of PDMS/Carboxen 1006 can be coated onto a fiber to increase analyte capacity. The shape of the pores in a carbon affect the rate of analyte adsorption and desorption. Carboxen particles have unique pores, compared with other porous carbons. The pores pass completely through the particle. Small analytes can be more rapidly desorbed, as they pass through the pore, rather than into it and back out. Molecules trapped in the mesopores must be desorbed by reversing the direction of movement. This requires a high desorption temperature, 300–320 °C, to reduce peak tailing. A comparison of three different SPME fibers used to extract hydrocarbons is shown in Table 7.

Table 7 *Comparison of SPME fibers for extracting C_2–C_6 hydrocarbons*

Analyte	100 μm PDMS	PDMS/DVB	PDMS/Carboxen
Ethane (C_2)	0	0	750
Propane (C_3)	0	0	20000
Butane (C_4)	0	340	72000
Pentane (C_5)	230	2150	108000
Hexane (C_6)	460	9300	106000

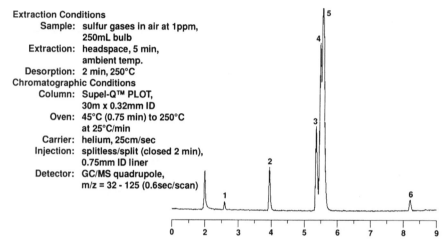

Extraction Conditions
    Sample: sulfur gases in air at 1ppm,
                       250mL bulb
    Extraction: headspace, 5 min,
                       ambient temp.
    Desorption: 2 min, 250°C
Chromatographic Conditions
    Column: Supel-Q™ PLOT,
                       30m x 0.32mm ID
    Oven: 45°C (0.75 min) to 250°C
                       at 25°C/min
    Carrier: helium, 25cm/sec
    Injection: splitless/split (closed 2 min),
                       0.75mm ID liner
    Detector: GC/MS quadrupole,
                       m/z = 32 - 125 (0.6sec/scan)

Figure 2 *Sulfur gases at 1 ppm by SPME: Carboxen™-PDMS fiber*

Sulfur gases were very difficult to extract. Figure 2 shows the extraction of sulfides using the PDMS/Carboxen fiber.

Carbowax/Templated Resin (CWITPR)

This fiber was introduced to extract polar analytes. The method of coating is the same as the general method for making other fibers consisting of porous particles and a liquid polymer. The templated resin used in this coating, SUPELCOGEL TPR-100, is a porous, high purity sphere that was developed to be an HPLC support. A monomer solution containing a mixture of hydrophilic monomers and divinylbenzene is forced into the pores of the starting material, silica particles. The solution is polymerized and hardened, then the silica is dissolved away at high pH, leaving behind a cast of the silica in the resin. What was once the silica's pore structure is now the backbone of the resin, and vice versa. The uniqueness of this approach is in the preliminary treatment of the silica and the composition of the monomer solution. The unique blend of hydrophobic and hydrophilic monomers gives the TPR-100

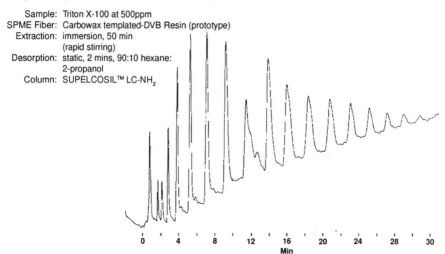

Sample: Triton X-100 at 500ppm
SPME Fiber: Carbowax templated-DVB Resin (prototype)
Extraction: immersion, 50 min
(rapid stirring)
Desorption: static, 2 mins, 90:10 hexane:
2-propanol
Column: SUPELCOSIL™ LC-NH₂

Figure 3   *Triton X-100 in water by SPME–HPLC*

resin unique selectivity, relative to other resins. The degree of crosslinking of the monomers is optimized to give the particle the highest possible efficiency without sacrificing mechanical strength. The method of preparing TPR-100 resin gives a tight pore size distribution and eliminates micropores less than 20 Å, and ensures a large surface area and large, mainly mesoporous pore volume. The very close particle size distribution around 5.5 μm also makes the SPME fiber highly reproducible. A unique application for this fiber is shown in Figure 3.

6   Coating Thickness

What are the advantages of having fibers that differ in coating thickness? We can ask the same question of capillary column technology. Column film thicknesses differ for analyses of volatiles and semivolatiles. The coating thickness affects the retention and separation of analytes—retention and separation increase with increasing film thickness. Various film thicknesses are used to increase analyte capacity, sharpen peaks, or more selectively separate specific components of a mixture. Similarly, different film thicknesses of the same phase have been introduced in SPME.[8]

In SPME, fiber coating thickness affects the following aspects of the extraction/desorption:

Selectivity for analytes
Extraction time
Sample capacity
Desorption time
Analyte carryover.

Table 8 *SPME phase coating thickness affects analyte recovery*

Analyte	PDMS film thickness/rel. recovery (%)		
	100 μm	30 μm	7 μm
Benzene	2	1	< 1
Toluene	5	1	< 1
Chlorobenzene	6	2	< 1
Ethylbenzene	3	4	1
1,3-Dichlorobenzene	17	5	2
1,4-Dichlorobenzene	15	5	1
1,2-Dichlorobenzene	15	4	1
Naphthalene	13	4	1
Acenaphthylene	19	8	3
Fluorene	29	18	8
Phenanthrene	37	27	16
Anthracene	49	38	32
Pyrene	69	54	47
Benzo[a]anthracene	105	91	96
Chrysene*	100	100	100
Benzo[b]fluoranthene	104	111	120
Benzo[k]fluoranthene	111	124	127
Benzo[a]pyrene	119	127	131
Indeno[1,2,3-cd]pyrene	61	140	148
Benzo[ghi]perylene	61	117	122

*Reference value.
SPME: fiber immersed in sample, 15 min, rapid stirring.

Table 8 shows the effect of extracting PAHs with SPME fibers having three different PDMS film thicknesses. From the table it is clear that small analytes are extracted more efficiently by a thicker coating, but extraction efficiency increases with decreasing film thickness.

7 Fiber Selection for SPME Analyses

Successful application of SPME depends primarily on the selection of a suitable fiber for a particular analysis. The efficiency of analyte extraction and desorption from the fiber depends on the following factors:

Molecular weight and size of analyte
Boiling point and vapor pressure of analyte
Polarity of analyte and fiber
Functional groups on analyte and fiber
Concentration range and detector type used.

The physical characteristics of the fiber (film thickness, phase polarity, porosity of the polymeric particle, if used) must be matched to those of the analytes. As a rule of thumb, 'like prefers like'—nonpolar phases will extract nonpolar analytes and more polar fibers should be used for extracting polar analytes. As polar analytes are highly water soluble, they are difficult to extract from water unless a suitable fiber is used. The polar polyacrylate fiber (polarity similar to Carbowax) is suitable for such extractions (*e.g.* phenol). Sample modifications, such as addition of salt, stirring, heating, and/or pH adjustment can help to optimize extraction efficiency.[3,4] As polyacrylate is solid at room temperature, analytes are absorbed and desorbed slowly unless the temperature is elevated.

7   New Generation of Fibers for SPME–HPLC

Initially, SPME fibers coated with polymeric phases were designed for SPME–GC applications, but the same fibers also were used for SPME–HPLC applications. PDMS-coated fibers perform very well under HPLC conditions, due to the flexibility provided by the coating. However, some of the more fragile fibers (*e.g.* PDMS/DVB, Carbowax/TPR) did not perform well with HPLC. The 110 μm diameter core support for GC fibers is fragile. In HPLC applications, the fibers are exposed to complicated mechanical maneuvering like passing through a ferrule, to higher pressures and flow rates, and to strong solvents. Some fibers were prone to breaking or stripping of the phase from the fused silica support. Sometimes the adhesive used to fix the fiber to the stainless steel rod dissolved in the solvent and introduced extraneous peaks into the chromatogram. It became necessary to improve the fibers in order to provide HPLC analysts with more reliable tools. Various techniques were tried to overcome the problems. Adhesive was replaced by mechanical fixing of the fiber to the rod. New fibers were made by coating new phases onto specially precoated fibers, for increased stability and flexibility. The new SPME–HPLC fibers have improved durability and afford greater confidence in consistency of results.

9   SPME Portable Field Sampler

The SPME portable field sampler (Figure 4) is a manual-style fiber holder in which, after sampling, the fiber can be stored by sealing it within a septum-sealed portion of the needle. Extracted compounds are safely sealed inside the needle, which is drawn into a replaceable sealing septum and locked in place. By then storing the sampler at subambient temperature, contamination and analyte breakdown and loss from the fiber can be reduced. Through proper conditioning and care, the fiber can be reused 50–100 times. The sampler is ideal for trace level analysis of volatile and semivolatile compounds. It is currently available with two types of coating.

10   New Fibers Under Development

Future application areas for SPME are vast. The future growth of SPME depends on the selectivity of fiber coatings. There is a need for new types of

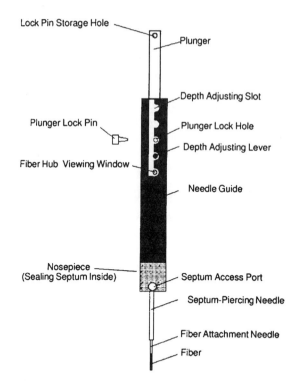

Figure 4   *SPME portable field sampler*

coatings for GC and HPLC analyses. New types of fibers with new selectivities can be used for biological applications.

There is also a need for design change, to make assemblies compatible with new technologies and to reduce the cost of using SPME. A recent design of SPME fibers will be compatible with the Merlin Microseal septum system, which requires a 23 gauge needle with a polished tip. The current SPME outer needle is a 24 gauge stainless steel needle. Because the Merlin Microseal eliminates the injection port septum, fiber damage will be reduced and extraneous peaks due to septum particles will be eliminated. This will also keep the inlet liner clean.

Another design change will be development of a 2 cm length fiber. The increase in length will be obtained by removing 1 cm of the inner tube. Doubling the fiber length will increase the capacity of the fiber.

In order to make unbreakable fibers for specific applications, it may be possible to make fibers with a stainless steel, tungsten or other metal core, in place of the fused silica core. For other unusual applications, like human breath analysis or other biological applications, the SPME device probably can be modified suitably.

Mixed bed coatings and coatings composed of differing layers (*e.g.* PDMS/Carboxen coated on fused silica, then coated with PDMS/DVB) will be interesting approaches to extracting different types of analytes at the same time. Such fibers should have the selectivities of both components. A PDMS/Carboxen–PDMS/DVB fiber will have a high efficiency for extracting odor compounds from water.

Derivatized α-, β- and γ-cyclodextrin coatings with various added functionalities will be suitable for extracting polar analytes.[9] Because cyclodextrins are capable of forming inclusion complexes with various analytes in the cavity of the cyclodextrin, cyclodextrin-based fibers will be very selective in extracting analytes.

Some work has been done on sol-gel particle coatings. Sol-gel is a porous silica with polarity similar to Tenax TA. Coatings made with this material have properties similar to the 100 μm PDMS coating, with slightly higher polarity. In this type of coating, as in cyclodextrin-based coatings, it is possible to add different functional groups to modify the capacity for extracting different analytes. Chong *et al.*[10] claimed to have developed a bonded sol-gel layer with PDMS that is thermally stable to more than 320 °C.

The sensitivity and success of SPME is governed mainly by the partition coefficient of the analyte between the coating and the matrix. Extraction selectivity can be achieved by using a suitable stationary phase that exhibits high affinity toward the target analyte. The possibility of incorporating antibodies or proteins onto the fiber will make SPME suitable for applications involving specific-molecules. This may be achieved by more advanced bonding technologies. The future of SPME technology and its applications will depend upon new fiber coatings for selective extraction of analytes.

There still are a few areas in SPME that can be improved. The breakage of fibers, stripping of coatings, and bending of the needle are common problems SPME users contend with. Fiber breakage can be reduced by careful handling of the fiber during sampling and when exposing the fiber in the GC injection port or SPME–HPLC interface. Using predrilled septa, correct septum nut tightness, and careful insertion of the fiber into the injection port will reduce this problem. A septumless injection port, such as the Merlin Microseal system, will be the best solution to overcome this problem.

Fiber swelling can be reduced by using lower concentrations of nonpolar organic solvents, or by using water as the mobile phase after using high concentrations of organic solvent.

It is also possible to expose the fiber in the sample headspace for a few minutes, after removing it from a sample with a high concentration of organic solvent, before retracting it into the inner tube. This will reduce swelling of the phase, which could cause stripping of the phase as the fiber is drawn back into the sheath.

References

1. R. Belardi and J. Pawliszyn, *Water Pollut. Res. J. Can.*, 1989, **24** 179.
2. C. Arthur, D. Potter, K. Buchholz, S. Motlagh and J. Pawliszyn, *J.LC-GC*, 1992, **10**, 656.

3. Application Note 17 Supelco, Bellefonte, PA, USA.
4. R.E. Shirey and V. Mani, Presentation at Pittcon 97.
5. T. Schumacher, *Supelco Reporter*, 1997, **16**, 8.
6. P. Popp and A. Daschko, *Chromatographia*, 1997, **46**, 419.
7. A. Boyd-Boland and J. Pawliszyn, *Anal. Chem.*, 1996, **68**, 1521.
8. W. Bertsch, W.G. Jennings and R.E. Kaiser, Recent Advances in Capillary Gas Chromatography, p. 115.
9. R.F. Dias and K.H. Freemann, *Anal. Chem.*, 1997, **69**, 944.
10. S.L. Chong, D. Wang, J.D. Hayes, B.W. Wilhite and A. Malik, *Anal. Chem.*, 1997, **69**, 3889.

CHAPTER 6

Sol-Gel Technology for Thermally Stable Coatings in SPME

ABDUL MALIK AND SAU L. CHONG

1 Introduction

Solid-phase microextraction (SPME) is a highly promising environmentally friendly sample preparation technique of great analytical potential.[1] The pioneering work of Pawliszyn and coworkers[2-4] in the late eighties and early nineties of the twentieth century led to the invention of this effective micro sample preparation technique. An important factor contributing to the rapid growth and popularity of SPME is its ability to perform sample extraction and preconcentration without requiring the use of hazardous organic solvents.[5] In SPME, a sorptive stationary phase coating, either on the outer surface of a fused silica fiber end (~ 1 cm segment)[1,6] or on the inner surface of a capillary,[7-8] actually serves as the extraction medium in which the analytes get preferentially sorbed and preconcentrated. It is quite evident that the stationary phase coating plays a vital role in SPME analysis. Because of the key role played by the stationary phase coating, future developments and further extension of SPME applications will greatly depend on new breakthroughs in the areas of coating technology.[9]

Polydimethylsiloxane (PDMS) is the most commonly used coating material in the current practice of solid-phase microextraction.[10] However, being non-polar in nature, PDMS coatings often show insufficient affinity for polar compounds. To overcome these difficulties, a number of polar SPME coatings were developed and evaluated. Most notable among these are polyacrylate coatings,[11-13] Carbowax/template resin and Carbowax/divinylbenzene coatings,[14] polyacrylic acid coatings,[15] Nafion perfluorinated resin coatings[16] *etc.*

For most commercial fibers, the exact methods and technical details for coating the fiber remain proprietary. However, from their low thermal stability characteristics it appears unlikely that such coatings are chemically bonded to the fiber substrates. It is more probable that such coatings are held on the fiber surface by physical forces of adhesion. SPME fibers with physically adhered stationary phase coatings suffer from a number of drawbacks that seriously limit the analytical potential of solid-phase microextraction. First, the physically coated stationary phase films are unstable at high temperatures. This limits the maximum analyte desorption temperature that can be used in SPME–GC. The upper limit of analyte desorption temperature, in turn, determines the range of analyte molecular weights amenable to SPME–GC analysis. Second, physically coated fibers also show limited stability toward organic solvents. This presents a serious hurdle to overcome if SPME is to couple with liquid-phase separation techniques that commonly use mobile phases rich in organic solvents. This also limits the use of SPME for direct extraction of analytes from nonaqueous liquid matrices.

From the above discussion it follows that the invention of an efficient technology for the creation of highly stable, relatively thick stationary phase coatings that can provide reliable SPME performance over a wide range of experimental conditions will have a major impact on the further development of the technique. In particular, these fiber coatings should possess high thermal and solvent stabilities, and also they should provide fast solute diffusion during extraction of analytes from the sample matrix and during desorption of the extracted analytes from the fiber coating into the GC injection port.

The first problem to be solved in this connection is to develop a suitable method for creating thick surface coatings on the fiber/capillary surface. Mention should be made that the static coating technique[17] which is widely used in capillary chromatographic column technology is not suitable for the creation of SPME fiber coatings. This technique is mostly suitable for surface coatings of sub-micrometer thickness similar to those used in GC (*e.g.* a typical stationary phase coating in a 250 μm i.d. GC column is 0.25 μm). Stationary phase coatings used on SPME fibers/capillaries are a few orders of magnitude thicker than those provided by the static coating technique (*e.g.* a typical coating thickness in SPME is 100 μm). Moreover, static coating technique is generally applied to the inner surface of the capillary, while in the most commonly used mode of SPME, coatings are to be created on the outer surface of solid fibers. Obviously, it is difficult to apply static coating technique for this purpose.

A number of different approaches have been reported in the literature to create thick coatings for SPME. Lee and co-workers[18] prepared porous-layer SPME coatings by gluing a layer of reversed-phase HPLC particles onto the fiber surface. The porous structure of the coating should significantly enhance the sorption–desorption kinetics during extraction and sample introduction steps of SPME–GC analysis. The authors demonstrated some interesting examples of extraction and chromatographic analysis using these porous-layer SPME coatings. However, the use of glue for fixing the particles on the fiber surface poses potential problems in terms of thermal stability. Such fibers can be expected to provide enhanced solvent stability since after curing the used glues are practically insoluble in common organic solvents. Because of this, such SPME fibers may be better suited for use in combination with liquid-phase separation techniques. Djozan and Assadi[19] described SPME fibers with a porous layer from activated charcoal created on the outer surface of fused silica rods. Analogous coatings from graphitized carbon black have been reported by Mangini and Cenciarini.[20] Porous structure of these fibers are likely to enhance sorption–desorption kinetics during extraction and sample introduction. This, in turn, should speed up SPME–GC analysis. However, technical details for the coating uniformity and performance stability, as well as preparation and preformance reproducibility for these fibers, remain to be explored.

Recently, we described a sol-gel chemistry-based approach to coating SPME fibers with a porous organic–inorganic hybrid polymer which is chemically bonded to the fused silica fiber.[9,10] In this approach, the coating is created on the fiber outer surface by dipping its bare end into a sol solution containing an alkoxide-based precursor, a hydroxy-terminated sol-gel active polymer and a surface derivatizing reagent dissolved in a suitable solvent system. Such coatings possess a porous structure, and are chemically bonded to the fiber outer surface. The coating thickness can be varied simply by controlling dipping time of the fiber end in the sol solution. This chapter focuses on demonstrating analytical potential of sol-gel coatings in solid-phase microextraction, highlighting universality of the approach, ease and versatility in fiber preparation, possibility of exploiting a wide range of material properties in analytical sample preparations and excellent thermal and solvent stability of the sol-gel coatings. Predictions are made regarding prospects and potentials of these coatings in SPME-hyphenated liquid-phase separation techniques including SPME–HPLC, SPME–CE and SPME–CEC.

2 Experimental

Chemicals and Materials

Fused silica fiber (200 μm) and fused silica capillary (250 μm i.d.) with protective polyimide coating were purchased from Polymicro Technologies, Inc. (Phoenix, AZ, USA). Tetramethoxysilane (TMOS), methyltrimethoxysilane (MTMOS), and hydroxy-terminated polydimethylsiloxane (PDMS) were procured from United Chemical Technologies, Inc. (Bristol, PA, USA). Polymethylhydrosiloxane (PMHS) and individual standards of polycyclic aromatic hydrocarbons (PAH), phenol derivatives, aniline derivatives and alkanes were purchased from Aldrich Chemical Co. (Milwaukee, WI, USA). HPLC-grade methylene chloride, methanol, and tetrahydrofuran (THF), graphite ferrules and PEEK tubing were purchased from Fisher Scientific (Pittsburgh, PA, USA). Trifluoroacetic acid (TFA) and individual standards of alcohols were obtained from Sigma Chemical Co. (St. Louis, MO, USA). Deionized water (18 MΩ) was obtained from a NanoPure Water Purification System. Eppendorf polypropylene microcentrifuge tubes were purchased from Brinkman Instruments, Inc. (Westbury, NY, USA). High-temperature silicone adhesive was obtained from Dow Corning Corp. (Midland, MI, USA).

Equipment

SPME–GC experiments were carried out on a Shimadzu Model 14A capillary gas chromatography system (Shimadzu Scientific Instruments, Inc., Columbia, MD, USA) equipped with a split-splitless injector and a flame ionization detector. A Barnstead Model 04741 Nanopure water purification system (Barnstead/Thermodyne: Dubuque, IA, USA) was used to obtain 18 MΩ deionized water. SPME–GC experimental data were collected and processed using Version 6.07 Chrome Perfect Software (Justice Innovations: Mountain Views, CA, USA) installed on a computer that was directly interfaced with the GC system. A Fisher Scientific Model G-560 Genie 2 vortex system was used for thorough mixing of sol solution ingredients. A Microcentaur Model APO 5760 centrifuge was used to separate the possible precipitates from the sol solution. A homemade syringe described in ref. 9 was used to facilitate fiber handling during extraction of analytes from the aqueous matrices and their subsequent desorption from the fiber into the GC injection port.

Sol-gel Coating of SPME Fibers

The details of creating sol-gel coatings on the outer surface of a fused silica fiber has been described in one of our recent publications.[9] Briefly, this was carried out as follows. A 15 cm piece of fused silica fiber was taken, and the polyimide coating was burnt off from the outer surface of approximately 1 cm long tip segment of the fiber using a cigarette lighter. The burned residues were then carefully removed from the fiber tip using a Kimwipe tissue paper soaked with methanol. The bare fiber segment was then kept submersed in a 0.1 M NaOH solution for 30 min. After this, the treated surface was thoroughly rinsed with deionized water and dried in air for 30 min. The bare end-segment of the fiber was then dipped into the sol solution and was kept submersed for a controlled period of time determined by the composition of the sol solution and the desired thickness of the coating to be created on the fiber. Typically, it ranged between 15 and 60 min. The coated surface was then dried in a flow of GC Grade helium and stored in a desiccator. To create thicker coatings, this procedure should be repeated for a few times using a fresh portion of the sol solution each time.

Fiber Conditioning

Sol-gel coated fibers were thermally conditioned before using them in SPME analysis. For this, the fiber was mounted on a specially designed syringe.[9] It was retracted into the interior of the syringe needle that served as the protective sheath for the fiber. The protected fiber was then introduced into the GC injection port by piercing its rubber septum with the needle. By depressing the syringe plunger, the coated fiber tip was exposed to the hot helium flow in the injection port. The conditioning of the fiber was carried out by step-wise increasing the conditioning temperature. In this work sol-gel SPME fibers were conditioned at temperatures up to 360 °C, starting from 250 °C using a step increment of 10 °C.

Sample Extraction

Aqueous samples of low-ppm and ppb level concentrations were prepared using chemical standards of *n*-alkanes, polycyclic aromatic hydrocarbons (PAHs), aliphatic alcohols, phenolic compounds and aromatic amines. For this, individual chemical standards were first dissolved in appropriate water-soluble organic solvents to prepare 100 ppm solutions. Appropriate volumes of these solutions were then added to deionized water to obtain aqueous samples of desired concentrations in the low-ppm or ppb range. Each extraction was carried out for 30 min using constant stirring. No salting-out or pH adjustments were used.

SPME–GC Analysis

SPME–GC analyses were performed using homemade open tubular columns with sol-gel polydimethylsiloxane (PDMS) or sol-gel Ucon stationary phase coatings. The sol-gel PDMS columns were prepared according to a procedure recently developed by us.[21] The sol-gel Ucon columns were prepared following a procedure analogous to the one that we originally developed for coating capillary electrophoresis columns.[22] The GC system was operated in the split-splitless injection mode using helium as the carrier gas. After sample extraction, the extracted analytes were desorbed from the fiber into the injection port that was maintained in the splitless mode for five minutes after the injection. The split valve was then opened providing a split ratio of 100:1. Depending on the boiling point and the polarity of the sample, the injector temperature was maintained in the range of 250–360 °C. After GC separation, analyte detection was carried out by a flame ionization detector (FID) whose temperature was set at a value 10 °C higher than the corresponding injector temperature. A computer loaded with Chrome Perfect software was used for on-line collection of the SPME–GC analysis data.

3 Results and Discussion

Sol-Gel Chemistry for SPME Coatings

The sol-gel approach to creating surface coatings has a number of advantageous features. These include simplicity and versatility in the creation of highly stable and efficient coatings, efficient molecular level mixing of sol-gel ingredients, possibility of creating hybrid organic–inorganic coatings with advanced material properties that will be difficult to achieve by conventional approaches, possibility of chemically binding the coatings onto the silica substrate as a result of the extension of the sol-gel condensation reactions with the surface silanol groups and commercial availability of extremely high purity grade precursors (*e.g.* 99.999% purity tetraethoxysilane can be obtained from Aldrich).

Sol-gel chemistry[23] provides a convenient pathway to create advanced material systems that can be effectively utilized to solve the SPME fiber technology problems mentioned in the introduction section of this chapter. It allows incorporation of organic ingredients into inorganic polymeric gel structure to obtain organic–inorganic hybrid copolymeric material systems that often possess superior stability characteristics, enhanced selectivity and mass transfer properties.[24] These advanced material systems can be created under extraordinarily mild thermal conditions (commonly at ambient temperature), using an appropriately designed sol solution.[25] Such material systems can be obtained in two different forms: (a) chemically bonded surface coatings [26,27] and (b) porous monolithic separation media.[28,29]

Sol-gel processes generally involve hydrolytic polycondensation reactions that are carried out in a sol solution. Recently we applied these reactions to prepare sol-gel coatings for SPME fibers[9] as well as GC[21] and CE[22] columns. For SPME coatings, we used a sol solution containing an alkoxide-based precursor, an organic polymer with sol-gel-active functional group(s), a surface derivatizing reagent, a porogenic agent and a sol-gel catalyst dissolved in a suitable solvent system containing a controlled amount of water (Table 1). Chemical reactions involved in sol-gel coating procedure for SPME fibers have been described in details in refs. 9–10. These reactions can be summarized as shown in Scheme 1. Hydrolysis of the alkoxysilane precursor (or its alkyl/aryl derivative) leads to the formation of reactive silanol-containing products that can further undergo polycondensation reactions with sol-gel-reactive species present in the sol solution to form a three-dimensional network of sol-gel polymer. As the polycondensation reactions proceed, the visco-elastic properties of the sol solution undergo progressive changes, and after a certain period of time it turns into a gel—a three-dimensional porous polymeric network with the solvent(s) trapped inside the pores. The chain of chemical events involved in this process can be effectively utilized to create surface-bonded stationary phase coatings in a simple way.

When the bare end of a fused silica fiber is immersed in such a sol solution, the sol-gel polymeric network evolving in the vicinity of the fiber outer surface may undergo condensation reaction with the silanol groups on the fiber. The bare fiber end is kept immersed inside the sol solution for a controlled period to allow the sol-gel reactions to proceed to a desired extent. The fiber is withdrawn from the sol solution after the controlled immersion period is over. As Scheme 1 shows, this simple procedure provides an effective way of creating a surface-bonded sol-gel coating on an SPME fiber.

The sol solution contains polymethylhydrosiloxane (PMHS) meant for deactivation of the coating. PMHS does not contain any sol-gel-active functionalities. However, it does contain reactive hydrogen atoms capable of derivatizing silanol groups at high temperatures. Sol-gel reactions are conducted at room temperature, and under these conditions PMHS molecules are not reactive to silanol moieties, and remain simply physically trapped in the sol-

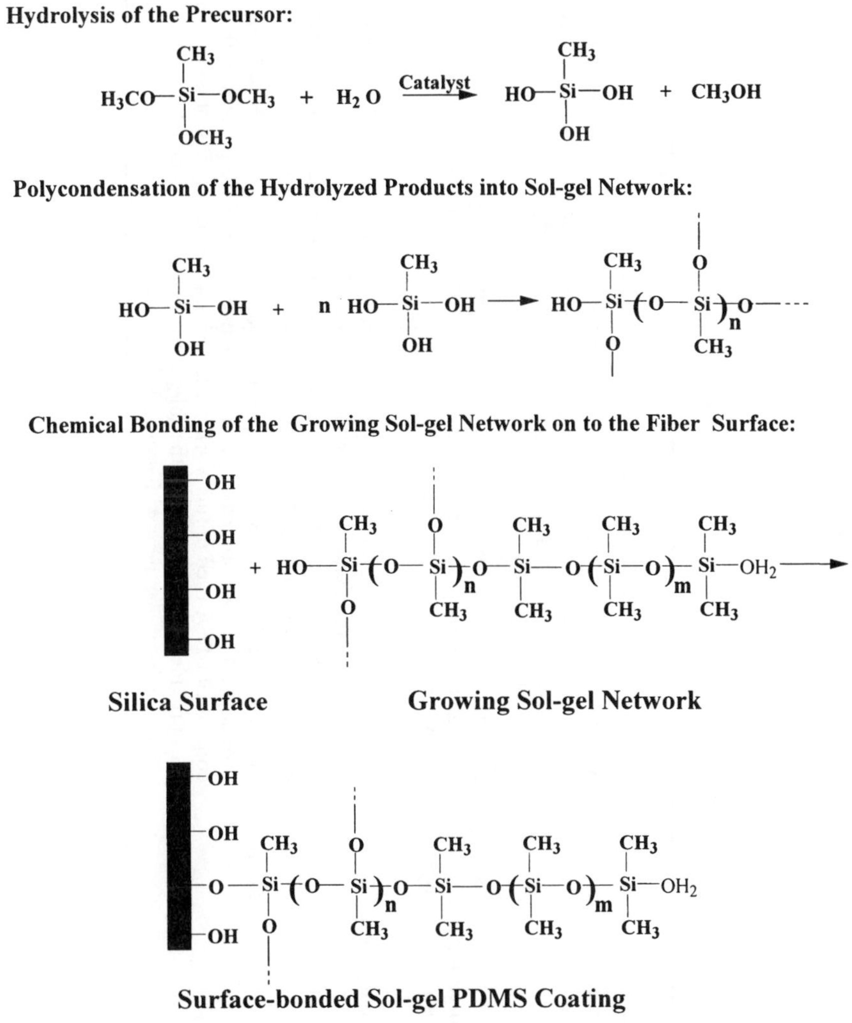

Hydrolysis of the Precursor:

$$H_3CO-\underset{\underset{OCH_3}{|}}{\overset{\overset{CH_3}{|}}{Si}}-OCH_3 \; + \; H_2O \; \xrightarrow{\text{Catalyst}} \; HO-\underset{\underset{OH}{|}}{\overset{\overset{CH_3}{|}}{Si}}-OH \; + \; CH_3OH$$

Polycondensation of the Hydrolyzed Products into Sol-gel Network:

$$HO-\underset{\underset{OH}{|}}{\overset{\overset{CH_3}{|}}{Si}}-OH \; + \; n \; HO-\underset{\underset{OH}{|}}{\overset{\overset{CH_3}{|}}{Si}}-OH \; \longrightarrow \; HO-\underset{\underset{O}{|}}{\overset{\overset{CH_3}{|}}{Si}}\left(O-\underset{\underset{CH_3}{|}}{\overset{\overset{O}{|}}{Si}}\right)_n O---$$

Chemical Bonding of the Growing Sol-gel Network on to the Fiber Surface:

Silica Surface   **Growing Sol-gel Network**

Surface-bonded Sol-gel PDMS Coating

Scheme 1 *Chemical reactions involved in sol-gel coating of fused silica SPME fiber surface*

gel network. However, during the high-temperature conditioning step, which is carried out after completion of the fiber coating procedure, PMHS molecules enter into derivatization reactions with the silanol groups in the sol-gel coating as described in ref. 9.

Sol-gel coatings comprise a unique stationary phase design and architecture suitable for the extraction of both polar and nonpolar compounds from aqueous or nonaqueous matrices. They should also allow for the simultaneous extraction of both types of analytes from such matrices. As was noted recently by Pawliszyn and coworkers,[16] these are the types of analytical problems that are difficult to address by the conventionally coated commercial fibers. One of the main reasons

Table 1 *Chemical ingredients of the sol-gel solution used in this study*

Sol-gel solution ingredient	Chemical name	Function
Sol-gel precursor	Methyltrimethoxysilane	Serves as a source for the inorganic component of the sol-gel organic–inorganic hybrid coating, and delivers it through hydrolytic polycondensation reactions. Also provides active silanol groups on the sol-gel coating to facilitate its chemical bonding to the fiber/capillary surface.
Organic polymer with sol-gel-active functional groups	(1) Hydroxy-terminated poly-dimethylsiloxane; (2) Poly(ethylene-propylene) glycol	Provides organic component for the sol-gel organic–inorganic hybrid coating through polycondensation reactions with the hydrolyzed products of the precursor.
Deactivation reagent	Poly(methylhydrosiloxane)	Provides desired level of high-temperature derivatization of silanol groups on the sol-gel coating, and thus allows for the control of the organic–inorganic hybrid coating polarity. The derivatization reaction takes place during thermal conditioning of the fiber after completion of the sol-gel coating procedure.
Sol-gel catalyst	Trifluoroacetic acid (TFA)	Plays a dual role in the sol-gel process: (a) as an acid-catalyst for the sol-gel process, and (b) as a source of water for the hydrolysis of the precursor (the commercial TFA used in this work contained approximately 5% of water). It thus provides a mechanism for the sol-gel process to be conducted in organic solvent-rich environment with controlled amount of water.

for such difficulties lies in the fact that conventional fibers have unimodal interaction sites on the coating. Most often such coatings are made out of polymers with a characteristic mode of interaction which inherently operates in the same fashion at all points on the coating. Such coatings fail to provide site-specific interactions with the analytes. Thus nonpolar fibers often fail to extract polar analytes from an aqueous environment because of their weaker solute affinity compared with water. For a similar reason, such fibers also fail to efficiently extract polar analytes from a nonpolar matrix. Fiber stability in the organic solvent environment is another issue of major concern in SPME.

Unlike conventional coatings, a sol-gel coating represents an advanced material system comprising a hybrid organic–inorganic copolymer. In the prepared sol-gel coatings, this hybrid copolymeric system consists of an *in situ* created silica-based inorganic component which is polar in nature, and a PDMS-based nonpolar component. These two building blocks are chemically bonded into the sol-gel network as integral parts of this advanced material system. Manipulation of the sol solution composition and deactivation of the coating with PMHS give effective ways of controlling the balance in the organic–inorganic compositions in this unique material system. Quite naturally, such a stationary phase should be able to efficiently extract both polar and nonpolar analytes, even from the same sample, as will be demonstrated by the examples that follow.

Figure 1 represents an example of SPME–GC analysis of C_{13}–C_{20} aliphatic hydrocarbons, illustrating the possibility of using sol-gel coated PDMS fibers for the analysis of nonpolar compounds. It can be assumed that in the direct extraction of such nonpolar analytes from aqueous matrices, the PDMS-based nonpolar sites on the sol-gel coating play an active role during extraction. This site-specific coating–analyte interaction serves as the key concept in the design of sol-gel coatings for SPME and chromatographic techniques. In principle, a wide range of polar and nonpolar moieties with sol-gel-active functionalities can be incorporated in the sol solution to create tailor-made sol-gel stationary phases.

Figure 2 represents an example of direct SPME–GC analysis of aliphatic alcohols from an aqueous environment—a problem which cannot be easily addressed by using conventional fibers.[16] Conventional polar coatings show higher affinity toward water matrix resulting in poor extraction characteristics for alcohols. The use of nonpolar coatings often fail do show enough affinity toward this polar class of compounds and do not provide satisfactory extraction results. For this class of compounds, satisfactory extraction results have only been achieved for methanol using headspace mode of SPME.[16] Sol-gel fibers, on the other hand, possessing a bimodal interaction site, can easily overcome these difficulties with the direct extraction of aliphatic alcohols and other polar compounds. The chromatogram presented in Figure 2 was obtained through extraction of C_{14}, C_{16}, C_{18} and C_{20} aliphatic alcohols from a 0.5 ppm aqueous solution. From the peak heights and baseline of the chromatogram, it can be assumed that sol-gel coated PDMS fibers will allow for the extraction of these alcohols from solutions of much lower concentrations. Mention should be made

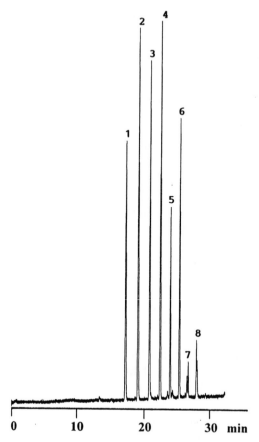

Figure 1 *SPME–GC analysis of aliphatic hydrocarbons from an aqueous matrix. Extrac-*
tion conditions. Fiber: 200 μm, fused silica; Stationary phase: sol-gel PDMS.
Coating thickness: 10 μm; direct 30 min extraction with stirring (no pH
adjustment and no salting out); Injection conditions. Injector temperature:
250 °C; Carrier gas: helium; Injection mode: split-splitless (First 5 min splitless,
followed by split with a ratio of 100 : 1); GC conditions. Column: 10 m × 250
μm i.d.; Stationary phase: sol-gel PDMS; Temperature programming: 40 °C
(5 min), 6 °C min^{-1}; Detector: FID, 300 °C. Peak identifications: (1) C_{13},
(2) C_{14}, (3) C_{15}, (4) C_{16}, (5) C_{17}, (6) C_{18}, (7) C_{19}, (8) C_{20} normal aliphatic
hydrocarbons

here that no salting out or pH adjustments were used in the extraction of above-
mentiond alcohols. Such measures are often necessary to achieve a satisfactory
extraction of polar compounds using conventionally coated fibers.

Because of direct chemical bonding to the fused silica fiber surface, sol-gel
SPME coatings provide substantially higher thermal stability than conventional
SPME fibers with physically held stationary phase coatings. Figure 3 illustrates
the thermal stability of a sol-gel-coated SPME fiber with PDMS-based sol-gel
coating using an example of SPME–GC analysis of polycyclic aromatic
hydrocarbons. In this example the analyte desorption temperature was 360 °C.

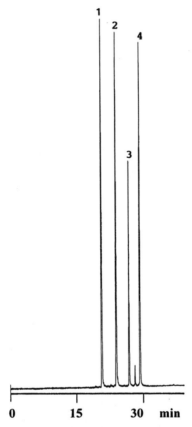

Figure 2 *Direct SPME–GC analysis of aliphatic alcohols from an aqueous sample matrix. Extraction conditions. Fiber: 200 μm, fused silica; Stationary phase: sol-gel PDMS; Coating thickness: 10 μm; direct 30 min extraction with stirring (no pH adjustment and no salting out); Injection conditions. Injector temperature: 250 °C; Carrier gas: helium; Injection mode: split-splitless (First 5 min splitless, followed by split with a ratio of 100:1); GC conditions. Column: 10 m × 250 μm i.d.; Stationary phase: sol-gel PDMS; Temperature programming: 40 °C (5 min), 6 °C min⁻¹; Detector: FID, 300 °C. Peak identifications: (1) C₁₂, (2) C₁₄, (3) C₁₆, (4) C₁₈ straight-chain aliphatic alcohols*

By comparison, conventionally coated PDMS fibers often start bleeding at desorption temperatures as low as 200 °C.[12] It should be mentioned that in our thermal stability evaluation experiments with sol-gel coated fibers we have not yet reached the upper temperature limit for the operation of such fibers. The upper temperature limit for sol-gel-coated fibers is likely to be significantly higher than 360 °C. In an on-going study on the thermal stability of sol-gel coated GC columns, we have established that PDMS-based sol-gel coatings in GC columns can easily withstand a conditioning temperature of 430 °C.[30] High thermal stability of sol-gel coatings in SPME and GC open new possibilities for extending the scope of SPME–GC analysis to higher boiling compounds

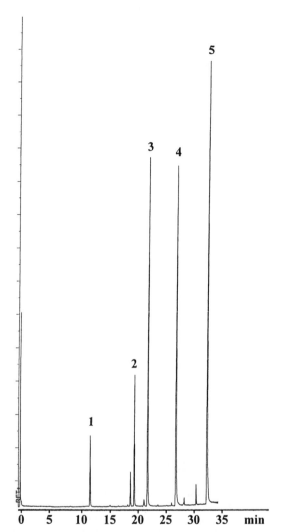

Figure 3   *Gas chromatogram of direct SPME–GC analysis of polycyclic aromatic hydro-*
carbons. Extraction conditions. Fiber: 200 μm, fused silica; Stationary phase:
sol-gel PDMS; Coating thickness: 10 μm; direct 30 min extraction with stirring
(no pH adjustment and no salting out); Injection conditions. Injector tempera-
ture: 340 °C; Carrier gas: helium; Injection mode: split-splitless (First 5 min
splitless, followed by split with a ratio of 100:1); GC conditions. Column: 10 m
× 250 μm i.d.; Stationary phase: sol-gel PDMS; Temperature programming:
40 °C (5 min), 6 °C min^{-1}; Detector: FID, 360 °C. Peak identifications:
(1) naphthalene, (2) acenaphthylene, (3) fluorene, (4) phenanthrene, (5)
fluoranthene

beyond the limit that is currently amenable to SPME–GC technology using
conventionally coated fibers.

Higher thermal stability of the sol-gel coated fibers also provides a versatile
solution for the carryover problem that is often encountered in SPME–GC

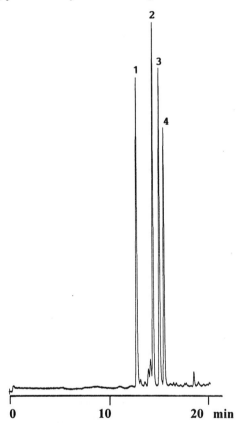

Figure 4 *Direct extraction by SPME from an aqueous sample matrix followed by GC analysis of dimethylphenol isomers. Extraction conditions. Fiber: 200 μm, fused silica; Stationary phase: sol-gel PDMS; Coating thickness: 10 μm; direct 60 min extraction with stirring (no pH adjustment and no salting out); Injection conditions. Injector temperature: 300 °C; Carrier gas: helium; Injection mode: split-splitless (5 min splitless, followed by split with a ratio of 100:1); GC conditions. Column: 10 m × 250 μm i.d.; Stationary phase: sol-gel PDMS; Temperature programming: 40 °C (5 min), 6 °C min⁻¹; Detector: FID, 300 °C. Peak identifications: (1) 2,6-dimethylphenol, (2) 2,5-dimethylphenol, (3) 2,3-dimethylphenol, (4) 3,4-dimethylphenol*

analyses of polar compounds.[11] Enhanced thermal stability of sol-gel coated fibers allows for efficient desorption of the extracted analytes into the GC injection port at significantly higher temperatures than are permissible with conventionally coated fibers without harming the stationary phase coating. Figure 4 presents an example to illustrate this point. An aqueous sample of dimethylphenol isomers was extracted and analyzed by SPME–GC with sol-gel-coated fibers and a sol-gel coated open tubular column. Carryover was not a problem for this phenolic sample, and no sample derivatization was necessary. With conventionally coated PDMS fibers derivatization is often necessary for phenolic compounds to reduce the carryover problem.[16]

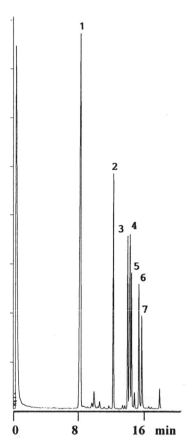

Figure 5 *Direct SPME–GC analysis of aniline derivatives from an aqueous matrix.*
Extraction conditions. Fiber: 200 μm, fused silica; Stationary phase: sol-gel
PDMS; Coating thickness: 10 μm; direct 60 min extraction with stirring (no pH
adjustment and no salting out); Injection conditions. Injector temperature:
250 °C; Carrier gas: helium; Injection mode: split-splitless (5 min splitless,
followed by split with a ratio of 100:1); GC conditions. Column: 10 m × 250 μm
i.d.; Stationary phase: sol-gel Ucon; Temperature programming: 40 °C (5 min),
6 °C min^{-1}; Detector: FID, 300 °C. Peak identifications: (1) N,N-dimethylani-
line, (2) N-methylaniline, (3) N-ethylaniline, (4) 2,6-dimethylaniline, (5) 2-
ethylaniline, (6) 4-ethylaniline, (7) 3-ethylaniline

Figure 5 represents another example of SPME–GC analysis of polar compounds with sol-gel-coated PDMS fiber. In this case, the analytes are aromatic
amines. As in the previous two examples, efficient extraction was achieved for
these closely-related aniline derivatives that were also efficiently analyzed on a
sol-gel-coated GC column. Figure 6 represents an important example illustrating the possibility of simultaneous extraction of both polar and nonpolar
compounds from the same sample using sol-gel coated fibers. Here an aqueous
sample containing trace concentrations of naphthalene, *N*-methylaniline and

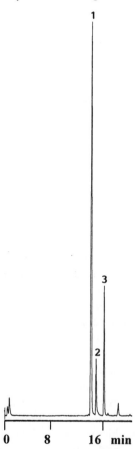

Figure 6 *SPME–GC analysis of an aqueous sample containing both polar and nonpolar compounds. Extraction conditions. Fiber: 200 µm, fused silica; Stationary phase: sol-gel PDMS; Coating thickness: 10 µm; direct 60 min extraction with stirring (no pH adjustment and no salting out). Injection conditions. Injector temperature: 250 °C; Carrier gas: helium. Injection mode: split-splitless (5 min splitless, followed by split with a ratio of 100:1); GC conditions. Column: 10 m × 250 µm i.d.; Stationary phase: Sol-gel PDMS; Temperature programming: 40 °C (5 min), 6 °C min⁻¹; Detector: FID, 300 °C. Peak identifications: (1) naphthalene, (2) N-methylaniline, (3) 2,3-dimethylphenol*

2,3-dimethylphenol was used for their extraction using a sol-gel-coated PDMS fiber. The used fiber showed, as is evident from the presented chromatogram, a higher affinity for the nonpolar analyte, naphthalene. However, appropriate adjustments in the composition of the coating sol solution should allow for the creation of coatings with higher affinity for polar compounds. As in the previous examples, here also no salting out or pH adjustments were used.

Scanning electron microscopic (SEM) investigation revealed that sol-gel coatings possess a porous structure[9,21] that provides enhanced surface area for the analytes to interact with the stationary phase. This should allow for the use of fibers with less coating thickness to achieve reasonable sample capacity for which conventionally prepared fibers would require a higher coating thickness. The use of thicker coatings essentially leads to slower mass transfer processes during extraction of analytes and their desorption from the fiber. Ultimately, this results in longer extraction times, and may also lead to sample carryover problems due to incomplete analyte desorption in the GC injection port. The porous structure of the sol-gel coatings provides fast sorption–desorption kinetics during extraction as well as sample introduction. Our experimental results on direct extraction of a number of polar and nonpolar analytes from aqueous matrices demonstrated[9] that sol-gel coated PDMS fibers required less than 10 min to reach extraction equilibria. Besides, fast desorption of the analytes from the porous sol-gel stationary phase coatings, in combination with their capability of withstanding high desorption temperatures, should make it possible to use shorter analyte desorption times for sol-gel coated fibers. This aspect of sol-gel coatings performance is currently being investigated in our laboratory.[31] Recently we demonstrated that this combination led to significant improvement in precision of SPME–GC analysis data.[9]

Sol-gel coatings are commonly prepared using tetraalkoxysilane precursors. However, it was shown by Mackenzie[24] that hydrolytic polycondensation of such precursors leads to sol-gel networks with very compact structures that are prone to cracking due to capillary thrusts generated during solvent evaporation from the pores.[32] Gel cracking is a thickness-dependent phenomenon, and may adversely affect the coating performance. According to Atkinson and Guppy[33] sol-gel coatings with thicknesses below 0.5 μm are practically crack-free. Since stationary phase coatings used in GC columns usually possess submicron thickness (*e.g.* a typical film thickness is 0.25 μm), cracking should not be a problem for high-efficiency performance of sol-gel coated GC columns. However, to achieve reasonable sample capacity, relatively high coating thicknesses (10–100 μm) are usually used on SPME fibers. For such coatings, cracking may have a significant effect on the fiber performance. Therefore, efficient application of sol-gel coatings to SPME fiber technology requires a satisfactory solution to this cracking problem.

The following equation describes the capillary forces that lead to the cracking of sol-gel coatings or monoliths during their drying.[34]

$$P = \frac{2\gamma \cos \theta}{r}$$

where P = capillary pressure generated in the pore, γ = surface tension of the pore liquid, θ = contact angle, r = radius of the cylindrical pore.

This equation suggests that during drying a pressure differential (capillary thrust) exists between two neighboring pores of different radii. This pressure differential works on the wall separating the two pores. The pore wall will crack if this capillary thrust exceeds the tensile strength of the wall material. A closer look into the above equation provides a number of possible solutions to this problem.[35] Since the pressure differential arises during drying due to a difference in pore diameters in the gel, sol-gel coatings containing identical (or very close) pore sizes will be subjected to no (or very little) capillary thrusts, and consequently will not suffer cracking. The use of drying control chemical additives (DCCA) leads to gel structures with uniform pore size, and may serve as a possible solution to the cracking problem.[36] Formamide is often used for this purpose.[37,38] Since the capillary thrust causing the gel cracking is directly proportional to the surface tension of the pore-liquid, the use of solvents with low surface tensions may provide another possible solution to the gel cracking problem. Supercritical fluids possess zero surface tension[39] and hence their use in the drying process may provide another alternative solution to the cracking problem.[40] A third possibility is to create a gel with more open structure that can efficiently relieve capillary strain generated during drying. Mackenzie *et al.*[24] showed that gels with such open structures can be prepared by using alkyl- or aryl-derivatized alkoxysilane precursors. In our research, this approach proved to be very effective in creating crack-free sol-gel thick coatings for solid-phase microextraction. Both in the present research and in our previous work,[9] methyltrimethoxysilane was used as the as the sol-gel precursor.

Because sol-gel coatings are chemically bonded to the fiber or capillary inner surface, such coatings provide enhanced stability to organic solvents. Whereas conventionally coated SPME fibers are not recommended to rinse with or expose to organic solvents,[41] in our laboratory we clean sol-gel coated fibers by rinsing with organic solvents. This outstanding solvent-stability of sol-gel coated fibers opens new possibilities for their application to direct extraction of polar compounds from organic solvent-rich matrices. Because of high solvent-stability, sol-gel coated fibers have great potential to become an appropriate tool that will facilitate effective and reliable interfacing of SPME with liquid- or fluid-based separation techniques. Currently, we are exploring these possibilities. Further development in this area can be expected in the coming years.

It should be noted that sol-gel SPME fiber technology is still in its early stages of development. Further studies are needed to explore the full potential and limitations of this technique. Of particular importance are: (1) characterization of the structural aspects of sol-gel coatings (including pore size, volume, their distribution, coating uniformity, *etc.*) and their influence on fiber performance, (2) developing practical methods for the control and fine-tuning of these structural characteristics, (3) conducting a detailed study on reproducibility aspects of sol-gel coating procedure, and performance of the prepared fibers, (4) exploring the influence of various parameters on structural characteristics, stability, and performance of sol-gel fibers, (5) investigating sample capacity of sol-gel coatings for various types of analytes and (6) comparing the performance of sol-gel coated fibers with that of commercially available coatings using various types of test solutes and matrices. The influence of moisture on the performance of sol-gel-coated fibers may be another aspect of further investigation.

Acknowldegements

Financial support for this research was provided, in part, by the University of South Florida (USF) Research and Creative Scholarship Grant Program under Grant #12-13-939-R0 and by the Faculty Development Program of USF College of Arts and Sciences.

Rererences

1. J. Pawliszyn, *Solid-Phase Microextraction. Theory and Practice*, Wiley-VCH, New York, 1997.
2. J. Pawliszyn and S. Liu, *Anal. Chem.*, 1987, **59**, 1475.
3. R. Berlardi and J. Pawliszyn, *Water Pollut. Res. J. Can.*, 1989, **24**, 179.
4. C.L. Arthur and J. Pawliszyn, *Anal. Chem.*, 1990, **62**, 2145.
5. Z.Y. Zhang, M.J. Yang and J. Pawliszyn, *Anal. Chem.*, 1994, **66**, A844.
6. H. Lakso and W.F. Ng, *Anal. Chem.*, 1997, **69**, 1866.
7. R. Eisert and J. Pawliszyn, *Anal. Chem.*, 1997, **69**, 3140.
8. H. Hartmann, J. Burhenne and M. Spiteller, *Fres. Environ. Bull.*, 1998, **7**, 96.
9. S.-L. Chong, D.-X. Wang, J.D. Hayes, B.W. Wilhite and A. Malik, *Anal. Chem.*, 1997, **69**, 3889.
10. S.-L. Chong, *Sol-Gel Coating Technologyfor the Preparation of Solid-Phase Microextraction Fibers of Enhanced Thermal Stability*, Masters Thesis, Department of Chemistry, University of South Florida, FL, 1997.
11. K.D. Buchholz and J. Pawliszyn, *Environ. Sci. Technol.*, 1993, **27**, 2844.
12. K.D. Buchholz and J. Pawliszyn, *Anal. Chem.*, 1994, **66**, 160.
13. W.H.J. Vaes, C. Hamwijk, E.U. Ramos, H.J.M. Verhaar and J.L.M. Hermens, *Anal. Chem.*, 1996, **68**, 4458.
14. A.A. Boyd-Boland and J. Pawliszyn, *Anal. Chem.*, 1996, **68**, 1521.
15. J.-L. Liao, C.-M. Zeng, S. Hjerten and J. Pawliszyn, *J. Microcol. Sep.*, 1996, **8**, 1.
16. T. Gorecki, P. Martos and J. Pawliszyn, *Anal. Chem.*, 1998, **70**, 19.
17. J. Bouche and J. Verzele, *J. Gas Chromatogr.*, 1968, **6**, 501.
18. Y. Liu, Y. Shen and M.L. Lee, *Anal. Chem.*, 1997, **69**, 190.
19. D. Djozan and Y. Assadi, *Chromatographia*, 1997, **45**, 183.
20. F. Mangini and R. Cenciarini, *Chromatographia*, 1995, **41**, 678.
21. D.-X. Wang, S.-L. Chong and A. Malik, *Anal. Chem.*, 1997, **69**, 4566.
22. J.D. Hayes and A. Malik, *J. Chromatogr. B*, 1997, **695**, 3.
23. C.J. Brinker and G.W. Scherer, *Sol-Gel Science. The Physics and Chemistry of Sol-Gel Processing*, Academic Press, San Diego, CA, 1990.
24. J.D. Mackenzie, in J.E. Mark, C.Y. Lee and P.A. Bianconi (eds.), *Hybrid Organic–Inorganic Composites* (ACS Symp. Ser. 585, American Chemical Society, Washington, DC., 1995), pp. 226–236.
25. J. Livage, M. Henry and C. Sanchez, *Solid State Chem.*, 1988, **18**, 259.
26. H. Dislich, in L.C. Klein (ed.), *Sol-Gel Technology for Thin Films, Fibers, Preforms, Electronics, and Specialty Shapes*, Noyes Publications: Park Ridge, NJ, 1988, pp. 50–79.
27. D.R. Uhlmann and G.P. Rajendran, in J.D. Mackenzie and D.R. Ulrich (eds.), *Ultrastructure Processing of Advanced Ceramics*, Wiley, New York, 1988, pp. 241–253.

28. A. Gupta, K. Biswas, A.B. Mallick and S. Mukherjee, *Bull. Mater. Sci.*, 1995, **18**, 497.
29. K. Nakanishi, H. Minakuchi, N. Soga and N. Tanaka, *Sol-Gel Sci. Technol.*, 1997, **8**, 547.
30. D.-X. Wang, F. Brignol and A. Malik, manuscript in preparation.
31. S.-L. Chong and A. Malik, study in progress.
32. G.W. Scherer, *J. Non-Cryst. Solids*, 1988, **100**, 77.
33. A. Atkinson and R.M. Guppy, in W.D. Nix, J.C. Bravman, E. Arzt and L.B. Freund (eds.), *Thin Films: Stresses and Mechanical Properties III*, Materials Research Society Symposium Proceedings, Vol. 230, MRS, Pittsburgh, PA. USA, 1992, pp. 553–559.
34. J. Zarzycki, M. Prassas and J. Phalippou, *J. Mater. Sci.*, 1982, **17**, 3371.
35. J.D.F. Ramsey, in D.J. Wedlock (ed.), *Controlled Particle, Droplet, and Bubble Formation*, Butterworth-Heineman, London, 1994, p. 30.
36. L.L. Hench, in L.L. Hench and D.R. Ulrich (eds.), *Science Ceramic Chemical Processing*, Wiley, New York, 1986, p. 52.
37. G. Orcel and L.L. Hench, *J. Non-Cryst. Solids*, 1986, **79**, 177.
38. N. Viart, D. Niznansky and J.L. Rehspringer, *J. Sol-Gel Sci. Technol.*, 1997, **8**, 183.
39. M. McHugh and V. Krukonis, *Supercritical Fluid Extraction: Principles and Practices*, Butterworth, Boston, 1986, p. 9.
40. S.S. Kistler, *J. Phys. Chem.*, 1932, **36**, 52.
41. Supelco 1998 Chromatography Products Catalog, p. 325.

CHAPTER 7

Solid versus Liquid Coatings

TADEUSZ GÓRECKI

1 Introduction

There are two distinct types of SPME coatings currently available from
Supelco. The most widely used poly(dimethylsiloxane) (PDMS) is a liquid
coating. Even though it looks like a solid, it is in fact a high-viscosity rubbery
liquid. Poly(acrylate) (PA) is a solid crystalline coating that turns into liquid at
desorption temperatures. Both PDMS and PA extract analytes via absorption.
The remaining coatings, including PDMS/DVB (divinylbenzene), Carbowax/
DVB, Carbowax/TR and Carboxen, are mixed coatings, in which the primary
extracting phase is a porous solid. Those coatings extract analytes *via* adsorption rather than absorption. Similarity of the names can be deceptive, since the
mechanisms of absorption and adsorption are different. The only common
feature is that both processes start with analyte molecules getting attached to
the surface of the coating. However, in absorption, analytes dissolve in the
coating and diffuse into the bulk of it during the extraction process, while in
adsorption they stay on the surface of the solid. It is the goal of this chapter to
explain the differences between the two processes and present the consequences
for quantitative analysis by SPME with the use of the two coating types.

2 Absorption *vs.* Adsorption[1,2]

Absorption is a process based on partitioning of a compound between two
immiscible phases, *e.g.* an organic solvent and water. The partitioning process is
determined by the relative fugacity of a compound in each phase, and at
equilibrium may be described for a two-phase system by a dimensionless
equilibrium constant (partition coefficient) K:

$$K = \frac{C_o^\infty}{C_w^\infty} \tag{1}$$

where C_o^∞ is the equilibrium concentration of a compound in the organic phase
(*e.g.* solvent, fibre coating), and C_w is the equilibrium concentration in water.

When the organic phase is a liquid coating on the SPME fibre, the partitioning coefficient is usually denoted as K_{fw} (see Chapter 1). The fugacity of a molecule in a given phase is determined by the chemical potential of this molecule in the phase. When fugacity f is high, the molecule has a tendency to 'escape' from the phase. The chemical potential μ_i of a compound in a solution is given by:

$$\mu_i = \mu^0 + RT \ln \gamma_i x_i \qquad (2)$$

where μ^0 is the chemical potential of the compound in the reference state (in this case pure liquid compound), R is gas constant, T is absolute temperature (K), γ_i is the activity coefficient of a compound in phase i, and x_i is the molar fraction of the compound in this phase. When two immiscible phases are contacted (*e.g.* organic solvent and water), the compound of interest is transferred between the two phases until equilibrium is reached, *i.e.* the chemical potentials of the compound in these two phases are equal:

$$RT \ln \gamma_o x_o = RT \ln \gamma_w x_w \qquad (3)$$

or

$$\gamma_o x_o = \gamma_w x_w \qquad (4)$$

where the indices o and w refer to organic solvent and water, respectively.

Since molar fractions are rarely used in analytical chemistry, it is more convenient to express equation 4 in terms of molar concentrations:

$$\gamma_o C_o V_{Mo} = \gamma_w C_w V_{Mw} \qquad (5)$$

where V_{Mo} and V_{Mw} are the molar volumes of the organic phase and water, respectively. Taking the above into account, the partitioning coefficient for a fibre extracting analytes by absorption can be expressed as:

$$K_{fw} = \frac{C_f^\infty}{C_w^\infty} = \frac{\gamma_w}{\gamma_f} \cdot \frac{V_{Mw}}{V_{Mf}} \qquad (6)$$

Molar volumes of the solvent (fibre coating) and water are unaffected by the presence of small amounts of organic molecules and can be assumed constant. The activity coefficients for nonpolar organic compounds in water can be quite large, ranging from 10^2 to 10^{11} (when pure liquid is used as the reference state). In contrast, the activity coefficients of such compounds in organic solvents (hence also in the fibre coating) are very small and often close to one, since the compound molecules experience very similar interactions to those in their pure liquid phase (for which $\gamma^0 = 1$).[1] Consequently, the magnitude of K_{fw} for nonpolar solutes is determined primarily by γ_w, the activity coefficient of a compound in water. Compounds with polar functional groups, which favour the aqueous phase and simultaneously disfavour very nonpolar organic phases, have higher activity coefficients in the coating and lower activity coefficients in water. As a result, K_{fw} for such compounds is usually much lower.

Several extremely important consequences for SPME stem from the above considerations. First of all, since the partition coefficient is determined primarily by the activity coefficient of an analyte in water, and this activity coefficient does not depend on the presence of other organic molecules in water (as long as their concentration is low), absorption is, in general, a non-competitive process. This means that the presence of other organic molecules in the matrix should not affect the uptake of the analyte of interest. The above statement is not true only when another organic compound is present in water in concentrations high enough to be treated as a co-solvent. In such a case, the activity coefficients of nonpolar organic compounds in water decrease, and consequently partition coefficients can decrease as well. On the other hand, when the concentration of an organic solvent in water is very high (which is possible only for polar solvents), or when a free nonpolar organic phase is present in water, the coating itself can also be affected. In the former case, the polarity of the coating may change since it becomes a mixed phase, and consequently the activity coefficients of nonpolar molecules in the coating also may change (increase for nonpolar compounds and decrease for polar compounds). In the latter case, the polarity of the coating remains almost unchanged, but its volume, and thus sorption capacity, increases. However, taking into account that SPME is geared towards trace analysis, in the vast majority of practical cases the assumption of non-competitiveness of the absorption process holds true.

Equation 6 points also to another very significant conclusion. Since the PDMS coating is a nonpolar liquid in which the activity coefficients of nonpolar organic molecules are close to one, and the partition coefficients of those molecules are determined almost exclusively by their activity coefficients in water, it cannot be expected that liquid coatings performing dramatically better for nonpolar molecules can be developed. The driving force for absorption, being the difference between the activity coefficients of analyte molecules in water and in the nonpolar coating, cannot be made much larger by decreasing the activity coefficients in the coating. There is some room for improvement in liquid coatings for polar analytes, but, realistically, no revolutionary break-through can be expected for those coatings either. For the activity coefficients in an SPME coating to be low for very polar compounds, the coating would have to be very polar as well, in which case it could become water-soluble. Also, even if the activity coefficients of polar analytes in the hypothetical liquid coating would be close to one, their activity coefficients in water would be relatively low as well ('like dissolves like'), and so would be the partition coefficients (see equation 6). An improvement over the currently available polar PA phase can be easily envisioned, yet one cannot expect very dramatic gains in sensitivity as long as pure absorption is responsible for the extraction process. Should the liquid coating contain functional groups that specifically interact with selected analytes, the sensitivity could be improved dramatically, but the extraction process in this case would be better described as chemisorption rather than absorption, which is outside the scope of this chapter.

Porous polymer-based solid extracting phases, including PDMS/DVB, Carbowax/DVB and Carboxen, constitute the second group of commercially available SPME fibre coatings. These coatings extract analytes via adsorption. The mechanisms of adsorption can be numerous. Molecules can be associated with surfaces *via* van der Waals, dipole–dipole, and other weak intermolecular forces. Poor solubility in water (hydrophobic interaction) drives the molecules out of it and onto solid surfaces. If the sorbate is ionizable in the aqueous phase, and the extracting phase exhibits the opposite charge, electrostatic interactions will occur. Finally, should the sorbate and sorbent exhibit mutually reactive moieties, some portion of the chemical may actually become bonded to the solid. Another important process that affects the sorption capacity of solids is capillary condensation. If the solid is porous, and the pore diameter is small enough (less than 100 Å), some molecules will tend to condense into small pores. When fluid condenses in a capillary, the radius of curvature of the fluid is usually negative, and as a result the vapour pressure of the fluid in a small pore is lower than the bulk vapour pressure of that fluid.[2] This process will continue until the level of the condensate in a pore reaches a point where the diameter of the pore becomes so large that the vapour pressure above the liquid becomes equal to the outside vapour pressure. If the pore diameter is uniform along its length, it can be nearly filled with liquid analyte at this stage, which explains the very high sorption capacity of some porous sorbents.

Not all of the above phenomena apply to commercially available solid SPME coatings. As of today, no charged coatings are available that would be able to extract ionized species. Chemisorption would be rather disadvantageous in SPME, since it would be very difficult to desorb the analytes for separation and final determination. Weak intermolecular interactions, as well as hydrophobic interactions when sampling from water, play the most important role for the three basic types of solid coatings available. Additionally, capillary condensation is important for Carboxen fibres, characterized by very high surface area and small pore sizes.

The fundamental difference between adsorption and absorption is that in adsorption molecules bind directly to the surface of a solid, while in absorption they dissolve into the bulk of the fluid. This difference can be illustrated by the following example. Let us assume that the extracting phase is a non-porous, large particle. Should the molecules be extracted by absorption, the amount extracted at equilibrium would not change if the particle was broken into many small pieces. On the other hand, the amount of molecules that adsorb onto the surface of a solid is proportional to the total surface area of the solid and not its volume or mass. Breaking up the extracting particle into many small pieces would significantly increase the total surface area, thus also the amount adsorbed.

Another crucial difference between absorption and adsorption is that in adsorption there are a limited number of surface sites where the process can take place. Two liquids can be miscible in any proportion, which means that no 'saturation' should ever occur in absorption, while in adsorption, when all active sites are occupied, no more analyte can be trapped (unless it can condense in small pores by the capillary condensation mechanism). This means that the dynamic range of solid, adsorption-type coatings cannot be as broad as for liquid coatings, and that the dependence between the concentration of an analyte in the sample and the amount extracted by adsorption-type SPME coating can be linear only in narrow concentration ranges. In addition, while absorption is a non-competitive process, adsorption is by definition competitive—molecules 'fight' for the available free sites on the surface of the solid. A molecule with higher affinity for the surface can replace a molecule with lower affinity. This means that the amount of an analyte extracted by the fibre from a sample can be affected by the matrix composition. A quantitative description of equilibrium absorption process is presented in Chapter 1. Following is a quantitative description of equilibrium extraction for adsorption-type solid coatings in which capillary condensation does not occur.

3   Theory of Analyte Extraction by Porous Polymer Coatings[3]

The following considerations pertain to PDMS/DVB and Carbowax/DVB fibres only. No equilibrium adsorption theory has been developed yet for the Carboxen fibre, for which capillary condensation plays an important role.

The dependence between the equilibrium concentration of a compound associated with the sorbent and its concentration in the solution is commonly referred to as adsorption isotherm. The term isotherm is used to indicate that one is considering sorption at a constant temperature. Depending on the dominating mechanisms, sorption isotherms may exhibit different shapes, as illustrated in Figure 1. Experimentally determined isotherms can commonly be fit with a relationship of the form:[2]

$$C_{ad} = K' C_s^n \tag{7}$$

where C_{ad} is the concentration of a compound on the solid surface, K' is a constant and n is the measure of nonlinearity involved. Equation 7 is known as the Freundlich isotherm. For SPME fibres, sorption isotherms similar to those for Case I are almost exclusively observed. Case II represents a situation when the 'attractiveness' of the solid surface for the sorbate remains the same for all levels of C_s. The linear isotherm observed in such a case would be ideal from the point of view of quantitative analysis by SPME. Unfortunately, such isotherms are observed only for relatively narrow ranges of C_s at low concentrations.

Mathematical description of the Freundlich isotherm given by equation 7 is purely phenomenological, *i.e.* it is not based on any physical model. While it is easy to fit experimental data to Freundlich isotherm, a much deeper insight into the adsorption process can be gained by applying a theory based on an actual model of the process. One of the oldest adsorption theories is that of Langmuir. While it has been refined many times to describe a broader range of specific cases, its basic assumptions remain generally unchanged. It will be illustrated

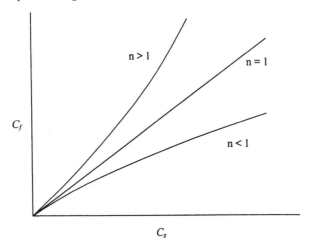

Figure 1   *Types of relationships between concentrations of sorbed and dissolved chemical (Freundlich isotherms)*

later in this chapter that the Langmuir adsorption isotherm well describes equilibrium analyte extraction by PDMS/DVB and Carbowax/DVB fibres; therefore it has been used to develop the theoretical description of the process that follows.

In the Langmuir model, the surface has a limited number of adsorption sites that can be occupied by the sorbate. The following assumptions apply: (1) the adsorbing molecule adsorbs into an immobile state; (2) all sites are equivalent; (3) each site can hold at most one molecule of the adsorbate; and (4) there are no interactions between adsorbate molecules on adjacent siles so that the equilibrium constant is independent of the coverage of the adsorbed species.[2] Assumption (3) means that a monolayer of the adsorbate can be formed at the surface at the most. Adsorption is treated as a reaction where a molecule A reacts with an empty site, S, to yield an adsorbed complex A_{ad}:

$$A + S \Leftrightarrow A_{ad} \tag{8}$$

It is assumed that adsorption and desorption are elementary processes where the rate of adsorption r_{ad} and desorption r_d are given by:

$$r_{ad} = k_{ad}[A][S] \tag{9}$$

$$r_d = k_d[A_{ad}] \tag{10}$$

where [A] is the concentration of A in the matrix, [S] is the concentration of unoccupied sites on the surface of the sorbent in mol cm^{-2}, $[A_{ad}]$ is the surface concentration of A in mol cm^{-2}, and k_{ad} and k_d are constants.

At equilibrium, the rate of adsorption equals the rate of desorption, hence

$$\frac{[A_{ad}]}{[A][S]} = \frac{k_{ad}}{k_d} = \Psi \tag{11}$$

where Ψ denotes equilibrium constant (this symbol will be used rather than the more common K to denote equilibrium constant to avoid confusion with the partition coefficient). When only a single compound undergoes adsorption, the site balance is:

$$[S] = [S_0] - [A_{ad}] \tag{12}$$

where $[S_0]$ is the total concentration of active sites on the surface ($=$ maximum concentration of the analyte) in mol cm^{-2}. Substituting equation 12 into equation 11 yields:

$$\frac{[A_{ad}]}{[A]\Psi} + [A_{ad}] = [S_0] \tag{13}$$

Hence

$$[A_{ad}] = [S_0]\frac{\Psi[A]}{1 + \Psi[A]} \tag{14}$$

It would be cumbersome to use surface concentration in mol cm^{-2} for the description of the SPME process. However, if we assume that the sorbent has a uniform pore size distribution and surface area throughout its bulk, surface concentrations can be replaced by bulk concentrations by multiplying both sides of equation 14 by the term Φ/V_f, where Φ is the area (in cm^2). We can now define the concentration of the analyte on the fibre C_{fA} and the maximum concentration of active sites on the coating $C_{f\,max}$ in the following way:

$$C_{fA} = [A_{ad}]\frac{\Phi}{V_f} \tag{15}$$

$$C_{f\,max} = [S_0]\frac{\Phi}{V_f} \tag{16}$$

We will also use the symbol C_{sA}^{∞} instead of $[A]$ to denote analyte concentration in the sample at equilibrium. From these, we can define the equilibrium concentration of an analyte on the fibre:

$$C_{fA}^{\infty} = \frac{C_{f\,max}\Psi C_{sA}^{\infty}}{1 + \Psi C_{sA}^{\infty}} \tag{17}$$

It is already evident that the concentration of an analyte on the fibre is *not* a linear function of its concentration in the sample, except when the product ΨC_{sA}^{∞} is much smaller than one. This may happen when either the affinity of an analyte towards the coating is low, or its concentration in the sample is very low. The reciprocal of this equation yields:

$$\frac{1}{C_{fA}^{\infty}} = \frac{1}{C_{f\max}} + \frac{1}{C_{f\max}\Psi C_{sA}^{\infty}} \tag{18}$$

Therefore the plot of $1/C_{fA}^{\infty}$ vs. $1/C_{sA}^{\infty}$ should be a straight line with a slope of $1/C_{f\max}\Psi$ and an intercept of $1/C_{f\max}$.

Even though equation 17 allows the determination of analyte concentration on the fibre at equilibrium (and thus the amount extracted by the fibre), it is difficult to use in practice, since it requires the knowledge of the analyte concentration in the sample at equilibrium. It seems much more interesting to determine the dependence between the *initial* concentration of the analyte in the sample (C_{0A}) and the amount extracted. Mass balance can be used for this purpose:

$$C_{0A}V_s = C_{sA}^{\infty}V_s + C_f^{\infty}V_f \tag{19}$$

From equation 17, equilibrium concentration of the analyte is:

$$C_{sA}^{\infty} = \frac{C_f^{\infty}}{\Psi(C_{f\max} - C_{fA}^{\infty})} \tag{20}$$

By combining equations 19 and 20, after a few rearrangements one gets:

$$n = C_{fA}^{\infty}V_f = \frac{\Psi C_{0A}V_s V_f(C_{f\max} - C_{fA}^{\infty})}{V_s + \Psi V_f(C_{f\max} - C_{fA}^{\infty})} \tag{21}$$

Equation 21 is an iterative dependence, since analyte concentration in the fibre at equilibrium appears on both its sides. Nevertheless, it gives a valuable insight into the nature of analyte extraction with porous polymer coatings. The form of the dependence is very similar to that of the equation for n when *liquid* coatings extracting analytes by absorption rather than adsorption are used (equation 5, Chapter 1):

$$n = \frac{KC_0 V_s V_f}{V_s + KV_f} \tag{22}$$

The main difference between equations 21 and 22 is the presence of the fibre concentration term ($C_{f\max} - C_{fA}^{\infty}$) in the numerator and denominator of equation 21. For very low analyte concentrations on the fibre, it can be assumed that $C_{f\max} \gg C_{fA}^{\infty}$. For this condition to be fulfilled, the analyte's concentration in the sample and/or its affinity for the fibre must be very low. When these requirements are met, a linear dependence should be expected. If, however, the amount of an analyte on the fibre is not negligible compared to the total number of active sites, the dependence ceases to be linear.

Equation 21 is in fact a quadratic equation, which can be solved analytically. Of the two roots obtained, only one has a physical meaning (the other one produces n values that are higher than the original amount of the analyte in the sample). The solution is:

$$n = \frac{C_{f\max}\Psi V_f + C_{0A}V_s\Psi + V_s - \sqrt{\Psi^2(C_{f\max}V_f - C_{0A}V_s)^2 + 2\Psi V_s(C_{f\max}V_f + C_{0A}V_s) + V_s^2}}{2\Psi}$$

(23)

A discussion of this dependence is presented in Section 4 of this chapter.

In real life situations, one can very rarely assume that only one compound will be extracted by the coating. Since adsorption is a competitive process, the presence of other compounds must affect the amount of analyte A extracted by the fibre (n_A). In the following derivation, only one competing compound is taken into account. The same reasoning can be applied, however, to any number of compounds present in the sample.

The concentration of analyte A on the fibre in the presence of a competing compound B is given by the following equation:[2,4]

$$C_{fA}^\infty = \frac{C_{f\max}\Psi_A C_{sA}^\infty}{1 + \Psi_A C_{sA}^\infty + \Psi_B C_{sB}^\infty}$$

(24)

where Ψ_A and Ψ_B are the adsorption equilibrium constants for compounds A and B, respectively, and C_{sB}^∞ is the equilibrium concentration of B in the sample. If more then two compounds were present in the sample, the denominator would contain additional $\Psi_i C_{s_i}^\infty$ terms. The mass balance for A is again described by equation 19. A derivation similar to that described above yields the following relationship:

$$n_a = C_{fA}^\infty V_f = \frac{\Psi_A C_{0A} V_s V_f(C_{f\max} - C_{fA}^\infty)}{(1 + \Psi_B C_{sB}^\infty)V_s + \Psi_A V_f(C_{f\max} - C_{fA}^\infty)}$$

(25)

It is clear from the above equation that n_A must be lower than n from equation 21, as there is an additional term in the denominator, which can only be greater than one. The difference does not have to be dramatic if the second term in the denominator is much larger than the first one, which can occur when the interfering compound is either present at a very low concentration, and/or is characterized by small affinity to the coating. In all other cases, one can expect that n_A will be significantly lower than n.

What is less obvious is the fact that adsorption of interfering compounds affects also the linear range of extraction. The term C_{sB}^∞ is the *equilibrium* concentration of B. Unless the volume of the sample is very large, in which case the equilibrium concentration of B is equal to its initial concentration, C_{sB}^∞

depends on the initial concentration of B and A in the same complex way in which C_{sA}^{∞} depends on C_{0A} and C_{sB}^{∞} (thus C_{0B}). Incorporating this dependence into equation 25 would make it very complex. Instead, we can picture this dependence in the following way: when B adsorbs on the surface of the coating, it effectively reduces the number of adsorption sites available for A. This means that $C_{f\,max}$ is lower, hence the non-linear range starts at lower concentrations of A compared with the case when the sample contains no interfering compounds. Equation 25 can be solved in the same way as equation 21. The only root with a physical meaning has the form:

$$n_A = C_{fA}^{\infty} V_f = \frac{C_{f\,max}\Psi_A V_f + C_{0A} V_s \Psi_A + V_s(1 + \Psi_B C_{sB}^{\infty})}{2\Psi_A}$$

$$- \frac{\sqrt{\Psi_A^2(C_{f\,max}V_f - C_{0A}V_s)^2 + 2\Psi_A V_s(1 + \Psi_B C_{sB}^{\infty})(C_{f\,max}V_f + C_{0A}V_s) + V_s^2(1 + \Psi_B C_{sB}^{\infty})^2}}{2\Psi_A}$$

$$(26)$$

Even though this dependence seems very complex, it can give a valuable insight into the extraction process, as will be illustrated in Section 4 of this chapter.

SPME extraction can be carried out by immersing the fibre in the sample (direct extraction), or by exposing it to the sample headspace. In fact, when volatile compounds are analyzed, headspace extraction is the preferred mode of operation (see Chapter 1). Mathematical description of the headspace extraction process is more complex than that of direct extraction, as in headspace extraction one has to deal with equilibria involving three phases: the sample, its headspace, and the fibre coating. Nevertheless, a similar procedure can be used to derive equations for the amount of analyte extracted from sample headspace both for a single analyte, and for multiple analytes. A detailed derivation can be found elsewhere.[3] General conclusions derived from those equations are the same as for two-phase systems, and therefore they will not be discussed separately.

4 Discussion

Figure 2 presents calibration curves obtained for *i*-propanol in the presence of methyl isobutyl ketone (MIBK).[5] Sampling was carried out from headspace with a PDMS/DVB fibre (for details, see ref. 5). The affinity of MIBK for the fibre coating was much higher than the affinity of *i*-propanol. As long as MIBK concentration remained low (10 times lower than the concentration of *i*-propanol; points represented by filled squares), the calibration curve remained linear up to ~ 75 μg L^{-1}, and the deviation from linearity at higher concentrations was not very significant. However, when MIBK concentration in the samples was equal to that of *i*-propanol (filled circles), the dependence could be approximated by a straight line only up to ~ 25 μg L^{-1}. Moreover, at higher MIBK concentrations, displacement of *i*-propanol was evident. The amount of *i*-propanol extracted from the sample at 150 μg L^{-1} was lower by almost 50%

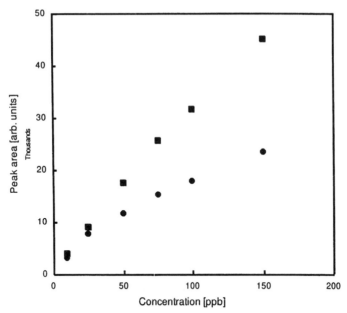

Figure 2 *Calibration curves for i-propanol in the presence of methyl isobutyl ketone (MIBK) (PDMS/DVB fibre, headspace sampling).* ■: *MIBK concentration 10 × lower than i-propanol concentration;* ●: *MIBK concentration equal to i-propanol concentration*

when MIBK concentration was also 150 μg L^{-1}, compared to the case when it was 15 μg L^{-1}. Figure 2 illustrates therefore that the presence of interfering compounds can affect both the amount extracted and the linear range of the method for porous polymer fibres.

Equation 18 predicts that the plot of $1/C_{fA}^{\infty}$ vs. $1/C_{sA}^{\infty}$ should be a straight line with a slope of $1/C_{fmax}\Psi_A$ and an intercept of $1/C_{fmax}$. Figure 3 presents a plot of this kind obtained for data reported by Górecki *et al.*[5] for headspace extraction of methyl isobutyl ketone (MIBK) with a PDMS/DVB fibre. The amount of analyte extracted by the fibre n (in arbitrary units of peak area) was used to create this plot rather than the concentration of the analyte in the fibre coating, due to the fact that the exact volume of the PDMS/DVB coating was unknown. Experimental details are given in ref. 5. The dependence is linear with a linear correlation coefficient of 0.99734, which indicates that within the concentration range examined the Langmuir isotherm model can be successfully applied for the description of analyte adsorption on selected porous polymer fibres. Since this plot was created based on data obtained for headspace extraction, the slope of the line is proportional to $1/C_{fmax}\Psi'_A$ rather than $1/C_{fmax}\Psi_A$, where $\Psi'_A = \Psi_{Ah}K_{Ha}$, Ψ_{Ah} is the adsorption equilibrium constant for analyte adsorption from the headspace over a liquid sample, and K_{HA} is dimensionless Henry's constant for analyte A.

Figure 4 illustrates the predicted dependence of the amount of analyte

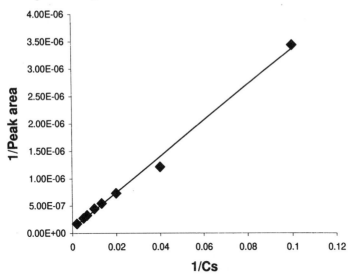

Figure 3 *Plot of $1/n$ vs. $1/C_{sA}^{\infty}$ for headspace MIBK extraction with PDMS/DVB fibre (for details see ref. 5)*

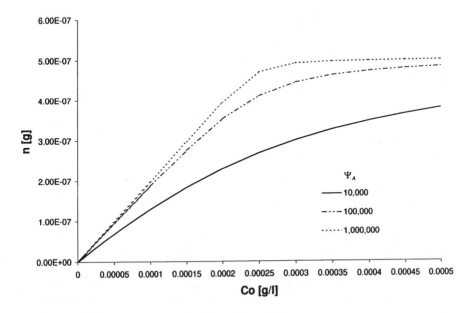

Figure 4 *Amount of analyte extracted by the fibre vs. initial concentration of the analyte in the sample for a single analyte and direct extraction, for three different equilibrium constants. Assumptions: $C_{fmax} = 1.0$ g l^{-1}, $V_f = 0.5$ μL, $V_s = 2$ mL*

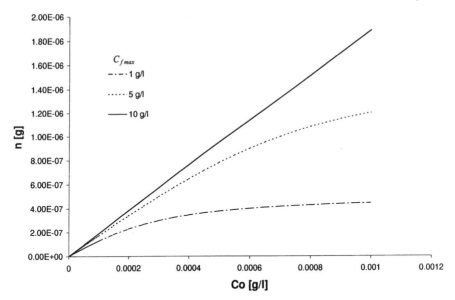

Figure 5   *Amount of analyte extracted by the fibre* vs. *initial concentration of the analyte in the sample for a single analyte and direct extraction, for three different C_{fmax} values. Assumptions:* $\Psi_A = 10\,000$, $V_f = 0.5\ \mu L$, $V_s = 2\ mL$

extracted by the fibre *vs.* the initial concentration of the analyte in the sample for direct extraction when a single analyte is present in the sample, for three different equilibrium constant (Ψ_A) values. The plots were determined using equation 23. At low analyte concentrations, the dependencies can be approximated by straight lines. At higher concentrations, they cease to be linear, and finally they level off when all the active sites on the fibre surface are occupied by the analyte molecules. The shape of the isotherms, and particularly their linear range, depends strongly on the Ψ_A value. When it is large (see the curve for $\Psi_A = 1\,000\,000$), the response remains practically linear until the fibre becomes saturated with the analyte. After this point, the curve levels off rather abruptly. When Ψ_A is low (see the curve for $\Psi_A = 10\,000$), the initial quasi-linear range is narrower, but *n* remains proportional (in a non-linear fashion) to the initial analyte concentration C_{0A} in a broader concentration range.

Figure 5 presents the theoretical dependence of the amount of analyte *n* extracted by the fibre on the initial concentration of the analyte in the sample for a single analyte and direct extraction, for three different $C_{f\,max}$ values. It is clear that the concentration of active sites on the fibre has a profound effect on linearity of the response. The higher the number of active sites, the broader is the linear range of the isotherm. This is obvious when looking at equation 21. When $C_{f\,max}$ is high, the value of the difference ($C_{f\,max} - C_{fA}^{\infty}$) is very close to $C_{f\,max}$ for a broader range of C_{fA}^{∞} values than when $C_{f\,max}$ is low.

Figure 6 presents the theoretical relationship between the amount of analyte A extracted by the fibre and the initial concentration of the analyte in the sample

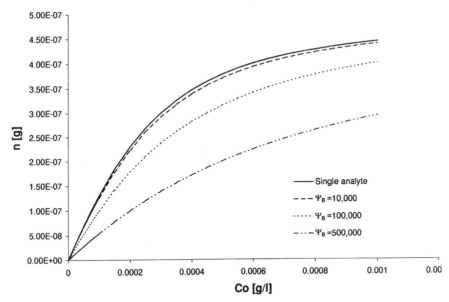

Figure 6 *Amount of analyte A extracted by the fibre vs. initial concentration of the analyte in the sample when two compounds are present in the sample (direct extraction), for three different Ψ_B values. Assumptions: $C_{fmax} = 1.0 \text{ g L}^{-1}$, $\Psi_A = 10\,000$, $V_f = 0.5 \,\mu\text{L}$, $V_s = 2 \text{ mL}$, $C_{sB}^{\infty} = 10 \,\mu\text{g L}^{-1}$*

when two compounds are present in the sample (direct extraction), for three different Ψ_B values. It is intuitively obvious that when the interfering compound has a high affinity for the fibre coating, the displacement effects are more pronounced. Indeed, Figure 6 illustrates that for the same C_{0A}, the higher the Ψ_B value is, the lower is the amount of extracted n_A. When the affinity for the fibre coating is similar for both compounds, the displacement effect is not very significant, especially when the concentration of the interfering compound(s) is not very high (see the curve for $\Psi_B = 10\,000$). On the other hand, when Ψ_B is high, displacement is significant. It should be emphasized that, as already mentioned in the discussion following equation 25, the curves in Figure 6 illustrate the effect of *equilibrium* concentration of the interfering compound B in the sample on the amount of analyte A extracted by the fibre. When all other parameters (especially sample volume V_s) are constant, equilibrium concentration of B for the same initial concentration C_{0B} is the lower, the higher is the Ψ_B value (*i.e.* more compound is extracted when the affinity for the coating is higher). In order for C_{sB}^{∞} to be the same for the three curves in Figure 6 corresponding to the case of two compounds that undergo extraction, the initial concentration of B would have to be the higher, the higher is the Ψ_B value. The curves do not illustrate therefore what is the effect of an interfering compound(s) when Ψ_B changes, while C_{0B} remains constant.

Figure 7 illustrates the predicted dependence of the amount of analyte A extracted by the fibre on initial concentration of this analyte in the sample when

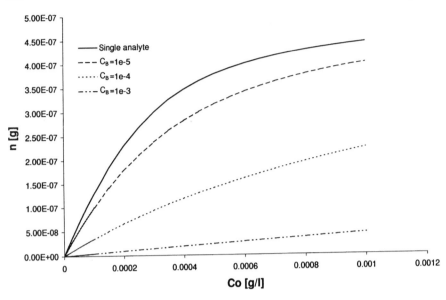

Figure 7 *Amount of analyte A extracted by the fibre* vs. *initial concentration of the analyte in the sample when two compounds are present in the sample (direct extraction), for three different C_{sB}^{∞} values. Assumptions: $C_{fmax} = 1.0 \text{ g L}^{-1}$, $\Psi_A = 10\,000$, $V_f = 0.5 \,\mu\text{L}$, $V_s = 2 \text{ mL}$, $\Psi_B = 100\,000$*

two compounds are present in the sample (direct extraction). In this Figure, Ψ_B is constant ($= 100\,000$), while C_{sB}^{∞} varies. It should be noted that in this case C_{fA}^{∞} is proportional to C_{0B}. It is clear from Figure 7 that the higher the concentration of the interfering compound, the lower is the amount of analyte A extracted by the fibre. This is intuitively obvious, since at higher concentrations of interfering compound(s), a larger fraction of adsorption sites is occupied and therefore fewer sites are available for analyte A. The courses of the relationships between the amount of analyte A extracted by the fibre and the initial concentration of the analyte in the sample when extraction is carried out from sample headspace are similar to those presented in Figure 4 for direct sampling and therefore will not be discussed here.

5   Conclusions

Figure 8 illustrates the initial and steady-state stages of the extraction process for absorption- and adsorption-type SPME coatings. At the beginning, independently of the nature of the coating, analyte molecules get attached to its surface. Whether they migrate to the bulk of the coating or remain on its surface depends on the diffusion coefficient of an analyte in the coating. Diffusion coefficients of organic molecules in PDMS are close to those in organic solvents and therefore this coating extracts analytes *via* absorption. Diffusion coefficients in poly(acrylate) are lower by about an order of magnitude, but still large

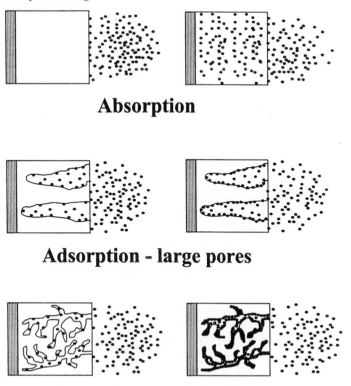

Absorption

Adsorption - large pores

Adsorption - small pores

Figure 8 *Comparison of absorption and adsorption extraction mechanisms. Diagrams on the left illustrate the initial stages of the processes. Diagrams on the right illustrate the steady-state condition*

enough for absorption to be the primary extraction mechanism. On the other hand, diffusion coefficients of organic molecules in divinylbenzene and Carboxen are so small that within the time frame of SPME analysis essentially all the molecules remain on the surface of the coating. Should the organic molecules remain there for a very long time (measured in days or weeks rather than hours), they still might diffuse into the bulk of the coating over short distances. This would manifest itself as persistent carryover, difficult to eliminate even after repeated desorptions.

This chapter illustrates the fact that, from the practical point of view, there are very significant differences between the two types of SPME coatings. Absorption is a non-competitive process and therefore quantitative SPME analysis with the use of liquid coatings is usually unaffected by matrix composition. Also, the linear range of analyte uptake is typically very broad. On the other hand, adsorption is a competitive process and therefore matrix composition, as well as extraction conditions, affects the amount of analyte extracted by the fibre. Additionally, there is only a relatively narrow quasi-linear range of analyte uptake, and in general this dependence is nonlinear. This makes quantitative analysis using solid coatings more difficult. The equilibrium theory developed for selected porous polymer coatings (PDMS/DVB, Carbowax/DVB) sheds light on the effect of several experimental variables on the amount of analyte extracted by the fibre coating. In general, porous polymer coatings can be expected to perform well for relatively clean matrices or matrices of constant composition, provided that the concentration of the analyte of interest is low (otherwise, the quasi-linear range of the calibration curve can be easily exceeded). Special strategies can be applied when interfering compounds with high affinity to the coating are present in the sample (see ref. 5).

It should be remembered that, in most practical cases, one has to deal with systems where more than one compound undergoes adsorption on the fibre coating. For example, porous polymer fibres are not entirely hydrophobic, and hence they always extract some water from samples. Fortunately, the affinity of water to porous polymers is not very high. On the other hand, other compounds present in samples may be extracted in significant amounts and adversely affect quantitation.

Carboxen coating is a special case. It extracts analytes *via* adsorption, therefore the general description of the extraction process is similar to that for porous polymer coatings. The main difference is that the pores in Carboxen are small enough to cause capillary condensation to occur. Consequently, one cannot talk about reaching equilibrium when Carboxen fibres are used. Obviously, one of the basic assumptions of the Langmuir isotherm model, stating that at a maximum the surface of the adsorbent can be covered by a monomolecular layer only is not fulfilled. Therefore the model presented herein is not directly applicable to Carboxen coatings. Capillary condensation can dramatically increase the sorption capacity of a coating for some analytes, but the fundamentals remain the same, and replacement effects can still occur. However, when concentrations of all the compounds that are extracted by Carboxen-coated fibre are low, this is not a very significant problem.

References

1. R.P. Schwarzenbach, P.M. Gschwend and D.M. Imboden, *Environmental Organic Chemistry*, John Wiley & Sons, New York, 1993.
2. R.I. Masel, *Principles of Adsorption and Reaction on Solid Surfaces*, John Wiley & Sons, Inc., New York, 1996.
3. T. Górecki, X. Yu and J. Pawliszyn, submitted to *The Analyst*.
4. W. Stumm and J.J. Morgan, *Aquatic Chemistry. Chemical Equilibria and Rates in Natural Waters*, John Wiley & Sons, New York, 1996.
5. T. Górecki, P. Martos and J. Pawliszyn, *Anal. Chem.*, 1998, **70**, 19.

Physicochemical Applications

CHAPTER 8

Application of SPME to Study Sorption Phenomena on Dissolved Humic Organic Matter

FRANK-DIETER KOPINKE, JÜRGEN PÖRSCHMANN
AND ANETT GEORGI

1 Introduction

Transport, bioavailability and finally the fate of hydrophobic organic compounds (HOCs) in the environment are strongly affected by their interactions with humic organic matter, which is ubiquitous in the natural world. Sorption of HOCs onto particulate organic matter in soils and sediments gives rise to transport retardation, whereas complexation with dissolved organic matter in surface and groundwater leads to their mobilization. Owing to the environmental importance of these processes they have been the subject of numerous scientific studies.[1,2]

For measuring the sorption of HOCs from an aqueous phase onto particulate sorbents, a simple phase separation, *e.g.* by sedimentation or centrifugation, is sufficient. For dissolved sorbents such a phase separation is more complicated and may affect the sorption equilibrium. The currently available techniques for measuring sorption on colloids are:[3]

- the fluorescence quenching technique (FQT)[4,5]
- fast solid phase extraction (SPE) or reversed phase (RP) method[6,7]
- the flocculation method[8]
- the solubility enhancement method[9-11]
- the dialysis method[12,13]
- the gas purging or headspace partitioning method.[14,15]

Three of these techniques will be considered in more detail.

The most frequently used method is the FQT. It is a true *in situ* method, based on the phenomenon that many fluorophores (*e.g.* PAHs) lose their fluorescence activity when they interact with dissolved polymers. The assumption of a 'dark complex' is, however, not undisputed.[16] According to this thesis, the fluorescence intensity is considered as a measure of the concentration of the freely-dissolved fluorophore. Problems may result from the absorption of excitation and fluorescence light by the polymer (the inner filter effect), which must be corrected for. The applicability of the method is limited to fluorescence-active compounds. It is a single component method, *i.e.* mixtures of analytes in a sample cannot be measured. However, the method is very sensitive. It can be used for time-resolved measurements, which may be useful for following the kinetics of sorption processes.

Very recently, Laor and Rebhun[8] proposed a simple flocculation method to determine binding coefficients of organic compounds to dissolved humic substances (DHS). The method is based on the precipitation of the humic substances by adding a coagulant, *e.g.* aluminium sulfate. After the flocculation is complete, the fraction of the analyte remaining in the clarified supernatant can be determined by conventional methods. The authors claim that the DHS–HOC complexes are sufficiently stable to survive the flocculation procedure. Nevertheless, it may strongly affect the DHS conformation and charge. It is well known from many previous studies that changing the pH value or the ionic strength in the aqueous solution may result in significant changes of the binding coefficients.[5,17] Therefore, it can hardly be assumed that such a flocculation does not disturb the original complexation equilibrium.

The fast SPE technique uses the selective adsorption of freely-dissolved HOCs on a reversed phase column, whereas the more hydrophilic DHS–HOC complex is permitted to pass the column. Again, the separation step is a significant interference with the sorption equilibrium. The removal of the freely-dissolved analyte fraction causes an equilibrium shift towards desorption. If the desorption kinetics are in the time scale of the column passage (some minutes), this gives rise to an underestimation of the sorption coefficients.

The remaining three techniques—solubility enhancement, dialysis and gas purging—also have various shortcomings, but will not be discussed here in detail. To summarize the present state of knowledge, none of the available methods is able to meet all the demands simultaneously. The essential criterion which has to be fulfilled by any experimental technique for measuring sorption coefficients is not to interfere with the original sorption equilibrium. A new technique makes this possible for the first time: solid-phase microextraction (SPME). The application of SPME for measuring sorption coefficients on DHS was first described by Kopinke *et al.*[18] in 1995 and later by Pörschmann *et al.*[19,20] The present paper is based on comprehensive experimental studies by Georgi.[21]

In a recent paper[22] SPME in the presence of dissolved humic acid (HA, Aldrich, $C_{HA} \leqslant 500$ ppm) is mentioned. The analytes investigated, however, were not sufficiently hydrophobic (log $K_{ow} \leqslant 3.15$) to observe significant interactions with the HA. The measurement of free concentrations of analytes and partition coefficients using SPME is addressed in Chapter 9 of this book.[23]

2 Methodological Aspects of SPME in the Presence of Dissolved Humic Substances

The Choice of SPME Sampling Mode

As described in Chapter 1, SPME sampling can be performed in three basic modes. Two of them, direct extraction and headspace extraction, were applied for DHS studies in our laboratory (Figure 1). In the direct mode, the fibre comes into contact with the DHS. Therefore, possible adverse effects have to be carefully checked. Indeed, with relatively hydrophobic dissolved polymers from a wastewater dump we observed a fouling of the fibre, leading to discoloration of the fibre coating and poor reproducibility. This is due to sorption and subsequent thermal decomposition of non-volatile compounds on the fibre coating. With most of the humic substances, more than 100 extractions per fibre could be performed without such fouling effects being observed. They can also in principle be avoided by headspace extraction. However, this mode fails for two groups of analytes:

- those with a very low Henry coefficient (*e.g.* PAHs with more than four rings) and
- those with a very high Henry coefficient (*e.g.* alkanes).

The latter analytes are highly enriched in the headspace volume. Consequently, changes in the concentration of the freely-dissolved fraction are not sufficiently reflected by the headspace concentration. Moreover, for less volatile analytes

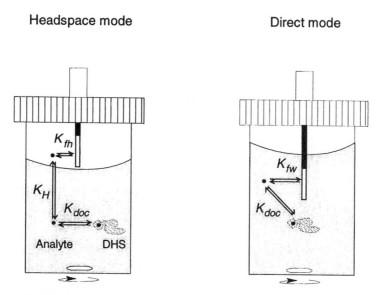

Figure 1 *SPME modes and sorption equilibria in the presence of dissolved humic substances (DHS)*

the headspace mode needs more time to reach extraction equilibrium. These limitations can make the direct extraction mode desirable, provided that the interactions between the fibre coating and the DHS are not significant. In order to minimize these interactions, the most hydrophobic coating material, poly(dimethylsiloxane) (PDMS), was used. The experimental data obtained in this study give ample evidence that the two SPME modes produce identical results for the great majority of investigated HOCs and DHS.

Fibre Coating–DHS Interactions

In order to recognize possible fibre coating–DHS interactions, we carried out some special experiments with several DHSs and 1-methylnaphthalene as the target analyte. First, a saturated aqueous solution of 1-methylnaphthalene was prepared by adding an analyte-filled silicone tube to deionized water. After equilibration, the solution was extracted with a 100 μm PDMS fibre for 20 min. Then, the same saturation procedure was applied to six DHS solutions, containing 100 and 500 ppm of three different humic substances, two hydrophobic commercial humic acids (from Aldrich and from Roth Ltd.) and a hydrophilic polymer (prepared by polymerization of hydroquinone). The substances were chosen to cover a broad range of DHS properties, particularly their hydrophobicities. The solutions were analyzed again by direct SPME, and the results are compiled in Figure 2.

Within the range of experimental error, the uptakes from all the solutions are equal, despite the different total concentrations of 1-methylnaphthalene (from

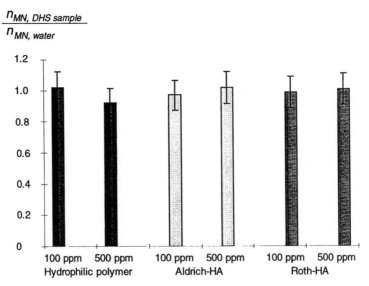

Figure 2   *SPME uptakes from solutions saturated with 1-methylnaphthalene (MN) in the presence of DHS (data are normalized to the uptake from a saturated solution free of DHS)*

27 up to 110 mg L^{-1}). This proves that the presence of DHS does not affect the equilibrium between the SPME-extracted and the freely-dissolved analyte, either by coadsorption of humic substances, or by blocking of the coating surface.

Kinetics of SPME in the Presence of DHS

An optimal approach to SPME analysis is to allow the extraction equilibrium to become established, at least to 95% of the final uptake. When equilibration times are excessively long, shorter extraction times may be desirable. In principle, quantitative measurements are also possible under nonequilibrium conditions. This requires identical mass transfer conditions for the two samples, the analytical sample and the external calibration sample, which may be achieved for many samples by strictly controlling the agitation intensity (*e.g.* speed of the magnetic stirrer) and the extraction time. However, DHSs are known to have surfactant-like properties. Therefore, they may affect the mass transfer through the boundary layer around the fibre. As an example, Figure 3 shows a comparison of extraction kinetics of fluoranthene from two samples, which differ only in their content of DHS. The ordinate is the fibre uptake normalized to the equilibrium value.

Surprisingly, the extraction from the humic acid (commercial source, Roth) solution proceeds faster than that from pure water. There seems to be a general tendency in the series of PAHs: for highly hydrophobic PAHs (four or more rings) the dissolved HA accelerates the extraction, for less hydrophobic compounds (naphthalene, biphenyl) it is retarded, and for medium-hydrophobic PAHs (three rings) the extraction kinetics is not significantly affected. In ref. 20 we found a retardation of the SPME of tetrabutyl tin in the presence of a fulvic and a humic acid. Apparently, DHS can affect the

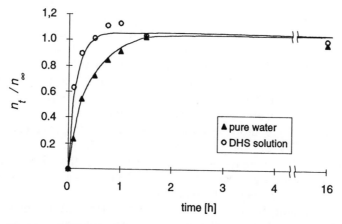

Figure 3 *Kinetics of SPME of fluoranthene from pure water and from a humic acid solution (direct extraction mode, 7 μm PDMS fibre, 0.2 ppm fluoranthene, 200 ppm HA, pH = 5.6, IS = 0.01 M, 700 rpm magnetic stirrer)*

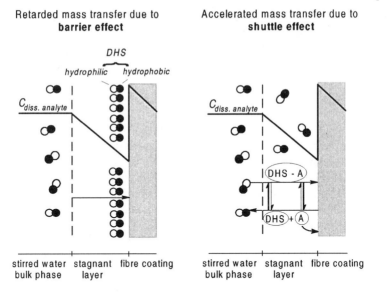

Figure 4 *Schematic presentation of possible mechanisms for the influence of DHS on SPME kinetics: the barrier effect and the shuttle effect (A: analyte, DHS-A: DHS–analyte complex)*

SPME kinetics in an opposite manner. Figure 4 illustrates two possible mechanisms. The first mechanism is the formation of a relatively rigid barrier of DHS molecules around the PDMS coating, which hampers the mass transfer. The second mechanism is based on the well known 'shuttle effect' of DHS molecules. DHS analyte complexes diffuse from the water bulk phase into the static layer around the fibre, where the freely-dissolved analyte is depleted due to the extraction. There, the complexes can increase the analyte concentration by desorption. If the sorption–desorption kinetics are fast compared with the residence time of DHS molecules inside the static boundary layer (about 0.1 s), a local equilibrium will always be established. In this case, there is no net contribution of the shuttle effect to the mass transfer rate. If, however, the sorption kinetics are in the time scale of the DHS diffusion (which is plausible), the contribution of the transport *into* the boundary layer dominates over the back-transport of analyte. The reason is simply that the DHS molecules which enter the static layer carry a high equilibrium loading of analytes from the water bulk phase, but leave this layer with a depleted analyte concentration. The higher the DHS loading with analyte is, the more significant the shuttle effect will be. This prediction is consistent with experimental observations.

The possible contribution of bound species to the diffusive mass transfer through the Nernst diffusion film around an extraction phase is discussed in detail by Jeannot and Cantwell for the analogous case of microextraction with *n*-octane.[24] The authors claimed large association and dissociation rates compared with the mass transfer rate as the precondition of such a contribution. Clearly, this is in contrast to our considerations discussed above.

Independent of the soundness of our hypothesis, the practical conclusion for SPME in DHS-containing solutions is clear: the kinetics of the SPME can be affected. Therefore, quantitative measurements should be performed under equilibrium conditions.

The time necessary for equilibrium SPME of a cocktail of different analytes is determined by the component with the highest fibre–water partition coefficient K_{fw}. In our PAH-cocktail this was benz[*a*]anthracene (log K_{fw} = 5.01). Its equilibration time ($n_t/n_\infty \geqslant 0.95$) with the 7 μm PDMS fibre in the direct extraction mode was about 5 h.

In contrast to SPME in the direct extraction mode, the rate-limiting step in the headspace mode may be different.[25,26] For highly volatile analytes, headspace SPME was found to be faster than direct SPME,[27] whereas for the semivolatile analytes this relation is reversed. For fluoranthene, about 10 h were necessary for equilibrium extraction with headspace SPME (compared with 2 h in the direct mode). 16 h of extraction time were found to be sufficient for reaching headspace equilibrium for PAHs up to pyrene (log K_{fw} = 4.57). For the least volatile PAH in our cocktail, benz[*a*]anthracene, even after 40 h the equilibrium was not achieved. Apart from benzanthracene, the thermodynamic criterion, that the fibre uptake has to be independent of the extraction mode under equilibrium conditions, was found to be fulfilled for all PAHs.

SPME of alkanes could not be performed in the headspace mode, because of their very high Henry coefficients (*e.g.* $K_H \approx 300$ for *n*-undecane[28]).

Determination of Fibre–Water Partition Coefficients

The calibration of SPME analysis by means of an external standard does not require the knowledge of fibre-water partition coefficients. Nevertheless, this knowledge can be helpful for validation of the method. In our study, K_{fw} values were determined for a series of PAHs and alkanes with the 7 μm PDMS fibre (V_f = 26 nL) from aqueous solution (pH = 7, IS = 0.007 M as NaCl and NaN$_3$, C_{HOC} = 2–100 ppb per component). The experimental problem of determining K_{fw} values lies in the absolute calibration of the GC detector. We used splitless injection of 0.5 μL of standard solutions into the hot split-splitless injector (280 °C) of the GC. It is well known, however, that even injections with closed split may give rise to discrimination of high boiling components. Although we did not observe indications of discrimination effects (FID response factors of naphthalene and benz[*a*]anthracene were identical), the use of this procedure introduces some uncertainty into the resulting K_{fw} values.

Figure 5 shows the correlation of log K_{fw} *versus* log K_{ow} over three orders of magnitude in the octanol-water partition coefficient. The correlations are fairly linear (for arenes and PAHs log K_{fw} = 0.72 log K_{ow} + 0.82 with r^2 = 0.966 and for alkanes log K_{fw} = 0.81 log K_{ow} + 0.20 with r^2 = 0.985) up to very hydrophobic analytes (log K_{ow} = 6.7 for *n*-undecane). The log K_{fw} values (RSD = 0.1) determined in the present study coincide well with those published

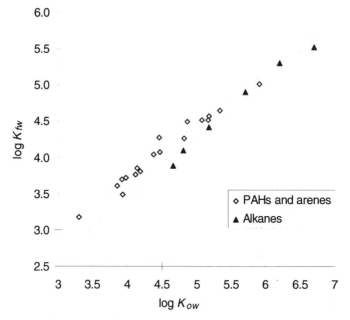

Figure 5 *Correlation between PDMS–water and octanol–water partition coefficients of a series of PAHs (naphthalene to benz[a]anthracene) and alkanes (n-heptane to n-undecane)*

by Potter and Pawliszyn[29] for a 15 μm PDMS coating. They contrast, however, with the observations of Langenfeld *et al.*[30] who found a stagnancy of the fibre uptake for analytes with log K_{ow} > 5.0.

The standard deviation of SPME analyses with the same fibre was in the range of 4–10% (average 7%); interfibre variation was 23%. The latter value includes differences between various fibres in the volumes of the PDMS coating. The relation between GC signal and analyte concentration was found to be linear (r^2 > 0.99) over the range of concentrations used (*e.g.* 20–300 ppb for naphthalene and 0.5–8 ppb for benz[a]anthracene). For very low PAH concentrations (*e.g.* <0.5 ppb of pyrene, corresponding to 0.5 ng fibre uptake) a deviation from linearity was observed, indicating a few preferred sorption sites on the fibre.

Sorption Coefficients—Theory

In Chapter 1 of this book the thermodynamic fundamentals of SPME have been discussed in detail. Three phases are involved: the fibre coating (V_f), the gas phase or headspace (V_h), and the pure water phase (V_w). Differing from the terminology of Chapter 1, we prefer to use here the index w (water) instead of s (sample), because the liquid sample considered in this chapter contains an additional phase: the sorbent phase (m_{doc}), which is colloidally dissolved in the water phase. The amount of analyte extracted onto the fibre coating is given then by

$$n = \frac{K_{fw} V_f C_0 V_w}{K_{fw} V_f + K_H V_h + V_w + K_{doc} m_{doc}} \tag{1}$$

where m_{doc} is the mass of the sorbent in the sample in terms of dissolved organic carbon and K_{doc} (mL g^{-1}) is the sorption coefficient of the analyte defined by equation 2:

$$K_{doc} = \frac{n_{sorb.analyte}}{m_{doc}} \frac{V_w}{n_{diss.analyte}} \tag{2}$$

$n_{sorb.analyte}$ is the amount of analyte in the bound state (analyte–DHS complex) and $n_{diss.analyte}$ is its amount in the freely-dissolved state. In most of our experiments, the conditions were adjusted in such a way that the capacities of the fibre coating and the headspace are small related to the capacity of the liquid phase. Therefore, equation 1 can be rewritten as:

$$n = \frac{K_{fw} V_f C_0 V_w}{V_w + K_{doc} m_{doc}} \tag{3}$$

If two identical samples are extracted, one containing DHS and the other not (*i.e.* the external standard), the corresponding fibre uptakes (peak areas) n_1 and n_2 can be used to calculate the sorption coefficient K_{doc}:

$$K_{doc} = \left(\frac{n_2}{n_1} \frac{K_{fw,1}}{K_{fw,2}} - 1 \right) \frac{V_w}{m_{doc}} \tag{4}$$

Assuming that $K_{fw,1} \equiv K_{fw,2}$ one comes to the modified sorption coefficient K_{doc}^{SPME}:

$$K_{doc}^{SPME} = \left(\frac{n_2}{n_1} - 1 \right) \frac{1}{C_{doc}} \tag{5}$$

This experimentally obtained sorption coefficient K_{doc}^{SPME} is identical with the conventional K_{doc} if the partition coefficient $K_{fw,1}$ between the fibre coating and the water phase is not affected by the presence of DHS. For low concentrations of DHS (<100 mg L^{-1}) this seems to be a reasonable approximation. Nevertheless, a more rigorous and general treatment of the multiphase partitioning equilibria must consider activities instead of concentrations of the analytes.[31] The fibre partition coefficient must then be defined as $K_{fw}^* = C_f / a_w$ where C_f and a_w are the concentration of the analyte in the fibre coating and its activity in the water bulk phase, respectively. The corresponding activity coefficient γ_w in $a_w = C_w \gamma_w$ is defined as $\gamma_w = 1.0$ for an infinitely diluted aqueous solution ($x_{water} \rightarrow 1.0$). The concentration of the extracted analyte C_f does not have to be transformed into the corresponding activity (a_f), because the fibre coating is unaffected by the sample composition. Equation 3 can now be transformed into equation 6, which shows that the fibre uptake is proportional to the activity of the analyte.

$$n = \frac{K_{fw}^* V_f C_0 \gamma_w V_w}{V_w + K_{doc} m_{doc}}$$
(6)

The activity-based fibre coefficient K_{fw}^* is characteristic of the polymer material, but not of the sample composition. This property is different from the concentration based fibre coefficient K_{fw}. Therefore, it strictly holds that $K_{fw,1}^* = K_{fw,2}^*$. Using $K_{fw,1}/K_{fw,2} = (K_{fw,1}^*/K_{fk,2}^*)(\gamma_{w,2}/\gamma_{w,1})$ and with $\gamma_{w,2} = 1.0$, equation 4 yields equation 7:

$$K_{doc} = \left(\frac{n_2}{n_1}\frac{1}{\gamma_{w,1}} - 1\right)\frac{1}{C_{doc}}$$
(7)

Considering the case of strong complexation of the analyte in the DHS sample ($n_1 \ll n_2$), equations 4 and 7 can be combined to give equation 8:

$$K_{doc}^{SPME} = \gamma_{w,1} K_{doc} = \frac{C_{sorb.analyte}}{a_{diss.analyte}} = K_{doc}^*$$
(8)

From equation 8 it is evident that the sorption coefficient K_{doc}^{SPME} calculated from the experimental SPME data is an activity-based coefficient K_{doc}^* under certain circumstances. A high degree of complexation of the analyte in the DHS-containing sample can normally be achieved either by selecting a strongly sorbing analyte or by increasing the DHS concentration.

What is the practical conclusion of this theoretical treatment? Let us consider the effect of any additive to the DHS-containing sample. This might be a salt or a cosolvent. The degree of sorption of the analyte and consequently K_{doc} will probably be affected for two reasons: (i) a change of the analyte activity in the water phase (*e.g.* salting-out effects) and (ii) possible changes of the DHS properties due to the additive (*e.g.* rearrangement of dissolved polymers). The change of K_{doc} reflects both effects. In contrast, K_{doc}^* reflects only the second effect. This enables us to separate the two effects. In other words, K_{doc}^* is a value characteristic of the sorption properties of the DHS, as it behaves under the specific aqueous conditions (such as pH value and ionic strength) in the sample. K_{doc} is a more complex parameter which describes the influence of the conditions on both participants, the sorbent and the sorbate.

It may be of interest to determine which type of sorption coefficient the other experimental techniques measure. The most popular method—FQT—measures the conventional, concentration-based K_{doc} value. This is because the freely-dissolved analyte fraction is detected by a concentration-proportional signal, the fluorescence intensity. Therefore, the combined application of FQT and

SPME can extend our knowledge about the driving forces of hydrophobic sorption.

Experimental Conditions

In order to simplify the quantitative treatment of the SPME data, we used conditions (7 μm PDMS fibre (Supelco), $\geqslant 100$ mL sample volume) where the depletion of the analytes from the sample and their gaseous fractions were insignificant ($[K_{fw}V_f + K_H V_h]/[K_{fw}V_f + V_w(1 + C_{doc}K_{doc}) + K_H V_h] < 0.05$).

The method was calibrated with external standards, containing all the analytes at the same ionic strength and pH value as the DHS samples, but without the DHS.

The reproducibility of absolute GC peak areas from SPME analyses was better than $\pm 10\%$ (RSD), in many cases better than $\pm 5\%$.

When applying the external calibration method, the same fibre must always be used for pairing results of the analytical sample and the corresponding standard sample. To avoid this limitation and the need for strictly stable GC conditions, the analysis may be calibrated by means of an internal standard. This can be a medium polarity compound, which is hydrophilic enough to interact only weakly with DHS and which is hydrophobic enough to be sufficiently extracted by the PDMS coating. We tested methyl benzoate as an internal standard. The results obtained were correct, but less reproducible than with the external calibration.

As analytes we investigated 16 PAHs, four nonanellated arenes and six alkanes (*cf.* Table 1). The analytes were spiked as ethanolic solutions to the various aqueous samples. Their final concentrations were in the range of 3 ppb (for benz[*a*]anthracene) to 70 ppb (for naphthalene).

The GC analyses were performed with several instruments (from Dani and Shimadzu) equipped with FID or QMS as detectors. An important step in the SPME analysis is the thermodesorption in the GC injector. We applied splitless injection at 290 °C for 3 min. The fibre was preconditioned (to avoid memory effects) at 290 °C for 15 min before each new extraction.

The DHS solutions for sorption experiments were prepared by dissolving the solid humic substance in diluted NaOH. After further dilution with deionized water and pH-adjustment (0.1 M HCl), NaN_3 (200 mg L^{-1}) was added to inhibit microbial activity. The sorption and SPME experiments were conducted in 100 mL Erlenmeyer flasks equipped with PTFE-lined septa and glass-coated magnetic stirrers. The solutions were purged with N_2 (10 min) before spiking with the ethanolic solution of the analytes. The final ethanol concentration was <0.2 vol.%. The solutions were protected from light (black paper) and stirred for several hours ($\geqslant 3$ h) at 25 ± 2 °C. Then the SPME was performed, either in the direct (6 h) or in the headspace mode (18 h).

3 Sorption Coefficients—Results

Sorption Coefficients of PAHs, Arenes and Alkanes

Table 1 compiles sorption coefficients of 26 HOCs, which were determined by means of SPME (direct mode) under equilibrium conditions for both processes, the sorption ($\geqslant 24$ h contact time between HOCs and DHS) and the extraction ($\geqslant 6$ h, for the alkane series 96 h sampling time). In most cases data from headspace SPME were also available; the results were identical with the two extraction modes.

In order to better understand the relationship between physical and chemical properties of HOCs and their sorption tendency, a lot of correlations of the type $\log K_{oc} = a \log K_{ow} + b$ were established.[21,32] They are linear free energy relationships (LFERs), based on the same driving force (solute hydrophobicity)

Table 1 *Sorption coefficients K_{doc} of organic compounds with a commercial humic acid (Roth, $C_{HA} = 200$ ppm, $C_{HOC} = 10$–700 ppb, pH = 7, IS = 0.01 M, $25 \pm 2\,^{\circ}C$)*

	HOC	$\log K_{ow}$	$\log K_{doc} \pm 2RSD(\log K_{doc})$
PAHs:	Naphthalene	3.30	3.29 ± 0.37
	Fluorenone	3.58	3.74 ± 0.17
	Carbazole	3.72	3.64 ± 0.20
	2-Methylnaphthalene	3.86	3.52 ± 0.24
	Acenaphthene	3.92	3.67 ± 0.19
	Acenaphthylene	3.94	3.71 ± 0.18
	Dibenzofuran	4.12	3.73 ± 0.17
	Fluorene	4.18	3.86 ± 0.14
	Dibenzothiophene	4.38	4.21 ± 0.10
	Anthracene	4.45	4.42 ± 0.08
	Phenanthrene	4.46	4.29 ± 0.09
	2-Methylphenanthrene	4.86	4.59 ± 0.08
	9-Methylanthracene	5.07	4.86 ± 0.07
	Fluoranthene	5.16	4.86 ± 0.07
	Pyrene	5.18	4.96 ± 0.07
	Benz[*a*]anthracene	5.91	5.49 ± 0.06
Arenes:	Biphenyl	3.98	3.48 ± 0.26
	Diphenylmethane	4.14	3.56 ± 0.22
	trans-Stilbene	4.81	4.29 ± 0.09
	4,4′-Dichlorobiphenyl	5.33	4.71 ± 0.07
Alkanes:	*n*-Heptane	4.66	4.02 ± 0.22
	iso-Octane	4.8	4.02 ± 0.25
	n-Octane	5.18	4.34 ± 0.13
	n-Nonane	5.7	4.85 ± 0.15
	n-Decane	6.2	5.30 ± 0.15
	n-Undecane	6.7	5.58 ± 0.28

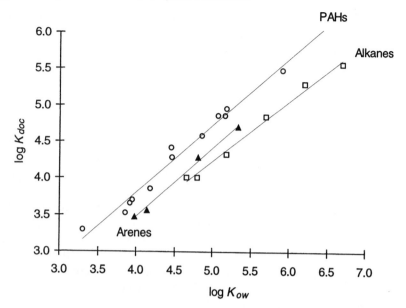

Figure 6 *Correlations between the sorption coefficients K_{doc} and the octanol–water partition coefficients K_{ow} of various PAHs, arenes and alkanes (data base is Table 1)*

of two partitioning processes. In Figure 6 such correlations for the three classes of PAHs, nonannelated arenes and alkanes are presented.

The following linear correlations were obtained:

- for PAHS: $\log K_{doc} = 0.91 \log K_{ow} + 0.16$ $(r^2 = 0.976)$
- for arenes: $\log K_{doc} = 0.94 \log K_{ow} - 0.29$ $(r^2 = 0.995)$
- for alkanes: $\log K_{doc} = 0.80 \log K_{ow} + 0.24$ $(r^2 = 0.987)$

The parameters a and b are empirical and have no theoretically founded meaning. It is worth mentioning that they are significantly different for the various classes of HOCs. Nevertheless, they permit a reliable estimation of sorption coefficients from easily available data (K_{ow}), without the necessity of experimental measurements.

The next step in generalizing the concept is to introduce parameters which describe the sorption potential of different DHS, under the specific conditions in the investigated sample. We have proposed such an approach, based on the solubility parameters δ of the HOC and the DHS.[33] Its application to the present data base is the subject of a forthcoming paper.[34]

The potential of SPME for measuring free concentrations of analytes in the presence of HA (Aldrich) is described very recently by Ramos *et al.*[35] The authors did not observe any significant interference in the extraction by the dissolved HA (20 mg L^{-1}, pH = 8.0) in the direct extraction mode. This finding is consistent with our results. In contrast to the experimental procedure applied in our study, Ramos *et al.* extracted under nonequilibrium conditions (*e.g.* 20 min in the direct extraction mode for DDT). This approach relies on identical extraction kinetics independent of the presence of a matrix. As we have shown above, this need not generally to be the case. In another paper, Vaes and Ramos *et al.* themselves pointed to the kinetic matrix effect.[23]

Kinetics of Sorption

As pointed out above, for quantitative analyses in the presence of DHS, SPME should be conducted under equilibrium conditions. This limits its range of application to steady states or at least to samples with relatively slow dynamics. Pörschmann *et al.*[19] attempted to overcome this limitation by performing a fast SPME (100 μm PDMS fibre, 10 s extraction time) combined with a fast GC analysis (10 s desorption time, 3 m $\times$ 0.1 mm deactivated fused silica capillary, isothermal at 270 °C). In this way, it was possible to complete an SPME analysis in less than one minute. The poorly reproducible fibre uptakes under these extraction conditions were compensated by applying deuterated surrogates and measuring peak ratios instead of absolute peak areas (*e.g.* the ion traces of $m/z = 178$ and 188 for nondeuterated and deuterated anthracene, respectively). By using this method, the authors measured the kinetics of complexation of some PAHs (naphthalene through pyrene, 5 ppb each) with DHS (100 ppm) from a wastewater pond. The sorption equilibrium was established within one minute with these DHS. In another study[20] the same authors found a significant fraction of tetrabutyltin with measurable sorption kinetics ($t_e \approx 5$ min). However, in these experiments the DHS was spiked into the aqueous solution, which may give rise to relatively slow changes in their conformation or aggregation.

   In the present study a similar approach was applied for various types of DHS. The analytes, 9-methylanthracene (MA) and pyrene (60 and 120 ppb, respectively) were equilibrated in DHS solutions (50 and 100 ppm) for 24 h. Then, the same amounts of deuterated PAHs ([^{2}H$_3$]MA and [^{2}H$_{10}$]pyrene) were added under vigorous stirring. The moment of this addition is the zero point on the time scale in Figure 7. The solutions were extracted (7 μm PDMS fibre) for 30 s each. A number of fibres were at our disposal, such that sufficient time for GC analyses (with a QMS as detector) was possible. The results for pyrene and a commercial HA (Roth) are shown in Figure 7. The ordinate displays the signal ratio for the ion traces $m/z = 202$ and 212 ([^{2}H$_0$]pyrene and [^{2}H$_{10}$] pyrene), normalized to that of a standard solution with equal amounts of both isotopomers. Values below 1.0 express a surplus of deuterated pyrene, which is a measure of the deviation from the sorption equilibrium. In contrast to our previous results,[19] this specific HA exhibits a significant fraction of slow sorption kinetics. Treatment of the data with pseudo first-order kinetics ($dC_{[^2H_{10}]pyrene}/dt = k'_{pyrene}C_{HA}C_{[^2H_{10}]pyrene} = k_{pyrene}C_{[^2H_{10}]pyrene}$) led to the following rate coefficients: $k_{pyrene} = 1.5 \pm 0.1$ h^{-1} and $k_{MA} = 2.0 \pm 0.2$ h^{-1}. The extrapolation of the data back to the starting point ($t = 0$) reveals PAH

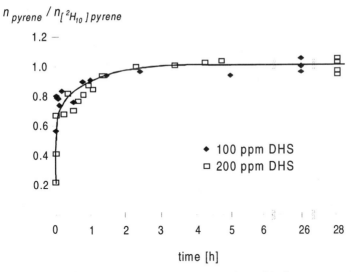

Figure 7 *Kinetics of sorption of pyrene on a humic acid (Roth), data are signal ratios of $[^2H_0]pyrene$ and $[^2H_{10}]pyrene$ (pH = 7, IS = 0.01 M)*

fractions of about $34 \pm 3\%$ (pyrene) and $29 \pm 4\%$ (MA), which follow slow sorption kinetics, whereas the remainder of the PAHs are bound to HA almost immediately. Surprisingly, the two HA concentrations (differing by a factor of two) yield equal pseudo first order rate coefficients. This indicates that the HA is not involved in the rate-determining step of the slow sorption. We explain this finding with the following reaction scheme:

$$\textbf{DHS + PAH} \underset{k'}{\overset{\text{fast}}{\rightleftharpoons}} \textbf{[DHS} \cdots \textbf{PAH]} \underset{k}{\overset{\text{slow}}{\rightleftharpoons}} \textbf{[DHS} \cdots \textbf{PAH]}$$

Scheme 1

The freely-dissolved PAH and the HA interact very quickly with each other (time scale: some seconds). The result is a loosely bound complex, which is instantaneously in equilibrium with the freely-dissolved fractions. This complex undergoes a rearrangement. Thereby, the sorbed PAH molecules occupy more favourable (hydrophobic) sites inside the humic polymer, *e.g.* 34% of the sorbed pyrene is bound at such sites. This is the relatively slow process observable by SPME. It can be understood as an unimolecular 'reaction'. Therefore, its half life time and rate coefficient do not depend on the HA concentration. The SPME is, of course, not able to directly detect intramolecular rearrangements. What we observe is the re-establishment of the fast sorption equilibrium, giving rise to further depletion of the freely-dissolved PAH fraction. Following this hypothesis, the rearranged complex has a higher sorption strength than the primary, labile complex. The experimentally determined sorption coefficient K_{doc} includes both sorbate fractions, the labile fraction (most likely sorbed on the outer sphere of the polymer) and the resistant fraction (possibly sorbed in the inner sphere of the humic polymer or its aggregates).

With an aquatic fulvic acid (FA) sample we did not observe such slow sorption kinetics within the time resolution of the method. The equilibrium was established after less than 5 min. This observation is consistent with our explanation: FAs are smaller and less hydrophobic than HAs. Accordingly, they do not have available this type of preferred sorption sites for the large PAH molecules.

4 Conclusions

SPME has been shown to be a versatile method for quantitative speciation of hydrophobic organic compounds in aqueous solutions in the presence of dissolved humic substances. Thereby, the labile complexation equilibrium is not disturbed. Both sampling modes, direct and headspace extraction, yield identical results. This and other findings prove the view that interactions between the humic substances and the PDMS coating do not significantly affect the fibre uptake. However, the extraction kinetics may be influenced by the presence of the humic substances. The tendency of the influence, retardation or acceleration of the mass transfer, depends on the analyte. Special attention is paid to the character of sorption coefficients K_{doc}^{SPME} calculated from SPME data. They are activity-based coefficients and may be different from the conventional concentration-based K_{doc} values. By measuring isotopomer ratios using SPME under dynamic (nonequilibrium) conditions, the method was utilized also to trace fast sorption processes, for which SPME under equilibrium conditions is not suitable.

Acknowledgements

This work was supported by the Deutsche Forschungsgemeinschaft (Project Ko 1334/2-1 to -4) and the Deutsche Bundesstiftung Umwelt (fellowship program).

Glossary

a, b	Empirical correlation parameters in the relation $\log K_{oc} = a \log K_{ow} + b$
a_f	Activity of the analyte in the fibre coating
a_w	Activity of the analyte in the water phase
C_{doc}	Concentration of dissolved organic carbon [g L^{-1}] ($C_{doc} \approx 0.5 C_{dhs}$)
C_{HA}, C_{pyrene}	Concentrations of HA and freely-dissolved pyrene in the water phase, respectively
DHS	Dissolved humic substances
DOC	Dissolved organic carbon, the carbon fraction of DHS
FA	Fulvic acid
FQT	Fluorescence quenching technique

HA	Humic acid
HOC	Hydrophobic organic compound
k_{MA}, k_{pyrene}	Pseudo first-order rate coefficients of sorption of 9-methylanthracene and pyrene, respectively
K_{doc}	Concentration based sorption coefficient ($K_{doc} = C_{\text{analyte sorbed on DOC}}/C_{\text{analyte dissolved in water}}$[ml g^{-1}])
K^*_{doc}	Activity based sorption coefficient ($K_{doc} = C_{\text{analyte sorbed on DOC}}/a_{\text{analyte dissolved in water}}$[ml g^{-1}])
K^{SPME}_{doc}	Experimental sorption coefficient, determined by using SPME
$K_{fw,1}$	Fibre–water distribution constant in the presence of DHS
$K_{fw,2}$	Fibre–water distribution constant in the absence of DHS ($K_{fw,2} = K_{fw}$)
K_{ow}	Octanol–water partition coefficient
m_{doc}	Amount of DOC in the sample ($m_{doc} \approx 0.5 m_{dhs}$)
$n_{diss.\ analyte}$	Amount of analyte in the freely-disolved state
$n_{sorb.analyte}$	Amount of analyte in the sorbed state
n_1, n_2	Extracted amounts of analyte (fibre uptakes) from a DHS containing sample and from a pure water sample, respectively
x_{water}	Molar fraction of water in a sample
γ_w, $\gamma_{w,2}$	Activity coefficient of the analyte in the water phase ($\gamma_w = 1.0$ for $x_{water} = 1.0$)
$\gamma_{w,1}$	Activity coefficient of the analyte in the water phase in the presence of DHS.

References

1. F.H. Frimmel and R.F. Christman, *Humic Substances and Their Role in the Environment*, John Wiley & Sons, Chichester, 1988.
2. A. Piccolo, *Humic Substances in Terrestrial Ecosystems*, Elsevier, Amsterdam, 1996.
3. M. Rebhun, F. De Smedt and J. Rwetabula, *Water Res.*, 1996, **30**, 2027.
4. T.D. Gauthier, E.C. Shane, W.F. Guerin, W.R. Seitz and C.L. Grant, *Environ. Sci. Technol.*, 1986, **20**, 1162.
5. M.U. Kumke, H.-G. Loehmannsroeben and T. Roch, *Analyst*, 1994, **119**, 997.
6. P.F. Landrum, S.R. Nihart, B.J. Eadie and W.S. Gardner, *Environ. Sci. Technol.*, 1984, **18**, 187.
7. C.R. Maxin and I. Koegel-Knabner, *Europ. J. Soil Sci.*, 1995, **46**, 193.
8. Y. Laor and M. Rebhun, *Environ. Sci. Technol.*, 1997, **31**, 3558.
9. C.T. Chiou, R.C. Malcom, T.I. Brinton and D.E. Kile, *Environ. Sci. Technol.*, 1986, **20**, 502.
10. S. Tanaka, K. Oba, M. Fukushima, K. Nakayasu and K. Hasebe, *Anal. Chim. Acta*, 1997, **337**, 351.
11. F.-D. Kopinke, A. Georgi and K. Mackenzie, *Anal. Chim. Acta*, 1997, **355**, 101.
12. J.F. McCarthy and B.D. Jimenez, *Environ. Sci. Technol.*, 1985, **19**, 1072.
13. F. De Paolis and J. Kukkonen, *Chemosphere*, 1997, **34**, 1693.
14. F. Lueers and Th.E.M. Ten Hulscher, *Chemosphere*, 1996, **33**, 643.
15. J.P. Hassett and E. Milicic, *Environ. Sci. Technol.*, 1985, **19**, 638.

16. V.A. Ganaye, K. Keiding, T.M. Vogel, M.-L. Viriot and J.-C. Block, *Environ. Sci. Technol.*, 1997, **31**, 2701.
17. M.A. Schlautman and J.J. Morgan, *Environ. Sci. Technol.*, 1993, **27**, 961.
18. F.-D. Kopinke and M. Remmler, *Naturwiss.*, 1995, **82**, 28.
19. J. Pörschmann, Zh. Zhang, F.-D. Kopinke and J. Pawliszyn, *Anal. Chem.*, 1997, **69**, 597.
20. J. Pörschmann, F.-D. Kopinke and J. Pawliszyn, *Environ. Sci. Technol.*, 1997, **31**, 3629.
21. A. Georgi, Ph.D. Thesis, University of Leipzig, 1998.
22. J. Dewulf, H. Van Langenhove and M. Everaert, *J. Chromatogr. A*, 1997, **761**, 205.
23. W.H.J. Vaes, E.U. Ramos, H.J.M. Verhaar, W. Seinen and J.L.M. Hermens, *Anal. Chem.*, 1996, **68**, 4463.
24. M.A. Jeannot and F.F. Cantwell, *Anal. Chem.*, 1997, **69**, 2935.
25. Z. Zhang and J. Pawliszyn, *Anal. Chem.*, 1993, **65**, 1843.
26. J. Ai, *Anal. Chem.*, 1997, **69**, 3260.
27. T. Nilsson, F. Pelusio, L. Montanarella, B. Larsen, S. Facchetti and J.O. Madsen, *J. High Resol. Chromatogr.*, 1995, **18**, 617.
28. D.R. Lide, Jr., *J. Phys. Chem. Ref. Data*, 1981, **10**, 1175.
29. D.W. Potter and J. Pawliszyn, *Environ. Sci. Technol.*, 1994, **28**, 298.
30. J.J. Langenfeld, S.B. Hawthorne and D.J. Miller, *Anal. Chem.*, 1996, **68**, 144.
31. J. Pörschmann, F.-D. Kopinke and J. Pawliszyn, *J. Chromatogr.*, in press.
32. J.R. Baker, J.R. Mihelcic, D.C. Luehrs and J.P. Hickey, *Water Environment Res.*, 1997, **69**, 136.
33. F.-D. Kopinke, J. Pörschmann and U. Stottmeister, *Environ. Sci. Technol.*, 1995, **27**, 941.
34. F.-D. Kopinke and A. Georgi, *Environ. Sci. Technol.*, in preparation.
35. E.U. Ramos, S.N. Meijer, W.H.J. Vaes, H.J.M. Verhaar and J.L.M. Hermens, *Environ. Sci. Technol.*, submitted for publication.

CHAPTER 9

The Use of SPME to Measure Free Concentrations and Phospholipid/Water and Protein/Water Partition Coefficients

WOUTER H.J. VAES, EÑAUT URRESTARAZU RAMOS, KARIN C.H.M. LEGIERSE AND JOOP L.M. HERMENS

SPME is mostly used to measure total concentrations of chemicals by maximising extraction efficiency to achieve high sensitivities. In addition, SPME enables measurement of the free concentration, if certain system requirements are met. For many applications, especially in pharmacology, toxicology and environmental sciences, the free concentration is of more interest than the total concentration.

In many natural as well as artificial systems, a matrix is present that binds the chemical of interest and thereby lowers its free concentration. By extracting only a minor fraction of the free concentration, the equilibrium between the matrix and aqueous dissolved chemical will not be perturbed, and only the freely dissolved chemical partitions to the fibre. This is referred to as negligible depletion SPME.

In this chapter, the requirements for measurement of the free concentration are described. Examples will be presented for measurements of free concentrations in an in vitro system that is in common use in toxicology and pharmacology. Additionally, the measurement of partition coefficients to matrices that are difficult to separate from the aqueous phase, *e.g.* phospholipid vesicles, and proteins, will be described.

1 Introduction

The concentration of chemicals in aqueous systems is not always well defined. Many compounds bind to organic materials like humic acids in surface waters, proteins in blood or cellular materials in many *in vitro* test systems. The total concentration is mostly measured using solvent or solid phase extraction procedures. The aim of these procedures is to extract as much of the analyte as possible to obtain good sensitivities. Nevertheless, on many occasions the free concentration of the analyte is of greater interest than the total concentration.

In this chapter, we define the free or bioavailable concentration as the concentration of the chemical that is freely dissolved in the aqueous phase and in no way bound to any matrix or system component. The free concentration is thus not only dependent on the total concentration of the analyte, but also on the concentration and capacity of matrix components, as well as the affinity of the chemical for the matrix.

The difference between the free concentration and the total concentration can be easily shown with the example in Figure 1. The system depicted in Figure 1 consists of an aqueous phase in which the analyte is dissolved, a matrix to which the analyte can bind and a receptor protein. Binding of the analyte to the receptor can cause a toxic or pharmacological effect, which is the parameter that is of main interest in this example. Clearly, a high affinity for the matrix decreases the concentration of the analyte that is available for binding to the receptor. Measurement of the total concentration would give an erroneous result, since it does not reflect the concentration at which a certain effect is observed. Therefore, the use of total concentrations should be avoided, especially when interpreting data quantitatively.

Consider the situation where different compounds are tested in the system of

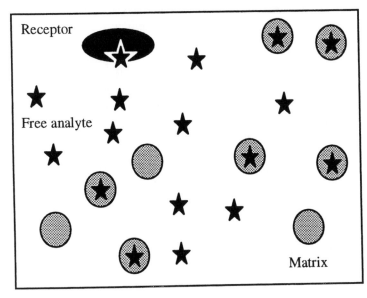

Total analyte concentration

Figure 1   *The free analyte is available for partitioning to the receptor as well as to the matrix, while analytes that are bound to matrix components are not available for the receptor*

Figure 1, to obtain a ranking in pharmacological activity. For this purpose, the concentration of the chemical at which it exhibits a certain response, for example the median inhibition concentration, *i.e.* IC_{50}, is determined. A ranking based on the free concentration would reflect the binding affinity to this receptor, while a ranking based on the total concentration would be disturbed by the affinity for the matrix. Thus, when comparing potencies of chemicals, the free concentration should be used to avoid bias due to matrix effects.

2   Measurement of Free Concentrations Using SPME

In this section, the technique to measure free concentrations of organic chemicals is described. Theory of negligible SPME will be discussed, followed by some practical considerations. Finally, an example of the applicability of the technique win be given.

Theoretical Aspects to Measure Free Concentrations Using SPME

Conventional Methods

Free concentrations are currently measured using equilibrium dialysis, micro-dialysis, semi-permeable membrane devices (SPMD), high flow rate LC or headspace analysis. A detailed description of these methods is given in ref. 1. Dialysis techniques use semi-permeable membranes that allow free passage of the analyte but not of the matrix. Equilibration times, although highly dependent on the system, are relatively long (in the order of hours or even days) and the semi-permeable membrane may adsorb chemicals, in particular the more hydrophobic ones. Headspace analysis is only suitable for volatile chemicals. In this chapter we describe a method, using SPME, to measure these free concentrations in a simple manner for chemicals that can be analysed using either GC or HPLC.

Negligible Depletion SPME

Free concentrations can be measured using a negligible depletion extraction.[2] Actually, the principles are remarkably simple. Since there exists an equilibrium between the compound dissolved in the aqueous phase and bound to the matrix, the main purpose of the extraction is not to perturb this equilibrium. The extraction of only a small percentage of the free concentration fulfils this requirement while still enabling the measurement of the analyte.

In general, solvent extraction techniques interfere in two ways with the equilibrium between the compound dissolved in water and bound to the matrix. First, during solvent extraction, a small amount of the solvent dissolves into the aqueous phase and binds to the matrix, thereby influencing the partitioning behaviour of the chemical and often disrupting the structural features of the matrix. Second, during the extraction the aqueous phase is depleted from the analyte, thereby forcing the equilibrium between the dissolved compound and the matrix in the direction of the aqueous phase, eventually causing a nearly complete extraction and thus depletion of the aqueous sample.

Thus, solvent extraction techniques are difficult to combine with free concentration measurements. The introduction of SPME by the group of Pawliszyn[3] made some changes in this respect. The possibilities of SPME to measure free concentrations was recognised a few years later.[4] Later, the first negligible depletion SPME extraction was described.[2] Herein, the assumption that, analogous to receptor–drug interactions, only the free concentration is available for partitioning to the fibre was confirmed by a combination of negligible depletion SPME and dialysis. In this paper one criterion was described that is fundamental to negligible depletion extraction (see equation 1). This equation was derived not only to comply with non-perturbation of the matrix/water equilibrium, but rather to obtain similar kinetics in samples with and without a matrix (for example, the calibration samples).[2]

$$K_{fw} \times \frac{V_f}{V_s} \ll 1 \tag{1}$$

The physical meaning of equation 1, where K_{fw}, V_f and V_s are the fibre/water partition coefficient, the volume of the fibre coating and the sample, respectively, is that the extraction efficiency, at equilibrium, should be as low as possible. Usually the maximum extraction efficiency is set at 5%. Therefore, we redefine equation 1:

$$K_{fw} \times \frac{V_f}{V_s} \le 0.05 \tag{2}$$

Although K_{fw} in equation 2 is defined at equilibrium, and equation 2 needs to be satisfied at all times, there is no need to perform the free concentration experiments at equilibrium. The measurement of the free concentration when equilibrium has not been reached causes an even lower depletion than the 5%. For very hydrophobic chemicals it might take hours or even days to reach a steady-state situation, which would be impractical. During the absorption of the analyte to the fibre, there still exists an unambiguous relationship between the free concentration in the aqueous phase and the concentration in the fibre coating. Nevertheless, there is an additional requirement when free concentrations are measured in the kinetic phase of the absorption profile. The depletion in the boundary layer surrounding the fibre can cause locally a shift in the equilibrium between the chemical in the aqueous phase and bound to the matrix, as suggested by He and Lee.[5] Therefore, the diffusion through the aqueous phase should be preferably the rate limiting step when measuring in the kinetic phase of the absorption profile. Thus, this leaves the diffusion rate in the coating to be the necessary rate limiting step in the absorption process. An easy way to determine in which phase the diffusion is limiting the absorption kinetics is given in ref. 6.

Practical Aspects to Measure Free Concentrations Using SPME

System Requirements

The procedure to measure the free concentration in aqueous matrices can not start until the partition coefficient to the fibre coating has been determined (or calculated). After all, equation 2 needs to be satisfied, and thus K_{fw} should be known (V_f and V_s are known system parameters).

Equation 2 also gives an indication of possible changes that can be made to the system in case of non-compliance. There are basically three possibilities to modify the system to lower the extraction efficiency below the 5% limit. First, lowering the affinity for the fibre coating by exchanging the polymer phase, thereby lowering the extraction efficiency. Second, lowering the volume of the fibre coating, either by taking a thinner film or by cutting the fibre to shorter length also reduces the extraction efficiency. Last, increasing the sample volume does not change the amount that will be extracted, but does again decrease the extraction efficiency. Thus, changing the system gives sufficient possibilities to comply with equation 2.

Procedure to Measure Free Concentrations

The procedure to measure free concentrations in aqueous matrices, which is given in Figure 2, is not really different from ordinary concentration measurements. The measurement consists of comparing the signal from samples for

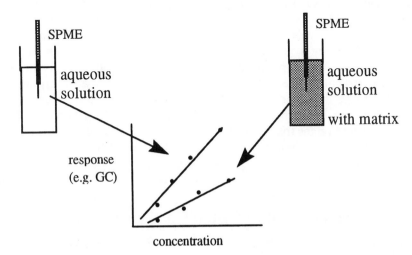

Figure 2   *The procedure to measure free concentrations is similar to other concentration measurements. The signal from samples with a matrix is compared to the signal from calibration samples*

which the concentration should be determined with the signal from calibration standards.

Calibration series for SPME are prepared in aqueous solutions, which is not always a straightforward procedure. For example, PCBs are difficult to dissolve and on many occasions the non-final concentration is significantly higher than the solvent extractable concentration. This discrepancy is not caused by low extraction recoveries, but mainly by volatilisation and binding to system components, *e.g.* glass walls. Therefore, it is good practice to calibrate the aqueous calibration standards to calibration samples that are prepared in an organic solvent, especially for volatile, degradable or extremely hydrophobic chemicals.

Example: Measurement of the free concentration in an in vitro assay. Currently, *in vitro* test systems are used to estimate the activity of pharmaceuticals, or the toxicity of chemicals to humans and other species. It is common practice to expose a monolayer of cells to chemicals that are dissolved in the culture medium. The effect on enzyme induction, biotransformation rates or viability is monitored as a function of time at different exposure concentrations. Unfortunately, the nominal exposure concentration is most frequently used to report the effect levels. Since the cells that are exposed in the *in vitro* system consist of organic materials like phospholipids, and proteins, it is likely, for relatively hydrophobic chemicals, that a significant amount of the chemicals will be absorbed to these materials. In addition, some *in vitro* test systems are supplemented with serum to improve cell longevity. Proteins that are present in this serum will bind the chemicals as well. Altogether this causes a lower free concentration.

In Figure 3, results are shown from an *in vitro* experiment where hepatocytes are exposed to four chemicals with varying hydrophobicity, in medium without serum. Data are taken from ref. 7. After an exposure of 10 minutes, the medium was isolated and analysed using negligible depletion SPME. Clearly, the most hydrophobic chemicals in this set, which can still be regarded as moderately hydrophobic, show a greater decrease in free concentration than the more hydrophilic ones. Hydrophobic chemicals are known to have a high affinity for lipophilic structures like cellular membranes, and thus the absorption was expected to be dependent on this property. The free concentrations of 4-*n*-pentylphenol and 4-chloro-3-methylphenol are reduced by a factor of almost three. The effect on the free concentration for highly hydrophobic chemicals like PCBs or dioxines can be expected to be ten- to thousand-fold worse. Thus, the toxicity of these compounds might be highly underestimated and the comparison of toxicities can be misinterpreted due to the effect of hydrophobic binding.

3   Determination of Partition Coefficients using SPME

Partition coefficients, which are defined as the ratio of the concentration of a chemical in two different phases at equilibrium, are of major interest in many disciplines. Partition coefficients, air–water, water–organic matter, and water–soil, are used in modelling the fate and distribution of chemicals in the

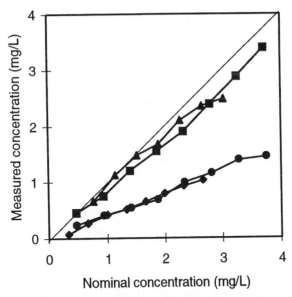

Figure 3   *The measured free concentration is plotted versus the nominal concentration in rat hepatocytes in primary culture for aniline (■), nitrobenzene (▲), 4-chloro-3-methylphenol (◆), and 4-n-pentylphenol (●). The drawn line represents the situation of complete bioavailability*

environment. Phospholipid–water partition coefficients are applied to model the absorption, distribution, and toxicity of drugs, and protein–water (which can be regarded as partition coefficients when the concentration of the chemical is much lower than the concentration of the protein) as a measure of binding affinity to transport or receptor proteins. In this section we propose a procedure to measure partition coefficients by negligible depletion SPME. This procedure is described in more detail in ref. 7.

Theoretical Aspects to Determine Partition Coefficients using SPME

In general, the partitioning of chemical X between an aqueous and organic phase is an equilibrium process:

$$X \text{ (aq)} \underset{\longleftarrow}{\overset{\longrightarrow}{}} X \text{ (organic)}$$

The partition coefficient K_{oa} is then defined at equilibrium as

$$K_{oa}(X) = \frac{[X]_{organic}}{[X]_{aq}} \tag{3}$$

Equation 3 immediately shows that by determining the concentration of a chemical in both the organic and the aqueous phase, the partition coefficient can be derived. Nevertheless, this approach requires a physical separation between those two phases. This poses experimental problems, especially when the organic phase is difficult to separate from the aqueous phase.

A different approach is to determine the difference in the free concentration in samples with and without the organic matrix. After all, the decrease in the free concentration can be subscribed to partitioning to the organic phase. Thus, assuming a mass balance, the concentration in the organic phase ($[X]_{organic}$) can be directly derived from the decrease in free concentration in the aqueous phase ($\Delta[X]_{aq}$), according to equation 4.

$$[X]_{organic} = \Delta[X]_{aq} \times \frac{V_{aq}}{V_{organic}} \tag{4}$$

where $V_{organic}$ and V_{aq} are the volumes of the organic and the aqueous phases, respectively. The partition coefficient can than be calculated by measuring the free concentration in a sample with and without an organic phase.

Practical Aspects to Determine Partition Coefficients Using SPME

Measurement of Phospholipid Vesicle–Water Partition Coefficients

The interest in phospholipid–water partition coefficients lies mainly in the fact that chemicals need to be absorbed over bilayer membranes to cause the eventual pharmacological or toxic effect, and these absorption rates are a non-linear function of the membrane–water partition coefficient. Since membrane–water partition coefficients are not available for many compounds, the octanol–water partition coefficient (K_{ow}) is mostly used as a surrogate. To test the predictive quality of K_{ow}, the partition coefficient between phospholipids (small unilamellar vesicles of L-α-dimyristoyl phosphatidylcholine) and water ($K_{DMPC,water}$) have been determined using negligible depletion SPME for a set of twenty-eight compounds.

For these measurements, a calibration series in water was prepared. Additionally, five samples were prepared with a known concentration of the chemical (which was equal to the highest standard) to which a known amount of vesicles was added. All samples were analysed by measuring their free concentration. Subsequently, the partition coefficient was calculated using equation 5.

$$K_{DMPC,water} = \frac{([X]_0 - [X]_a) \times \dfrac{V_a}{V_{DMPC}}}{[X]_a} \tag{5}$$

where $[X]_0$ is the nominal concentration, $[X]_a$ is the freely available concentration in the solution containing the vesicles and V_a and V_{DMPC} are the volumes of the aqueous and the phospholipid phase, respectively.[7]

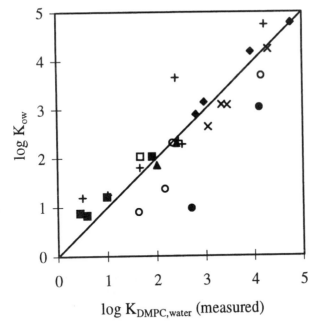

Figure 4 *A comparison of measured log $K_{DMPC,water}$ and log K_{ow} values. Data are shown for benzenes ($\blacklozenge$), alcohols ($\blacksquare$), anilines ($\bigcirc$), phenols ($\times$), nitrobenzenes ($\blacktriangle$), quinoline ($\square$), esters ($+$), and amines ($\bullet$)*

The results presented in Figure 4 show that K_{ow} should be used with caution, especially with more polar or even charged chemicals. Nevertheless, for apolar chemicals similar results were obtained as in other studies, which is a good validation for the method.[7] Thus, the determination of the membrane–water partition coefficient itself does not pose any experimental problem, especially since SPME can be completely automated.

Measurement of Protein–Water Partition Coefficients

In the aquatic environment, organic chemicals are absorbed mostly by the lipid phases in aquatic organisms. Usually, among fish, fat contents vary from roughly 3 to 10% of the total body weight. For species with low fat contents, other organic phases might contribute significantly to the bioconcentration behaviour of organic chemicals. Recently, Legierse *et al.*[8] found that bioconcentration factors (defined as the ratio of the concentration in the organism, and in the water at equilibrium) for the pond snail (*Lymnaea stagnalis*), which has a fat content of only 0.5%, did show the usual hydrophobicity dependent behaviour for nonpolar chemicals, but not for a polar chemical like the organophosphorous insecticide chlorthion. Therefore, the contribution of binding to snail protein was investigated.

Several snails were homogenised using a Potter–Elvehjem homogeniser.

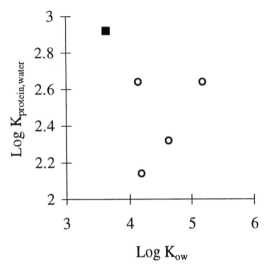

Figure 5   *Binding constants to proteins from Lymnaea stagnalis for chlorobenzenes (○) and chlorthion (■) versus their respective hydrophobicities (log K_{ow})*

Subsequently, the homogenate was centrifuged to isolate only cytosolar proteins. Binding to these proteins was studied with varying concentrations of the compound of interest. The decrease in the free concentration was measured using negligible depletion SPME. Partition coefficients were determined by calculating the free fraction and using equation 6.

In Figure 5, the partition coefficient to protein is plotted as a function of the hydrophobicity. Clearly, chlorthion shows a higher partition coefficient than would be expected based on the data for non-polar chemicals. As a result, the accumulation of chlorthion in the protein phase of the pond snail contributes significantly to its total bioconcentration.[9]

4   General Remarks

SPME has proven to be, already shortly after its introduction, a very versatile technique for analytical chemistry. In this chapter, SPME is presented to have more possibilities than other sampling techniques. Using SPME for free concentration measurements makes it an even wider applicable tool than most conventional methodologies.

References

1. H. Suffet, C.T. Javfert, J. Kukkonen, M.R. Servos, A. Spacie, L.L. Williams and J.A. Noblet, *Bioavailability*, CRC Press, Boca Raton, FL, USA, 1994.
2. W.H.J. Vaes, E. Urrestarazu Ramos, H.J.M. Verhaar, W. Seinen and J.L.M. Hermens, *Anal. Chem.*, 1996, **68**, 4463.
3. C.L. Arthur and J. Pawliszyn, *Anal. Chem.*, 1990, **62**, 2145.

4. F.-D. Kopinke, J. Pörschmann and M. Remmler, *Naturwissenschaften*, 1995, **82**, 28.
5. Y. He and H.K. Lee, *Anal. Chem.*, 1997, **69**, 4627.
6. W.H.J. Vaes, C. Hamwijk, E. Urrestarazu Ramos, H.J.M. Verhaar and J.L.M. Hermens, *Anal. Chem.*, 1996, **68**, 4458.
7. W.H.J. Vaes, E. Urrestarazu Ramos, C. Hamwijk, I. Van Holsteijn, B.J. Blaauboer, W. Seinen, H.J.M. Verhaar and J.L.M. Hermens, *Chem. Res. Toxicol.*, 1997, **10**, 1067.
8. K.C.H.M. Legierse, D.T.H.M. Sijm, C.J. Van Leeuwen, W. Seinen and J.L.M. Hermens, *Aquat. Toxicol.*, 1998, in press.
9. K.C.H.M. Legierse, W.H.J. Vaes, T. Sinnige and J.L.M. Hermens, submitted 1998.

CHAPTER 10

Estimation of Hydrophobicity of Organic Compounds

ALBRECHT PASCHKE AND PETER POPP

1 Introduction

In this chapter we report on a special physicochemical application of SPME. After a short theoretical section and a review on the established techniques for the determination of octanol–water partition coefficient (K_{ow}) as a common measure of hydrophobicity, we compare published SPME fibre-coating–water distribution constants (K_{fw}) of different substance classes with their K_{ow} values. Then we present actual results of long-term direct extraction of BTEX compounds, selected halogenated benzenes, chlorinated pesticides and poly-chlorinated biphenyls with the commercially available SPME fibre types and a new C_8-coating. The chapter ends with a discussion on the possibilities and limitations of SPME for the estimation of K_{ow} values.

It is necessary in many scientific disciplines such as organic chemistry, bio-chemistry, pharmaceutics, or environmental science to consider and quantify the hydrophobicity of substances. The simplest procedure to characterize the hydrophobic nature of a chemical is to dissolve it in a nonpolar liquid that is immiscible with water and then to determine the amout of it that could be extracted by water. Based on the pioneering work of Hansch and co-workers,[1-3] n-octanol–water has become the standard two-phase system and the partition coefficient of a substance in this system is a widely used physicochemical property for hydrophobicity characterization, e.g. in drug design[4,5] or in modelling of environmental fate of chemicals.[6,7] The octanol–water partition coefficient (K_{ow}) defined as an equilibrium concentration ratio, can be converted to a ratio of solute activity coefficients (γ) in both liquid phases:

$$K_{ow} = \frac{C_o^\infty}{C_w^\infty} = \frac{\gamma_w}{\gamma_o}\left(\frac{v_w}{v_o}\right) \tag{1}$$

Since the molar volumes (v) of both phases are constants and most substances behave nearly ideally in the octanol phase ($\gamma_o \approx 1$–3),[8] it becomes clear that the variations in K_{ow} are mainly caused by the value of γ_w which can span several orders of magnitude.[9] Hence the octanol–water partition coefficient measures mainly the relative non-ideality of a chemical in water and thus its hydrophobicity.

The rational basis for the use of a single standard two-phase system for hydrophobicity studies of chemicals, *i.e.* for the possible transformation of distribution constants from one solvent–water system to another, is a linear free-energy relationship (LFER)* first proposed by Collander[10] (*cf.* also refs. 2–4):

$$\log K_{sw} = A + B \log K_{ow} \tag{2}$$

Equation 2 can be derived from thermodynamic considerations assuming that the chemical potential of solute (in each of the liquid phases) is an additive quantity of its functional groups and thus the free energy of phase transfer can be expressed as a linear combination of the contributions of all structural groups of the solute.

There exists already a great number of experimental K_{ow} values summarized in several reviews,[11,12] handbooks,[6,7,13] and databases.[14,15] Various estimation methods have also been developed[4,6] including the well-known CLOGP software.[4,14] The octanol–water partition coefficients of organic compounds vary over at least ten orders of magnitude (from 10^{-2} to 10^8). Uncertainties still exist, especially for very hydrophobic substances ($K_{ow} \geqslant 10^5$). This is illustrated by the greater differences in published experimental results (*e.g.* K_{ow} data collections for PCB congeners[7,15]). But even for well-investigated more water-soluble pesticides one can find very different K_{ow} values.[12,16]

The experimental determination of K_{ow} can be done directly by batch equilibration techniques. The conventional shake-flask procedure[4,17] is useful for substances with $K_{ow} \leqslant 10^4$ whereas the slow-stirring method[18,19] can be used for more hydrophobic substances. Both methods are relatively time-consuming and laborious if the determination of solute concentration in both phases cannot be done *via* UV/Vis spectrometry. A time- and substance-sparing possibility of indirect determination of $K_{ow} \leqslant 10^6$ provides the RP-HPLC.[20,21] Here, the measured HPLC capacity factors of reference substances are directly correlated with their octanol–water partition coefficient to obtain a calibration line which can be inverted to calculate the K_{ow} of the test chemicals. The inherent assumption in this procedure is again the existence of a LFER according to equation 2 between the distribution constant K_{HPLC} ($= C_{stat}/C_{mob}$) and K_{ow}. The practicability of the RP-HPLC method has been demonstrated for different stationary-phase/mobile-phase pairs although the molecular mechanism of partitioning in an alkane- or phenyl-coated silica column is probably different from that in the octanol–water system. The incompleteness of a parallel comparison becomes obvious in the problems which are associated

---

*The partition coefficient (as equbrium constant) of a substance in the two-phase system organic solvent–water and the Gibbs free energy of phase transfer (ΔG_{sw}) are related in the following way: $\Delta G_{sw} = - RT [\ln K_{sw} - \ln (v_w/v_s)]$.

with the correct choice of the standard substances for calibrating the method and in the dependence of the measured capacity factors on the mobile-phase composition.[22] Additionally, with very hydrophobic substances, practical problems arise due to the long retention times.

By considering the above mentioned limitations of existing measuring techniques it seemed very interesting to use SPME as an alternative method for the indirect determination of K_{ow} values, the more so as some correlations between SPME distribution constants of different analytes and the associated octanol–water partition coefficients had already been reported by Pawliszyn's group.[23–25] Considering these data, it was not unrealistic to suggest that the K_{fw} values for other substance classes can also correlate with their K_{ow} values. The first papers focusing especially on the use of SPME for K_{ow} estimation were published in 1996.[26,27]

2 Description of $\log K_{fw}$–$\log K_{ow}$ Correlations for Reported Data

We originally tested the 100 μm PDMS fibre with some hydrophobic chlorinated substances and obtained only a poor correlation between $\log K_{ow}$ and $\log K_{fw}$ calculated from experiments with fibre exposure times from 30 to 60 min.[19] Therefore we have collected reported K_{fw} values to see how well these data correlate with K_{ow}. In Table 1 we have summarized data sets[23–31] for which a significant $\log K_{fw}$–$\log K_{ow}$ correlation exists. But one can find in the literature also K_{fw} values such as that from Chai *et al.*[32] for volatile C_1–C_3 halogenated organics or from Magdic and Pawliszyn[33] for chlorinated pesticides which do not correlate with K_{ow}. Figure 1 shows reported K_{fw} values for 97 substances *vs.* their K_{ow} values.

Table 1 *Linear relationships $\log K_{fw} = A + B \log K_{ow}$ for reported K_{fw} values (n = number of individual substances in the data set, r = correlation coefficient). The required $\log K_{ow}$ values were taken from different sources (mainly from refs. 11, 12 and 14)*

Substance group (n)	Fibre type	A	B	r	Ref.
BTEX (5)	100 μm PDMS	0.2275	0.9729	0.998	23
BTEX (5)	100 μm PDMS	0.4986	0.6229	0.995	28
BTEX (5)	7 μm PDMS	0.7729	0.6026	0.933	28
Alkylbenzenes (15)	PDMS[a]	0.2090	0.8023	0.969	29
PAH (4)	15 μm PDMS	1.0188	0.6392	0.971	24
PAH (5)	monomeric C_{18}	1.3006	0.2727	0.900	30
PAH (5)	polymeric C_{18}	1.1788	0.3396	0.908	30
Substituted Phenols (11)	95 μm PA	−0.5371	0.5935	0.958	25
Substituted Phenols (6)	85 μm PA	0.5953	0.7592	0.927	26
Iodoalkanes (9)	70 μm PA	0.4945	0.7375	0.983	31
Miscellaneous SVOC (19)	85 μm PA	−0.0855	0.8342	0.867	27

[a] Determined indirectly by a $\log K_{fw}$–LTPRI relationship.[29]

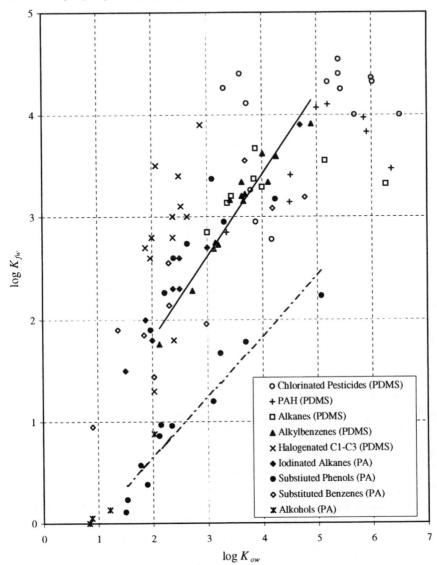

Figure 1 *Logarithmic scaled plot of reported K_{fw} values of 97 substances vs. their K_{ow}. The K_{wf} values were taken from refs. 25–29, 31–33; the $\log K_{ow}$ from refs. 7, 11, 12, 14, 19, 26, 27. The regression lines are drawn for the alkylbenzenes[29] (solid line) and the substituted phenols[25] (dashed–dotted line)*

Although there is a clear overall trend to higher K_{fw} values with increasing hydrophobicity of substances, good correlations over several $\log K_{ow}$ orders are only seen for iodinated alkanes,[31] alkylbenzenes[29] and substituted phenols.[25] The latter data set shows a remarkable different intercept to the data from ref. 26, but does not scatter as much from the straight line (*cf.* also Table 1). Furthermore, in Figure 1 one can see a decrease of $\log K_{fw}$ at $\log K_{ow} > 5$. For the more hydrophobic substances (chlorinated pesticides, polyaromatic hydro-carbons, higher alkanes) the distribution equilibrium was probably not reached with the SPME and thus too low K_{fw} values were determined. We have performed long-term experiments for the K_{fw} determination with the commer-cially available SPME fibre types and have included in our investigation a new poly(octylmethylsiloxane) coated fibre (C8) which was expected to have a similar extraction behavior as the reversed phases in HPLC.

3   Determination of K_{fw} in Long-term Experiments

Materials and Methods

Investigated Substances

For the K_{fw} determination we have selected three groups of substances: first the BTEX compounds, secondly some more volatile halogenated benzenes (further mentioned as halobenzenes, see Table 3 for details) and thirdly a group of semivolatile chlorinated benzenes, pesticides and biphenyls (later named as SVOCs, *cf.* Table 3). The chemicals were purchased dissolved in methanol from different producers.

Standard and Test Solutions

A stock solution in methanol was prepared using the standard mixtures. An aliquot of the stock solution was dissolved in bidistilled water to get an aqueous test solution for SPME with an initial concentration of individual substance of 1 mg L^{-1} for the BTEX compounds and the halobenzenes group, and of 0.5 μg L^{-1} for the chlorinated semivolatiles (SVOC group).

SPME Fibres

We used the following SPME fibre assemblies (purchased form Supelco): 100 μm PDMS, 65 μm PDMS-DVB, 85 μm PA, 65 μm CW-DVB. In addition we tested two fibre prototypes, a 65 μm Carboxen-PDMS fibre which is now commercially available in a film thickness of 75 μm, and a 65 μm poly(octyl-methylsiloxane) fibre (C8).*

Extraction Procedure

The fibres were conditioned as recommended by Supelco and were checked with a blank desorption run directly before use. The microextraction was carried out in most cases with 8 mL vials filled with 5 mL water sample. The sample agitation was performed by magnetic stirring with glass-coated mini-impellers at 1000 r.p.m. The digital stirrers used (Heidolph M 3000) ensured a constant stirring speed and did not change the temperature of the vial during long-term extractions.

*We thank G. Moskopp from Supelco, Germany for making these fibres available for test purposes.

Our experiments[34] and the great number of reported investigations on direct SPME of BTEX compounds (*e.g.* refs 23–25) have shown that the extraction equilibrium is reached with fast magnetic stirring (1000 r.p.m.) after 90–120 min. In earlier work[35] on the optimization of the direct extraction of the halobenzenes with the 100 μm PDMS fibre, we found that 60 min is sufficient to reach equilibrium for these compounds. To ensure equilibrium conditions, we have selected an extraction time of 180 min. For the SVOC group it was expected that the equilibration time would be much longer than for the BTEX aromatics and halobenzenes. Because of the lack of detailed information on this problem we have performed some tests with the CW/DVB fibre from six to 48 hours and have decided then to carry out SPME over 48, 72 and 120 h with the SVOC mix.

Desorption and Gas Chromatographic Analysis

The quantification of extracted analytes was carried out with different GC systems but always with a desorption time of 2 min in splitless injection mode. The special GC conditions for the different analytes are summarized in Table 2.

Calibration via Liquid Injection

The quantification of amount of analyte extracted onto the fibre, n_f, was done with liquid standard solutions (injected *via* autosampler/1 min splitless). For the BTEX and halobenzenes, single mid-point standard injections have been used for n_f calculation due to the signal linearity of FID. For the SVOC group we have performed a 12-point calibration of the ECD signal. The validity of this calibration was checked daily using a lower and higher calibration standard. n_f values were calculated from the ECD response curve using a smoothed spline function.[36]

Calculation of K_{fw} Values and their Variance

In the calculation of fibre-coating–water distribution constant it is necessary to consider also the evaporated amount (n_h) and the adsorbed amount (n_a) of test substance. The final equation for K_{fw} is:

$$K_{fw} = \frac{C_f^\infty}{C_w^\infty} = \frac{n_f/V_f}{(n_{w(0)} - n_f - n_h - n_a)/V_w} \tag{3}$$

The inital amount of substance ($n_{w(0)}$) is known and the extracted amount (n_f) will be determined. The evaporated amount of analyte (n_h) can be estimated using tabulated Henry's Law constants (Table 3). This inevitable change in water-phase concentration cannot be neglected for the substances investigated as the calculated volatilization losses under the actual experimental conditions show (see Table 3, where additionally the logK_{ow} values of the substances are listed).

Table 2 Specification of the GC conditions for the different substance groups

Substance group	GC system[a]	Capillary column	Injector temp. (°C)	Oven temperature program	Detector temp. (°C)	Make-up gas (N₂)
BTEX	HP 5890 II with FID[b]	CP-SIL 5 CB (25 m × 0.32 mm; 5 μm)	250[c]	30 °C (4 min) → 10 °/min → 150 °C	300	45 mL min⁻¹
Halobenzenes	HP 5890 II with FID[b]	CP-SIL 8 CB (25 m × 0.32 mm; 0.25 μm)	250	60 °C (1 min) → 12 °/min → 250 °C	300	60 mL min⁻¹
SVOC	CP 9001 with ECD	CP-SIL 8 CB (25 m × 0.32 mm; 0.25 μm)	250	80 °C (8 min) → 6 °/min → 250 °C	300	60 mL min⁻¹

[a] Carrier gas (N₂): 0.8–1 mL min⁻¹.
[b] Hydrogen: 35 mL min⁻¹; air: 260 mL min⁻¹.
[c] For C8 fibre 230 °C; for Carboxen/PDMS fibre 280 °C.

Table 3 *Henry's Law constants, K_H, and estimated volatilization losses, Δ_8, of the substances under investigation at 25 °C in an 8 mL vial (with 5 mL aqueous sample and 2.9 mL headspace) and their $logK_{ow}$ values*

Substance	K_H	$\Delta_8\ [\%]$[e]	$logK_{ow}$
Benzene	0.225[a]	11.5	2.13[f]
Toluene	0.274[a]	13.7	2.73[f]
Ethylbenzene	0.358[a]	17.2	3.15[f]
o-Xylene	0.228[a]	11.7	3.12[f]
m-Xylene	0.294[a]	14.9	3.20[f]
Chlorobenzene	0.153[a]	8.2	2.98[g]
o-Dichlorobenzene	0.098[a]	5.4	3.44[g]
p-Dichlorobenzene	0.065[a]	3.6	3.44[h]
Bromobenzene	0.085[a]	4.7	2.99[h]
o-Dibromobenzene	0.036[b]	2.0	3.79[h]
p-Dibromobenzene	0.028[b]	1.6	3.64[h]
Hexachlorobenzene (HCB)	0.053[a]	3.0	5.84[g]
α-Hexachlorocyclohexane (α-HCH)	0.00035[c]	0.02	3.79[i]
β-Hexachlorocyclohexane (β-HCH)	0.00005[c]	0.00	3.88[i]
γ-Hexachlorocyclohexane (γ-HCH)	0.00005[c]	0.00	3.72[i]
δ-Hexachlorocyclohexane (δ-HCH)	0.00003[c]	0.00	4.17[i]
1,1,1-Trichloro-2,2-bis(4-chlorophenyl)ethane (p,p'-DDT)	0.00095[c]	0.06	6.00[g]
1,1,-Dichloro-2,2-bis(4-chlorophenyl)ethane (p,p'-DDD)	0.00026[c]	0.01	6.02[h]
1,1-Dichloro-2,2-bis(4-chlorophenyl)ethylene (p,p'-DDE)	0.00321[c]	0.19	5.69[h]
2,4,4'-Trichlorobiphenyl (PCB 28)	0.0081[d]	0.47	5.62[h]
2,2',5,5'-Tetrachlorobiphenyl (PCB 52)	0.0081[d]	0.47	6.09[h]
2,2',4,5,5'-Pentachlorobiphenyl (PCB 101)	0.014[a]	0.82	6.85[h]
2,2',3,4,4',5'-Hexachlorobiphenyl (PCB 138)	0.00085[d]	0.05	7.25[h]
2,2',4,4',5,5'-Hexachlorobiphenyl (PCB 153)	0.00093[d]	0.05	7.16[h]
2,2',3,4,4',5,5'-Heptachlorobiphenyl (PCB 180)	0.00040[d]	0.02	8.04[h]

[a] Ref. 13.
[b] For the estimation of the dibromobenzene values from vapour pressure and aqueous solubility data we thank Dr. Ralph Kühne (UFZ).
[c] L.R. Sunito *et al.*, *Rev. Environ. Contam. Tox.*, 1988, **103**, 1.
[d] S. Brunner *et al.*, *Environ. Sci. Technol.*, 1990, **24**, 1751.
[e] $\Delta = |(C_{w(0)} - C_w^{\infty})/C_{w(0)}| \times 100$.
[f] Ref. 11.
[g] Ref. 19.
[h] Ref. 14.
[i] A. Paschke and G. Schüürmann, *Fresenius Environ. Bull.*, 1998, **7**, 258.

The amount of substance adsorbed at the glass walls and septum during fibre exposure (n_a) is difficult to quantify and thus we treat it as an experimental uncertainty. By introducing a 'headspace-corrected' water concentration ($C_w = C_{w(0)} - n_h/V_w$) we obtain the following simplified equation for calculating K_{fw}:

$$K_{fw} = \frac{n_f/V_f}{C_w - n_f/V_w} \qquad (4)$$

From equation 4 for K_{fw} we obtain the following equation for its variance s_K^2 by using the law of error propagation:

$$s_K^2 = \left(\frac{\partial K_{fw}}{\partial n_f}s_{n_f}\right)^2 + \left(\frac{\partial K_{fw}}{\partial V_f}s_{V_f}\right)^2 + \left(\frac{\partial K_{fw}}{\partial C_w}s_{C_w}\right)^2 + \left(\frac{\partial K_{fw}}{\partial V_w}s_{V_w}\right)^2 \qquad (5)$$

We assumed the following realistic estimates for the standard deviation of parameters describing the fibre geometry: 5 μm for a, 10 μm for b, 0.1 mm for h. For C_w, V_w, the directly injected amount of analyte, and the peak areas after fibre desorption and calibration we calculated with a relative standard error of 5%.

Results

Extraction Time Profiles with Selected Fibres

Figure 2 shows the ECD response of the CW-DVB fibre desorption after different extraction times (6–120 h) in the SVOC spiked water sample. The GC signal for all SVOCs goes through a maximum at 24 h and reaches a nearly constant level at longer extraction times. The heights of response signals depend on the ECD sensitivity for the compounds. Except C8, all fibre types tested show a similar extraction time profile for the SVOCs. An extraction maximum and a following decrease to a definite level has already been observed in other studies. This fact can be caused by a non-equilibrium situation where the uptake of analyte in the coating dominates the overall process due to the steep concentration gradient towards the coating.

Figure 3 gives the extraction *vs.* time profile (1–120 h) with the C8 fibre. Here the maximum at 24 h is not distinctly marked (with exception of HCB) and some compounds (*e.g.* DDE and PCB 180) seem to have a very slow uptake kinetics into the fibre coating.

Fibre-coating–Water Distribution Constants

For the BTEX aromatics and the halobenzenes we have used the results of a 3 h fibre exposure for the calculation of K_{fw} according to equation 4. The obtained values are listed in Table 4. As already shown in a former study[34] the Carboxen-PDMS fibre extracts the BTEX compounds much more efficiently than any

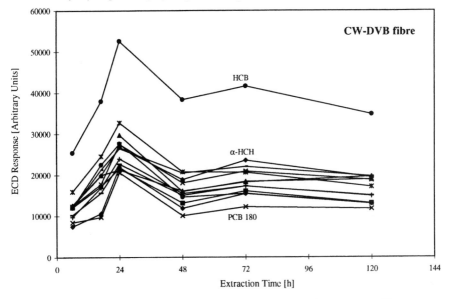

Figure 2 *Extraction* vs. *time profiles of selected SVOCs with the CW-DVB fibre*

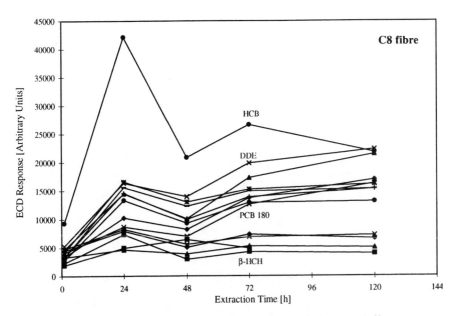

Figure 3 *Extraction* vs. *time profiles of selected SVOCs with the new C8 fibre prototype*

Table 4 Fibre/coating distribution constants, K_{fw} (calc. according equation 4) and their estimated standard deviations (equation 5) for BTEX and the halobenzene group (chloro- to p-dibromobenzene) after an extraction time of 3 h and for the SVOC group (HCB to PCB 180) after 120 h

Substance	PDMS (100 µm)	PA	C8	CW-DVB	PDMS-DVB	Carboxen-PDMS
Benzene	88 ± 9	114 ± 12	469 ± 49	183 ± 19	73 ± 7	8564 ± 1534
Toluene	231 ± 25	258 ± 27	980 ± 105	322 ± 33	333 ± 34	10 288 ± 2005
Ethylbenzene	541 ± 60	509 ± 55	509 ± 53	537 ± 56	940 ± 101	8438 ± 1502
o-Xylene	486 ± 51	526 ± 56	450 ± 47	528 ± 55	843 ± 90	9164 ± 1691
m-Xylene	546 ± 61	541 ± 58	621 ± 65	585 ± 61	996 ± 107	8654 ± 1558
Chlorobenzene	519 ± 57	747 ± 82	138 ± 14	1396 ± 154	1578 ± 176	13 568 ± 2980
o-Dichlorobenzene	1147 ± 136	2356 ± 297	511 ± 53	3589 ± 448	5306 ± 724	43 578 ± 21 093
p-Dichlorobenzene	1181 ± 140	2134 ± 264	437 ± 45	3658 ± 459	5354 ± 733	87 476 ± 76 257
Bromobenzene	581 ± 65	1034 ± 117	198 ± 20	2002 ± 229	2552 ± 301	21 528 ± 6235
o-Dibromobenzene	1566 ± 193	4910 ± 746	966 ± 104	7039 ± 1044	12 774 ± 2393	75 555 ± 57 907
p-Dibromobenzene	1755 ± 221	4825 ± 729	1074 ± 116	7536 ± 1143	15 218 ± 3104	71 222 ± 51 859
HCB	1897 ± 176	4242 ± 437	3657 ± 326	8861 ± 1012	12 307 ± 1612	1416 ± 114
α-HCH	782 ± 65	3312 ± 320	1280 ± 100	11 624 ± 1484	34 672 ± 8350	4093 ± 394
β-HCH	93 ± 10	23 916 ± 5821	1376 ± 108	20 696 ± 3561	79 803 ± 37 012	2270 ± 194
γ-HCH	674 ± 56	3360 ± 325	1136 ± 88	28 476 ± 5989	22 116 ± 3960	3815 ± 361
δ-HCH	750 ± 62	4190 ± 430	3103 ± 268	46 921 ± 14 136	38 309 ± 13 871	2771 ± 245
p,p'-DDD	3947 ± 436	14 255 ± 2480	28 996 ± 6172	24 724 ± 4743	20 530 ± 3516	1086 ± 86
p,p'-DDE	3473 ± 369	68 684 ± 38 975	91 937 ± 48 158	21 795 ± 3868	13 077 ± 1762	823 ± 64
PCB 28	3780 ± 412	18 264 ± 3702	13 435 ± 1834	17 769 ± 2802	99 354 ± 55 688	950 ± 74
PCB 52	3411 ± 361	16 809 ± 3232	26 217 ± 5222	25 152 ± 4878	15 844 ± 2349	1063 ± 84
PCB 101	3549 ± 380	19 086 ± 3982	45 891 ± 13 592	34 734 ± 8376	11 561 ± 1473	927 ± 72
PCB 138	15 687 ± 3387	90 486 ± 65 656	(609 722)	52 357 ± 17 179	23 825 ± 4466	613 ± 47
PCB 153	2562 ± 252	45 380 ± 18 087	83 377 ± 40 143	14 093 ± 1969	7383 ± 791	663 ± 51
PCB 180	1489 ± 133	8330 ± 1097	19 002 ± 3112	5411 ± 528	3007 ± 259	463 ± 35

other fibre type. The same relations are found for the halobenzenes, but the DVB blended coatings also yield relatively high K_{fw} values for these substance groups. The new C8 fibre is comparable to the latter fibre types for the BTEX compounds but has the lowest extraction efficiency for the halobenzenes.

For the SVOC we have calculated the fibre-coating–water distribution constants based on an extraction time of 120 h. Table 4 contains these K_{fw} values. For HCB and the HCH isomers CW-DVB and PDMS-DVB yield the highest distribution constants, whereas the picture is very different for DDD, DDE and the PCBs. The smallest K_{fw} values were obtained with Carboxen-PDMS followed by 100 μm PDMS. In the case of Carboxen-PDMS it seems to exhibit a completely different analyte uptake mechanism in the coating compared with the other fibres. Carboxen-PDMS is porous and discriminates larger molecules.[37] Interestingly polyacrylate (PA) shows very much higher K_{fw} than 100 μm PDMS. Also the new C8 coating seems to be a valuable tool for the direct SPME of SVOCs.

For some of the tabulated K_{fw} values we have calculated very high standard deviations (*e.g.* the Carboxen-PDMS values for BTEX and halobenzenes). Such high standard deviations are obtained if the amount of analyte initially in the sample is nearly completely extracted onto the fibre. This was already demon-strated by Pawliszyn.[29] But in considering the standard deviations K_{fw} in Table 4 one must bear in mind that we have assumed relatively high estimates for uncertainty of primary experimental parameters (see above). The detailed error analysis based on equation 5 shows that the errors in n_f and in C_w contribute to a much greater extent to the K_{fw} variance than the errors in V_f and V_w. (Under the actual experimental conditions and assumptions the ratio of the four summands in equation 5 is approximately 10–1:0.1:3–1:0.001).

4   Discussion

Our final goal was to study whether SPME can be used to estimate hydro-phobicity of organic compounds. We have correlated the logarithms of the determined K_{fw} values and the logK_{ow} at first for the three substance groups separately and then for all substances together. The significant linear relation-ships are summarized in Table 5.

For the BTEX compounds the results are as good as reported in literature (*cf.* Table 1). Exceptions here are Carboxen-PDMS and C8 which do not differ-entiate between the five components. For the halobenzenes all fibre types yield strong linear logK_{fw}–logK_{ow} relationships but for the more hydrophobic SVOC group we obtain a positive correlation only with the C8 fibre. The apparent decrease of K_{fw} values with increasing hydrophobicity for Carboxen-PDMS can be explained with the size-exclusion or pore-diffusion effect in this porous fibre coating.

Figure 4 shows a logarithmic plot of all measured K_{fw} values *vs.* K_{ow}. Only K_{fw} values resulting from the use of the C8 fibre show a remarkable linear increase from moderate to very hydrophobic substances. The outliers from the regression line in the upper region can be explained. For DDE and HCB we

Table 5 *Linear relationships* $\log K_{fw} = A + B \log K_{ow}$ *for determined* K_{fw} *values*
(n = number of individual substances in the data set, r = correlation
coefficient; for $\log K_{ow}$ *values see Table 3)*

Substance group (n)	Fibre type	A	B	r
BTEX (5)	PDMS (100 μm)	0.3169	0.7593	0.998
	PA	0.6635	0.6498	0.997
	CW-DVB	1.2523	0.4700	0.998
	PDMS-DVB	−0.4114	1.0705	1.000
Halobenzenes (6)	PDMS (100 μm)	0.8331	0.6436	0.962
	PA	−0.0109	0.9876	0.972
	C8	−0.9059	1.0472	0.963
	CW-DVB	0.7035	0.8416	0.954
	PDMS-DVB	0.0078	1.0996	0.937
	Carboxen-PDMS	1.6226	0.8897	0.851
SVOCs (13)	C8	1.4887	0.4676	0.806
	Carboxen-PDMS	4.2433	−0.1999	−0.952
All substances (24)	PDMS (100 μm)	1.9646	0.2373	0.764
	PA	1.9163	0.3734	0.812
	C8	1.1300	0.5197	0.902
	Carboxen-PDMS	5.1778	−0.3345	−0.794

suppose a special solubility behavior due to molecular structure. Interestingly the K_{fw} of both substances are nearly identical with PA and C8. The PCB 138 value is relatively uncertain (maybe much too high) and for PCB 180 the calculated K_{fw} can be too low because extraction equilibrium is not reached.

The selection of a fibre coating hydrophobicity can be rationalized. If the definition of distribution constants in terms of the activity coefficients (equation 1) is introduced in equation 2, one obtains a linear relationship between the ratios of the analyte's activity coefficients in the considered phases. This relationship can be reduced to a direct proportionality between the activity coefficients in the compared organic solvents (γ_s, that is here γ_f, and γ_o) under the assumption that the activity coefficients in the different aqueuos phases are nearly identical. In the actual case we equate the more or less solvent-free water matrix where the fibres are exposed with octanol-saturated water. Based on solubility measurements,[38,39] it can be concluded that the above made assumption is held and we obtain the following equation:

$$\log \gamma_f = A' + B \log \gamma_o \tag{7}$$

Thus, we need to find a SPME fibre coating for which the activity coefficient of the target analyte rises in a similar way as in octanol. The parameter B should not be too much below 1 to secure a high sensitivity of the method. γ_f can be

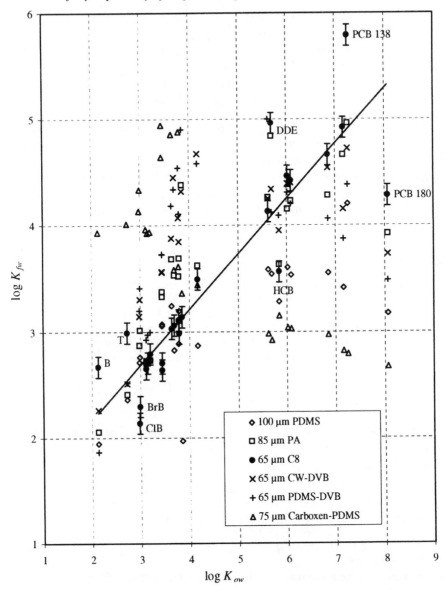

Figure 4 *Logarithmic scaled plot of determined K_{fw} values of BTEX, halobenzenes, and SVOCs vs. their K_{ow} (K_{fw} values from Table 4, $logK_{ow}$ values from Table 3). The line is obtained by linear regression of all C8 data points. The error bars of the C8 point of PCB 138 are drawn too small*

calculated from retention time measurements of the substance on a GC column with a stationary phase corresponding to the fibre coating.[40,41] Zhang and Pawliszyn[42] have demonstrated with some McReynolds probe substances that the use of headspace SPME can yield equivalent results to the GC method.

In addition to the problem with the selection of the appropriate fibre coating, the time-consuming determination of extraction equilibrium, especially if $\log K_{ow} > 5$ is expected, is a serious limitation of the SPME application for estimation of hydrophobicity. Additional problems can arise at long exposure times if hydrolysis of the substance under investigation cannot be neglected. In such cases it seems to be ingenious to combine the conventional octanol–water partitioning methods with SPME as an advantageous analytical tool. We have demonstrated this approach recently for chlorinated organic substances including the very hydrophobic PCB 209[19] but have also obtained reasonable results for selected polyaromatic hydrocarbons.[43]

Independently of the use for the estimation of hydrophobicity, the actual determined K_{fw} values can help to optimize the experimental conditions for direct SPME of HCH isomers, DDT metabolites and PCBs.

References

1. T. Fujita, J. Iwasa and C. Hansch, *J. Am. Chem. Soc.*, 1964, **86**, 5175.
2. A. Leo and C. Hansch, *J. Org. Chem.*, 1971, **36**, 1539.
3. A. Leo, C. Hansch and D. Elkins, *Chem. Rev.*, 1971, **71**, 525.
4. C. Hansch and A. Leo, *Exploring QSAR. Fundamentals and Applications in Chemistry and Biology*, American Chemical Society, Washington, DC, 1995.
5. H. Kubinyi, *QSAR: Hansch Analysis and Related Approaches*, VCH, Weinheim, 1993.
6. W.J. Lyman, W.F. Reehl and D.H. Rosenblatt, *Handbook of Chemical Property Estimation Methods – Environmental Behavior of Organic Compounds*, American Chemical Society, Washington, DC, 1990.
7. D. Mackay, W.Y. Shiu and K.C. Ma, *Illustrated Handbook of Physical-Chemical Properties and Environmental Fate for Organic Chemicals*, Lewis Publishers, Boca Raton, FL, 1992.
8. S.R. Bhatia and S.I. Sandler, *J. Chem. Eng. Data*, 1995, **40**, 1196.
9. D.A. Wright, S.I. Sandler and D. DeVoll, *Environ. Sci. Technol.*, 1992, **26**, 1828.
10. R. Collander, *Acta Chem. Scand.*, 1951, **5**, 771.
11. J. Sangster, *J. Phys. Chem. Ref. Data*, 1989, **18**, 1111.
12. A. Noble, *J. Chromatogr.*, 1993, **642**, 3.
13. D.R. Lide (ed.), *Handbook of Chemistry and Physics*, 78th Edn., CRC Press, Boca Raton, FL, 1997.
14. A. Leo, MedChem Database, Daylight Chemical Information Systems Inc., Irvine, CA, 1997.
15. J.A. Beauman and P.H. Howard, Physprop Database. Syracuse Research Corp., Syracuse, NY, 1995.
16. A. Finizio, M. Vighi and D. Sandroni, *Chemosphere*, 1997, **34**, 131.
17. OECD Guideline for Testing of Chemicals No. 107, *Partition Coefficient (n-octanol/water), Shake Flask Method*, OECD, Paris, 1995.
18. D.N. Brooke, A. Dobbs and N. Williams, *Ecotoxicol. Environ. Safety*, 1986, **11**, 251.
19. A. Paschke, P. Popp and G. Schüürmann, *Fresenius' J. Anal. Chem.*, 1998, **360**, 52.
20. J.G. Dorsey and M.G. Khaledi, *J. Chromatogr. A*, 1993, **656**, 485.
21. OECD Guideline for Testing of Chemicals No. 117, *Partition Coefficient (n-octanol/water), HPLC Method*, OECD, Paris, 1989.

22. K. Valkó, L.R. Snyder and J.L. Glajch, *J. Chromatogr. A*, 1993, **656**, 501.
23. D.D. Potter and J. Pawliszyn, *J. Chromatogr.*, 1992, **625**, 247.
24. D.D. Potter and J. Pawliszyn, *Environ. Sci. Technol.*, 1994, **28**, 298.
25. K.D. Buchholz and J. Pawliszyn, *Environ. Sci. Technol.*, 1993, **27**, 2844.
26. J.R. Dean, W.R. Tomlinson, V. Makovskaya, R. Cummings, M. Hetheridge and M. Comber, *Anal. Chem.*, 1996, **68**, 130.
27. W.H.J. Vaes, C. Hamwijk, E.U. Ramos, H.J.M. Verhaar and J.L.M. Hermens, *Anal. Chem.*, 1996, **68**, 4458.
28. J.J. Langenfeld, S.B. Hawthorne and D.J. Miller, *Anal. Chem.*, 1996, **68**, 144.
29. J. Pawliszyn, *Solid Phase Microextraction – Theory and Practice*, Wiley-VCH, New York, 1997.
30. Y. Li, M.L. Lee, K.L. Hageman, Y. Yang and S.B. Hawthorne, *Anal. Chem.*, 1997, **69**, 5001.
31. P.A. Frazey, R.M. Barkley and R.E. Sievers, *Anal. Chem.*, 1998, **70**, 638.
32. M. Chai, C. Arthur, J. Pawliszyn, R. Belardi and K. Pratt, *Analyst*, 1993, **118**, A1501.
33. S. Magdic and J. Pawliszyn, *J. Chromatogr. A*, 1996, **723**, 111.
34. P. Popp and A. Paschke, *Chromatographia*, 1997, **46**, 419.
35. P. Popp, A. Paschke, U. Schröter and G. Oppermann, *Chem. Anal. (Warsaw)*, 1995, **40**, 897.
36. K. Ebert, H. Ederer and T.L. Isenhour, *Computer Applications in Chemistry*, VCH, Weinheim, 1989.
37. R.E. Shirey and V. Mani, Pittcon Conference, Atlanta, 1997.
38. M.M. Miller, S.P. Wasik, G.L. Huang, W.Y. Shiu and D. Mackay, *Environ. Sci. Technol.*, 1985, **19**, 522.
39. A. Li and A.W. Andren, *Environ. Sci. Technol.*, 1994, **28**, 47.
40. R.E. Pecsar and J.J. Martin, *Anal. Chem.*, 1966, **38**, 1661.
41. K.L. Mallik, *An Introduction to Nonanalytical Applications of Gas Chromatography*, Peacock Press, New Delhi, 1976.
42. Z. Zhang and J. Pawliszyn, *J. Phys. Chem.*, 1996, **100**, 17648.
43. A. Paschke, P. Popp and G. Schüürmann, *Fresenius' J. Anal. Chem.*, 1999, in press.

Environmental Applications

CHAPTER 11

Air Sampling with SPME

PERRY A. MARTOS

1 Introduction

This chapter describes sampling airborne compounds with SPME. The sampling of two groups of analytes is presented: hydrocarbons and aldehydes. Various theoretical and practical considerations are provided with a strong emphasis on the latter. Both liquid injection and permeation tube standard gas generation techniques were used to calibrate SPME for each group of compounds. Finally, data from field studies comparing SPME air sampling to standard air sampling methods are presented.

2 Hydrocarbons

It is known that the sampling of airborne hydrocarbons occurs because they are prevalent and as such there is concern for their point of origin, their distribution, and fate in the environment.[1] The hydrocarbons also act as air quality markers[1] and therefore appropriate sampling methods for them are required.[2,3] The group of hydrocarbons chosen for air sampling with SPME is summarized in Table 1. These hydrocarbons include compounds with a broad range of boiling points and vapour pressures and they are most familiar to environmental scientists, *e.g.* terpenes, aliphatic hydrocarbons, mono aromatics and alkyl substituted aromatics.

Standard Gas Mixtures for Calibrating SPME: the PDMS Coating

Two basic methods can be used to calibrate SPME for air sampling: static gases and dynamic gases. The former suffers from a number of problems that are obviated with dynamic gas standards. Two dynamic standard gas generation systems are presented in this chapter. The first dynamic standard gas system used to calibrate PDMS for the sampling of hydrocarbons is presented in Figure 1, based on the 'liquid injection' of target analyte mixtures into a dilution gas.[4,5] The second is based on calibrated permeation tubes, presented below (Figure 2). The 'liquid injection' standard gas system (Figure 1) features

Table 1 *Summary of test hydrocarbons*

Analyte	Formula/MW	B.P. (°C)	D_{AB} (cm^2 s^{-1})	V. P. (mmHg)
Pentane (n-)	C_5H_{12}/72	36	0.092	420
Hexane (n-)	C_6H_{14}/86	69	0.084	124
Benzene	C_6H_6/78	80	0.090	75
Toluene	C_7H_8/92	111	0.082	21
Ethylbenzene	C_8H_{10}/106	136	0.076	7
Xylene (p-)	C_8H_{10}/106	138	0.076	9
Xylene (o-)	C_8H_{10}/106	144	0.076	7
Pinene (α-)	$C_{10}H_{16}$/136	155	0.065	3
Mesitylene	C_9H_{12}/120	165	0.071	2
Limonene (d-)	$C_{10}H_{16}$/136	176	0.065	1
Undecane (n-)	$C_{11}H_{24}$/156	195	0.061	<1

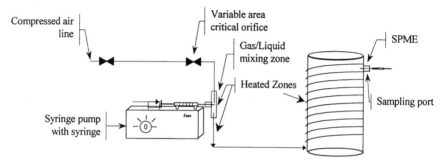

Figure 1   *A dynamically generated standard gas system. The analyte mixture is placed in a syringe and injected at a constant flow into a heated zone where its vapours are diluted with a flowing stream of air. The sampling chamber is maintained at constant temperature*

tremendous flexibility in generating standard gases of widely varying concentration and composition, as described by equation 1.

$$C_{g(T)} = \left[\frac{(m/t)_{mix}}{(F)_g} \times \left(\frac{\% Analyte}{Total} \right) \right] \tag{1}$$

where $C_{g(T)}$ is the gas concentration of analyte in *mass/volume* at a given temperature, m/t is the mass delivery rate of the liquid mixture, F is the dilution gas flow rate, and %*Analyte/Total* is the ratio of an individual analyte to the total. A common error normally encountered is when the mass/volume, *e.g.* μg L^{-1}, is converted to ppm. This conversion requires temperature, pressure and analyte molecular weight, as shown in equation 2. The mass of analyte must be converted to the *volume* it occupies relative to the total volume of the system.

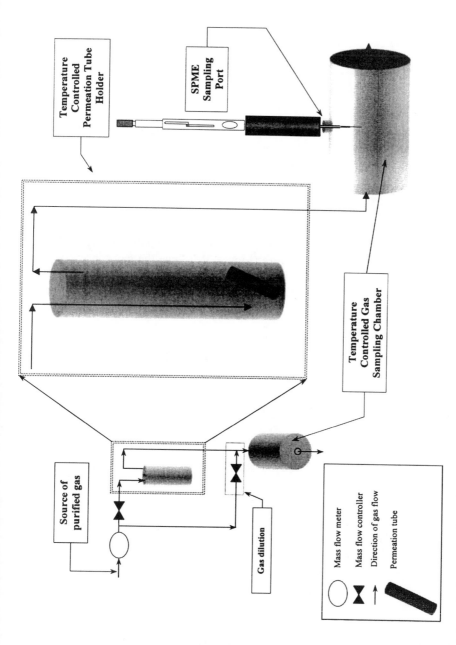

Figure 2 *A dynamically generated standard gas system based on permeation tubes and mass flow controlled gas dilution*

For example, for 1 μg L^{-1} (1 mg m^{-3}) benzene gas concentration (25 °C and 1 atm), the conversion to 1 ppmv (1 part/10^6 parts) is as follows:

$$C_{g(ppmv)}\left(\frac{\mu L}{L}\right) =$$
$$C_{g(\mu g/L)}\left(\frac{mg}{m^3} \times \frac{10^{-3}g}{mg} \times \frac{10^{-3}m^3}{L}\right)\left(\frac{0.0821\ L \times atm \times mol^{-1} \times K^{-1} \times 298 K}{1\ atm \times 78\ g \times mol^{-1}}\right)$$

$$(2)$$

Neglecting this can result in *serious* gas concentration errors. For example, it could be thought that 1 μg L^{-1} is 1 ppb in the gas phase. This is an error of more than 430 fold for pentane and more than 300 fold for benzene.

Similarity between SPME Gas Extractions and Standard Methods

The 100 μm thick PDMS fiber coating was used to sample a standard gas mixture of the test hydrocarbons (Table 1) and was compared to charcoal tube extractions of the same gas mixture (Figure 3).[4] Since the gas concentrations (mass vol^{-1}) of each analyte are equal, the mass loadings of analyte on the charcoal tubes are almost identical. This was observed by comparing the peak

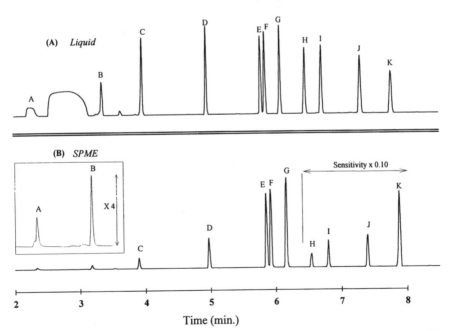

Figure 3 (A) *Chromatograms of charcoal tube and* (B) *PDMS SPME extraction of the standard gas (34 μg L^{-1}, 298 K) from the standard gas generating device. Peak identification, A: n-pentane, B: n-hexane, C: benzene, D: toluene, E: ethylbenzene, F: p-xylene, G: o-xylene, H: α-pinene, I: 1,3,5-trimethylbenzene (mesitylene), J: d-limonene, K: n-undecane*

areas, but not the peak heights (Figure 3A). By comparison, the SPME extraction of the same gas mixture resulted in lower mass loading of compounds such as pentane relative to *n*-undecane. This phenomenon is easily explained with equation 3 (refer to Chapter 2 of this book).

$$n_f^\infty = K_{fg} V_f C_g^\infty \tag{3}$$

The equation indicates that as the affinity of the analyte increases for the fiber coating, so does its mass at equilibrium. It is known that the affinity of *n*-undecane is more than 310 times greater than that of *n*-pentane, and as such, when both are present at equal concentration and extracted by the PDMS coating, they will follow this ratio. This is not a disadvantage. In fact, this phenomenon has led to the simple determination of distribution coefficients from chromatographic parameters, the retention index system,[6] with a simplified description shown here in equation 4. Here, if one knows the retention index value of a compound (LTPRI), then the K_{fg} (25 °C) can be easily calculated.

$$\log_{10} K_{fg} = 0.00415(\text{LTPRI}) - 0.188 \tag{4}$$

Effect of Sampling Temperature

The effect of temperature is important for the PDMS coating, but can be easily compensated by equation 4.[4] Here, when the PDMS fiber has been calibrated at one temperature, yet sampled at another temperature, the effect of temperature can be corrected.[4]

$$\log K_{fg} = a \times \left(\frac{1}{T}\right) + b \tag{5}$$

where the terms are described in detail elsewhere.[4] It is emphasized that the sampling temperature must be known and corrected if it differs from the calibration temperature. More than 10% error is possible with a 1 °C difference between the two.[4]

3 Formaldehyde

Formaldehyde (HCHO) is a ubiquitous airborne contaminant in our environment as a result of its use in a considerable number of products, and locations,[7-15] is a probable human carcinogen and is a known animal carcinogen.[16,17] Typical HCHO air concentrations are significantly less than 200 ppbv, which presents difficulties for its sampling and determination. SPME, however, can be used to sample HCHO with on-fiber derivatization.[18] Various concentrations of HCHO were generated using a permeation tube source of HCHO[18] (Figure 2). The generation of formaldehyde gas from this permeation tube occurs following the thermal decomposition of paraformaldehyde into formaldehyde (*e.g.* 80 °C). Concurrent to the steady-state release of formaldehyde from the permeation tube is its constant dilution with a mass flow controlled gas dilution system (Figure 2). This process can be described with equation 6.

$$C_{gas} = \frac{KR}{F} \times 10^6 \tag{6}$$

where C_{gas} is the analyte component concentration (ppmv), K is the inverse density of the permeant in (cm^3 g^{-1}), R is the permeation rate through the membrane (g min^{-1}), and F is the dilution gas flow rate (cm^3 min^{-1}). The 'R' is established gravimetrically, *i.e.* by weighing the permeation tube over a period of time. The 'K' is calculated using equation 7,

$$K = \frac{V_m}{M} \times \frac{760}{P} \times \frac{T}{273.15} \tag{7}$$

where the volume of one mole of gas V_m is 22.414 at STP, M is the molecular weight of the permeating gas, T is temperature (K), and P is the pressure (mmHg).

Figure 4 shows the derivatization reagent used and the product of the reaction following exposure to HCHO. Here, the derivatization reagent, PFBHA, is first loaded onto PDMS/DVB fiber coatings, then exposed to various concentrations of HCHO (Figure 5) for 10 s and 300 s. The result is that with longer exposure times larger amounts of product are formed on the coating. The method detection limit for the 300 s exposure time is ~5 ppbv HCHO! On the other hand, shorter exposure times also yield excellent calibration curves, with lower sensitivity, yet the dynamic linear range is extensive. Typical precision for this sampling system was ~10% RSD. Alternatively, the empirically determined first order rate uptake can be used to quantify the HCHO gas concentration.[18] In this approach, no calibration curves are required. The only criterion is that the consumption of derivatization reagent must remain negligible, which is simple to satisfy for short exposures (10 s) to

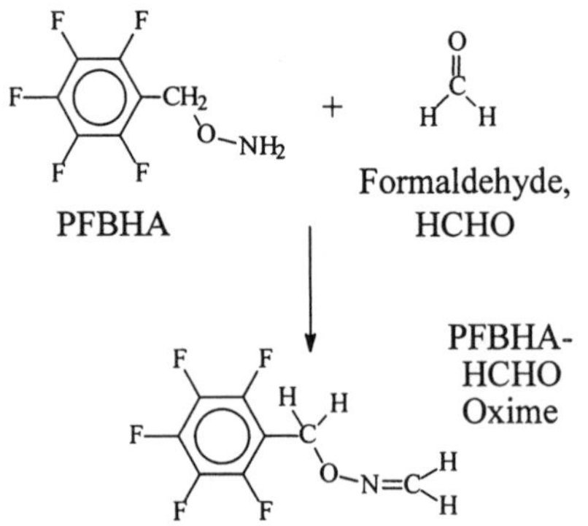

Figure 4 *On-fiber reaction between the derivatization reagent, PFBHA, and formaldehyde[18]*

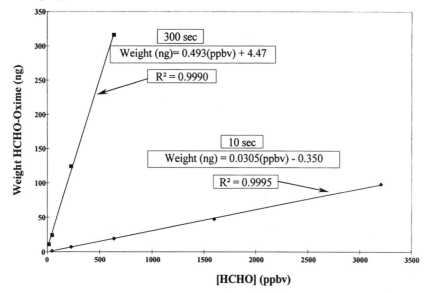

Figure 5 *Amount of PFBHA-HCHO oxime formed at 10 and 300 seconds sampling times as a function of various [HCHO] (ppb$_v$). The equations for the curves are presented. Note that the slope for the 300 second sampling time is approximately 16 times larger than that observed for the 10 second sampling time*

large concentrations of HCHO and long exposures (300 s) to low HCHO concentrations.

Formaldehyde in Cosmetics

Figure 6 shows the chromatograms obtained from the determination of HCHO from commercially available hair gel.[18] This sampling method, as demonstrated by Figure 6, was used to sample HCHO, and other aldehydes, from particleboard, biological material, and coffee grounds.[18,19]

4   Field Sampling Data

Hydrocarbons with PDMS

Table 2 summarizes the observed styrene concentrations at an industrial site using SPME, and two recognized air sampling methods. These SPME samples were acquired *via* TWA sampling[4,5] (Figure 7) (see Chapter 2), *i.e.* the fiber coating was retracted into the needle housing during sampling.[20]

Formaldehyde with PFBHA on PDMS/DVB

Table 3 summarizes the HCHO gas concentrations at a facility that stored formalin preserved tissue. Again, these samples were acquired with the fiber in

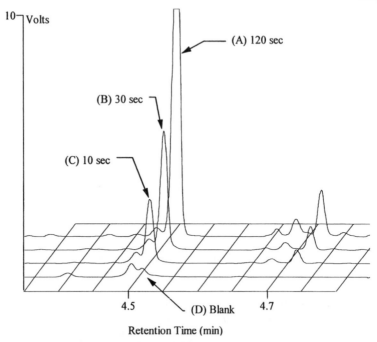

Figure 6 *Exposure of PFBHA loaded PDMS/DVB fibers to the headspace of approximately 3.5 g hair gel. The various times presented are different sampling times*

Table 2 *Industrial concentrations of styrene [μg L^{-1} (23 °C and 25% relative humidity)] using two different methods of field sampling, comparing SPME with PDMS to active sampling with charcoal and passive badge sampling*

Sampling time (min)	SPME 100 μm PDMS	Charcoal Tube (NIOSH Method 1500/1501)	Passive Badge (3M OV-3500)
30	56	54	72

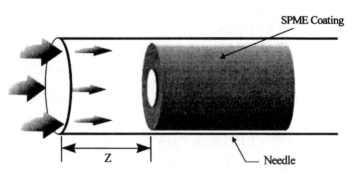

Figure 7 *The position of SPME to acquire TWA samples (Chapter 2)*

Table 3 *TWA (7 h) (ppb$_v$) gas phase concentrations of formaldehyde at a pathology storage facility using three different sampling and analysis methods. TWA SPME of formaldehyde uses PFBHA-PDMS/DVB fibers. Other methods were 3M passive badges (3721, colorimetric assay) and NIOSH Method 2541(HMP/XAD-2, GC/FID)*

Location	3-M (ppb$_v$)	SPME (ppb$_v$)	2541 (ppb$_v$)
1	88	109	102
2	149	152	160
3	38	57	51

the retracted position (Figure 7) for 420 min. The SPME data were compared with two completely different conventional sampling and analysis methods, and were found to agree extremely well with them.

6 Summary

SPME can be used as an air sampler for a broad range of analytes. It shows tremendous promise to be used as an air-sampling tool, which is in many ways superior to conventional air sampling systems. For example, its small size and high sampling sensitivity lend itself to sampling in areas not possible with other methods. Its flexibility extends to both rapid (spot) sampling and time-weighted average sampling.

References

1. M.L. Davis and D.A. Cornwell, *Introduction to Environmental Engineering*, McGraw-Hill, Inc., 1991, Ch. 6.
2. US EPA Method T01, *Method for the Determination of Volatile Organic Compounds in Ambient Air Using Tenax Adsorption and GCMS*, 1984.
3. US EPA Method T011, *Determination of Volatile Organic Compounds in Ambient Air Using SUMMA Passivated Canister Sampling and GC Analysis*, 1988.
4. P.A. Martos and J. Pawlisyzn, *Anal. Chem.*, 1997, **69**, 206.
5. J. Pawliszyn, *Solid Phase Microextraction: Theory and Practice*, Wiley-VCH, 1997.
6. P.A. Martos, A. Saraullo and J. Pawlisyzn, *Anal. Chem.*, 1997, **69**, 402.
7. B.A. Tichenor and M.A. Mason, *J. Air Pollut. Control. Assoc.*, 1988, **38**, 264.
8. M.F. Boeniger, *Am. Ind. Hyg. Assoc. J.*, 1995, **56**, 590.
9. F. Lipari, J.M. Dasch and W.F. Scruggs, *Atmos. Environ.*, 1984, **18**, 326.
10. T. Godish, *Am. J. Public Health*, 1989, **79**, 1044.
11. C.H. Risner, *J. Chrom. Sci.*, 1995, **33**, 168.
12. D.K. Milton, M.D. Walters, K. Hammond and J.S. Evans, *Am. Ind. Hyg. Assoc. J.*, 1996, **57**, 889.
13. J.S. Bennett, C.E. Feigley, D.W. Underhill, W. Drane, T.A. Payne, P.A. Stewart,

R.F. Herrick, D.F. Utterback and R.B. Hayes, *Am. Ind. Hyg. Assoc. J.*, 1996, **57**, 599.

14. P. Carlier, H. Hannachi and G. Mouvier, *Atmos. Envir.*, 1986, **20**, 2079.
15. R. Otsen and P. Felin, *Gaseous Pollutants: Characterization and Cycling*, John Wiley & Sons, 1992, pp. 345–421.
16. *Threshold Limit Values for Chemical Substances and Physical Agents*, American Conference of Governmental Industrial Hygienists, 1996, p. 23.
17. R.S. Bernstein, L.T. Stayner, L.J. Elliott, R. Kimbrough, H. Falk and L. Blade, *Am. Ind. Hyg. Assoc.*, 1984, **45**, 778.
18. P.A. Martos and J. Pawliszyn, *Anal. Chem.*, 1998, **70**, 2311.
19. P.A. Martos, Ph.D. Thesis, University of Waterloo, 1997, Ch. 4.
20. P.A. Martos and J. Pawliszyn, accepted in *Anal. Chem.*

CHAPTER 12

The Application of SPME in Water Analysis

CHRISTOPH GROTE AND KARSTEN LEVSEN

1 Introduction

Water is our most important food. Extensive analytical monitoring (in particular of those water supplies which are used for drinking water production) is required to secure and maintain constant high quality of this basic food.

While sophisticated instruments (such as GC–MS and HPLC–MS) have been designed in the past for reliable identification and quantification of trace organics in aqueous samples, less progress has been achieved in the development of advanced and automated methods for sample extraction and clean-up. Thus, even today many standardized methods in water analysis still rely on *liquid/ liquid extraction* (LLE), a very laborious technique for which large amounts of (in part) toxic solvents are used. The introduction of *solid-phase extraction* (SPE) in water analysis has reduced the amount of solvents necessary for extraction, but also the manpower involved in routine analysis of aqueous samples, as automatic extraction units are available today. No solvents are necessary for *headspace analysis*, a technique which again may be readily automated and which has therefore found widespread application in water analysis, in particular for volatile compounds.

The introduction of SPME in water analysis represents a further important step forward towards the development of simple, efficient, solvent-free extraction methods. This is a technique which again may be readily automated.

Indeed, water was the matrix to which SPME was applied first and even today most SPME studies are still reported on water analysis. Over 100 publications (including two reviews)[1,2] have demonstrated the usefulness of this new extraction method for the analysis of many different compound classes in aqueous matrices. These studies will not be covered exhaustively in this chapter, but rather only those reports will be discussed which highlight the use of SPME in water analysis, *e.g.* in which the parameters influencing SPME of analytes from aqueous samples, basic aspects and also quantification by SPME are discussed in more detail. Moreover, the basic aspects in water analysis and

169

the parameters influencing the extraction will be illustrated in this chapter using waste water as an example, an aqueous matrix with extreme properties.

In addition, this chapter will provide a general guideline for the development of SPME methods in water analysis.

The last section of the chapter is devoted to a specific application of SPME in water analysis, *i.e.* the development of a new automatic system for semi-continuous monitoring of surface, ground and (industrial) waste water.

2 Basic Considerations

The theoretical aspects of solid-phase microextraction were discussed in detail in the first chapter of this book (see also a recent monograph).[3] In that chapter, the practical procedures for both manual and automatic SPME (using a modified autosampler) are also reported.

SPME is an equilibrium method. Thus, usually only a small fraction of the analytes present in the sample is extracted. This is a major difference compared with SPE where exhaustive extraction of a preselected sample volume is desirable. The maximum amount of a given compound is extracted once its equilibrium distribution between the fibre coating and the sample matrix has been reached. As shown in Chapter 1, the amount of analyte extracted by SPME is independent of the sample volume, if the latter is very large. Thus, direct sampling of large water reservoirs such as lakes is possible. However, exhaustive extraction by SPME is also possible, in principle, if the sample is very small and the distribution constant very large. The effect of sample volume on the extraction yield was described by Górecki and Pawliszyn.[4]

In water analysis, SPME is performed by exposing the fibre coated with a suitable polymer directly to the aqueous sample or to the headspace above the sample.

In both modes, the extraction will be accelerated by agitating the sample to allow a rapid mass transfer of the analytes. In the case of direct SPME, agitation is mainly achieved by magnetic stirring,[5,6] by vibration of the needle in which the fibre is mounted[7,8] or by sonication.[5,6] In headspace SPME, magnetic stirring is the method of choice.[9-11] Only in flowing aqueous systems is agitation not necessary or less important.[8,12]

Headspace SPME is particularly suited for complex and dirty aqueous matrices, if the analytes are sufficiently volatile. Technical aspects of this method have been described,[9-11,13,14] including the use of a cooled fibre to enhance the extraction yield.[15] In general, the extraction time for volatile compounds is shorter for headspace SPME than for direct extraction (see also Chapter 1). This has been shown for BTEX compounds,[9,16] where, *e.g.* with polydimethylsiloxane/Carboxen fibres, an equilibrium is reached after ∼25 min in the headspace mode, but within 120–180 min in the direct mode.[16]

Analysis by SPME is very easy, if combined with gas chromatography (GC) or gas chromatography coupled to mass spectrometry (GC–MS). The fibre is simply transferred to the GC inlet where thermal desorption occurs. With this equilibrium method, only a small fraction of the analytes is extracted (see above). However, this entire fraction is transferred into the GC, leading to low detection limits (see Section 5).[17,18] Moreover, this 'in-line' method reduces sample contamination during the sample preparation step.

A necessary prerequisite for SPME–GC analysis is that the compound is sufficiently volatile and stable to be amenable to GC. In water analysis, there are numerous compound classes which are so polar or thermally unstable that extensive peak tailing (as with phenols) or even thermal degradation (as with certain pesticide classes or explosives) occurs.

Derivatization of the analytes allows this shortcoming to be overcome at least partially. Derivatization with suitable agents can occur inside the fibre coating, in the GC injector port, in the headspace of the sample or directly in the aqueous phase, where the latter approach is the most straightforward. Derivatization of fatty acids, amines, phenols and several metal ions for SPME–GC analysis was reported.[19–29]

The range of compounds amenable to SPME analysis can be extended to include polar and thermally labile compounds, if SPME is combined with HPLC or HPLC/MS (see Chapters 23 and 24). This approach has been applied to the analysis of PAHs,[30] surfactants in water[31] and to the determination of thermally labile pesticides in aqueous samples.[32–34] However, at present, detection limits with this method are too high to verify the limits set in the European Drinking Water Guideline (0.1 μg L^{-1}).[33,34] An automated SPME–HPLC device has also been introduced.[33]

Other chromatographic methods such as CE[35] and SFC[36,37] have also been combined with SPME in water analysis. Coupling of SPME to spectroscopic detectors such as AED,[27,38] ICP–MS,[39] PED,[40] UV spectroscopy[41,42] and Raman spectroscopy[43] have been described.

SPME can be applied to all aqueous matrices, such as drinking, surface, ground and waste water. A new instrument for the quasi-continuous monitoring of industrial waste water based on SPME is described in detail below.

3 Parameters Determining the Extraction

General Remarks

If all other parameters are kept constant, the maximum extractable amount under equilibrium conditions* is determined by the *distribution constant*. Depending on the coating and the individual compounds, these distribution constants may vary substantially, leading to varying extraction yields, as shown in Table 1 for two polymeric phases and 24 compounds typically found in the waste water of one German chemical plant. These differences in extraction yield are a disadvantage in water analysis.

For headspace analysis, the distribution constants can be estimated from retention indices,[44–47] which is also useful for quantification in SPME (see also Section 4).

---

* Note that with headspace SPME the equilibrium comprises at least three phases, with direct extraction at least two phases.

Table 1 *Comparison between 85 μm PA and 100 μm PDMS fibres and precision*
(85 μm PA fibre)

Compound order: elution on PTE-5 extraction time: 60 min *: peak area 2-bromophenol (PDMS) = 1	(a) Relative peak area* PA	PDMS	(b) Relative standard deviation (%RSD, n=8) PA
Fluorobenzene	37	40	3.7
1-Bromobutane	33	60	3.1
1,1,2-Trichloroethane	12	9	1.1
2-Fluorotoluene	115	148	2.1
Chlorobenzene	107	95	1.2
o-Xylene	189	270	1.3
2-Bromothiophene	62	37	2.8
Ethyl-2-bromoisobutyrate	9	16	2.8
Bromoacetaldehyde diethyl acetal	4	3	6.6
4-Bromotoluene	331	300	1.1
4-Fluoroacetophenone	10	4	1.8
2-Bromophenol	34	1	1.4
3-Bromochlorobenzene	288	232	1.2
2-Fluoropropiophenone	41	27	1.4
2,5-Dibromothiophene	243	152	1.3
2,3-Dimethylphenol	51	4	1.1
3-Chlorophenol	55	2	1.2
4-Bromo-2-chlorophenol	154	3	3.3
1,4-Bromoethoxybenzene	311	208	1.2
3,5-Dibromotoluene	585	620	2.7
3-Chloro-4′-fluoropropiophenone	27	12	6.3
4-Bromo-2,6-dimethylphenol	271	13	1.3
2-Bromo-4-nitrotoluene	179	64	1.8
4,4′-Difluorobenzophenone	386	131	2.1

Polymeric Coating

For any given analyte, very different distribution constants and thus very different extraction yields are observed for different polymeric coatings. Today, SPME fibres with various polymeric phases are commercially available and suitable for water analysis. Coatings recommended for GC analysis are polydimethylsiloxane (PDMS), polyacrylate (PA), polydimethylsiloxane/divinylbenzene (PDMS/DVB), polydimethylsiloxane/Carboxen (PDMS/Carboxen) and Carbowax/divinylbenzene (CW/DVB).[48] The polymeric coatings differ in thickness, but in particular in their chemical composition and polarity. Whereas the first two 'liquid' coatings (PDMS, PA) extract by *absorption*, the latter three (PDMS/DVB, PDMS/Carboxen, CW/DVB) can be considered as 'solids' which extract by *adsorption*. The properties of these coatings are described in Chapters 5 and 6. For samples containing analytes in concentrations covering several orders of magnitude, the PDMS and PA polymers are the coatings of choice, since displacement of trace analytes by other analytes of higher affinity are of minor concern (see also below). The different extraction ability of two selected coatings is demonstrated in Table 1 for 24 compounds found in industrial waste water. Although all compounds were present in the same concentration, the extracted amount differs significantly for the 100 μm PDMS and the 85 μm PA fibre due to different K values. (Note that the equilibrium was not reached in all cases.)

Extraction Time

The dependence of the extracted amount of analyte on the extraction time (often termed extraction time profile) gives valuable information for SPME method development and allows the experimental determination of the equilibration time. Figure 1 shows this time dependence for five compounds (found in industrial waste water) and two polymeric coatings. It is apparent that the extraction time profile (absorption time profile) depends on the individual analyte, but in particular on the polymeric phase. While for the PDMS fibre an equilibrium for all selected compounds is already reached after 30 min, much longer equilibration times are necessary if a PA fibre is used. Although in principle it is desirable to continue extraction until equilibrium is achieved, in routine analysis there is often insufficient time to do so. At non-equilibrium conditions (see also Chapter 2), the extraction yield strongly depends on the extraction time, as shown in Figure 1. Precise determinations are still possible at shorter times, if the extraction time is kept very precisely constant.

Temperature

Extraction by SPME is an exothermic process. Thus, lowering of the temperature increases the distribution constant and the extraction yield at equilibrium. However, in practical applications when the extraction is stopped before reaching the equilibrium, not only thermodynamic but also kinetic aspects become important. This is demonstrated in Figure 2, where the extraction yield (expressed as peak area) is plotted against the temperature. It is apparent that the extraction yield passes through a maximum. Before this maximum, the extraction yield increases with increasing temperature due to the enhanced mass transfer (kinetics), while at temperatures above this maximum, uptake decreases due to thermodynamic reasons (decreasing distribution constant).

Matrix Effects

The distribution constant, and thus the extracted amount, strongly depends on the matrix. The most important parameters influencing the extraction are described below.

a)

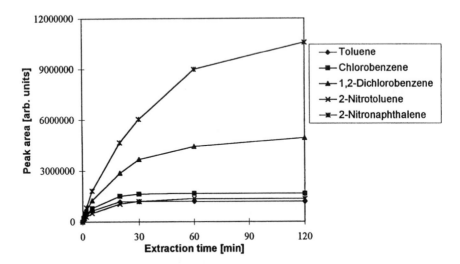

b)

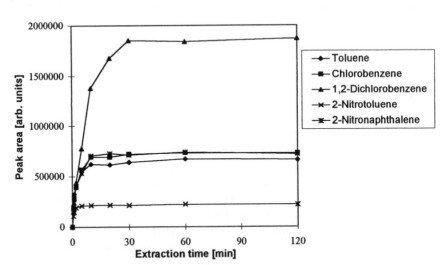

Figure 1   *Absorption time profiles for selected compounds* (a) 85 μm *PA*, (b) 100 μm
PDMS fibre

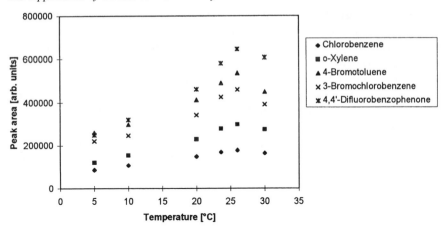

Figure 2 *Effect of temperature on the extraction yield (non-equilibrium conditions)* *(85 μm PA fibre,* 30 min *extraction,* 0.1 ppm)*

pH

A strong dependence of the extraction yield on the pH value is observed for acidic and basic compounds (see also Chapter 1).[21,22,49–51] Such compounds may only be extracted quantitatively by SPME, if they are present in the neutral form. For instance, the waste water contaminants, listed in Table 1, comprise several phenolic compounds, which are best extracted at low pH.

Salt

The effect of salt addition to enhance the extracted amount of an analyte by SPME was studied in detail by several authors.[17,18,47,49,51] Salting (in general by the addition of sodium chloride) is well known to improve extraction of organics from aqueous solution. Although salt addition usually increases the amount extracted, the opposite behaviour is also observed.[17,18] In general, the effect of salt addition increases with the polarity of the compound.[18] In particular, industrial waste water may contain higher amounts of inorganic salts (up to a few percent). While such concentrations generally have little impact on the extraction, this has to be verified in each instance.

Organic Solvents, Lower Alcohols

Usually, water samples do not contain high amounts of organic solvents. However, if for method validation aqueous samples are spiked with reference compounds dissolved in polar, volatile solvents, the concentration of these additional solvents should be sufficiently low in order not to influence the SPME process.

It was demonstrated for a series of aromatic hydrocarbons that the analyte uptake is not affected by methanol concentrations up to 1%.[52] Other authors report that a methanol content of up to 5% does not lower the extraction yield of selected triazines by more than 10%.[53] In contrast, ethanol present in aqueous solution has a strong effect on the extraction of several pesticides even at low concentrations ($\leqslant 5\%$).[54] This demonstrates that the effect of organic solvents is compound-dependent. In view of the fact that the waste water from many chemical companies has a variable content of lower alcohols, the influence of these on the extraction of the analytes has to be verified by additional experiments. However, in general, small amounts of organic solvents do not significantly affect the extraction yield.

Compounds in Excess

In many aqueous samples, pollutants are present in a wide range of concentrations. For instance, in the above-mentioned industrial waste water, the concentration of the various contaminants may differ by several orders of magnitude. It is at least conceivable that compounds present in large excess displace minor components when absorbed by the fibre coating. Several authors have studied the effect of excess compounds.[18,55,56] In all instances, displacement processes were not observed, *i.e.* the concentration of a trace compound was not influenced by the concentration of a compound in excess. Such studies have usually been carried out using PDMS and PA fibres, where analytes are mainly absorbed. The situation may differ for fibres, where adsorption processes prevail (see also Chapter 5).

Organic Matter

Humic compounds are often present in ground water in concentrations > 1 mg L^{-1}, *i.e.* in greater excess than anthropogenic pollutants. Again, displacement reactions during the absorption of analytes may be anticipated. However, studies on pesticides and other compounds did not show a pronounced effect of humic compounds on the extraction of other pollutants, provided their concentration did not exceed 10 mg L^{-1}.[18,57,58] Dewulf *et al.* investigated the effect of humic acids on the extraction of halomethane and haloethane compounds. They found that even levels of 100 mg L^{-1} of humic acids did not significantly affect the extraction yield.[47]

Other authors investigated the matrix effects by spiking different types of waters (Milli-Q, tap, sea, waste water) with a constant amount of reference compounds.[59] They found that the natural organic matter content in waste water with a DOC level of 212 mg L^{-1} can reduce the extraction yield by more than 20% compared to spiked Milli-Q water.

In summary, the SPME process may be affected by several matrix parameters. Thus, the matrix properties in waste water analysis by SPME should be known.

4 Quantification by SPME

With SPME, the quantification of analytes can be carried out by using the following techniques:[3]

(1) External standard calibration
(2) Internal standard calibration (and isotopic dilution)
(3) Standard addition
(4) Application of a known distribution constant

Each technique has its particular advantages and disadvantages. The method of choice strongly depends on the accuracy which has to be reached and the characteristics of the matrix. All above-mentioned methods are described in more detail in Chapter 1.

Whilst the matrix may have a significant effect on the extraction yield, many quantifications are performed using external calibration,[17,49,59] although the use of internal calibration is also described.[60,61] As a result of the large differences in K values observed for many analytes, the choice of appropriate internal standards is difficult (see also Section 6). Isotopically labeled compounds are best suited, but require the use of a mass spectrometer. Moreover, these compounds are often very expensive and sometimes not available. When variations in matrix are expected, standard addition is the method of choice.[17,25,62]

Finally, known distribution constants (K values) of the analyte may be used for quantification which may be calculated by determining the amount of analyte extracted at equilibrium (see Chapter 1). However, this procedure is very time-consuming.

When sampling gaseous matrices (*i.e.* in headspace analysis), a prediction of distribution constants based on physical chemical properties of the coating is possible (see Chapter 8).[44–47] The method is based on the linear relationship observed when plotting log K against the retention index obtained in a linear temperature programmed GC separation. Since only a single GC run is needed to determine the retention index, this method is easily performed. It has to be kept in mind that the fibre and column coating must be of the same material.

Originally, the SPME technique was considered by many scientists to be a screening method. However, it has been shown, particularly in water analysis, that reliable quantitation of analytes using SPME down to ppt level is possible.[63–65]

Interlaboratory comparisons have been organized between several laboratories in Europe and North America. Eleven laboratories took part in the first test, where 12 pesticides (organochlorine, organonitrogen and organophosphorus) were analyzed by GC–MS at low ppb level using a 100 μm PDMS fibre.[63] The second test was performed with selected VOCs at low ppb levels using a 100 μm PDMS coating and GC with MS, FID or ECD detection.[64] In a third interlaboratory comparison, triazine herbicides and some of their degradation products were determined at ppt level using the Carbowax/DVB fibre and MS or NPD detectors.[65] The results demonstrate that accurate quantification is possible by SPME even at trace levels (see also Chapter 15).

5 Development of an SPME–GC Method in Water Analysis

Development of an SPME method in combination with GC (MS) involves the following steps:

(a) Optimization of the chromatographic separation, selection of a suitable detector

(b) Selection of SPME coating

A fibre coating should be selected which provides optimum, but also very similar extraction yields for most target compounds. As mentioned in Chapter 1, a simple rule holds for liquid polymers: like dissolves like. If compounds differing markedly in their physicochemical properties, in particular their polarity and volatility, are to be analyzed, extraction yields will vary substantially for individual compounds, leading to greatly differing detection limits (see Section 3). Compounds which cannot be extracted at all should be clearly defined. Although time-consuming, it is advisable to determine distribution constants of selected reference compounds (see also Chapter 17 where different fibre coatings are evaluated for the analysis of environmental samples and detection limits are summarized for many industrial pollutants).

(c) Selection of the sampling mode (in particular direct SPME–headspace SPME)

If the target compounds are sufficiently volatile, the headspace method should be preferred as matrix interferences are minimized.

(d) Determination of extraction time profiles

Extraction time profiles are to be determined for several *representative* reference compounds to define the extraction time. If no equilibrium conditions are reached, the method extraction time will be a compromise between short overall analysis time, acceptable detection limit and precision.

Note that, the steeper the slope of the extraction time profile at a given time, the larger the possible error due to small variations in extraction time. This may be very important when SPME is performed manually, but is less critical for automated systems.

(e) Determination of the desorption conditions

The desorption conditions (time and temperature) should be selected to realize not only quantitative desorption, but also a minimum of carry-over of analytes.[22,49,57,66-68] The maximum desorption temperature is limited by the stability of the polymeric coating (reported by the manufacturer). Harsh desorption conditions may affect the life span of the fibre coating, and thus the lowest possible temperature should be applied.

It has to be kept in mind that the manual SPME device allows the adjustment of the fibre depth within the GC insert. The device should be adjusted in such a way that the fibre is exposed to the hottest part of the injector. Since no solvent is evaporated during the SPME process, low volume inserts (specially designed for SPME) should always be used. This leads to narrow peaks.[69]

(f) Optimization of further parameters
For development of an SPME method, it is not always necessary to optimize all parameters discussed in Section 3, as much information is available from the literature. However, the influence of these parameters on the SPME process should be kept in mind. In water analysis, adjustment of the pH is of particular importance, if basic and acidic compounds are present. As mentioned above, salt addition may enhance the extraction yield of polar compounds and often leads to more equal extraction yields.[18]

(g) Selection of the calibration method (see Section 4)

(h) General method validation
The method validation will include the determination of the *linear range, the limit of detection and quantification*, the *precision* and *accuracy, selectivity* and *ruggedness*.

If an FID or an MS is used as detector, SPME methods are characterized by a *large linear range*. For example, for all compounds shown in Table 1 (with the exception of 1,1,2-trichloroethane) a linear range over four orders of magnitude ($R^2 \leqslant 0.99$) was observed.

SPME methods in water analysis usually show very good *detection limits*, if selective detectors (nitrogen–phosphorus detector, electron capture detector, mass spectrometer) are employed. For thermally stable pesticides, detection limits in the low ppt range have been reported (depending on distribution constant and detector response), which readily allow the verification of the limits set in the European Drinking Water Guideline for this compound class. Using headspace SPME with an 80 μm PDMS/Carboxen fibre detection limits ranging from 0.1 to 0.5 ng L^{-1} have been reported for many halomethanes and haloethanes.[16]

The *accuracy* of the SPME method for the determination of pesticides and chlorinated hydrocarbons in aqueous samples has been demonstrated by interlaboratory comparison (see Section 4). An adequate *selectivity* in trace analysis of organics in water samples is usually only achieved with MS detection.

Many factors contribute to the ruggedness of an analytical method. For SPME analysis of aqueous samples, the stability of polymeric coating is one important criterion. While developing an instrumental set-up for a quasi-continuous on-site monitoring of industrial waste water (see Section 6), the stability of the PA fibre employed was investigated using real waste water which was extracted repeatedly after adjusting the pH to 10 (initially < 1). Using an 85 μm PA fibre, 55 extraction/desorption cycles were performed, the results of which are presented in Figure 3. The relative standard deviation of one late

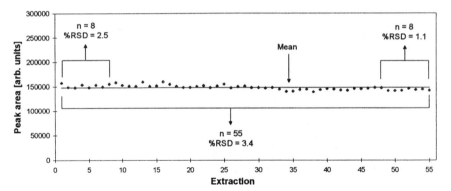

Figure 3 *Fibre stability tested with a real waste water sample (55 extraction/desorption cycles, PA fibre)*

eluting peak chosen as indicator of the precision was 3.4% ($n = 55$). Neither a downwards trend in the extracted amount nor any visible deterioration of the polymeric coating was observed. Moreover, the precision at the end of this study was even better than at the beginning of the study. Thus, the method appears to be sufficiently robust for on-site monitoring of waste water.

6 Automated Quasi-continuous On-site Monitoring of Water

General Aspects

SPME in water analysis may be performed manually or by use of a GC autosampler with integrated SPME fibre. With the latter method, the vials still have to be filled and placed in the autosampler carousel manually.

However, there is a need for fully automated instrumental set-ups (including sampling and sample pretreatment) which may operate continuously or quasi-continuously to monitor, *e.g.* surface or waste water.

Consider industrial waste water as an example: The central waste water plant of a chemical company usually carries out a biological treatment step the performance of which can be reduced by a high input of toxic chemicals to such an extent that an additional load of pollutants is introduced into the environment.

Continuous on-site monitoring of the influent or effluent of this central waste water plant for individual organic compounds provides an early warning of such accidental discharges of chemicals. Several instrumental set-ups for continuous monitoring, *e.g.* by on-line SPE–GC or SPE–HPLC, have been proposed.[70,71]

In the following, a relatively simple instrumental set-up for a quasi-continuous on-site monitoring of organic pollutants in water based on SPME is described.[72] The focus is on water analysis.

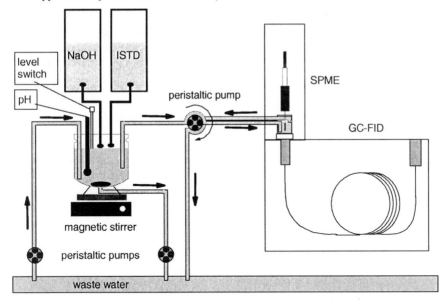

Figure 4 *System for fully automated monitoring of organic compounds in industrial waste water*

The Instrumental Set-up

Design

The instrumental set-up is based on a stop-flow method. All the above-mentioned aspects of the SPME technique have been taken into account and have led to the general scheme presented in Figure 4, which shows the application of the on-line system to the analysis of industrial waste water.

The fully automated procedure for the analysis of organic compounds in water consists of the following steps:

- Sampling
- Sample preparation (pH adjustment, addition of salt, derivatization)
- Addition of standards
- Extraction by SPME
- Chromatographic separation and detection (GC–FID)
- Quantification, comparison with critical limits
- Data transmission and alarm

These steps will be described in more detail below.

Sampling

A representative water sample is taken and transferred to a 250 mL vessel by means of a peristaltic pump. The vessel is equipped with a magnetic stirrer and a level switch based on two electrodes which allows automatic filling to a constant volume. A second pump is connected to the bottom of the vessel so that several rinsing cycles can be performed. This is especially important when the pH value is adjusted from acidic to basic conditions because of the possible precipitation of hydroxides which have to be removed.

Sample Preparation (pH Adjustment, Addition of Salt, Derivatization)

Since the pH of industrial waste water may have extreme values leading to a potential deterioration of the SPME fibre by the agressive matrix, the pH value has to be adjusted prior to the extraction. pH adjustment also becomes necessary when basic or acidic compounds like amines or phenols have to be analyzed because these compounds are extracted by SPME only in their neutral form (see Section 3). The reagent (acid or base) is added using a programmable burette. Depending on the compounds of interest, the pH value can be set automatically to a constant value. Besides the pH value, the salt content has a marked effect on the extraction process. In general, addition of salt enhances the extraction efficiency. Therefore, it might be useful to add salt prior to the SPME step. Moreover, a derivatizing agent can also be added via another burette prior to extraction to convert analytes into compounds amenable to GC analysis or compounds with higher affinity to the fibre. Introduction of suitable functional groups (e.g. halogen-containing groups if an electron capture detector is used) will also increase the sensitivity.

Addition of Standard(s)

A second burette is used to add one or more reference compound(s). These compounds can be used as internal standards for quantification but also for monitoring of the overall performance of the instrument. External calibration can also be performed in this way.

Extraction by SPME

A two-channel peristaltic pump is used to deliver the sample, prepared as described above, into a 10 mL extraction vessel where the SPME extraction is performed. Since the sample volume has to be kept constant to assure reproducible extraction, special attention has been paid to the filling technique.

This second vessel is designed in such a way that the volume remains constant during extraction, as shown in Figure 5. A magnetic stirrer is integrated into this set-up to achieve fast equilibration times.

The extraction vessel is suited for both direct (Figure 5a) and headspace SPME (Figure 5b). Since the lowest end of the sample outlet tube determines the water level in the cell, switching between direct and headspace SPME is easily possible by varying the length of this tube. In addition, the extraction vessel may be thermostated at higher temperatures by pumping a suitable preheated liquid

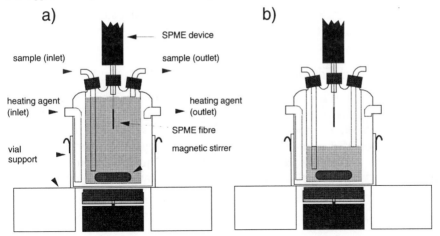

Figure 5 *Heatable extraction vessel with integrated magnetic stirrer: (a) direct SPME; (b) headspace SPME*

through the jacket placed around the inner vessel. This is important for headspace SPME where a higher temperature normally enhances the extraction efficiency due to the better mass transfer between the matrix and the headspace. Moreover, since the extraction process is exothermal,[3] it is necessary for direct SPME to maintain a constant temperature of the sample to ensure reproducible extraction (see Section 3).

A commercially available autosampler which incorporates the SPME fibre is modified to accommodate both the extraction vessel and a magnetic stirrer. Software designed by us is used to control the three pumps, the level switch and the burette. Once the extraction is under way, the next sample preparation is performed, resulting in short overall analysis times and high sample through-put. Analytical cycle times of less than 60 min can be achieved, which is usually sufficient for a quasi-continuous monitoring of surface or waste water.

Since the sample flow into the vessel is stopped during extraction, methods developed by manual SPME can be transferred directly to the automated system.

Chromatographic Separation and Detection (GC–FID)

After the desorption of analytes into the hot injector of the GC, the separation and detection are carried out automatically. Although all common GC detectors can be used (AED, ECD, FID, NPD, MS), the FID is advantageous for routine on-site analysis of target compounds because of its large linear dynamic range, similar response to most organic compounds, ruggedness and low cost. However, compound confirmation is not possible. To partly overcome this drawback, separation can be performed simultaneously on two columns with different polarity.

A mass spectrometer, although more expensive, is better suited for non-target analysis. With this detector, compound identification is possible, in principle.

Quantification, Comparison with Critical Limits

As external calibration is time-consuming, internal standard calibration is preferred with this instrument.

Since a constant amount of internal standard is added to each sample, monitoring of its peak area can serve as quality control to check the overall performance of the entire analytical system. After quantification, the concentrations of the analytes of interest are compared with the critical limits.

Data Transmission and Alarm

Data transmission from the on-site location to a remote laboratory is possible by using a modem. The GC software is modified to generate an electronic signal, if the peak area of a given compound surpasses a pre-selected level. This signal is then used to trigger an optical, acoustical or electronic alarm to initiate appropriate safety measures in time.

Validation of the On-line System

To validate the procedure, 24 reference compounds typically found in industrial waste water of one chemical plant in Germany have been selected. To check the precision of the automated system, a solution of 1% NaCl was pumped from a reservoir into the 250 mL vessel where a methanolic standard solution containing the reference compounds (0.1 ppm each) was added automatically by means of a burette. After two rinsing cycles, the prepared solution was pumped into the extraction vessel, where the SPME extraction was performed. During the chromatographic separation, the next sample preparation step was started. The entire procedure was repeated 16 times. The relative standard deviations ranged between 1.1 and 7.1%. Similar results were obtained in the headspace mode. The method is linear over four orders of magnitude with $r^2 > 0.99$ (using an FID). Detection limits are in the range <1–10 μg L^{-1} (depending on the K value of the compound) which is sufficient for waste water analysis.

To sum up, the procedure is characterized by long maintenance intervals. These are only limited by the stability of the SPME fibre and the volumes of the reagent containers. Unattended operation of at least one week is anticipated.

Running Costs

The complete automation of the system saves valuable operator time and thus keeps running costs to a minimum. Besides the normal costs required to run a GC, only a few additional costs are necessary for SPME fibres, internal standards and reagents.

Outlook

Field tests will be starting soon at an industrial waste water plant. In addition to waste water analysis, which is described in detail above, the on-site system is also suitable for the analysis of drinking water, ground water and all types of surface water. Besides industrial waste water plants, biological treatment systems of municipal water systems are also prone to be affected by toxic chemicals. The monitoring system proposed here will also be of value in this case. Other potential applications include process control applying the head-space technique.

Combining the automated system with the rapid GC technique would allow an even higher sample throughput, thus reducing the cycle time and thus the time until analytical information is available.

7 Conclusions

Recent research on SPME performed by many groups working in the field of water analysis clearly demonstrates that SPME is an effective and reliable tool for qualitative and quantitative analysis of many different compounds. The SPME technique has major advantages compared to other currently used sampling techniques such as LLE and SPE. The method is robust, easy-to-handle, economical and there is no need of toxic organic solvents. It is particularly useful if only small aqueous samples are available. The response is linear over several orders of magnitude and the data obtained for precision and detection limits are satisfying. The selectivity and sensitivity will be further improved as more coatings become available. Several international interlaboratory comparisons demonstrated that the SPME technique provides accurate data even at trace levels. Thus, it is time to implement standardized procedures for water analysis based on SPME.

Using the fully automated on-line system described in this chapter, unattended round-the-clock operation for about one week is possible. The system is suitable for both modes of operation, direct and headspace SPME, and manual methods can be directly transferred to the automated system. Due to its user-friendliness and its cost-saving potential, it is expected that the instrument will find its applications in routine water analysis.

References

1. R. Eisert and K. Levsen, *J. Chromatogr. A*, 1996, **733**, 143.
2. R. Eisert and J. Pawliszyn, *Crit. Rev. Anal. Chem.*, 1997, **27**, 103.
3. J. Pawliszyn, *Solid Phase Microextraction: Theory and Practice*, 1997, Wiley-VCH, New York.
4. T. Górecki and J. Pawliszyn, *Analyst*, 1997, **122**, 1079.
5. D. Louch, S. Motlagh and J. Pawliszyn, *Anal. Chem.*, 1992, **64**, 1187.
6. S. Motlagh and J. Pawliszyn, *Anal. Chim. Acta*, 1993, **284**, 265.
7. Z. Penton, H. Geppert and V. Betz, *GIT Spez. Chromatogr.*, 1996, **2**, 112,
8. R. Eisert and J. Pawliszyn, *J. Chromatogr. A*, 1997, **776**, 293.

9. Z.Y. Zhang and J. Pawliszyn, *Anal. Chem.*, 1993, **65**, 1843.
10. B.D. Page and G. Lacroix, *J. Chromatogr. A*, 1997, **757**, 173.
11. J. Ai, *Anal. Chem.*, 1997, **69**, 3260.
12. R. Eisert and K. Levsen, *J. Chromatogr. A*, 1996, **737**, 59.
13. Z. Zhang and J. Pawliszyn, *J. High Resolut. Chromatogr.*, 1993, **16**, 689.
14. B. MacGillivray, J. Pawliszyn, P. Fowlie and C. Sagara, *J. Chromatogr. Sci.*, 1994, **32**, 317.
15. Z. Zhang and J. Pawliszyn, *Anal. Chem.*, 1995, **67**, 34.
16. P. Popp and A. Paschke, *Chromatographia*, 1997, **46**, 419.
17. A.A. Boyd-Boland, S. Magdic and J. Pawliszyn, *Analyst*, 1996, **121**, 929.
18. R. Eisert and K. Levsen, *J. Am. Soc. Mass Spectrom.*, 1995, **6**, 1119.
19. L. Pan, M. Adams and J. Pawliszyn, *Anal. Chem.*, 1995, **67**, 4396.
20. L. Pan and J. Pawliszyn, *Anal. Chem.*, 1997, **69**, 196.
21. P. Bartak and L. Cap, *J. Chromatogr. A*, 1997, **767**, 171.
22. K.D. Buchholz and J. Pawliszyn, *Anal. Chem.*, 1994, **66**, 160.
23. L. Pan, J.M. Chong and J. Pawliszyn, *J. Chromatogr. A*, 1997, **773**, 249.
24. R. Eisert and K. Levsen, *GIT. Fachz. Lab.*, 1996, **40**, 581, 586–588.
25. Y. Cai and J. M. Bayona, *J. Chromatogr. A*, 1995, **696**, 113.
26. T. Górecki and J. Pawliszyn, *Anal. Chem.*, 1996, **68**, 3008.
27. S. Tutschku, S. Mothes and R. Wennrich, *Fresenius' J. Anal. Chem.*, 1996, **354**, 587.
28. Y. Morcillo, Y. Cai and J.M. Bayona, *J. High Resolut. Chromatogr.*, 1995, **18**, 767.
29. M. Guidotti and M. Vitali, *Ann. Chim.*, 1997, **87**, 497.
30. J.A. Chen and J. Pawliszyn, *Anal. Chem.*, 1995, **67**, 2530.
31. A.A. Boyd-Boland and J. Pawliszyn, *Anal. Chem.*, 1996, **68**, 1521.
32. K. Jinno, T. Muramatsu, Y. Saito, Y. Kiso, S. Magdic and J. Pawliszyn, *J. Chromatogr. A*, 1996, **754**, 137.
33. R. Eisert and J. Pawliszyn, *Anal. Chem.*, 1997, **69**, 3140.
34. M. Möder and P. Popp, to be published.
35. A.L. Nguyen and J.H.T. Luong, *Anal. Chem.*, 1997, **69**, 1726.
36. Y. Hirata and J. Pawliszyn, *J. Microcolumn Sep.*, 1994, **6**, 443.
37. A. Medvedovici, P. Sandra and F. David, *J. High Resolut. Chromatogr.*, 1997, **20**, 619.
38. R. Eisert, K. Levsen and G. Wünsch, *J. Chromatogr. A*, 1994, **683**, 175.
39. L. Moens, T. De Smaele, R. Dams, P. Van Den Broeck and P. Sandra, *Anal. Chem.*, 1997, **69**, 1604.
40. B. Rosenkranz, J. Bettmer, W. Buscher, C. Breer and K. Cammann, *Appl. Organometal. Chem.*, 1997, **11**, 721.
41. B.L. Wittkamp, S.B. Hawthorne and D.C. Tilotta, *Anal. Chem.*, 1997, **69**, 1197.
42. B.L. Wittkamp, S.B. Hawthorne and D.C. Tilotta, *Anal. Chem.*, 1997, **69**, 1204.
43. B.L. Wittkamp and D.C. Tilotta, *Anal. Chem.*, 1995, **67**, 600.
44. B. Schäfer, P. Hennig and W. Engewald, *J. High Resolut. Chromatogr.*, 1997, **20**, 217.
45. A. Saraullo, P.A. Martos and J. Pawliszyn, *Anal. Chem.*, 1997, **69**, 1992.
46. R.J. Bartelt, *Anal. Chem.*, 1997, **69**, 364.
47. J. Dewulf, H. Van Langenhove and M. Everaert, *J. Chromatogr. A*, 1997, **761**, 205.
48. Supelco chromatography products catalog 1998, Supelco, Bellefonte, PA, USA.
49. M. Möder, S. Schrader, U. Franck and P. Popp, *Fresenius' J. Anal. Chem.*, 1997, **357**, 326.
50. B. Schäfer and W. Engewald, *Fresenius' J. Anal. Chem.*, 1995, **352**, 535.
51. L. Müller, E. Fattore and E. Benfenati, *J. Chromatogr. A*, 1997, **791**, 221.

52. C.L. Arthur, L.M. Killam, K.D. Buchholz, J. Pawliszyn and J.R. Berg, *Anal. Chem.*, 1992, **64**, 1960.
53. R. Eisert, K. Levsen and G. Wünsch, *Vom Wasser*, 1996, **86**, 1.
54. L. Urruty and M. Montury, *J. Agric. Food Chem.*, 1996, **44**, 3871.
55. P. Popp, S. Mothes and L. Brüggemann, *Vom Wasser*, 1995, **85**, 229.
56. C.L. Arthur and J. Pawliszyn, *Anal. Chem.*, 1990, **62**, 2145.
57. K.K. Chee, M.K. Wong and H.K. Lee, *J. Microcolumn Sep.*, 1996, **8**, 131.
58. S.D. Huang, C.P. Cheng and Y.H. Sung, *Anal. Chim. Acta*, 1997, **343**, 101.
59. I. Valor, J.C. Molto, D. Apraiz and G. Font, *J. Chromatogr. A*, 1997, **767**, 195.
60. F.J. Santos, M.T. Galceran and D. Fraisse, *J. Chromatogr. A*, 1996, **742**, 181.
61. J. Ritter, V.K. Stromquist, H.T. Mayfield, M.V. Henley and B.K. Lavine, *Microchem. J.*, 1996, **54**, 59.
62. K.J. James and M.A. Stack, *Fresenius' J. Anal. Chem.*, 1997, **358**, 833.
63. T. Górecki, R. Mindrup and J. Pawliszyn, *Analyst*, 1996, **121**, 1381.
64. T. Nilsson, R. Ferrari and S. Facchetti, *Anal. Chim. Acta*, 1997, **356**, 113.
65. R. Ferrari, T. Nilsson, R. Arena, P. Arlati, G. Bartolucci, R. Basla, F. Cioni, G. Del Carlo, P. Dellavedova, E. Fattore, M. Fungi, C. Grote, S. Guidotti, S. Morgillo, L. Müller and L. Volante, *J. Chromatogr. A*, 1998, **795**, 371.
66. R. Young, V. Lopezavila and W.F. Beckert, *J. High Resolut. Chromatogr.*, 1996, **19**, 247.
67. T. Nilsson, F. Pelusio, L. Montanarella, B. Larsen, S. Facchetti and J.O. Madsen, *J. High Resolut. Chromatogr.*, 1995, **18**, 617.
68. D.W. Potter and J. Pawliszyn, *J. Chromatogr. A*, 1992, **625**, 247.
69. J.J. Langenfeld, S.B. Hawthorne and D.J. Miller, *J. Chromatogr. A*, 1996, **740** 139.
70. H. Mol and U.A.T. Brinkman, *J. Chromatogr. A*, 1995, **703**, 277.
71. H. Bagheri, E.R. Brouwer, R.T. Ghijsen and U.A.T. Brinkman, *J. Chromatogr. A*, 1993, **647**, 121.
72. C. Grote, K. Levsen and G. Wünsch, *Anal. Chem.*, submitted.

CHAPTER 13

The Application of SPME to Pesticide Residue Analysis

STEPHEN J. CROOK

1 Introduction

It is generally accepted that the continued use of pesticides is necessary in order that the ever increasing world population can be fed. However, industry does not ignore the perceived risks attributed to the use of pesticides and as a consequence generates a substantial quantity of data to support the safe use of pesticides. Also it provides appropriate methodology to allow monitoring of these products in the environment as required by regulatory bodies.[1-3]

As a result of this, the registration and development process for a modern pesticide or herbicide requires extensive field studies to assess the magnitude of residues in crop commodities in order that risk assessment can be made and tolerances can be set. In addition, it is important that the environmental fate of the pesticide must be considered in terms of leaching potential to groundwater and persistence in soil.

Consequently, during the course of the registration process, many hundreds of crop, soil and water samples requiring quantitative residue analysis for trace levels of the parent compound and metabolites are generated. It is, therefore, essential that the modern pesticide residue laboratory can provide a high sample throughput whilst keeping control of analytical costs. In addition, any methodology produced must be readily transferable to regulatory and government laboratories in order to allow routine monitoring of registered products and the level at which they and their metabolites occur in raw and processed foodstuffs and in the environment.

Safety issues are also of increasing importance and need consideration when developing any analytical method or process; pesticide analytical methodology is no exception. Procedures that use large volumes of organic solvent can often be time consuming and expensive to run, and have the added drawback of exposing the analyst to the hazard of the solvent. Also they are likely to have a negative impact on the environment as the disposal of large quantities of organic solvent must be addressed.

Pesticide residue analysis is, therefore, an on-going challenge to the analytical chemist. Clearly the above issues can make the development of suitable analytical methodology a complex and time consuming process. Consequently, it is essential in analytical method development to use the most prudent technology for the delivery of accurate and informative data in an expedient manner.

Solid phase microextraction (SPME) has been applied to the analysis of a range of pesticides[4-7] and can offer solutions to the above issues at a low cost. The main advantages of SPME for pesticide residue analysis are that it is a solvent free technique and that it is capable of giving high sensitivity for many analytes. Full automation is possible and SPME can produce extremely good quantitative data. Whilst SPME has been widely used for the analysis of pesticides in water, very little published data exists on its application to analysis in other matrices such as soil, sediment and particularly crop, although, some information is available on other analytes.[8,9]

2   Objectives

Conventional pesticide residue methodology employs procedures such as solvent extraction, liquid–liquid partition and solid phase extraction.[10-12] These techniques require the use of large volumes of organic solvent, are often time consuming and can be difficult to automate. The aim of the work described here was to apply the technique of automated SPME to routine, quantitative, pesticide residue analysis. In addition the results should prove that analytical data comparative to conventional methodology can be generated in a much shorter time and at a much reduced cost. Also it was important to demonstrate that SPME could be applied to the analysis of a wide range of matrices, namely environmental water, crop and soil samples.

Four triazole fungicides, diclobutrazol, flutriafol, hexaconazole and paclobutrazol (Figure 1), were chosen as examples of analytes typical of those requiring residue analysis on a routine basis. These compounds have been in existence for a number of years and have been in regular use in Europe in recent years.

3   Conventional Analytical Methodology

Traditionally, the analysis of the four chosen triazoles has been carried out using a range of techniques, dependent on the matrix from which analysis was required.

For water analysis, analytes are retained from the aqueous phase on to a C_{18} solid phase cartridge which is subsequently dried and eluted with an organic solvent. Samples may then be concentrated further prior to final determination by gas liquid chromatography (GC) using either thermionic nitrogen specific

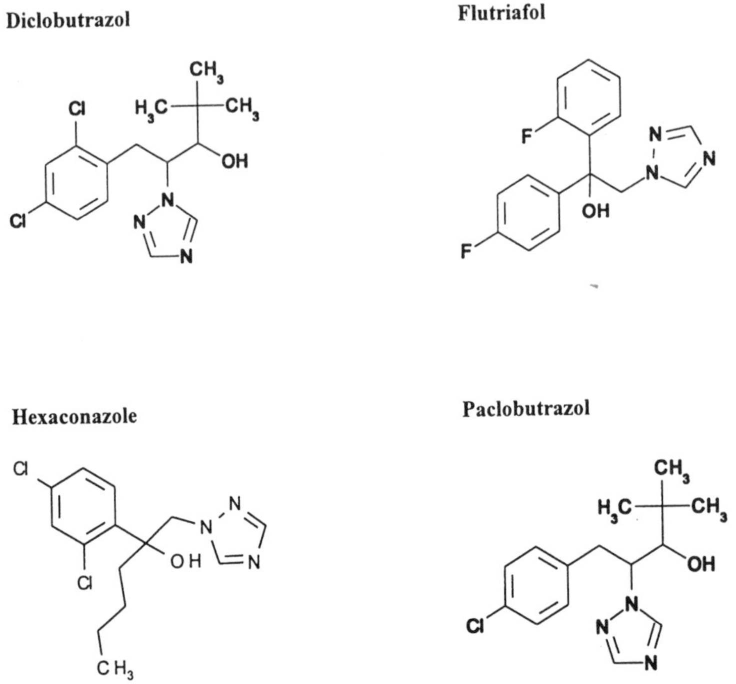

Diclobutrazol

Flutriafol

Hexaconazole

Paclobutrazol

Figure 1 *Structures of triazole fungicides*

(TSD) detection or mass selective detection (MSD). Approximately 500 mL of water is required to achieve the required limit of quantification (LOQ) of 0.1 μg L^{-1}.

Soil analysis involves extraction of the analytes from soil using mixtures of organic solvent and water, either employing sample agitation and in some cases reflux or soxhlet extraction. Analytes are then often subjected to complex clean up procedures which include techniques such as liquid/liquid partition and solid phase extraction (SPE). Often, further sample concentration is then required to achieve the required LOQ of 0.01 mg kg^{-1} prior to final determination again by GC using either TSD or MSD.

Crop analysis involves extraction of the analytes from the matrix using mixtures of organic solvent and water, employing high speed maceration and in the case of hexaconazole a reflux in 0.1 M sodium hydroxide solution in methanol. Analytes are then subjected to complex clean up procedures, as in the soil methodology, which again include liquid/liquid partition, solid phase extraction (SPE) and sample concentration stages in order to achieve the required LOQ of 0.01 mg kg^{-1}. Final determination is again by GC using either TSD or MSD.

Using the methodologies described above an analyst could typically analyse between 10 and 15 samples in a seven hour working day prior to running the samples overnight on the GC for the final determination. Whilst these times are typical for routine residue analysis, it was envisaged that a well designed SPME method would be capable of eliminating much of the sample work-up procedures and could thus greatly improve sample throughput.

4 Experimental

Materials

Diclobutrazol, flutriafol, hexaconazole, and paclobutrazol analytical standards of purity >90%.

Ultra pure water as obtained from an Elga Maxima laboratory water purification system available from Elga Ltd., High Street, Lane End, High Wycombe, Bucks HP14 3JH, UK.

Sodium chloride, Analar grade, available from Merck Ltd., Hunter Boulevard, Magna Park, Lutterworth, Leicestershire LE17 4XN, UK.

Acetonitrile, super purity grade, available from Romil Ltd., The Source, Convent Drive, Waterbeach, Cambridge CB5 9QT, UK.

Instruments/Apparatus

GC Conditions

A Hewlett Packard 5890 series 2 GC fitted with an HP 5972 Mass Selective Detector operated in the selected ion monitoring (SIM) mode was used. A Varian 8200 autosampler with SPME capability incorporating constant sample agitation was mounted on to the GC to allow automated SPME to be carried out. Connection was by means of an adapter kit supplied by Varian Instruments.

Ions monitored:
Diclobutrazol	m/z	= 270
Flutriafol	m/z	= 219
Hexaconazole	m/z	= 256
Paclobutrazol	m/z	= 236

Splitless injection was used using a standard 4 mm internal diameter pre-silanised glass liner available from Thames Chromatography, Fairacres Industrial Centre, Dedworth Road, Windsor, Berkshire SL4 4LE.

The GC column used was a 30 m Hewlett Packard HP-5 crosslinked 5% phenylmethylsilicone with a 0.25 mm internal diameter and a 0.25 μm film thickness. The GC column temperature programme used was to hold the temperature initially at 60 °C for 6 minutes (note that in the desorption experiments the initial hold time was kept equivalent to the desorption time). The column was then programmed at a rate of 20 °C min^{-1} to a final temperature of 280 °C. The column was then held at this temperature for 8 minutes.

SPME parameters: 85 μm polyacrylate fibre for use with Varian 8100 or 8200
                    autosampler available from Supelco.
                    Equilibrium time = 45 mins with constant agitation.
                    Desorption time = 6 mins.
                    Desorption temperature = 300 °C.
                    All SPME was made directly from the aqueous sample.
                    0.4 g salt dissolved in 1.2 ml aqueous sample.

5   Methods

Fibre Selection

The most appropriate fibre for the analysis of the triazoles has proven to be the
85 μm polyacrylate. The configuration used was that designed for use with the
Varian 8200 autosampler using constant sample agitation; in all cases this was
used for the described work.

Absorption Time

A time profile of the absorption of the triazoles on to the 85 μm polyacrylate
was determined in order to assess the optimum SPME sampling period. 1.2 mL
aliquots of 0.001 μg mL^{-1} standard solutions of the four triazoles in ultra pure
water were analysed by SPME using mass selective detection with sampling
periods of 10, 20, 30, 60 and 120 minutes. Desorption time was set to 10 minutes
at 275 °C for these experiments. 0.4 g sodium chloride was added to each vial.

Desorption Time

A desorption time profile of the four triazoles was established to prevent carryover of analyte from one analysis to the next. 1.2 mL aliquots of 0.001 μg mL^{-1}
standard solutions of the four triazoles in ultra pure water were analysed by
SPME using mass selective detection with a sampling period of 45 minutes; 0.4 g
salt was added to each vial. Desorption times of 2, 4, 6, 8, and 10 minutes were
investigated. Subsequent to each sample injection, a blank injection of the fibre
was made in order to determine if any carry-over was occurring.

Sodium Chloride Addition

The addition of sodium chloride to the aqueous sample has regularly been used
as an aid to improve the extraction of analytes from solution in SPME.[13–15]
    1.2 mL aliquots of standard solutions of the triazole fungicides in ultra pure
water were dispensed into GC autosampler vials containing different amounts
of sodium chloride from zero to 0.6 g (saturated). Samples were analysed by
SPME using a 45 minute absorption time and a 6 minute desorption at 275 °C,
and the response obtained was compared.

Quantitative Residue Analysis

Once appropriate SPME conditions had been established from the above experiments, the technique could be applied to the quantitative analysis of samples. To establish the scope to which SPME could be applied analysis was attempted from environmental water, soil, and crop samples.

Water Analysis by SPME

For the purposes of these experiments water sampled from the River Thames was used as an example of a typical environmental water sample. 5 mL aliquots of Thames river water were fortified with the four triazole fungicides at levels ranging from 0.1 ppb to 0.5 ppb. 1.2 mL of the fortified samples were transferred to a GC autosampler vial containing 0.4 g sodium chloride. Vials were shaken until complete dissolution of the sodium chloride occurred. Fortified samples were than analysed by GC using mass selective detection in the SIM mode. Samples were compared against standard solutions prepared in ultra pure water. 0.4 g sodium chloride was dissolved in 1.2 mL of standard solutions to give direct comparison with the fortified samples.

Soil Analysis by SPME

Control soil sampled from local farmland was used for the purposes of these experiments. Prior to analysis the soil was air dried for 24 hours and then finely sieved to give a homogenous sample.

Very importantly, the extraction system used in the conventional methodology had to be retained as this is well established and has been proven to extract the pesticide residue in extensive laboratory tests.

10 g aliquots of prepared soil were fortified at levels ranging from the desired limit of quantification (0.01 mg kg^{-1}) to 0.1 mg kg^{-1}. Residues were extracted by shaking for 30 minutes using an acetonitrile/water mixture (70/30 v/v, 20 mL). Samples were then centrifuged at a speed of 2500 rpm for 5 minutes to obtain separation of the supernatant liquid from the particulate soil. Clearly the extracted sample contains a high percentage of organic solvent at this stage and it is proven that the content of organic solvents has a significant effect on the partition characteristics of analytes.[15] Therefore, to eliminate this problem 0.2 mL aliquots of the supernatant liquid were taken and diluted to a final volume of 4.0 mL to give a final concentration of 0.02 g mL^{-1}. At the required limit of quantification of 0.01 mg kg^{-1} this gives an actual concentration of analyte in the final sample vial of 0.0002 μg mL^{-1}. Whilst this is a low concentration in conventional terms, SPME when combined with a suitable detection technique can give superior detectability in comparison, as all analyte that is absorbed on the fibre enters the GC and is detected. After dilution, 1.2 mL of sample was transferred to a suitable autosampler vial containing 0.4 g sodium chloride. Vials were shaken until complete dissolution occurred.

Samples were analysed using GC–MSD and compared with standard solution in ultra pure water.

Crop Analysis by SPME

Maize forage and fodder were chosen as crop types in which it can be difficult to obtain adequate sample clean up by conventional methodologies. Again, as for the soil experiments it was important that the sample extraction system for the crop samples be retained as for the conventional methodology.

10 g aliquots of maize forage were fortified with known amounts of hexa-conazole and diclobutrazol at levels between 0.01 and 0.1 mg kg^{-1}. Samples were subjected to high speed maceration with a mixture of acetonitrile and water (35/65 v/v, 25 mL). Samples were then centrifuged at a speed of 2500 rpm for 5 minutes to obtain separation of the supernatant liquid from the particulate crop material. As in the example of the soil analysis the extracted sample contains a high percentage of organic solvent at this stage. So, as above, elimination of the solvent affect was crucial. Therefore, 0.1 mL aliquots of the supernatant liquid were taken and diluted to a final volume of 4.0 mL to give a final sample concentration of 0.01 g mL^{-1}. At the required limit of quantifica-tion of 0.01 mg kg^{-1} this gives an actual concentration of analyte in the final sample vial of 0.0001 μg mL^{-1}. This is comparable to the levels used in the soil experiments. After dilution, 1.2 mL of sample was transferred to a suitable autosampler vial containing 0.4 g sodium chloride. Vials were shaken until complete dissolution occurred. Samples were analysed using GC–MSD in the SIM mode and compared to standard solutions in ultra pure water.

6 Results and Discussion

Absorption Time

A graph of the response obtained (amount of analyte absorbed on to the fibre) against the SPME sampling period is displayed in Figure 2 and clearly shows that the equilibrium profiles for all four of the triazoles are extremely similar, and that in all cases equilibrium is still not reached after 2 hours even with constant sample agitation. For the purposes of running samples in a routine laboratory this sort of equilibrium time is unacceptable. This equates to analysing only seven or eight samples on the GC overnight which negates any benefits obtained by the rapid sample work up. Therefore, to give the technique parity with the conventional method a 45 minute absorption time was chosen allowing around 20 samples to be run on the GC overnight. Fortunately the Varian SPME autosampler can function in the 'prep-ahead' mode which means that the next sample can be sampled while the previous one is being run on the GC. It should be noted that, although the systems are not at equilibrium, quantitative analysis was still feasible. As long as each sampling period was identical the variation seen due to this fact would be minimal.

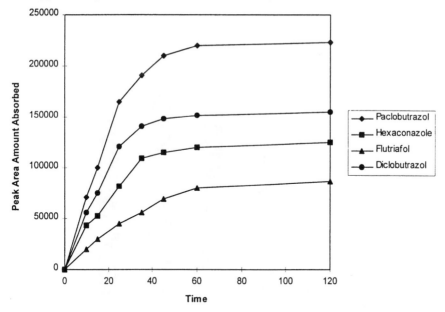

Figure 2   *SPME absorption profiles of diclobutrazol, flutriafol, hexaconazole and paclobutrazol*

Desorption Time

The relationship between the response obtained (amount of analyte absorbed
on to the fibre) against the SPME desorption time for flutriafol is shown in
Figure 3 and clearly shows that the maximum response is obtained after 6
minutes and, thereafter no benefit is seen for longer desorption periods. Profiles
of the other analytes were similar and no carry over in a subsequent blank
injection was seen after this time. Therefore 6 minutes was chosen as the
optimum desorption time for the analysis of all four triazoles.

Effect of Sodium Chloride

As expected, the addition of sodium chloride to the samples had a significant
influence on the amount of analyte absorbed on to the fibre. Results for
flutriafol are displayed graphically in Figure 4. Results for the other analytes
were similar.

   In all cases, the extraction of analyte was improved by the addition of sodium
chloride; however, the extent of the increase in the amount extracted was
compound dependent. The extraction of the most polar of the analytes,
flutriafol, was clearly the most improved by salt addition. This agrees with
previous findings where as a general rule it would appear that the addition of
sodium chloride is beneficial for the extraction of pesticides by SPME. Certainly
for the triazines, chloroacetanilides, thiocarbamates, 2,6-dinitroanilines and
uracils significant improvements are seen when sodium chloride is added.[5] As a

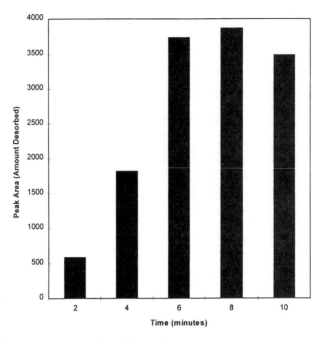

Figure 3   *Desorption time profile of flutriafol*

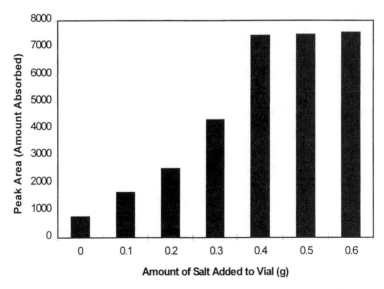

Figure 4   *The effect of sodium chloride concentration on amount of flutriafol extracted by SPME fibre*

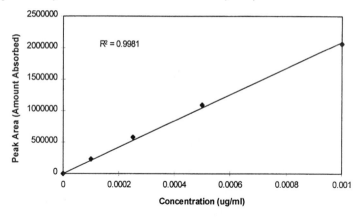

Figure 5 *Linearity plot for hexaconazole*

rule the extraction of more polar hydrophilic pesticides, as with flutriafol, is improved to a greater extent than the non-polar pesticides. However, for certain non-polar compounds such as organophosphorus pesticides, no advantage is obtained, indeed at high concentrations of sodium chloride a decrease in the amount of analyte may be observed.[4]

For quantitative experimental work it was decided that the optimum concentration was to use 0.4 g salt dissolved in 1.2 mL of sample solution. Saturated salt solution was not considered as a viable option as problems had been observed with salt crystals forming in the hollow barrel of the SPME syringe causing blockage and damage to the syringe mechanism at these concentrations. Owing to the same problem it was found beneficial to insert a vial containing water containing no sodium chloride, after every 8–10 sample injections, as a wash to remove the build up of sodium chloride around the fibre.

Quantitative Data

Results were assessed for the percentage recovery obtained in comparison to the fortification level and also how reproducible they were in terms of the relative standard deviation (% RSD). Prior to quantitative analysis the linearity of response of the mass selective detector was proven as displayed for hexaconazole in Figure 5. Linearity plots were similar for the other analytes.

Chromatograms were also assessed for any interference caused by co-extracted material. A summary of the quantitative recovery data generated from the analysis of environmental water, soil and crop samples is presented in Table 1.

The quantitative results obtained were extremely good. Mean recovery values were at an acceptable level as were % RSD values and were comparable to those obtained in the course of conventional analytical procedures. Note that the EPA guidelines[16] specify pesticide residue methods to produce a mean recovery between 70–110% with a % RSD of $\leqslant 20\%$. The slightly high recovery level

Table 1 *Summary of quantitative SPME residue analysis*

Matrix	Analyte	n	Mean recovery (%)	%RSD	LOD (ppb)
River water	Paclobutrazol	13	99	12	0.05
	Flutriafol	6	88	4	0.05
	Hexaconazole	10	92	18	0.05
	Diclobutrazol	10	92	13	0.05
Soil	Paclobutrazol	6	101	14	10
	Diclobutrazol	6	75	12	10
Maize forage	Hexaconazole	5	91	10	10
	Diclobutrazol	6	87	12	10
Maize fodder	Hexaconazole	5	114	10	10
	Diclobutrazol	5	93	12	10

seen in the hexaconazole fodder analysis may be attributed to a slight matrix enhancement, this could be corrected by addition of a suitable internal standard prior to GC analysis.

Of particular note are the results obtained for the maize fodder and forage analysis. Despite the absence of any form of sample clean up, apart from sample dilution, the results obtained do not appear to be affected by the presence of crop matrix except perhaps for the slight enhancement seen in the hexaconazole fodder results. Also it should be noted that fibre performance did not deteriorate after up to 40 multiple injections of crop matrix samples. These are extremely encouraging results as the sample clean up and manipulation required for the analysis of such samples by conventional means can be extremely time and labour intensive. Chromatograms were free of interference in all cases which allowed accurate integration of all peak data. Typical chromatograms are displayed in Figure 6.

7 Conclusions

The technique of SPME has been applied to the routine residue analysis of four triazole fungicides in environmental water samples, soil extracts and crop extracts. The automated version of SPME with constant sample agitation has proven to be a viable alternative to the conventional analytical methodology. Analysis in all cases was achieved in significantly reduced timescales. It is estimated that an approximate fourfold increase in sample throughput was achieved using automated SPME over the conventional methodology. Also a significant reduction in the quantity of consumable items and organic solvents required for analysis was made.

0.2 ppb Diclobutrazol in River Water

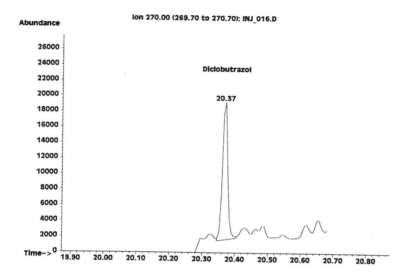

0.01 mg kg^{-1} Hexaconazole in Soil

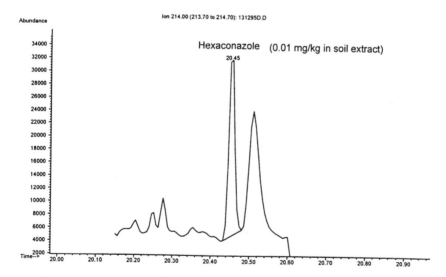

Figure 6 *Example chromatograms*

Very importantly, quantitative analysis of the analytes investigated was carried out to the desired limits of quantification of 0.1 ppb in environmental water samples and to 0.01 mg kg^{-1} in crop and soil samples. It is also of note that these results were achieved with the system not at equilibrium. The selectivity of SPME combined with mass selective detection allowed analysis with no observable interference from coextracted material and without the requirement for any further sample preparation or clean-up.

Whilst SPME has been applied successfully in these cases it should be noted that applications which involve the determination from sample matrices with high lipid content will create problems for SPME analysis direct from the sample extract. Other forms of sample pre-treatment will be required for successful SPME in these cases. Also where multiple analyte determinations are required and one or more of the analytes is of a polar nature, this may preclude the use of SPME if the analytes can be analysed simultaneously using conventional methods such as SPE.

What is abundantly clear is that SPME is ideally suited to certain pesticide residue analytical applications and in these cases significant reductions in analytical costs both in terms of manpower and reagents can be achieved. In the modern laboratory where the demands of processing ever increasing sample numbers at a low cost are paramount, a technique such as SPME can be an invaluable weapon in the analyst's armoury. The advent of new fibre coatings and the eventual automation of SPME–HPLC may also help to extend the range of SPME applications in the field of pesticide analysis.

The increasing pressure from regulatory authorities and increased public awareness of issues surrounding pesticide usage can only mean that the requirement for analysis of these compounds will increase, just as the required detection limits can only get lower. SPME will undoubtedly help to address some of these issues and its value to the analyst can surely only increase.

References

1. Liste I de la Directive 76/464/CEE du Conseil. JO No. 1129 du 18/5 1976, 23.
2. A.R. Agg Ba and T.F. Zabel, *J. Inst. Water Environ. Manag.*, 1990, **4**, 44.
3. Directive 80/778/EEC relating to the quality of water intended for human consumption (95/C 131/03).
4. R. Eisert and K. Levsen, *J. Am. Soc. Mass Spectrom.*, 1995, **6**, 1119.
5. A.A. Boyd-Boland and J.B. Pawliszyn, *J. Chromatogr. A.*, 1995, **704**, 163.
6. R. Eisert and K. Levsen, *Fresenius' J. Anal. Chem.*, 1995, **351**, 555.
7. M.J. Redondo, M.J. Ruiz, R. Boluda and G. Font, *Chromatographia*, 1993, **36**, 187.
8. G. Durand, S. Chiron, V. Bouvot and D. Barcelo, *Int. J. Environ. Anal. Chem.*, 1992, **49**, 31.
9. E. Bolygó and N.C. Atreya, *Fresenius' J. Anal. Chem.*, 1991, **339**, 423.
10. A. Junker-Buchheit and M. Witzenbacher, *J. Chromatogr. A.*, 1996, **737**, 67.
11. B.D. Page and G. Lacroix, *J. Chromatogr. A.*, 1997, **757**, 173.
12. I. Valor, J.C. Moltó, D. Apraiz and G. Font, *J. Chromatogr. A.*, 1997, **767**, 195.
13. M.T. Sng, F.K. Lee and H.Å. Lakso, *J. Chromatogr. A.*, 1997, **759**, 225.
14. B.D. Page and G. Lacroix, *J. Chromatogr. A.*, 1993, **648**, 199.
15. H. Verhoeven, T. Beuerle and W. Schwab, *Chromatographia*, 1997, **46**, 63.
16. EPA Residue Chemistry Test Guidelines, OPPTS 860.1000, 1996.

CHAPTER 14

Inter-laboratory Validation of SPME for the Quantitative Analysis of Aqueous Samples

TORBEN NILSSON AND TADEUSZ GÓRECKI

1 Introduction

A large number of successful quantitative as well as qualitative applications of SPME have been reported.[1] However, in order for SPME to gain further acceptance and to be considered by standardisation organisations its performance should be validated in inter-laboratory studies. In this chapter, the results of three inter-laboratory studies concerning the accuracy and precision of SPME are summarised. The statistical data treatment was performed in accordance with the ISO standard 5725[2,3] in all cases. A brief description of the statistics for the inter-laboratory validation of a method on the basis of a reference material is given below. In one study a standardised method for the SPME analysis of twelve pesticides representing the main pesticide groups (organochlorine, organonitrogen and organophosphate) in aqueous samples was tested at low μg L^{-1} level by eleven European and North American laboratories.[4] The two other studies were related to the testing of standard methods for the SPME analysis of volatile organic compounds[5] and triazine herbicides,[6] respectively, in aqueous samples. The performance of the standard method for the analysis of volatile organic compounds was tested at low μg L^{-1} level in a study including twenty Italian laboratories.[7] Two different SPME approaches were examined, and SPME was compared to the traditional techniques, purge-and-trap and static headspace. The performance of the standard method for the analysis of triazine herbicides was examined by ten European laboratories at concentrations around the European limit of 100 ng L^{-1} for individual pesticides in drinking water.[8] On the basis of the positive results, both standard methods were presented to the Italian Standardisation Organisation, Unichim, and their presentation at European level is in progress.

2   Statistics for Inter-laboratory Validation

The statistical data treatment in the validation studies was performed in accordance with the ISO standard 5725 which describes inter-laboratory statistics based on the analysis of variance (ANOVA) technique.[2,3] Throughout this presentation of the statistics for the validation of a method on the basis of a reference material the following index notation will be used: i is the laboratory number, varying from 1 to p. In order for the inter-laboratory study to be reliable p should be at least 8. j is the replicate number, varying from 1 to n_i. n_i is the total number of replicates for laboratory i and should be at least 2 for all of the laboratories. A set of n_i results is called a cell. The statistics are valid only when the following assumptions are true:

- The data distribution within a given cell follows the normal distribution law.
- The laboratory mean distribution follows the normal distribution law.
- The intra-laboratory variances are equal.

Therefore, initially data which do not fulfil these assumptions are eliminated by Cochran's test of the within-laboratory variabilities and Grubb's two-sided tests of the laboratory mean distribution, *i.e.* the between-laboratory variability. If the difference between the test statistics and the critical value is significant at the 95% confidence level, the data point is called a straggler, while it is defined as an outlier if the difference is significant at the 99% confidence level. Stragglers are left in the set, while the entire cell is excluded from the further calculations if the variance or the average of a cell is an outlier. If more than two out of nine cells are outliers, the inter-laboratory study is not considered reliable.

In Cochran's test, C is calculated from equation 1 and compared with the corresponding critical maximum values. When an outlier is detected, the test is repeated with the remaining data until no more outliers occur.

$$C = \frac{s_{max}^2}{\sum_{i=1}^{p} s_i^2} \tag{1}$$

$$s_i^2 = \frac{\sum_{j=1}^{n_i} x_{ij}^2 - \frac{\left(\sum_{j=1}^{n_i} x_{ij}\right)^2}{n_i}}{n_i - 1} \tag{2}$$

s_i^2 is the intra-laboratory variance, *i.e.* s_i is the standard deviation of the results from laboratory i, and s_{max}^2 is the highest intra-laboratory variance in the set.

In Grubb's test, the cell average or intra-laboratory mean, $\bar{x}_i$, is calculated from equation 3. G_p for the highest value and G_1 for the lowest value are calculated using equations 4 and 5, respectively, and compared with the corresponding critical maximum values. For clarity the laboratory means have been denoted simply by x_i in these equations.

$$\bar{x}_i = \frac{\sum_{j=1}^{n_i} x_{ij}}{n_i} \tag{3}$$

$$G_p = \frac{x_p - \bar{x}}{s} \tag{4}$$

$$G_1 = \frac{\bar{x} - x_1}{s} \tag{5}$$

$$\bar{x} = \frac{1}{p} \sum_{i=1}^{p} x_i \tag{6}$$

$$s = \sqrt{\frac{1}{p-1} \sum_{i=1}^{p} (x_i - \bar{x})^2} \tag{7}$$

If no single outliers are found, Grubb's test for the two highest and the two lowest values is applied. G is calculated using equations 8 and 9, respectively, and compared with the corresponding critical minimum values.

$$G = \frac{s_{p-1,p}^2}{s_0^2} \tag{8}$$

$$G = \frac{s_{1,2}^2}{s_0^2} \tag{9}$$

$$s_0^2 = \sum_{i=1}^{p} (x_i - \bar{x})^2 \tag{10}$$

$$s_{p-1,p}^2 = \sum_{i=1}^{p-2} (x_i - \bar{x}_{p-1,p})^2 \tag{11}$$

$$\bar{x}_{p-1,p} = \frac{1}{p-2} \sum_{i=1}^{p-2} x_i \tag{12}$$

$$s_{1,2}^2 = \sum_{i=3}^{p} (x_i - \bar{x}_{1,2})^2 \tag{13}$$

$$\bar{x}_{1,2} = \frac{1}{p-2} \sum_{i=3}^{p} x_i \tag{14}$$

After rejection of outliers, the statistical analysis is performed using the ANOVA technique, which gives estimates of the gross average, the inter-laboratory variance, and the intra-laboratory or repeatability variance on the basis of equation 15. This equation is often rewritten in abridged form (equation 16), where SS_t is the total sum of squares to gross average, SS_L is the factorial or inter-laboratory sum of squares, and SS_r is the residual or intra-laboratory sum of squares.

$$\sum_{i=1}^{p}\sum_{j=1}^{n_i}(x_{ij} - \bar{\bar{x}})^2 = \sum_{i=1}^{p}(\bar{x}_i - \bar{\bar{x}})^2 + \sum_{i=1}^{p}\sum_{j=1}^{n_i}(x_{ij} - \bar{x}_i)^2 \tag{15}$$

$$SS_t = SS_L + SS_r \tag{16}$$

The gross average $\bar{\bar{x}}$ can be calculated from equation 17.

$$\bar{\bar{x}} = \frac{\sum_{i=1}^{p}\sum_{j=1}^{n_i} x_{ij}}{N} \tag{17}$$

$$N = \sum_{i=1}^{p} n_i \tag{18}$$

Formulas can be developed from equation 16 for calculation of repeatability variance s_r^2 (equation 19), inter-laboratory variance s_L^2 (equation 21) and reproducibility variance s_R^2 (equation 25).

$$s_r^2 = \frac{SS_r}{N - p} \tag{19}$$

$$SS_r = \sum_{i=1}^{p}\left[\sum_{j=1}^{n_i} x_{ij}^2 - \frac{\left(\sum_{j=1}^{n_i} x_{ij}\right)^2}{n_i}\right] \tag{20}$$

$$s_L^2 = \frac{(p-1)\left(\dfrac{SS_L}{p-1} - s_r^2\right)}{N'} \tag{21}$$

$$SS_L = SS_t - SS_r \tag{22}$$

$$SS_t = \sum_{i=1}^{p}\sum_{j=1}^{n_i} x_{ij}^2 - \frac{\left(\sum_{i=1}^{p}\sum_{j=1}^{n_i} x_{ij}\right)^2}{N} \tag{23}$$

$$N' = N - \frac{\sum_{i=1}^{p} n_i^2}{N} \tag{24}$$

$$s_R^2 = s_L^2 + s_r^2 \tag{25}$$

Repeatability and reproducibility are defined as the closeness of agreement between measurements obtained under repeatability and reproducibility conditions, respectively. The standard deviation of the difference between two measurements is $s\sqrt{2}$ when each measurement has a standard deviation of s. The critical difference or confidence limit is determined by multiplication of $s\sqrt{2}$ by a critical range factor f, which depends on the confidence level and the distribution of data. For a normal distribution at a 95% confidence level f is 1.96. Thus, the repeatability limit r, which is an estimate of the reliability of a method, and the reproducibility limit R can be determined from equations 26 and 27, respectively.

$$r = 2.8s_r \qquad (26)$$

$$R = 2.8s_R \qquad (27)$$

Accuracy of a method is defined as the closeness of the test results to the true value. In order to claim that a method is accurate, the confidence interval CI determined in an inter-laboratory study (equation 28) must include the true value of the reference material. Preferably, the gross average should be near this value.

$$\text{CI} = \bar{\bar{x}} \pm r \qquad (28)$$

The ISO standard 5725 describes also method validation by comparison to a reference method and a comparison of two alternate methods. These statistics were presented and applied in one of the inter-laboratory studies only.[7]

3 Volatile Organic Compounds

Volatile organic compounds (VOCs) are present in the aquatic ecosystem owing to their industrial applications and the decomposition of higher molecular organic compounds by the oxidising agents used in wastewater treatment. Many VOCs, including the BTEX compounds (benzene, toluene, ethylbenzene and xylenes), are known to be harmful, and thus they are measured on a routine basis in drinking and ground water.

Standard Method

A SPME standard method was established as a rapid, simple and solventless alternative to the traditional methods for quantitative analysis of VOCs in aqueous samples.[5] The method can be automated easily for the purpose of routine analysis. The optimisation study has been described previously.[9] The recommended analytical conditions were 30 minute extraction at room temperature from the headspace over a stirred sample using the 100 μm poly(dimethylsiloxane) coating followed by 5 minute desorption at 220 °C and

GC–MS analysis. Furthermore, a number of specific suggestions were given for special classes of compounds.

Validation

The results of the inter-laboratory validation of SPME on the basis of a reference material are presented in Table 1. The number of outliers never exceeded two out of nine; thus the inter-laboratory study can be considered reliable. The relative number of outliers in the purge-and-trap and static headspace analyses were approximately the same. The 95% confidence intervals included the true value in all cases but chloroform, and the gross average was typically very near the true value, *i.e.* the SPME method proved to be accurate. The average gross average was 3.10 for both SPME approaches. For a comparison, this value was 2.96 for purge-and-trap and 2.95 for static headspace analyses. No significant differences between the accuracies of SPME and the traditional techniques were seen in statistical calculations considering the latter methods as reference methods. Also, a statistical comparison of the two SPME approaches revealed no significant differences in the accuracy. The average relative repeatability standard deviation with SPME was around 10% in this inter-laboratory study, while lower values have been reported previously.[9] Better precision was achieved by HS-SPME than with the fibre immersed in the aqueous phase. This was especially true for the reproducibility. The repeatability and reproducibility of purge-and-trap and static headspace were comparable to those of HS-SPME. Detection limits were in the low ng L^{-1} range with MS detection, and the linearity of the response was good in all cases. The average r^2 value of the SPME calibration curves was 0.994. Thus, SPME proved to be a valid alternative technique for the quantitative analysis of VOCs at trace level in aqueous samples.

4 Pesticides

Pesticides are widely present in the environment owing to their use in agriculture. Many pesticides are harmful; therefore limits have been imposed on their highest acceptable concentrations in drinking water, and routine analyses are performed. Traditional extraction techniques for this purpose are liquid–liquid extraction and solid-phase extraction using cartridges. Recently, also analysis by SPME–GC–MS has proven successful for a large number of pesticides.[1] The results of the two inter-laboratory studies on the application of SPME for the analysis of organochlorine, organonitrogen and organo-phosphate pesticides, as well as triazine herbicides, are given below. They confirmed that SPME–GC–MS is a reliable technique for the quantitative analysis of such pesticides at trace level in aqueous samples.

Table 1 *Results of the inter-laboratory validation of SPME for the analysis of volatile organic compounds in aqueous samples[7]*

Compound	Technique	Data material		Accuracy ($\mu g\ L^{-1}$)			Precision (%)	
		p	Outliers	True value	$\bar{\bar{x}}$	CI	Repeatability s_r	Reproducibility s_R
Chloroform	HS-SPME	11	1	3	3.96	3.18–4.74	7.1	29
	SPME	10	0	3	4.52	3.27–5.77	9.7	37
1,1,1-Trichloroethane	HS-SPME	11	1	3	3.21	2.31–4.11	10	49
	SPME	12	2	3	3.04	2.03–4.05	12	36
Carbon tetrachloride	HS-SPME	11	1	3	3.02	2.22–3.82	9.3	42
	SPME	12	2	3	3.14	2.44–3.84	8.0	48
Benzene	HS-SPME	8	0	3	3.10	2.25–3.95	9.6	20
	SPME	8	1	3	3.47	2.15–4.79	14	36
Trichloroethene	HS-SPME	9	1	3	2.98	2.39–3.57	7.0	28
	SPME	10	1	3	2.88	1.99–3.77	11	41
Bromodichloromethane	HS-SPME	10	1	3	2.93	2.46–3.40	5.8	33
	SPME	10	0	3	3.23	2.48–3.98	8.4	41
Toluene	HS-SPME	8	0	3	2.71	1.30–4.12	18	18
	SPME	10	0	3	2.61	1.65–3.57	13	49
Tetrachloroethene	HS-SPME	11	1	3	3.32	2.33–4.31	11	36
	SPME	12	1	3	3.05	2.00–4.10	12	39
Ethylbenzene	HS-SPME	8	0	3	3.26	2.44–4.08	8.9	25
	SPME	10	0	3	3.09	1.77–4.41	15	56
p-Xylene	HS-SPME	8	1	3	3.09	2.39–3.79	8.1	27
	SPME	11	1	3	2.69	1.64–3.74	14	58
Bromoform	HS-SPME	10	2	3	2.95	2.08–3.82	11	15
	SPME	11	1	3	2.95	2.10–3.80	10	33
1,2-dichlorobenzene	HS-SPME	9	2	3	2.70	2.26–3.14	5.9	22
	SPME	10	1	3	2.57	1.79–3.35	11	39

Organochlorine, Organonitrogen and Organophosphate Pesticides

Standardised SPME conditions for the analysis of organochlorine, organo-nitrogen and organophosphate pesticides in aqueous samples were 45 minute extraction at ambient temperature from a stirred sample using the 100 μm poly(dimethylsiloxane) coating, followed by 5 minute desorption at 250 °C and GC–MS analysis.[4] In another study, the 85 μm polyacrylate coating was used, and the ionic strength of the sample was increased by salt addition, thereby improving the sensitivity of the method and reaching detection limits in the low ng L^{-1} range.[10] The results of the inter-laboratory validation are presented in Table 2. It follows from this table that only very few outliers were found. The gross averages were generally very near, though slightly below, the true values. The true values were within the 95% confidence intervals in all cases. Thus, the accuracy of the standardised method was satisfactory. Also, precision of the method was good, and except for a few cases it was comparable to that of the standard method for the analysis of VOCs. The average linear correlation coefficient was 0.996. Taking into account the diversity of the test compounds, it is very likely that the use of SPME can be extended to reliable quantitative analysis of many other classes of semi-volatile compounds.

Triazine Herbicides

Optimisation of SPME conditions aiming at obtaining maximum sensitivity was described previously.[8] The standard method for the analysis of triazine herbicides in aqueous samples called for a 30 minute extraction with the 65 μm Carbowax-divinylbenzene coating at ambient temperature and neutral pH from a rapidly stirred sample containing 0.3 g mL^{-1} NaCl.[6] Five minute desorption was performed directly in the injection port of the gas chromatograph at 240 °C. The results of the inter-laboratory validation are presented in Table 3. The new task in this inter-laboratory study as compared with the inter-laboratory study of organochlorine, organonitrogen and organophosphate pesticides was to perform quantitative analyses at very low concentrations, close to the European limit of 100 ng L^{-1} for individual pesticides in drinking water. The average detection limits of the standard method were in the range from 4 to 24 ng L^{-1} for the triazine herbicides, and somewhat higher for the more polar degradation product desethylatrazine. These detection limits were below typical values reported earlier and below the method detection limits of the EPA standard methods for triazine herbicides. The lowest detection limits were obtained with nitrogen–phosphorus detection. The average r^2 values of the calibration curves were above 0.99 for all of the analytes. The number of participants was slightly below the requirements of the ISO standard 5725. However, practically no outliers were found, so the results of the inter-laboratory study were considered to be reliable anyway. The determined concentrations of the reference sample compared well with the true values, thus proving the good accuracy of the method. Precision was also satisfactory. Thus, it was concluded that reliable

Table 2 *Results of the inter-laboratory validation of SPME for the analysis of organochlorine, organonitrogen and organophosphate pesticides in aqueous samples[4]*

Pesticide	Data material		Accuracy ($\mu g\ L^{-1}$)			Precision (%)	
	p	Outliers	True value	$\bar{\bar{x}}$	CI	Repeatability s_r	Reproducibility s_R
Dichlorvos	10	0	25	27.3	21.5–33.1	7.5	20
EPTC	11	0	10	9.9	8.3–11.5	5.7	17
Ethoprofos	11	1	17	15.5	13.2–17.8	5.3	31
Trifluralin	11	0	2	1.6	0.8–2.4	17	39
Simazine	9	0	25	23.6	17.0–30.2	9.9	18
Propazine	11	0	10	9.5	6.1–12.9	13	25
Diazinon	11	0	10	8.2	6.4–10.0	7.7	27
Methyl chlorpyriphos	11	1	2	1.6	1.3–2.0	7.5	21
Heptachlor	11	0	10	8.9	3.1–14.7	23	40
Aldrin	11	1	2	2.0	0.5–3.5	27	46
Metolachlor	10	1	17	15.7	13.6–17.8	4.6	19
Endrin	11	1	10	8.8	6.3–11.3	9.9	36

Table 3 Results of the inter-laboratory validation of SPME for the analysis of triazine herbicides in aqueous samples[8]

| Triazine herbicide | Data material | | Accuracy (ng L^{-1}) | | | Precision (%) | |
	p	Outliers	True value	$\bar{\bar{x}}$	CI	Repeatability s_r	Reproducibility s_R
Desethylatrazine	5	1	100	98	76–120	8.0	17
Simazine	6	1	50	54	34–73	13	13
Atrazine	6	0	70	78	61–96	8.1	9.9
Propazine	6	0	60	63	38–87	14	16
Terbuthylazine	7	0	90	96	69–122	9.9	9.9
Ametryn	7	0	110	123	93–152	8.1	12
Prometryn	8	0	120	134	100–167	9.0	14
Cyanazine	5	0	80	87	71–102	6.3	17

quantitation of triazine herbicides can be performed by SPME at concentrations below the European limit for individual pesticides in drinking water.

References

1. J. Pawliszyn, *Solid Phase Microextraction. Theory and Practice*, Wiley-VCH, New York, 1997.
2. ISO 5725-1994, *Accuracy (Trueness and Precision) of Measurement Methods and Results*, International Organisation for Standardisation, Geneva, 1994.
3. M. Feinberg, *Trends Anal. Chem.*, 1995, **14**, 4550.
4. T. Górecki, R. Mindrup and J. Pawliszyn, *Analyst*, 1996, **121**, 1381.
5. UNICHIM 1210-1997, *Qualità dell'acqua, Determinazione di idrocarburi e idrocarburi alogenati volatili, Metodo per microestrazione su fase solida (SPME) e gascromatografia capillare*, Unichim, Milano, 1997.
6. UNICHIM 1211-1997, *Qualità dell'acqua, Determinazione di erbicidi triazinici, Metodo per microestrazione su fase solida (SPME) e gascromatografia capillare*, Unichim, Milano, 1997.
7. T. Nilsson, R. Ferrari and S. Facchetti, *Anal. Chim. Acta.*, 1997, **356**, 113.
8. R. Ferrari, T. Nilsson, R. Arena, P. Arlati, G. Bartolucci, R. Basla, F. Cioni, G. Del Carlo, P. Dellavedova, E. Fattore, M. Fungi, C. Grote, M. Guidotti, S. Morgillo, L. Müller and M. Volante, *J. Chromatogr. A.*, 1998, **795**, 371.
9. T. Nilsson, F. Pelusio, L. Montanarella, B. Larsen, S. Facchetti and J. Ø. Madsen, *J. High Resolut. Chromatogr.*, 1995, **18**, 617.
10. R. Eisert and K. Levsen, *J. Chromatogr. A.*, 1996, **733**, 143.

CHAPTER 15

SPME for the Determination of Organochlorine Pesticides in Natural Waters

KOK KAY CHEE, MING KEONG WONG AND
HIAN KEE LEE

A solid-phase microextraction (SPME) method was developed for the extraction and determination of seventeen organochlorine pesticides (OCPs), and two surrogate standards, in water. SPME parameters including gas chromatograph injector port temperature, initial gas chromatograph oven temperature, and duration of desorption were studied for the effects on the thermal desorption of the OCPs in the gas chromatograph inlet. The optimisation of these parameters was carried out using a three-level orthogonal array design (OAD) with an $OA_9(3^4)$ matrix. The effects of pH salt and dissolved humic substances on the extraction efficiency were also investigated. The method is capable of achieving detection limits in the low ng L^{-1} region, with a %RSD of 3.0–10.0, depending on the compound concerned. The optimised SPME approach developed is suitable for OCP determination in natural waters.

1 Introduction

Contamination of natural waters is a global problem since no international boundaries exist for pollutants. The world's oceans are the main environmental reservoirs of environmental pollutants including the noxious, toxic and persistent organochlorine pesticides (OCPs).[1] With stricter environmental regulations, allowable limits for these compounds may be lower than parts per trillion. Thus, it is necessary to develop more sensitive analytical techniques for OCPs that do not, at the same time, compromise on cost-effectiveness, simplicity, rapidity, precision and suitability for routine and field analyses. Sample preparation is the most important and time-consuming step of an analytical technique, particularly when the matrix in which the analytes of interest are present is an environmental sample of appreciable complexity. Significant analyte losses often occur during sample preparation. Thus, more accurate and precise results can be obtained by combining a sensitive analytical technique with an appropriate sample preparation procedure that reduces analyte loss from adsorption to glassware, sample reconstitution and other factors.[2,3]

There are two widely-used methods for the extraction of the semi-volatile OCPs from water: liquid–liquid extraction (LLE) and solid-phase extraction (SPE). The former method is time-consuming and requires large volumes of solvent.[4] SPE methods, with sorbents packed into cartridges[5,6] or impregnated into membranous disks[7,8] have been employed for the extraction of these compounds. Although they do not have the same problems as LLE, clogging of the cartridges or disks does occur for less-clean natural water samples, compromising their extraction efficiency.

Solid-phase microextraction (SPME) is a solvent-less technique developed by Belardi and Pawliszyn[9] for the extraction of organic pollutants from water samples. This technique, in the form of either direct or headspace analysis, has been applied for the determination of BETX and aromatic compounds, volatile halogenated hydrocarbons, polychlorinated biphenyl congeners, polyaromatic hydrocarbons, organochlorine pesticides and caffeine in water and food.[10–18] In the case of polar compounds, a relatively polar poly(acrylate)-coated fibre has been used for the extraction of phenolic compounds in water.[19] The dynamics of sorption and optimization of SPME conditions have been studied by Pawliszyn and co-workers.[20–22] In the present paper, the use of a relatively non-polar poly(dimethylsiloxane) (PDMS)-coated fibre is described. The partition coefficient (K) for the PDMS fibre towards a particular analyte has been found to correspond to the octanol–water partition coefficient (K_{ow}) values available in the literature.[22] Octanol can be used to extract both polar and non-polar compounds because it contains a non-polar saturated hydrocarbon chain as well as a polar hydroxyl end group. PDMS is relatively non-polar and can easily extract the generally non-polar OCPs. Its behaviour is similar to that of hexane used in LLE.

In SPME, thermal desorption is needed to transfer analytes from the fibre into the gas chromatographic column for analysis. This takes place in the injector port of the instrument itself. It is important that thermal desorption conditions be optimised in order to ensure a more accurate and precise determination of the analytes.

An example of an optimisation strategy is orthogonal array design (OAD). A description of OAD and its use as a chemometric method for optimising analytical procedures have been reported previously.[23–29] In the present work, a three-level OAD approach was considered, in view of the fact that not all relationships between the factor and output can be described as linear; advantages of such an approach have been published.[25]

An evaluation of the SPME technique coupled to gas chromatography (GC) with electron-capture detection (ECD) for OCPs is reported in this paper. The optimum conditions for extracting 19 OCPs in natural waters were established using a three-level OAD with an $OA_9(3^4)$ matrix. Finally, a preliminary survey of the level of contamination by OCPs of coastal sea waters collected from the

Straits of Johore, which lies between Singapore and the southern tip of peninsular Malaysia, was carried out, using the OAD-optimised SPME conditions.

2 Experimental

Instrumentation and Reagents

A Supelco (Bellefonte, PA, USA) PDMS-coated fibre (100 μm thickness) for SPME was used for OCP extraction. Before use, the fibre was conditioned in a gas chromatograph injector port at 210 °C for 4 hours under a flow of purified N_2. OCP analysis was performed on a Hewlett-Packard (Palo Alto, CA, USA) Series II gas chromatograph equipped with a ^{63}Ni electron-capture detector. Analytes were separated on a Hewlett-Packard Ultra-2 capillary column (30 m × 0.32 min i.d.; phase thickness 0.25 μm). The split-splitless injector was operated in splitless mode.

For syringe injection, the injector port and detector temperatures were held at 270 °C and 360 °C, respectively. The oven temperature was initially held at 100 °C for 1 minute, increased to 170 °C at 25 °C/minute, ramped to 270 °C at 2 °C/minute, and finally increased to 290 °C at 25 °C/minute, with the final temperature held for 3 minutes. For introduction of the fibre, the initial oven and injector port temperatures were set according to the OAD given in Table 1. The initial oven temperature was held for 1.5 minutes and the 'purge-on' duration was similar to the analyte desorption time for the fibre.

All pesticide-grade solvents were purchased from Fisher Scientific (Fair Lawn, NJ, USA). Purified water was obtained from a Milli-Q system (Millipore,

Table 1 *Assignment of the factors and levels of SPME experiments using an $OA_9(3^4)$ matrix with the results of selected variables on the responses*

Trial No.	Column No.[a]				Response
	1	*2*	*3*	*4*	*NAR (%)*[b]
1	210	50	10		100.0
2	210	60	7.5		72.3
3	210	70	5		74.0
4	190	50	7.5		62.3
5	190	60	5		64.2
6	190	70	10		60.2
7	170	50	5		60.1
8	170	60	10		81.5
9	170	70	7.5		59.9

[a] 1 = injector port temperature (°C); 2 = initial oven temperature (°C); 3 = desorption time (min); 4 = unassigned column.
[b] NAR = normalised average recovery.

Milford, MA, USA). A standard stock solution containing 17 OCPs (α-BHC, β-BHC, δ-BHC, γ-BHC, Heptachlor, Aldrin, Heptachlor epoxide, Endosulfan I, Dieldrin, *p,p'*-DDE, Endrin, Endosulfan II, *p,p'*-DDD, Endrin aldehyde, Endosulfan sulphate, *p,p'*-DDT and Methoxychlor) was prepared in 1:1 (v/v) hexane:toluene at a concentration of 250 mg L^{-1}. A working standard solution of 1 mg L^{-1} was prepared by dilution of the standard stock solution with methanol and ethyl acetate. A surrogate standards mixture containing 2,4,5,6-tetrachloro-*m*-xylene and dibutyl chlorendate was prepared in 1:1 (v/v) hexane:toluene at a concentration of 2000 mg L^{-1} each. Further dilution was carried out with methanol and ethyl acetate. All OCPs and the surrogate standards were purchased from Ultra-Scientific (North Kingston, RI, USA). The humic acid mixture was purchased from Tokyo Kasei Kyogo (Tokyo, Japan).

SPME Procedure

Direct sampling of OCPs spiked (500 ng L^{-1}) into 40 ml of distilled water was carried out using the PDMS fibre. Samples were stirred and extracted until equilibrium was attained. Three SPME parameters (factors) were selected for optimisation: (1) GC injector port temperature (factor A), (2) initial GC oven temperature (factor B), and (3) duration of desorption (factor C). A three-level OAD with an $OA_9(3^4)$ matrix was employed to assign the variables considered. Selection of variables was based on previous experience and knowledge of SPME conditions. The assignments of the variables and their levels are given in Table 1.

An equilibration study was performed to determine the most suitable extraction time for all the 19 analytes. The SPME fibre was exposed to a spiked solution containing 500 ng L^{-1} each of the OCPs and surrogate standards for 5, 10, 15, 20, 30 and 45 minutes, and the GC peak areas for each compound was plotted against extraction time. A carryover study was also carried out. The fibre was exposed to three injector port temperatures (170 °C, 190 °C and 210 °C) for 5 minutes each. The effect of pH, salinity and humic acid on the extraction efficiency was investigated. The pH of the solutions were adjusted to 3.5 or 8.5 with 0.1 M HCl or NaOH. The procedure for preparing dissolved humic acid compounds in water has been described.[30] The same method was used for this work. A comparison between OCP recoveries from spiked into sea water and in those from spiked Milli-Q water was carried out.

Environmental Analysis

Employing the optimised conditions, a survey of sea water with respect to OCP contamination was carried out. The water was collected from the Straits of Johore, which lies between peninsular Malaysia and Singapore island.

Table 2 *ANOVA table of experimental results*

Source	Sum of squares	Degrees of freedom	Mean square	F ratio	Significance
Injector port temp. (A)	1314.1	2	657.1	24.7	**$P < 0.001$
Initial oven temp. (B)	311.4	2	155.7	5.9	**$P < 0.025$
Desorption time (C)	1611.7	2	805.9	30.3	**$P < 0.001$
Pooled error[a]	292.5	11	26.6		
Total	3629.6	17			

The critical F value is 13.81 (**$P < 0.001$) and 5.26 (**$P < 0.025$).
[a] Pooled error result from unassigned column effect (column 4, Table 1).

3   Results and Discussion

Optimisation of SPME Conditions

The normalised average recoveries (NAR) for the 19 OCPs and surrogate standards from nine pre-designed experimental trials are tabulated (Table 1). When applying the optimisation procedure, some two-variable interactions could be ignored owing to previous experience and knowledge of SPME conditions. Thus, as an indication of whether any important interactions have been missed, the error variance of this optimisation scheme could be calculated by pooling the sum of squares from the unassigned column 4. From the analysis of variance (ANOVA) (Table 2), three variables have been identified to be significant as far as extraction efficiency is concerned, at a 97.55% level of confidence.

From the ANOVA results given in Table 2, optimum SPME conditions were desorption time: 10 minutes; injector port temperature: 210 °C; and initial oven temperature: 50 °C. The use of a longer desorption time permitted the reduction of the injector port temperature. The latter reduced the possibility of bleeding of the fibre material and also prolonged the lifetime of the fibre. Establishing the appropriate injector port temperature for thermal desorption is dependent on the thermal properties of the analytes. Normally, too high an injector port temperature causes decomposition of the thermally labile species and will also affect the cryofocusing of the analytes in the column. However, due to the affinity of some OCPs for the PDMS-coated fibre, a higher injector port temperature (220 °C) is required to effect complete desorption. A relatively low initial oven temperature is required for the reconcentration of the analytes in the GC column.

Equilibration

Equilibration profiles for the OCPs are given in Figure 1(a)–(c). Most of the extractions reached equilibrium in under 45 minutes [Figure 1 (a)] while some OCPs reached equilibrium with the fibre in 15 minutes, *e.g.* TCMX, β-BHC, δ-BHC, γ-BHC, Heptachlor, Aldrin, Heptachlor epoxide, and Endrin aldehyde.

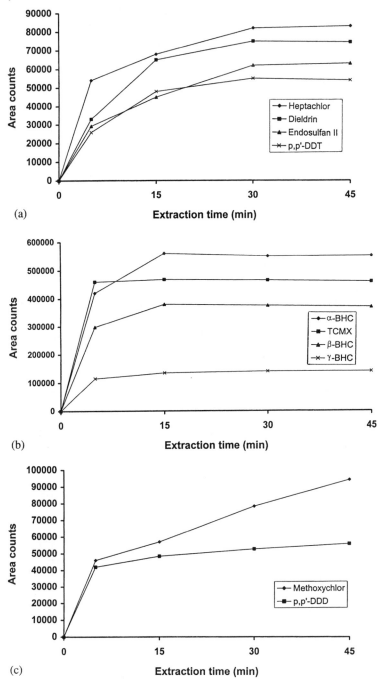

Figure 1 *Equilibration profiles of OCPs during SPME. (a) Heptachlor, Dieldrin, Endo-sulfan II and p,p'-DDT. Equilibration time: 30 min. (b) TCMX, α-BHC, β-BHC, δ-BHC and γ-BHC. Equilibration time: 15 min; (c) p,p'-DDD and Methoxychlor. Equilibration time: 45 min*

After 60 minute extractions, the absorbed masses of some OCPs were only *ca.* 70–85% of those adsorbed after 45 minutes.

During extraction, it is assumed that the only equilibrium occurring was between the analytes in the aqueous phase and the PDMS fibre. However, with prolonged extraction, OCPs were likely to partition between the aqueous phase and the walls of the sample vial, resulting in substantial losses of the analytes from solution owing to their high hydrophobicities. Although it was reported[18] that much longer equilibration times were required for some OCPs (*i.e.* Dieldrin, Methoxychlor, *etc.*), it was important also to take into consideration other factors concerning sample preparation time, loss of analyte during extraction, and sensitivity of the SPME method when deciding on the final applied equilibration time. Each GC run could be completed within 45 minutes.

Carryover

In SPME, the properties of analytes can affect the efficiency with which they are extracted. Analytes with higher boiling points can cause carryover problems, which may be the most significant disadvantage of the technique, if not properly addressed. The affinity of analytes towards a particular sorbent will also affect the extraction efficiency, resulting in carryover. Results of our investigation on carryover effects are summarised in Table 3. The data are for three successive

Table 3 *Carryover study of OCP extraction by SPME*

| OCP | Temperature (°C) | % Carryover[a] | | |
		First	Second	Third
TCMX[b]	170	0.7	–	–
	190	0.2	–	–
	210	–	–	–
α-BHC	170	0.4	–	–
	190	–	–	–
	210	–	–	–
β-BHC	170	–	–	–
	190	–	–	–
	210	–	–	–
δ-BHC	170	2.2	–	–
	190	–	–	–
	210	–	–	–
γ-BHC	170	0.5	–	–
	190	–	–	–
	210	–	–	–
Heptachlor	170	20.5	–	–
	190	4.8	–	–
	210	–	–	–

(Continued)

Table 3 *(continued)*

OCP	Temperature (°C)	% Carryover[a] First	Second	Third
Aldrin	170	0.4	–	–
	190	–	–	–
	210	–	–	–
Heptachlor	170	5.2	1.0	0.5
epoxide	190	2.2	0.5	0.4
	210	0.9	0.2	<0.05
Endosulfan I	170	9.2	0.5	–
	190	1.9	–	–
	210	0.4	–	–
Dieldrin	170	11.2	0.7	–
	190	2.6	–	–
	210	0.4	–	–
p,p'-DDE	170	37.2	4.5	–
	190	11.8	–	–
	210	3.0	–	–
Endrin	170	13.6	1.3	0.5
	190	4.0	0.5	0.4
	210	1.8	0.2	<0.05
Endosulfan II	170	15.3	1.3	–
	190	2.9	–	–
	210	0.3	–	–
p,p'-DDD	170	13.7	1.1	–
	190	4.4	–	–
	210	1.0	–	–
Endrin	170	13.4	1.5	–
aldehyde	190	3.4	–	–
	210	1.6	–	–
Endosulfan	170	15.3	1.6	–
sulphate	190	3.9	–	–
	210	0.7	–	–
p,p'-DDT	170	17.8	2.2	–
	190	4.6	–	–
	210	0.4	–	–
Methoxy-	170	31.5	8.7	1.7
chlor	190	12.6	0.9	–
	210	2.2	–	–
Dibutyl	170	44.2	16.0	6.3
chlorendate[b]	190	25.7	5.0	1.4
	210	9.8	1.5	0.6

[a] Expressed as percentage of peak area resulting from first desorption.
[b] Surrogate standard.

carryover blanks and expressed as percentage areas of the first desorption. Carryover was most severe for late-eluting, and therefore the relatively less volatile, OCPs, as expected. The effect of carryover can be minimized by retaining the fibre in the injector port for a further 15 minutes after the first desorption.

ESV, Retention Times, Linearity, Detection Limits and Precision

The affinities of the OCPs for the PDMS coating determine the extent of absorption and concentration of the OCPs in the aqueous sample. Buchholz and Pawliszyn[22] have described the phenomena quantitatively. Another way to represent affinity and PDMS extraction ability towards OCPs is by the equivalent sample volume (ESV).[31] For example, if the ESV = 2, it means the amount of analyte extracted by the fibre is equivalent to the amount present in 2 mL of the total volume of the sample. ESV is simple and convenient to operate, as it depends on the affinity of the analytes for the sorbent and is unaffected by the analyte concentration within the linear range, as long as the amount of analytes extracted are above the instrumental detection limits. The ESV values for the 19 analytes are provided in Table 4. ESV data were compared to some K_{ow} values and their solubility data in water (S_w,) in order to establish a correlation, if any, amongst all these parameters. Basically, OCPs

Table 4 *Equivalent sample volume (ESV), octanol–water partition coefficient (K_{ow}), and solubility data of OCPs*

OCP	Log K_{ow}[31]	Solubility in water (S_w) (μg L^{-1})[31]	ESV (mL)
TCMX[a]	–	–	0.69
α-BHC	3.46	1630	1.25
β-BHC	3.80	240	0.30
δ-BHC	2.80	–	0.33
γ-BHC	3.20	7300	0.79
Heptachlor	4.40	180	3.12
Aldrin	5.17	17	4.10
Heptachlor epoxide	3.65	275	1.50
Endosulfan I	3.55	530	1.43
Dieldrin	3.69	200	2.60
p,p'-DDE	5.69	65	4.36
Endrin	3.21	220	1.19
Endosulfan II	3.62	2802	1.45
p,p'-DDD	5.06	20	3.80
Endrin aldehyde	–	–	1.72
Endosulfan sulphate	3.66	117	2.00
p,p'-DDT	4.89	1.2	3.79
Methoxychlor	3.31	40	3.10
Dibutyl chlorendate[a]	–	–	2.65

[a] Surrogate standard.

with higher K_{ow} should have higher ESV values since both parameters describe the affinity of analytes towards the fibre coating. For example, *p,p'*-DDE, Aldrin and *p,p'*-DDD have relatively higher K_{ow}; similarly, their ESVs are also the highest. Generally, a trend could be observed in which OCPs with similar K_{ow} values have similar ESVs. There were some deviations from this trend, however. β-BHC, for example, should have an intermediate ESV since it has higher affinity for the fibre coating except for *p,p'*-DDE, Aldrin, *p,p'*-DDD, *p,p'*-DDT and Heptachlor. Yet, its ESV was the lowest. The reason could be that instrumental sensitivity, and hence its response, was lower for this OCP. When compared with S_w data of most of the OCPs, a reasonably good inverse relationship could be established with the respective ESVs (greater affinity for water instead of for the fibre translated to lower ESV values). In summary, the ESV approach can be effective in estimating the affinity of analytes towards the fibre coating, since it is simple and convenient to apply.

Retention time data for the OCPs and surrogate standards are provided in Table 5; these were obtained under the optimum SPME conditions described above. All 19 OCPs were completely resolved within 41 minutes. Table 5 also shows the linear range for OCPs with GC–ECD analysis. For most of these compounds, the range was linear over a 100-fold concentration window. Generally, the linear range for most analytes was between 50 and 5000 mg L^{-1}. The lower end of the range was normally limited by the GC–ECD detection limits.

Table 5 *Retention time data, SPME linear range, detection limits and precision*

OCP	Retention time (min)	SPME linear range (ng L^{-1})	SPME detection limits (ng L^{-1})	Precision (%)
TCMX[a]	11.51	50–10 000	1.00	5.9
α-BHC	13.03	50–10 000	0.50	3.2
β-BHC	14.17	20–5000	1.50	6.9
δ-BHC	15.61	10–5000	0.10	4.2
γ-BHC	14.47	10–5000	0.10	3.0
Heptachlor	16.61	50–10 000	0.80	6.0
Aldrin	18.24	10–5000	0.07	7.7
Heptachlor epoxide	20.34	50–10 000	1.50	4.7
Endosulfan I	23.32	10–5000	0.10	3.9
Dieldrin	25.56	10–5000	0.10	3.9
p,p'-DDE	27.46	10–5000	0.04	10.0
Endrin	29.07	20–10 000	0.10	6.1
Endosulfan II	29.85	10–5000	0.10	4.0
p,p'-DDD	30.83	10–5000	0.05	8.0
Endrin aldehyde	31.42	25–10 000	0.10	4.3
Endosulfan sulphate	33.34	10–5000	0.05	5.0
p,p'-DDT	33.84	10–5000	0.05	7.1
Methoxychlor	38.87	10–5000	0.10	5.5
Dibutyl chlorendate[a]	41.69	50–10 000	1.00	6.7

[a] Surrogate standard.

Detection limits were determined by comparing the S/N ($=3$) ratio to the lowest detectable concentrations, based on the individual calibration plots. Detection limits are shown in Table 5. For some OCPs, these can be as low as 0.04 ng L^{-1}. In general, analytes with relatively higher K_{ow} values would have lower detection limits—although this observation can be offset by lower instrumental response towards individual OCPs.

The precision of the technique for each OCP was determined by carrying out eight consecutive injections of solutions spiked at 0.5 μg L^{-1} concentrations. As Table 5 shows, for most of the OCPs the precision was generally below 10 %RSD.

Effect of pH, Salinity and Humic Acid

The effects of pH, salinity and humic acid substances on the average recoveries are given in Table 6. A matrix of pH 3.5 or 8.5 had no significant effect on the extraction of all OCPs. However, it is common to reduce the pH of aqueous samples to improve the extraction of the more polar compounds (like phenols).[19] The effect of salinity has previously been studied for the SPME of phenols;[19] the addition of NaCl to near saturation condition was suitable for the extraction of these compounds as their solubilities were much reduced, permitting greater partitioning to the SPME sorbent. In the present work, however, salinity had no significant impact on OCP extraction.

Humic acid substances present in water have been reported to reduce extraction efficiency by SPE, either by interaction with the analytes or by permeation of the SPE sorbent.[30,33,34] Our results indicated that SPME was not seriously affected by these materials at concentrations equivalent to 10 mg L^{-1} humic acid, except for some late-eluting OCPs (Endosulfan sulphate, Methoxychlor, 2,4-Dibutyl chlorendate). Losses of these OCPs were between 10 and 15%.

Table 6 *Effect of pH, salinity and dissolved humic acid on the average recoveries (AR) of OCPs*

Conditions	Level	AR (%)[a]
pH	3.5	90.5
	8.5	89.0
Salinity	0% (w/v) NaCl	100
	36% (w/v) NaCl[b]	90.3
Dissolved humic acid	0 mg L^{-1}	100
	10 mg L^{-1}	81.0

[a] Average of two determinations.
[b] Saturated solution.

Table 7 *Relative recoveries of OCPs from spiked sea water*
(at 0.5 ng L^{-1})

OCP	Relative recovery (%)[a]
TCMX[b]	80.5 ± 6.3
α-BHC	89.3 ± 3.1
β-BHC	79.5 ± 4.3
δ-BHC	89.1 ± 3.7
γ-BHC	81.3 ± 5.0
Heptachlor	78.5 ± 7.3
Aldrin	89.1 ± 8.1
Heptachlor epoxide	100 ± 5.6
Endosulfan I	98.5 ± 5.1
Dieldrin	76.4 ± 4.9
p,p′-DDE	87.7 ± 3.8
Endrin	87.8 ± 8.1
Endosulfan II	91.5 ± 7.9
p,p′-DDD	90.1 ± 8.0
Endrin aldehyde	83.2 ± 3.8
Endosulfan sulphate	85.6 ± 3.5
p,p′-DDT	86.1 ± 4.3
Methoxychlor	87.5 ± 5.0
Dibutyl chlorendate[b]	71.0 ± 7.9

[a] Mean ± %RSD ($n = 5$).
[b] Surrogate standard.

Spiked and Real Sample Analysis

Prior to extracting genuine water samples, the optimised SPME conditions were used to extract OCPs spiked into water at a concentration of 0.5 mg L^{-1}. Recoveries were in the range 71–100% with %RSD values 3.1–8.1, as shown in Table 7. A typical chromatogram of a spiked sea water extract is shown in Figure 2.

Sea water collected from the Straits of Johore was analysed by this procedure. A chromatogram of a sea water extract is shown in Fig. 3. Generally, water from the Straits was free of OCPs, or if present, they were below the detection limit. However, careful analysis of chromatographic data (represented typically by Figure 3) reveals that one OCP, α-BHC, might possibly be present. Its concentration was measured to be 0.42 μg L^{-1}.

4 Conclusion

The present study demonstrates the applicability of SPME with PDMS sorbent for the effective extraction of OCPs from aqueous samples, in conjunction with the OAD optimisation approach. Based on our observations, the optimum GC conditions for the procedure in relation to OCP analysis are: injector port

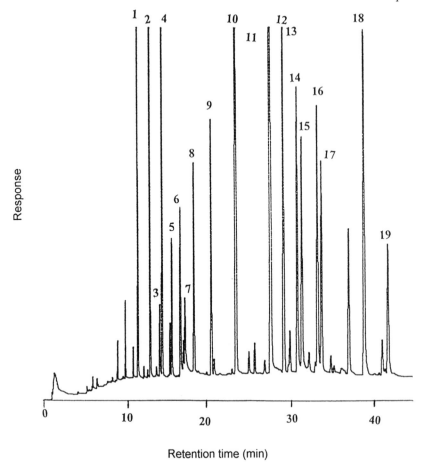

Figure 2 *Gas chromatogram of 19 OCPs and surrogate standards extracted by SPME from spiked water sample (at 0.5 ng L⁻¹). Peak identities: (1) TCMX (surrogate standard); (2) α-BHC; (3) β-BHC; (4) γ-BHC; (5) δ-BHC; (6) Heptachlor; (7) Aldrin; (8) Heptachlor epoxide; (9) Endosulfan I; (10) Dieldrin; (11) p,p'-DDE; (12) Endrin; (13) Endosulfan II; (14) p,p'-DDD; (15) Endrin aldehyde; (16) Endosulfan sulphate; (17) p,p-DDT; (18) Methoxychlor; (19) Dibutyl chlorendate (surrogate standard). Chromatographic conditions are given in the text*

temperature: 220 °C; initial oven temperature: 50 °C; duration of thermal desorption: 10 minutes.

Acknowledgments

The authors thank the Government of Canada for providing financial assistance under the ASEAN-Canada Cooperative Programme on Marine Science—Phase II. K.K. Chee is grateful to the National University of Singapore for the award of a research scholarship.

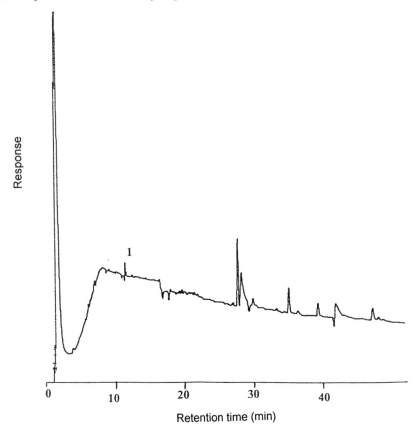

Figure 3   *Gas chromatogram of extract of sea water collected from the Straits of Johore. Peak 1 may possibly be due to α-BHC. Chromatographic conditions are given in the text*

References

1. I. Cruz, D.E. Wells and I.L. Marr, *Anal. Chim. Acta*, 1993, **283**, 260.
2. C. Markell, D.F. Hagen and V.A. Bunnelle, *LC-GC*, 1993, **9**, 332.
3. M.C. Hennion, *Trends Anal. Chem.*, 1991, **10**, 317.
4. US Environmental Protection Agency, *Method 3510A, Separatory Funnel Liquid–Liquid Extraction. Test Methods for Evaluating Solid Waste, 1B: Laboratory Manual of Physical/Chemical Methods*, SW-846, 3rd Edn., Washington, DC, USA, 1986.
5. G. Manes, Y. Pico, J.C. Molto and G. Font, *J. High Resolut. Chromatogr.*, 1990, **13**, 843.
6. L. Torreti, A. Simonella, A. Dossena and E. Toretti, *J. High Resolut. Chromatogr.*, 1992, **15**, 99.
7. B.A. Tomkins, R. Merriweather and R.A. Jenkins, *J. Assoc. Off. Anal. Chem., Int.*, 1992, **75**, 1091.
8. L.M. Davi, M. Baldi, L. Penazzi and M. Liboni, *Pestic. Sci.*, 1992, **35**, 63.
9. (a) R.P. Belardi and J.B. Pawliszyn, *Pollut. Res. J. Can.*, 1989, **23**, 179; (b) C.L.

Arthur, D.W. Potter, K.D. Buchholz, S. Motlagh and J.B. Pawliszyn, *LC-GC*, 1992, **10**, 656.

10. D.W. Potter and J.B. Pawliszyn, *J. Chromatogr.*, 1992, **625**, 247.

11. C.L. Arthur, K.D. Buchholz, J. Berg and J. Pawliszyn, *Anal. Chem.*, 1992, **64**, 1968.

12. C.L. Arthur, L.M. Killman, S. Motlagh, D.W. Potter and J. Pawliszyn, *Environ. Sci. Technol.*, 1992, **26**, 979.

13. C.L. Arthur, L.M. Killman, K.D. Buchholz, L. Berg and J. Pawliszyn, *J. High Resolut. Chromatogr.*, 1992, **15**, 741.

14. B.D. Page and G. Lacroix, *J. Chromatogr.*, 1993, **638**, 199.

15. C. Arthur and J. Pawliszyn, *Anal. Chem.*, 1990, **62**, 2145.

16. D.W. Potter and J. Pawliszyn, *Environ. Sci. Technol.*, 1994, **28**, 298.

17. S.B. Hawthorne, D.J. Miller, J. Pawliszyn and C.L. Arthur, *J. Chromatogr.*, 1992, **603**, 185.

18. S. Magdic and J. Pawlisyzn, *J. Chromatogr. A*, 1996, **723**, 111.

19. K.D. Buchholz and J. Pawliszyn, *Environ. Sci. Technol.*, 1993, **27**, 2844.

20. D. Louch, S. Motlagh, J. Pawliszyn and C.L. Arthur, *Anal. Chem.*, 1992, **64**, 1187.

21. C.L. Arthur, L.M. Killam, K.D. Buchholz and J. Pawliszyn, *Anal. Chem.*, 1992, **64**, 1960.

22. K.D. Buchholz and J. Pawliszyn, *Anal. Chem.*, 1994, **66**, 160.

23. W.G. Lan, M.K. Wong, N. Chen and Y.M. Sin, *Analyst*, 1994, **119**, 1659.

24. W.G. Lan, M.K. Wong, N. Chen and Y.M. Sin, *Analyst*, 1994, **119**, 1669.

25. H.B. Wan, W.G. Lan, M.K. Wong and C.Y. Mok, *Anal. Chim. Acta*, 1994, **289**, 371.

26. H.B. Wan, W.G. Lan, M.K. Wong, C.Y. Mok and Y.H. Poh, *J. Chromatogr. A*, 1994, **677**, 255.

27. W.G. Lan, M.K. Wong, N. Chen and Y.M. Sin, *Talanta*, 1994, **41**, 1917.

28. W.G. Lan, K.K. Chee, M.K. Wong and Y.M. Sin, *Analyst*, 1995, **120**, 273.

29. W.G. Lan, K.K. Chee, M.K. Wong and H.K. Lee, *Analyst*, 1995, **120**, 281.

30. W.E. Johnson, N.J. Fendinger and J.R. Plimer, *Anal. Chem.*, 1991, **63**, 1510.

31. H.B. Wan, H. Chi, M.K. Wong and C.Y. Mok, *Anal. Chim. Acta*, 1994, **298**, 219.

32. J. H. Montgomery, *Agrochemicals Desk Reference: Environmental Data*, Lewis Publishers, Boca Raton, FL, USA, 1993.

33. A. Di Corcia and R. Samperi, *Anal. Chem.*, 1993, **65**, 907.

34. I. Lisca, E.R. Brouwer, H. Lingeman and U.A.Th. Brinkman, *Chromatographia*, 1993, **37**, 13.

CHAPTER 16

Determination of Sulfur-containing Compounds in Wastewater

PETER POPP, MONIKA MÖDER AND IMELDA McCANN

1 Introduction

In the area of Leipzig–Halle–Bitterfeld, the pollution of rivers, lakes and in some cases also groundwater is a major environmental problem. Of particular interest for our work were wastewater deposits from coal briquette manufacturing, such as the 'Schwelvollert' lake situated in a disused open-cast lignite mine. Chemical analysis of this lake revealed that contamination was mainly caused by liquid fractions comprising large amounts of hydroxylated aromatic compounds as well as hydrocarbons and a variety of sulfur compounds. A number of organic sulfur compounds were also found in the groundwater around former chemical plants in Bitterfeld. Consequently we decided to develop methods for the determination of volatile and semi-volatile sulfur compounds using the sensitive, solvent-free SPME procedure.

The SPME extraction of substances with higher boiling points from water samples is described in many papers. Johansen and Pawliszyn[1] used SPME for the extraction of semi-volatile sulfur compounds from water. The authors describe the trace analysis of heteroaromatic compounds in water and polluted groundwater. The motivation for the study was that in Denmark the contamination of the groundwater by creosote (which contains compounds with nitrogen, sulfur or oxygen in the aromatic ring) from disused gasworks is a matter of increasing concern. Using polydimethylsiloxane (PDMS), poly-acrylate (PA) and Carbowax/divinylbenzene fibres, extraction yields and detection limits were also determined for the sulfuric compounds thiophene, benzothiophene and dibenzothiophene.

The SPME enrichment of low boiling substances is more difficult. The extraction and quantitative analysis of volatile and highly volatile compounds in water by SPME is described in several papers[2–8]. In most cases 100 μm PDMS and 85 μm PA fibres were used. Pelusio et al.[9] used headspace SPME to analyse volatile organic sulfur compounds in black and white truffle aroma. Seven volatile sulfur compounds including dimethyl sulfide, dimethyl disulfide and dimethyl trisulfide were found; however, quantitative analysis was not performed. Another paper[10] describes the application of automated SPME with a 75 μm Carboxen/PDMS fibre for the determination of highly volatile sulfur compounds in beer. Dimethyl sulfide, diethyl sulfide and di-*n*-propyl sulfide were identified. This work and other recent investigations show that the newly developed carbon-coated fibres are eminently suitable for analysing highly volatile compounds. Manganio and Cenciarni[11] developed a fused-silica fibre coated with graphitized carbon black Carbograph I (Alltech) with increased thermal stability, and reported extraction and calibration curves of VOCs in gaseous and aqueous samples. Popp and Paschke[12] showed that by using a 80 μm Carboxen/PDMS fibre (Supelco) in connection with GC and an electron-capture detector, detection limits of halocarbons in the range of a few ng L^{-1} can be achieved. Djozan and Assadi[13] used extra-fine powdered activated charcoal as a coating layer and headspace SPME. In combination with GC-FID, detection limits of the BTEX compounds in the range of 1.5–2 ng L^{-1} were calculated. The results of these investigations lead us to theorize that, if appropriate fibres are used, it should also be possible to attain low detection limits for highly volatile sulfur compounds. The rest of this chapter describes the development of methods for the determination of volatile and semi-volatile sulfur compounds with SPME–GC–MS and applications for wastewater analysis.

2 Materials and Methods

Preparation of Standard Solutions

The methods were optimised and validated with 12 sulfur compounds. The volatile compounds (dimethyl sulfide, ethyl methyl sulfide, thiophene, carbon disulfide, dimethyl disulfide) were available as pure liquids. Concentrated solutions were prepared in methanol. For SPME samples, 10 μL of these solutions were directly added to 4 mL water. Hence the content of methanol in any sample did not exceed 0.5% (v/v). The standard solution of the semi-volatile compounds (benzothiophene, 7-methylbenzothiophene, 3,5-dimethyl-benzothiophene, dibenzothiophene, naphtho[2,1-*b*]thiophene, phenanthro[4,5-*bcd*]thiophene and phenanthro[3,4-*b*]thiophene) was prepared in toluene with a concentration of 100 μg mL^{-1} for each compound. This solution was diluted in methanol and the SPME samples were prepared in the same way as the highly volatile compounds.

The analytes were placed in 5.0 mL vials with stir bars and sealed with Teflon-lined septa and hole caps. The headspace was about 2.0 mL. To increase the ionic strength of the spiked water samples and the groundwater or wastewater samples, 0.16 g NaCl was added and samples were adjusted to a pH of approximately 3. The latter procedure was necessary as the groundwater samples from the chemical plants in Bitterfeld had pHs between 2 and 3.

SPME Procedure

The SPME fibres coated with 100 μm PDMS, 65 μm polydimethylsiloxane/divinylbenzene (PDMS/DVB), 65 μm Carbowax/divinylbenzene (CW/DVB), 85 μm PA and 75 μm Carboxen/PDMS were obtained from Supelco. New fibres were conditioned for two hours under a nitrogen or helium stream at the desorption temperature recommended by the manufacturer. All extractions were carried out at room temperature. The headspace procedure was used to extract the highly volatile substances; the other substances were directly extracted under strong stirring (about 1000 rpm).

The desorption temperature of the Carboxen/PDMS fibre was 280 °C, whereas for all other fibres a desorption temperature of 250 °C was chosen. The desorption time was 2 min.

Instrumentation

Method development was performed on a Finnigan GCQ system and a HP 6890 with MSD. Initially only fibres with PA of 85 μm thickness and PDMS of 100 μm thickness were tested.[14] Helium with a flow rate of 40.0 cms^{-1} was used for all fibre injections. Separation was performed on a 25 m, 0.25 mm i.d., 0.25 μm film HP-5 capillary column. During desorption the column was maintained at 50 °C. The temperature was then increased at a rate of 7 K min^{-1} to a final temperature of 280 °C and then held for 10 minutes. The transfer line temperature was 285 °C and the ion source temperature of the direct coupled mass spectrometer was 180 °C.

The HP 6890/MSD system was equipped with a split-splitless injector and a SPME insert. For the determination of the volatile sulfur compounds, a 30 m, 0.32 mm i.d., 1.0 μm film CP SIL 8 CB capillary column (A) was used. The analysis of the higher boiling compounds was performed with a 30 m, 0.25 mm i.d., 0.25 μm HP 5 MS capillary column (B).

Helium at 40 mL min^{-1} served as a carrier gas. The temperature program for column A was 35 °C hold 5 min, 7 K min^{-1} to 80 °C, hold 1 min, 30 K min^{-1} to 280 °C, while the program for column B was 50 °C, hold 2 min, 7 K min^{-1} to 260 °C, 35 K min^{-1} to 280 °C hold 1 min.

The transfer line temperature was 280 °C. For all investigations (ITD and MSD) the single ion mode (SIM) was selected to detect the analytes. Target ions with the following m/z values were used for the selective detection of the sulfur compounds: 62, 76, 84, 94, 134, 148, 162, 184, 208 and 234.

3 Results and Discussion

Comparison of Fibre Coatings

Figure 1 compares the extraction efficiencies of the fibres used for highly volatile sulfur compounds. The extraction yields given in a logarithmic scale generally increase with the boiling points of the investigated compounds, the exception being carbon disulfide, which has a low boiling point but high extraction yields.

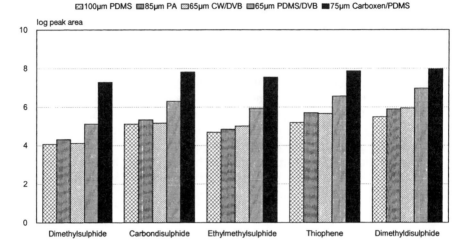

Figure 1 *Extraction efficiencies of various fibre coatings for sampling volatile sulfur compounds. Concentration of each compound:* 20 μg L^{-1}

It is shown that with the 65 μm PDMS/DVB fibre recommended for polar volatiles, tenfold higher extraction yields were found compared to the PDMS, PA and CW/DVB fibres, whereas using the 75 μm Carboxen/PDMS fibre the extraction yields increase to a factor of 100–1000. This fibre was used for the further investigations to extract the volatile compounds.

Figure 2 shows the extraction efficiencies for sulfur compounds with higher boiling points. In this case the differences between the five fibres are lower. In

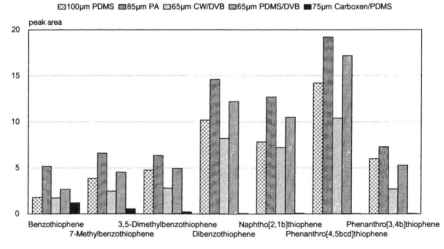

Figure 2 *Extraction efficiencies of various fibre coatings for sampling semi-volatile sulfur compounds. Concentration of each compound:* 2.5 μg L^{-1}

contrast to the extraction of the highly volatile compounds, the Carboxen/
PDMS fibre only provides acceptable extraction yields for benzothiophene.
This fibre is not suitable for the extraction of high boiling compounds. The
other fibres have much higher extraction efficiencies (given in peak area counts),
but the use of the 85 μm PA fibre is most propitious for the enrichment of semi-
volatile sulfur compounds. Consequently this fibre was chosen for all further
experiments.

Extraction Time Profiles

Figures 3 and 4 show the profiles of extraction amount *versus* time. In order to
extract the volatile compounds using the Carboxen/PDMS fibre, 40 min is
sufficient for dimethyl sulfide, ethyl methyl sulfide, thiophene and dimethyl
disulfide to reach equilibrium. In the case of carbon disulfide, a loss of substance
was observed after 15 min. This was probably caused by leakage through the
puncture of the SPME needle in the vial septum.

   The extraction profiles of the semi-volatile sulfur compounds with PA fibre
and direct extraction from water (Figure 4) show that the compounds with the
lowest boiling points (benzothiophene, 7-methylbenzothiophene) reach equilibrium after 40 min. The higher boiling substances have a high affinity to the
fibre and accordingly reaching the partition equilibrium takes longer. An
extraction time of 40 min was chosen for all further experiments.

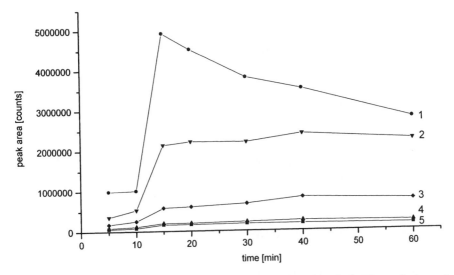

Figure 3   *Exposure time profiles for volatile sulfur compounds with the 75 μm Carboxen/
PDMS fibre: 1, Carbon disulfide (1.0 μg L^{-1}); 2, Thiophene (0.5 μg L^{-1});
3, Dimethyl disulfide (0.2 μg L^{-1}); 4, Ethyl methyl sulfide (0.2 μg L^{-1});
5, Dimethyl sulfide (0.2 μg L^{-1})*

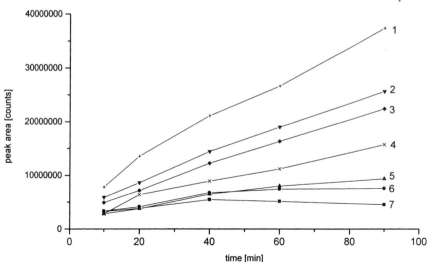

Figure 4 *Exposure time profiles for semi-volatile sulfur compounds with the 85 μm PA fibre: 1, Phenanthro[4,5-bcd]thiophene; 2, Dibenzothiophene; 3, Naphtho[2,1-b]thiophene; 4, Phenanthro[4,5-bcd]thiophene; 5, 3,5-Dimethylbenzothiophene; 6, Benzothiophene; 7, 7-Methylbenzothiophene. Concentration of each compound: 2.5 μg L⁻¹*

Carryover Profiles

In all experiments blanks were regularly run to ensure there was no contamination or carryover from fibre or water. At the highest concentration levels (20 μg L^{-1}) for the Carboxen/PDMS fibre, carryover between 0.1% (dimethyl sulfide) and 0.6% (dimethyl disulfide) was observed.

Carryover was also observed for the extraction of the semi-volatile sulfur compounds with the PA fibre. It was found that the compounds with the highest carryover were those with the highest extraction yields, which points to incomplete desorption at the selected conditions. The average carryover from seven extractions for dibenzothiophene was 0.6%, while the value for naphtho[2,1-*b*]thiophene was 0.9%. Phenanthro[4,5-*bcd*]thiophene had the highest carryover at 1.8%.

Linearity and Detection Limits

To determine linearity and LOD, solutions between 20 ng L^{-1} and 20 μg L^{-1} were prepared. LOD was calculated from a signal-to-noise ratio of about 3. Figures 5 and 6 show that the linearity of the combination of SPME (75 μm Carboxen/PDMS, 85 μm PA) and GC–MSD (SIM-mode) exceeds more than three orders of magnitude.

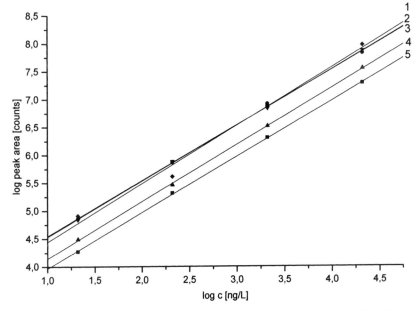

Figure 5 *Linearity of the Carboxen/PDMS–GC–MS method for volatile sulfur compounds: 1, Dimethyl disulfide; 2, Thiophene; 3, Carbon disulfide; 4, Ethyl methyl sulfide; 5, Dimethyl sulfide*

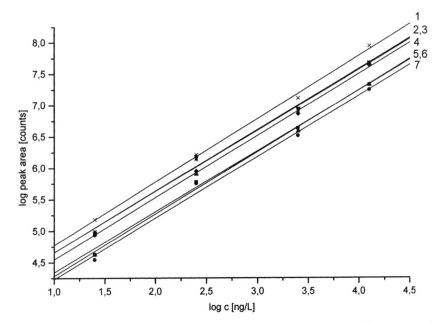

Figure 6 *Linearity of the PA–GC–MS method for semi-volatile sulfur compounds: 1, Phenanthro[4,5-bcd]thiophene; 2, Dibenzothiophene; 3, Naphtho[2,1-b]thiophene; 4, Phenanthro[3,4-b]thiophene; 5, 3,5-Dimethylbenzothiophene; 6, Benzothiophene; 7, 7-Methylbenzothiophene*

Table 1 *Detection limits (LOD) of volatile sulfur
compounds for headspace extraction with a
75 μm Carboxen/PDMS fibre*

Compound	LOD-MSD (ng L^{-1})
Dimethyl sulfide	5
Carbon disulfide	1
Ethyl methyl sulfide	3
Thiophene	1
Dimethyl disulfide	1

Table 2 *Detection limits (LOD) of semi-volatile sulfur compounds for direct extraction with a 85 μm PA fibre*

Compound	LOD-MSD (ng L^{-1})
Benzothiophene	1.2
7-Methylbenzothiophene	1.5
3,5-Dimethylbenzothiophene	1.2
Dibenzothiophene	0.6
Naphtho[2,1-*b*]thiophene	0.6
Phenanthro[4,5-*bcd*]thiophene	0.4
Phenanthro[3,4-*b*]thiophene	0.8

The detection limits are listed in Tables 1 and 2. The values range from 0.4 ng
L^{-1} to 5 ng L^{-1} and are very low. Johansen and Pawliszyn[1] also calculated a
low LOD of 20 ng L^{-1} for benzothiophene and dibenzothiophene using the
combination of PA fibre and mass spectrometric detection. However, the
detection limit of thiophene (1 μg L^{-1}) is three orders of magnitude higher
than the value we calculated from studies using the Carboxen/PDMS fibre.

A comparison of our values with octanol–water partition coefficients is not
useful. In the case of the Carboxen/PDMS fibre, the adsorption of the carbon in
the pores is probably the predominant process,[12,15] whereas in the case of the
extraction of the semi-volatile compounds with the PA fibre equilibrium was not
reached for most compounds.

Environmental Sample Analysis

The methods developed for the determination of volatile and semi-volatile
sulfur compounds were applied to the analysis of environmental samples. The
first sample (A) was a contaminated groundwater from Antonie landfill in
Bitterfeld. A second sample (B) was groundwater from Kanena landfill near

Halle. A highly contaminated lignite-derived wastewater sample (C) was taken from a disused industrial plant in Lauchhammer, and the water sample (D) was taken at a depth of 24 m from the Schwelvollert lake to the south-west of Leipzig. The analysis results are contained in Table 3.

The compounds dimethyl sulfide to dimethyl disulfide were determined using the 75 μm Carboxen/PDMS fibre while the semi-volatile compounds were analysed with the 85 μm PA fibre.

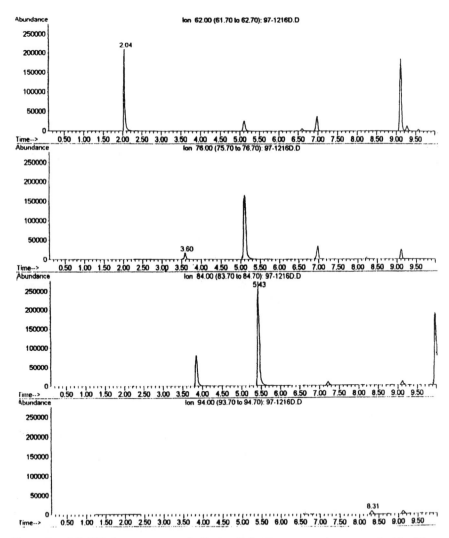

Figure 7 *GC–MS chromatogram of a lignite-derived water sample extracted with a 75 μm Carboxen/PDMS fibre. SIM plots of dimethyl sulfide at 2.04 min (m/z 62), ethyl methyl sulfide at 3.60 min (m/z 76), thiophene at 5.43 min (m/z 84) and dimethyl disulfide at 8.31 min (m/z 94).*

Table 3 *Concentrations of sulfur compounds in wastewater samples* (μg L^{-1})

Compound	Sample A	Sample B	Sample C	Sample D
Dimethyl sulfide	0.04	0.4	4.1	1.8
Carbon disulfide	19.1	0.7	n.d.	2.3
Ethyl methyl sulfide	0.1	0.1	0.4	0.1
Thiophene	0.2	0.3	3.2	0.8
Dimethyl disulfide	0.3	0.2	0.5	1.1
Benzothiophene	0.1	0.02	n.d.	1.0
7-Methylbenzothiophene	n.d.	n.d.	0.5	0.2
3,5-Dimethylbenzothiophene	n.q.	n.d.	n.d.	0.1
Dibenzothiophene	0.1	0.01	0.2	0.02
Naphtho[2,1-*b*]thiophene	n.q.	n.d.	n.q.	0.1
Phenanthro[4,5-*bcd*]thiophene	n.d.	0.02	n.d.	n.d.
Phenanthro[3,4-*b*]thiophene	n.d.	0.05	n.d.	n.d.

n.d.: not detectable.
n.q.: not quantifiable (co-elution with other peaks).

The RSD (n = 3) of the wastewater samples for the volatile substances ranged from 4 to 18% and for the semi-volatile compounds from 7 to 17%. To assess any possible matrix influences, one of the environmental samples (D) and HPLC water were spiked with the same standard solution of semi-volatile sulfur compounds. The sampling procedure was carried out under optimum conditions. It was found that the peak areas of the compounds were nearly identical, within the error of measurements.

Figure 7 shows the selective ion monitoring plots of the heavily contaminated sample C. These examples illustrate the suitability of the SPME–GC–MS procedure for the trace analysis of volatile and semi-volatile sulfur compounds in contaminated waters. The combination of a 75 μm Carboxen/PDMS fibre for the extraction of volatiles and a 85 μm PA fibre for the enrichment of semi-volatiles enables the determination of sulfur compounds with a wide range of boiling points with detection limits in the lower ng L^{-1} range.

References

1. S.S. Johansen and J. Pawliszyn, *J. High Resol. Chromatogr.*, 1996, **19**, 627.
2. C.L. Arthur, L.M. Killam, S. Motlagh, M. Lim, D.W. Potter and J. Pawliszyn, *Environ. Sci. Technol.*, 1992, **26**, 979.
3. M. Chai, C.L. Arthur, J. Pawliszyn, R.P. Belardi and K.F. Pratt, *Analyst*, 1993, **118**, 1501.
4. Z. Zhang and J. Pawliszyn, *J. High Resol. Chromatogr.*, 1993, **16**, 689.
5. T. Nilsson, F. Pelusio, L. Montanarella, B. Larsen, S. Facchetti and J.Ö. Madsen, *J. High Resol. Chromatogr.*, 1995, **18**, 617.
6. I. Valor, C. Cortada and J.C. Molto, *J. High Resol. Chromatogr.*, 1996, **19**, 472.
7. F.J. Santos, M.T. Galceran and D. Fraisse, *J. Chromatogr. A*, 1996, **742**, 181.
8. K.J. James and M.A. Stack, *Fresenius' J. Anal. Chem.*, 1997, **358**, 833.

9. F. Pelusio, T. Nilsson, L. Montanarella, R. Tilio, B. Larsen, S. Facchetti and J.Ö. Madsen, *J. Agric. Food Chem.*, 1995, **43**, 2138.
10. Z. Penton, *Varian SPME Application Note 16*, 1998.
11. F. Mangani and R. Cenciarni, *Chromatographia*, 1995, **41**, 678.
12. P. Popp and A. Paschke, *Chromatographia*, 1997, **46**, 419.
13. D. Djozan and Y. Assadi, *Chromatographia*, 1997, **45**, 183.
14. I. McCann, *Thesis*, The Queen's University of Belfast, October 1997.
15. R.E. Shirey, *PittCon'97, Book of Abstracts*, 305.

CHAPTER 17

Analysis of Creosote and Oil in Aqueous Contaminations by SPME

SYS STYBE JOHANSEN

1 Background

In many countries a major part of the drinking water resources originate from groundwater reservoirs and thus groundwater quality has received considerable attention during the last century. Investigations have shown that soil and groundwater in the vicinity of old gasworks, wood treatment facilities, asphalt factories or the oil industry are often contaminated with creosote and/or oil that typically has entered the subsurface as a result of accidental spills or leaks.[1] These contaminated sites are widespread throughout the world in high numbers.[2] Once the contamination has reached the groundwater level, it is spread further and therefore creosote and oil are a potentially significant contamination problem.

Creosote and oil are complex mixtures of an unknown number of organic compounds (Figure 1). Their chemical composition varies depending on the production process, temperature, and coal type *etc*. The major classes of compounds identified in creosote are up to 85% polyaromatic hydrocarbons (PAH) including naphthalenes, 1–10% phenols, 5–13% heteroaromatic compounds (NSO), and 0.5–3% monoaromatic hydrocarbons (MAH), while oil constitutes of about 13–30% alkanes, 45–55% cycloalkanes and 25–35% aromatic compounds involving MAH, PAH and NSO. Aromatic compounds of up to 3–4 ring systems are of interest because of their relatively high water solubility.[3,4] In risk assessments of groundwater at such contaminated sites it is therefore of interest to measure the content of aromatic compounds such as MAH, phenols, NSO, and PAH, and these will be the compounds reviewed here. For interest some studies have reported on the SPME analysis of alkanes and cycloalkenes in water.[5–7]

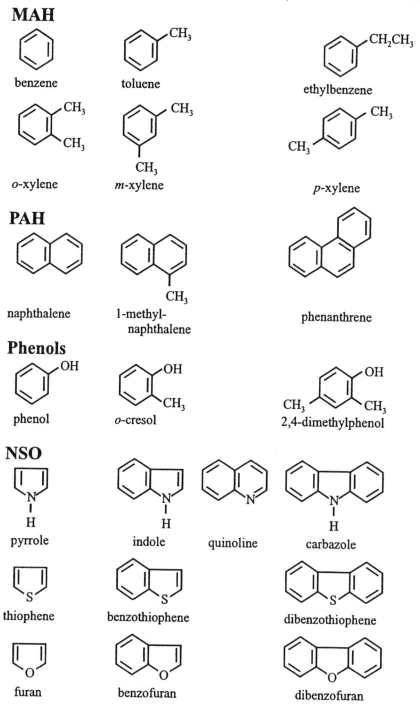

Figure 1 *Selected aromatic compounds originating from creosote and oil divided into organic classes*

2 Analysis

The choice of instrumental analysis for environmental samples of organic micropollutants such as aqueous contaminations of aromatic compounds is mostly capillary GC, preferably in combination with MS in order to achieve high resolution, selectivity and sensitivity. Applications of HPLC are less common owing to longer analysis times, lower separation capacity, and most importantly, for trace analysis, detection options are limited. Before instrumental analysis of environmental samples at trace level, extensive extraction and preconcentration are usually required. An analytical detection limit in the order of ng L^{-1} is commonly required to meet the acceptance criteria for organic contaminants in groundwater used for drinking water, which typically are 0.1–10 μg L^{-1} in the EU.[8] However, as a screening tool a level of μg L^{-1} is acceptable.

The water-soluble aromatic compounds are a heterogeneous group with a wide spectrum of physical and chemical properties ranging from very volatile, *e.g.* benzene/thiophene, and polar compounds, *e.g.* phenol/pyridine, to some non-polar and semi-volatile compounds, *e.g.* dibenzofuran/phenanthrene. These properties result in stringent requirements for the sample preparation technique. Several sample preparation techniques have been developed and are now used routinely for the trace analysis of aromatic compounds in aqueous samples, *e.g.* liquid–liquid extraction using various organic solvents, *e.g.* dichloromethane, diethyl ether,[3,9] and solid-phase extraction by different reversed-phase columns, *e.g.* octadecyl.[9,10]

Generally, the use of liquid–liquid extraction on samples with trace contamination levels is a very time-consuming and expensive technique involving large quantities of toxic solvents, which eventually must be condensed and disposed of. Simpler sample preparation techniques involving reduced solvent consumption are therefore of interest. Solid-phase extraction could be an alternative, but the technique suffers from insufficient retention and low extraction efficiency of some aromatic compounds, *e.g.* the volatile and the highly water-soluble compounds.[3,5,11] Alternatives are therefore still needed. Here the use of SPME will be presented as a potential candidate for determining aqueous contaminations of creosote and oil.

MAH

Table 1 shows SPME applications for analysis of the major constituents of creosote and oil in water. The analysis of MAH by SPME has been studied thoroughly, probably due to the excellent abilities of SPME for MAH. Table 1 shows that the desired detection limit of MAH in water is achieved either by HS or liquid extraction within a few minutes using the PDMS fiber and GC–FID or GC–ITMS measurements. Between 3–30% of each MAH is extracted by the 100 μm PDMS from aqueous solutions; the lowest extraction efficiency is achieved with benzene. Use of saturated salt conditions enhanced the sensitivity. At optimized conditions (saturated salt and stirring) more than 95% of

Table 1 SPME applications on creosote, oil and related compounds in aqueous samples. The sampling conditions and the analytical parameters are listed, e.g. limit of detection (LOD), linear range (LR), and relative standard deviation (RSD)

Class	Conditions	Abs. time	Analysis	LOD (ppb)	LR (ppb)	RSD (%)	Reference
MAH	Dir, 56 PDMS	2–5 min	GC–FID	1–3	15–3000	3–5	12
	Dir, uncoated fiber	2 min	GC–FID	50–200	200–1500	4–11 (50)	13
	Dir, 100 PDMS[a]	3½ min	GC–FID	–	35–850		
	Dir, 100 PDMS[a]	30 min	GC–FID	0.03–0.07 (0.2)	–	<20	14
	HS, 100 PDMS + salt	50 min	GC–FID	0.19–0.35 (0.7)	4–140	<20	15
	Dir, 100 PDMS	30 min	GC–ITMS	0.01–0.015	0.03–100	<7.5	16
	Dir, 100 PDMS	300 min	GC–FID	0.31–0.93 (3.6)	4–6 orders	<17	6
	Dir, 100 PDMS	10 min	GC–FID	0.03–0.1 (0.22)	1–2000	<9	17
	HS, 100 PDMS	10 min	GC–FID	0.08–0.41	1–2000	<11	
	HS, 100 Charcoal[b]	15 min	GC–FID	0.0015–0.002	0.005–50	<8	18
Phenols (alkylated)	Dir, PDMS	15 min	GC–ITMS	0.2–7.6	20–2000	<5	19 (20, 21)
	Dir, PA	40 min	GC–FID	2.1–30	6.7–670	<5	
	Dir, PA	40 min	GC–ITMS	0.02–0.8	2–3 orders	<5	
	Dir, PA + salt	40 min	GC–FID	0.42–5.5	2–3 orders	<5	
	Dir, PA + salt	40 min	GC–ITMS	0.005–0.15		<5	
	HS, PA + salt	60 min	GC–FID		100–20000	<9	22
PAH	Dir, 100 PDMS	300 min	GC–FID	0.03–0.42	3–5 orders	<12	6
	Dir, 100 PDMS[a]	30 min	GC–FID	0.09–0.15	2–3 orders	<20	14
	Dir, 15 PDMS	10 min	GC–ITMS	0.001–0.02		<10	
NSO	Dir, PA + salt	200 min	GC–FID	0.5–15	0.5–500	<15	23
	Dir, PA + salt	200 min	GC–ITMS	0.02–0.04 (–10)	0.02–500	5–14	24
Pyridines, thiazoles	Dir, 100 PDMS	10 min	GC–MS (SIM)	25	50–5000	<5	28
N-herbicides	Dir, CWDVB	10 min	GC–MS (SIM)	5	5–500	<5	29
	Dir, PA	50 min	GC–ITMS	0.00001–0.015	2–4 orders	<20	
	Dir, PA	50 min	GC–NPD (FID)	0.01–0.8 (0.2–15)	0.03–30	<20	
	Dir, PA	25 min	GC–NPD	0.01–0.09	3 orders	1–7	
n-Alkanes (C6–C10)	Dir, PA	25 min	GC–MS (SIM)	0.007–0.024		<10	
	Dir, PDMS	300 min	GC–FID	0.07–0.16	3–6 orders	<14	

Abbreviations: Abs. time: absorption time. Dir: direct sampling. HS: headspace sampling. + salt: addition of salt to the sample. PDMS: 15, 56 or 100 μm polydimethylsiloxane. PA: 85 μm polyacrylate. CW/DVB: 65 μm Carbowax/divinylbenzene. IT: ion trap. [a] Non-stirred SPME system. [b] Non-commercial fiber of 100 μm activated charcoal—see reference.

MAH were recovered using the experimental activated charcoal fiber.[18] The results of using activated charcoal as a fiber coating with GC–FID detection are comparable in LOD with the 100 μm PDMS coating and GC–ITMS detection. This indicates that a higher sensitivity is possible by using the charcoal fiber and GC–MS detection. Unfortunately, this fiber coating is not commercially available yet.

While fiber carry-over is not a problem for these very volatile compounds, volatilization of MAH, *e.g.* benzene, may occur during sampling and injection resulting in analyte loss, poor reproducibility and reduced sensitivity. A tight seal in the vial during sampling and a minimum of time between the end of sampling and injection is therefore required. Benzene is the most problematic MAH to analyse due to its high volatility and low K_{fw} value. If the goal is to measure these volatile compounds in complex samples, it is clearly advantageous to use HS sampling to limit the total number of compounds the fiber is exposed to. If the fiber is exposed directly to samples high in particular matter, material from the matrix could coat the solid phase and interfere with the extraction. Furthermore, the short time of equilibration for MAH and therefore small sampling time is also beneficial when complex samples are analysed. Valor *et al.*[17] compared the LOD between headspace (HS) and direct SPME sampling of MAH in water and found that the LOD obtained by direct SPME were lower in absolute values than those obtained by HS-SPME, with the exception of benzene. Another important aspect of analysing volatiles is that the SPME technique is readily portable and has good potential for field monitoring.

Although nearly all studies mentioned in Table 1 are performed with the PDMS coating, the PA coating extracts MAH efficiently as well. The PA coating has an affinity for polar (see phenols below) and non-polar compounds due to its structure, consisting of a hydrocarbon chain backbone with relatively polar ester side chains.[19]

PAH

Analysis of non-polar semi-volatile PAH, including naphthalenes, has been studied using the PDMS coating as illustrated in Table 1. The equilibration time of PAH is several hours, but usually a much shorter absorption time is used under well defined conditions. The extraction efficiency of PAH by a 100 μm PDMS is high, generally 15–50%. Naphthalene has the lowest sensitivity of the PAH because of its low K_{fw}. It was observed by Johansen and Pawliszyn[24] that the extraction efficiency of high molecular weight NSO, *e.g.* dibenzofuran, was reduced up to 40% by increasing the ionic strength of the solution. This phenomenon is probably due to their low water solubility resulting in precipation or adsorption to glass walls. Similar observations are therefore expected with the PAH

Carry-over is the major limiting factor for PAH analysis by SPME, especially for the four-ring systems and higher. To clean the fiber, very high desorption temperatures are required with the temperature being limited by the upper temperature limits of the coating. Interference may be prevented or significantly reduced by limiting the absorption time to the necessary minimum, making several consecutive desorptions of the fiber in an unoccupied injection port within each analysis and checking the background level of the fiber before analyses.[23] Chen and Pawliszyn[25] found another solution for carry-over problems by coupling SPME with HPLC and using solvent desorption. By this desorption procedure no carry-over was observed. Another coupling that might reduce carry-over problems was developed by Ngoyen and Loung,[26] who coupled SPME and cyclodextrin-modified capillary electrophoresis. They achieved LOD of between 8 and 75 μg L^{-1} of the 16 EPA priority PAH using a PDMS coating; however, the approach is still under development.

As for MAH, the PA coating is also able to extract PAH. This was observed in studies of complex samples where some PAH were detected; Johansen and Pawliszyn[24] observed excellent extraction efficiency for NSO by the PA coating, and similarities in extraction characteristics between NSO and PAH are expected, due to their similar physico-chemical properties, *e.g.* dibenzofuran and fluoranthene (see section on NSO).

Phenols

Phenols have been analysed by SPME with success in several studies (see Table 1). These studies showed that the amount of phenols extracted is enhanced in a low pH (pH = 2) and saturated salt environment (NaCl), but this coincides with an equilibrium time increase from 40 to 60 min. Importantly, only the extraction of phenols with pK_a below 7, such as the chlorinated phenols, is influenced by the pH of the environment. Therefore sample pH is not relevant for the phenols in creosote, which have pK_a of about 10. However, saturated salt conditions enhance the extraction of phenol and alkylated phenols, typically by a factor of 2–5. Clark and Bunch[21] observed a high affinity of the PA coating toward the alkylated phenols of cigarette smoke condensate, which are similar to those of creosote. An optimum extraction efficiency was obtained in the temperature range of 20–30 °C by the PA coating.[27] At optimum conditions about 1–6% of the phenols were extracted by the PA coating from aqueous solutions. The lowest recovery determined was for phenol itself, owing to its high water solubility and thereafter low K_{fw}. As illustrated in Table 1 it was possible to achieve the desired detection limit for phenols in water, using the PA fiber, and a 40 min extraction of a salt saturated sample, with GC–FID or GC–ITMS detection. Carry-over is only a problem for high molecular weight phenols, and can be overcome by using the PA coating that tolerates high desorption temperatures (300 °C).

In order to extract the polar phenols by the non-polar PDMS coating, an *in situ* derivatization where the hydroxyl group is replaced by an acetate group is necessary. This treatment increased the method sensitivity with the PDMS coating several times, but it is still inefficient compared with that achieved directly with the PA coating.[19] The derivatization of the phenols may improve their chromatography, however, with the commercial capillary GC columns of today, direct analysis of phenols is simple.

HS-SPME of phenols has a longer equilibrium time due to slow transfer of the phenols from the aqueous phase through the headspace to the fiber. Method sensitivity was lower for HS-SPME (high ppb level) compared with direct SPME when using an extraction time of 60 min.[19,22] HS-SPME of phenols in aqueous samples is therefore not recommended.

NSO

As illustrated in Table 1, SPME fibers are able to extract NSO by direct sampling from aqueous solutions. The extraction efficiency strongly depends on the fiber coating and the compound. In comparison between PDMS, PA and Carbowax coatings, the PA coating extracted the highest amounts of NSO, even for the relatively non-polar dibenzofuran and dibenzothiophene. The low molecular weight NSO, *e.g.* pyridine and pyrrole, had very low absolute recoveries (0.2–1.3%). Variation in pH between 7 and 10 did not influence the extraction of pH-sensitive NSO, *e.g.* pyridines; however under acidic pH the extracted amount of pyridines and quinolines is reduced. An increase in ionic strength of the sample enhanced the extraction efficiency by a factor of 3–10 for the majority of NSO except for the relatively non-polar dibenzofuran and dibenzothiophene, for which the extraction efficiency dropped nearly 40% at saturated conditions. These two compounds have high K_{fw} values, so the sensitivity at saturated conditions was still good. The optimum conditions for most NSO (using PA fiber in 4 mL salt saturated sample at pH 8) resulted in detection limits of 20–40 ng L^{-1} for the semi-volatile NSO, and 1–10 μg L^{-1} for the very polar and volatile NSO. Owing to the wide range of NSO physico-chemical characteristics, their equilibration times vary from 60 to 200 min, and their extraction efficiencies vary between 0.15 and 75%. Furthermore, loss of volatile NSO, *e.g.* thiophene and 1-methylpyrrole, was observed for extractions longer than 1 hour.[24]

Coleman[28] performed several experiments with the Carbowax/DVB and PDMS coatings, for other creosote related flavour compounds, *e.g.* pyridines and thiazoles. As shown in Table 1, a ppb limit of detection was possible with both fibers. This sensitivity could be improved by increasing the extraction time and the ionic strength of the samples.

Studies of other polar compounds reveal many similarities in methods between nitrogen-containing pesticides, *e.g.* triazines and NSO (see Table 1). PA is the preferred coating; they have high equilibration times (hours), there is a strong effect of salt addition and they have comparable detection limits.

Complex Aqueous Samples

Water analysis often involves problems caused by a high molecular weight matrix or other inorganic components that are able to interact with the SPME coating. These effects are due to the basis of SPME as an equilibration extraction. Experiments made by Müder *et al.*[20] have shown that matrix effects may result in incomplete extraction after a normal extraction time, but this could be improved by using a longer extraction time. The effect was found to be essentially an added polymer molecule causing a delay of the extraction, related to an increase in density and viscosity of the sample and a corresponding slower diffusion of the analytes (phenols) to the coating. They observed no other changes in the analysis, *e.g.* the LOD was similar under corrected sampling time. However, a large extension of the sampling period increases the risk of the coating surface being blocked by adsorbed polymers, *e.g.* humic acid particles.

Other studies observed that the RSD increased with the complexity of the sample,[17,24] but this was not observed consistently. Several studies have indicated that other organic analytes interfere with the recovery of the analytes.[19,20,31] However, analyses of groundwater samples for polar pesticides have not shown any interference and it was demonstrated that humic acid below 10 mg L^{-1} (normal level for groundwater) had no significant effect on the extraction. Even if one component with a high affinity toward the fiber is present at a much higher concentration than the other compounds, the relative extraction efficiency is not affected.[30]

In contrast, analysis of a water sample for VOC at ppb level by SPME gave comparable results with the classical method (purge and trap) while in heavily contaminated samples (mg L^{-1}) the results were only in the same order of magnitude.[31] This difference is probably caused by interactions and competitions between the high number of analyte in the fiber extraction. In such circumstances SPME could be applied as a screening tool.

Müder *et al.*[20] observed comparable concentrations of phenols in original wastewater (103 mg L^{-1} humic acid) loaded with coal-derived substances, by 45 min SPME with PA coating and liquid–liquid extraction using dichloromethane. The results obtained differed by a factor of 1.4–2.9 and a partial explanation is probably found in the differences of the techniques' basic principles. The organic solvent may to some degree extract analytes adsorbed on to the polymer matrix, while SPME accumulates only the analytes readily available for extraction. Langenfeld *et al.*[6] observed a good agreement between 45 min SPME with PDMS coating and dichloromethane extraction for the determination of PAH concentrations in creosote-contaminated water, and Eisert and Levsen[30] determined concentrations (ppt level) of polar pesticides in wastewater with SPME–GC–MS in close agreement with those obtained by conventional SPE–GC–MS.

These studies indicate that the matrix effects are mainly of importance in heavily contaminated samples, *e.g.* wastewater, while in analysis of drinking water the effects are insignificant. Matrix effects can be overcome to a large extent by normalization to similar conditions of pH, ionic strength *etc.* However, the use of an internal standard or standard addition is strongly recommended for quantitative analysis of complex samples. Additionally Nilsson *et al.*[31] observed that this variance was significantly reduced when the absolute response for VOC was normalized to that of the internal standard, and that the relative standard deviation of SPME in real samples is usually below 10%, and is generally better than for the conventional methods.[3,6,24]

3   Summary

The potential of SPME for determining individual organic compounds in aqueous mixtures is good. SPME is a more versatile technique for extraction of organics in water than conventional techniques. SPME is recommended for screening and monitoring investigations of aqueous contaminations with creosote and/or oil/gasoline. In screening analysis, SPME will give fast evaluation of the character, the level of contamination and the dominating species in each sample. The technique has great possibilities in field applications. A contamination plume is easily verified and characterized using SPME coupled to GC–MS. These studies indicate that SPME–GC–MS is an excellent method for trace analysis of aromatic compounds in water.

SPME of creosote- and oil-contaminated water requires a compromise in the sampling conditions. However, observations indicate that the optimal SPME set-up will probably involve a PA coated fiber, a salt saturated and stirred sample in neutral pH, a sampling time between 30 and 60 min and a high desorption temperature (*e.g.* 280 °C) in the GC injection port.

Future analysis of heavily contaminated samples must involve careful studies of the influences of accompanying substances prior to the set-up of SPME routines. Furthermore, statistical studies of the interferences (*e.g.* matrix effects) is necessary in order to evaluate the potentials of SPME as a quantitative method for complex samples.

References

1. S.S. Johansen, E. Arvin, A.B. Hansen and H. Mosbæk, *GWMR*, 1997, **1**, 106.
2. A.R. Lotimer, D.W. Belanger and R.B. Whiffin, in *Subsurface Contamination by Immiscible Fluids*, ed. Weyer, Balkema, Rotterdam, 1992, pp. 411–416.
3. S.S. Johansen, A.B. Hansen, H. Mosbæk and E. Arvin, *J. Chromatogr.*, 1996, **738**, 395.
4. K. Østgaard and A. Jensen, *Int. J. Environ. Anal. Chem.*, 1983, **14**, 55.
5. J. Ritter, V.K. Stromquist, H.T. Mayfiled, M.V. Henley and B.K. Lavine, *Microchem.*, 1996, **54**, 59.
6. J.J. Langenfeld, S.B. Hawthorne and D.J. Miller, *Anal. Chem.*, 1996, **68**, 144.
7. A. Saraullo, P.A. Martos and J. Pawliszyn, *Anal. Chem.*, 1997, **69**, 1992.
8. Danish Environmental Protection Agency guidelines (in Danish) 1992. No 4/1992 on sites contaminated with products for the pressure-impregnation of wood; No. 6/1992 on sites contaminated with tar/asphalt; No. 8 on microbiological decontamination of contaminated soil.
9. US EPA, Method for the determination of organic compounds in drinking water. Method 500-600, Environmental Monitoring Systems Laboratory, Office of Research and Development, US Environmental Protection Agency, Cincinnati, Ohio, 1988.
10. G.A. Junk and J.J. Richard, *J. Res. Nat. Bur. Stand.*, 1988, **93**, 274.
11. C.E. Rostad, W.E. Pereira and S.M. Ratcliff, *Anal. Chem.*, 1984, **56**, 2856.
12. C.L. Arthur, L.M. Killam, S. Motlage, M. Lim, D.W. Potter and J. Pawliszyn, *Environ. Sci. Technol.*, 1992, **26**, 979.

13. L.P. Sarna, G.P.B. Webster, M.R. Friesen-Fischer and R.S. Ranjan, *J. Chromatogr.*, 1994, **677**, 201.
14. M.M.E. van der Kooi and Th.H.M. Noij, Proceedings of the 16th International Symposium on Capillary Chromatography, Riva del Garda, Italy, 1994, pp. 1087–1098.
15. B. MacGillivray, J. Pawliszyn, P. Fowlie and C. Sagara, *J. Chromatogr. Sci.*, 1994, **32**, 317.
16. D.W. Potter and J. Pawliszyn, *J. Chromatogr.*, 1992, **625**, 247.
17. I. Valor, C. Cortada and J.C. Molto, *J. High Resol. Chromatogr.*, 1996, **19**, 472.
18. Dj. Djozan and Y. Assadi, *Chromatographia*, 1997, **45**, 183.
19. K.D. Buchholz and J. Pawliszyn, *Anal. Chem.*, 1994, **66**, 160.
20. M. Müder, S. Schrader, U. Franck and P. Popp, *Fresenius' J. Anal. Chem.*, 1997, **357**, 326.
21. T.J. Clark and J.E. Bunck, *J. Chromatogr. Sci.*, 1996, **34**, 272.
22. P. Bartak and L. Cap, *J. Chromatogr.*, 1997, **767**, 171.
23. D.W. Potter and J. Pawliszyn, *Environ. Sci. Technol.*, 1994, **28**, 298.
24. S.S. Johansen and J. Pawliszyn, *J. High Resol. Chromatogr.*, 1996, **11**, 627.
25. J. Chen and J. Pawliszyn, *Anal Chem.*, 1995, **67**, 2530.
26. A. Nguyen and J.H.T. Luong, *Anal. Chem.*, 1997, **69**, 1726.
27. S. Huang, C. Cheng and Y. Sung, *Anal. Chim. Acta*, 1997, **343**, 101.
28. W.M. Coleman, *J. Chromatogr. Sci.*, 1997, **35**, 245.
29. A.A. Boyd-Boland and J.B. Pawlizsyn, *J. Chromatogr.*, 1995, **704**, 163.
30. R. Eisert and K. Levsen, *Fresenius' J. Anal. Chem.*, 1995, **351**, 555.
31. T. Nilsson, F. Pelusio, L. Montanarella, R. Tilio, B.R. Larsen, S. Facchetti and J.Ø. Madsen, Proceedings of the 16th International Symposium on Capillary Chromatography, Riva del Garda, Italy, 1994, pp. 1148–1158.

CHAPTER 18

Direct Analysis of Solids Using SPME

JOHN R. DEAN AND P. HANCOCK

1 Introduction

Normally, the analysis of environmental solid materials (soil, sludge or related matter) prior to chromatographic separation and detection requires some form of extraction with an organic solvent (or solvent mixture). This has been traditionally done by either heating or agitation of the organic solvent–solid mixture. The former has utilised such techniques as Soxhlet extraction or soxtec extraction while the latter utilises sonication or shake-flask extraction. More recently other instrumental extraction techniques have been applied and these include supercritical fluid extraction, microwave-assisted extraction and accelerated solvent extraction. All these approaches are costly in terms of organic solvent usage (and disposal) or equipment costs. The aim of this chapter is to investigate whether solid phase microextraction (SPME) can be utilised for the direct environmental analysis of solids.

Chlorinated benzenes are used as industrial solvents, pesticides, deodorants and chemical intermediates.[1] For example, 1,2,4-trichlorobenzene has wide usage including a solvent in chemical manufacturing, a carrier to apply dyes to polyester materials, a dielectric fluid in transformers, a component in lubricants as a heat transfer medium and as an ingredient in insecticides, herbicides and wood preservatives.[2,3] The presence of chlorobenzenes in the environment is growing as a result of industrial atmospheric discharges[4] and release of solid/liquid effluents.[5] As much as $0.25 \, \text{mg m}^{-3}$ of 1,2,4-trichlorobenzene has been measured in the air of Los Angeles[6] and concentrations ranging from 0.007 to $275 \, \text{mg L}^{-1}$ have been measured in the drinking water supplies of US cities.[6] Of the USA Environmental Protection Agency's principal organic hazardous constituents, chlorobenzene is considered one of the most difficult compounds to incinerate.[7] This is primarily due to the strength of the C—Cl bond which is $95 \, \text{kcal mol}^{-1}$ as compared with more typical values of around $85 \, \text{kcal mol}^{-1}$.[8] This may in part account for their resistance to degradation in the environment. Also as the number of Cl atoms increases in the chlorobenzene series the vapour pressure and solubility decrease and boiling point and K_{oc} values increase.[9] The chlorobenzenes' general insolubility in water coupled with their resistance to degradation can lead to persistence in the environment[10] and their high K_{oc} values can lead to bioaccumulation, which may reach a level of toxicity detrimental to the well being of plants and animals.[11,12]

The problem of using SPME for the analysis of solid samples is how to calibrate the instrumental response. In this chapter two approaches are considered, both based on the sampling of the headspace above a solid soil sample placed in a sealed vial. The vial is heated in a water bath to speed up the release of contaminants from the soil. In the first approach the responses from the headspace sampling of a series of liquid standards are compared with the responses from a slurried soil sample. In the second approach separate quantities of soil are spiked with increasing concentrations of the analyte of interest prior to headspace sampling and analysis (the method of standard additions).

The aim of this study was to develop a method for quantifying the levels of chlorinated benzenes present in contaminated land samples. Two soils were supplied by the European Commission, Community Bureau of Reference (BCR); one was a sandy soil and the other a clay soil. The clay soil was known to be contaminated to a much higher degree than the sand.

2 Experimental

Quantification was carried out using a GC–MSD system (Hewlett Packard G1800A GCD). Separation was achieved using a DB-1 column of 30 m length × 0.25 mm i.d. × 0.25 μm film thickness. The GC was operated at an injector temperature of 280 °C with an initial column temperature of 50 °C which was held for 10 minutes during SPME desorption. The split valve was opened after 11 minutes. The column was temperature programmed at 10 °C min^{-1} up to 285 °C.

Solid phase microextraction was done using either an 85 μm coating thickness polyacrylate fibre or a 100 μm coating thickness polydimethylsiloxane fibre (Supelco, Poole, Dorset, UK). Fibres were conditioned in the injection port according to the manufacturer's instructions prior to extractions.

All standards (1,3,5-trichlorobenzene, 1,2,4-trichlorobenzene, 1,2,3-trichlorobenzene, 1,2,4,5-tetrachlorobenzene and pentachlorobenzene) and solvents used were of HPLC grade. Owing to the insoluble nature of chlorobenzenes in water a 1000 ppm stock solution was prepared in acetone. Serial dilutions into HPLC-grade water were prepared from the stock prior to extraction/analysis.

Contaminated soils were kindly donated by the European Commission, Community Bureau of Reference (BCR), reference material number 529 sandy soil and reference material number 530 clay soil. 2 ml screw top autosampler vials were supplied by Phase Separations Ltd. (Clwyd, Wales).

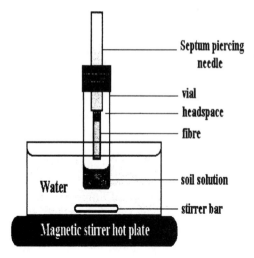

Figure 1 *Schematic diagram of manual headspace SPME experimental arrangement*

SPME Procedure

A manual SPME method was used for the extraction of chlorobenzenes from contaminated soil. The method developed involved two main steps:

(1) The soil ($\sim$0.02 g sandy soil or $\sim$0.002 g clay soil) was placed in a 2 ml crimp topped vial and maintained at a constant temperature in a water bath. The fibre needle was used to pierce the vial septum and the plunger lowered the fibre into the vial thus exposing the fibre to the headspace above the sample matrix to allow partitioning of the analytes between sample and fibre.

(2) After a certain length of time, the fibre was removed from the sample and introduced into the injection port of the GC where thermal desorption took place at a temperature of 280 °C. To afford instantaneous injection by SPME the analytes were focused at a low temperature at the head of the GC column.

Figure 1 shows the experimental arrangement used for manual headspace extraction.

3 Results and Discussion

Selection of SPME Fibre

Initial studies were carried out to quantify chlorobenzenes in aqueous solution and to compare the GC–MSD responses obtained for both the 85 μm poly-acrylate and the 100 μm polydimethylsiloxane fibre by direct immersion in

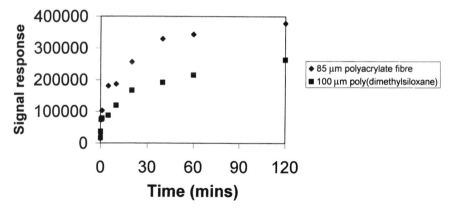

Figure 2 *Effect of SPME fibre on signal response for 1,2,4-trichlorobenzene*

aqueous solutions (1 ml). It was found that all the chlorobenzenes studied, except pentachlorobenzene, gave a higher signal response when using the polyacrylate fibre. Similar signal responses were obtained by both fibres for pentachlorobenzene. A typical response for 1,2,4-trichlorobenzene using the two different fibres is shown in Figure 2. The polyacrylate fibre was used for all subsequent work. A typical chromatogram of the chlorobenzene standard mixture using SPME–GC–MSD is shown in Figure 3.

Note: All initial optimisation studies for the headspace sampling of the contaminated soil were done using the sandy soil.

Influence of Soil Wetting and Extraction Temperature on Signal Response

The headspace extraction of dry and wet soil using a fixed SPME sorption time of 20 minutes at varying temperatures (20, 40, 60 and 80 °C) was evaluated. It is clearly seen from Figure 4, using 1,2,4-trichlorobenzene as an example, that signal response enhancement occurs when the soil is previously wetted (typically 10-fold). This response is raised further on heating. Similar responses were obtained for the other chlorobenzenes studied. In this case 0.1 ml of water is added to the soil prior to SPME. However, while the signal response enhancement is linear with respect to temperature for the dry soil, this is not the case for the wetted soil which showed increasing signal response with respect to increasing temperature. For example, the ratio of the signal response for the wet to dry soil at 24 °C was 11 for 1,2,3-trichlorobenzene and this increased to 15 at 80 °C. It was concluded that wetting the soil prior to SPME produced signal enhancement and that these benefits could be maximised at a temperature of 80 °C.

Peak identification: 1 = 1,3,5-trichlorobenzene; 2 = 1,2,4-trichlorobenzene; 3 = 1,2,3-
trichlorobenzene; 4 = 1,2,4,5-tetrachlorobenzene; peak 5 = pentachlorobenzene. The
concentration of 1,3,5-trichlorobenzene, 1,2,4-trichlorobenzene, 1,2,3-
trichlorobenzene, 1,2,4,5-tetrachlorobenzene and pentachlorobenzene was 0.04, 1.2,
0.6, 1 and 0.1 μg ml⁻¹, respectively.

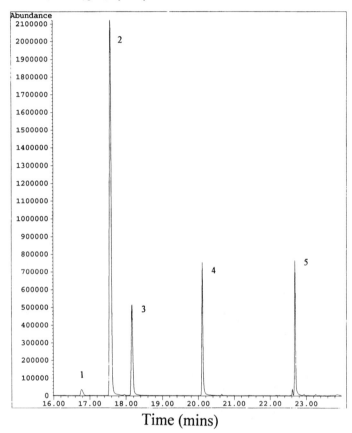

Figure 3   *Chromatographic separation of chlorobenzenes using SPME–GC–MSD*

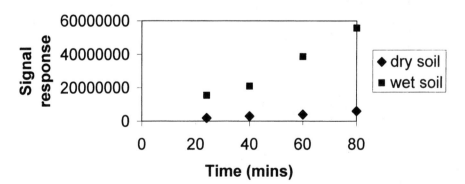

Figure 4   *Effect of soil wetting and temperature on SPME of 1,2,4-trichlorobenzene*

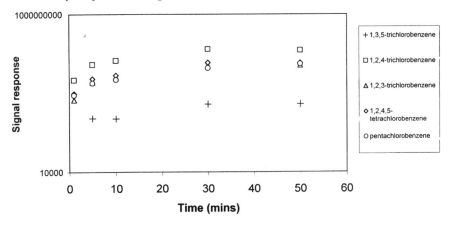

Figure 5 *SPME–time profiles for chlorobenzenes*

Effect of Extraction Time

The effects of sampling the headspace above the wetted soil at 80 °C for varying lengths of time (up to 50 min) were then assessed. The results for the five chlorobenzenes are shown in Figure 5. It is observed that a typical sorption (extraction) profile consists of an initial rapid partitioning followed by a slower prolonged uptake and finally a steady-state equilibrium between the fibre and the vapour phase of the analyte. Considerations have to be made concerning the expected precision of the method, when selecting an optimum time to carry out the sorption. Obviously taking a sorption time on the initial rapid uptake part of the system would lead to poorer precision, particularly for the manual approach used in this work, as slight deviations in the extraction time lead to larger variations in the amount extracted. However, this has to be also considered from the viewpoint of the total time taken for the entire process (extraction and analysis), the faster the analysis the more results can be generated. It was therefore considered appropriate to use a 20 min extraction time.

Effect of Desorption Time

The effect of desorption time in the GC injector was also investigated. Initial studies, so far, have used a 10 min desorption time. However, it was considered appropriate to investigate the effect of desorption time (0.5–40 min) for the five chlorobenzenes under investigation. A typical trace is shown in Figure 6 for 1,2,4-trichlorobenzene. The results, for all chlorobenzenes, indicate that a 10 min desorption time is the minimum time required for quantitative desorption of all analytes studied.

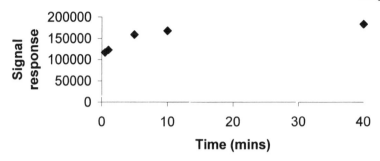

Figure 6   *Effect of desorption time on signal response for 1,2,4-trichlorobenzene*

Optimum SPME Conditions for Chlorobenzenes

It is concluded that the optimum conditions for the headspace SPME of chlorobenzenes from soil were as follows:

- SPME fibre: 85 μm polyacrylate
- Sample: wetted soil
- Extraction temperature: 80 °C
- Headspace sorption time: 20 min
- Desorption time in GC injector: 10 min

A typical chromatogram for the separation of the chlorobenzenes extracted from the sandy and clay soil samples, under the experimentally determined optimum conditions, are shown in Figure 7 (a and b). Comparing Figure 7 with Figure 3 (an SPME of a standard solution of the chlorobenzenes) it can be observed that other analytes are also extracted from the two soils. It is noted that the sandy soil is not as heavily contaminated as the clay soil.

Repeatability of SPME

An assessment of the repeatability of headspace SPME was done. It was found that the precision was better for the wetted sandy soil than for the wetted clay soil. The precisions for 1,3,5-trichlorobenzene, 1,2,4-trichlorobenzene, 1,2,3-trichlorobenzene, 1,2,4,5-tetrachlorobenzene and pentachlorobenzene from the sandy soil were 8.8, 3.3, 3.6, 7. 0 and 6.9%RSD ($n = 4$), respectively, while from the clay soil they were 8.9, 10.4, 22.7, 14.0 and 19.5%RSD ($n = 4$), respectively.

Quantitation of Levels of Chlorobenzenes in Soil Samples: Use of Aqueous Standards

As soil matrices vary greatly in character from different locations (*i.e.* organic carbon content, pH, cation exchange capacity), choosing an identical clean matrix to artificially spike and use as an external standard is virtually impossible. Adding water to the soil allows us the advantage of treating the

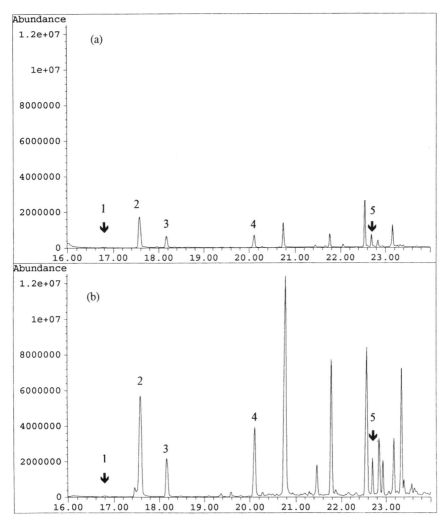

Time (mins)

Figure 7 *Headspace SPME–GC–MSD of* (a) *sandy soil (BCR 529) and* (b) *clay soil (BCR 530). Peak identification as in Figure 3*

soil as a solution (technically a slurry) and therefore an ability to use aqueous standards to quantify the levels of chlorobenzenes present in the soils.

Using the previously optimised conditions a calibration curve was plotted for each analyte using headspace SPME–GC–MSD. The graphs obtained showed linearity over three orders of magnitude, *e.g.* for 1,2,4-trichlorobenzene over the concentration range 25 ng ml^{-1} to 16 μg ml^{-1}, the R^2 value was 0.9736. The results using this approach for the sandy and clay soils are shown in Table 1.

Table 1 *Quantitation of chlorobenzenes in contaminated soils using headspace-SPME followed by GC–MSD. Results expressed in* mg kg^{-1}

| | Sandy soil (BCR 529)* | | | | Clay soil (BCR 530)* | | | |
| | Aqueous calibration (n = 7) | | Standard additions (n = 3) | | Aqueous calibration (n = 11) | | Standard additions (n = 3) | |
Compound	Mean	%RSD	Mean	%RSD	Mean	%RSD	Mean	%RSD
1,3,5-Trichlorobenzene	0.20	18.4	0.19	48.0	3.74	48.2	5.13	14.9
1,2,4-Trichlorobenzene	5.14	18.0	4.35	28.3	277.1	29.3	274	13.5
1,2,3-Trichlorobenzene	1.69	34.7	3.04	21.9	39.2	64.9	30.0	19.3
1,2,4,5-Tetrachlorobenzene	2.05	91.2	5.83	29.9	61.5	82.2	47.3	8.7
Pentachlorobenzene	0.63	36.3	0.72	40.3	51.5	26.1	53.9	42.6

* Soils obtained from the Community Bureau of Reference (BCR), Brussels, Belgium.

Quantitation of Levels of Chlorobenzenes in Soil Samples: Use of the Standard Additions Method

In order to negate the influence of the matrix (soil), separate quantities of soil were spiked with increasing concentrations of each chlorobenzene solution and subjected to headspace SPME–GC–MSD using the optimised conditions. The results shown in Table 1 demonstrate close agreement to the results obtained using aqueous standard solutions.

Results for the analysis of chlorobenzenes in BCR 529 (sandy soil) have recently been reported. Santos *et al.*[13] analysed the sandy soil for several chlorobenzenes including 1,2,3-trichlorobenzene and pentachlorobenzene using headspace SPME and Soxhlet extraction. Experimental data obtained by this group were also compared with data provided by BCR as part of an intercomparison exercise. The results are shown in Table 2. Excellent agreement is reported by this group between their experimental data (both approaches)

Table 2 *Chlorobenzenes in BCR 529 (sandy soil). Results adapted from Santos et al.*[13]

| | Headspace SPME | | Soxhlet | | Intercomparison exercise | |
Compound	Mean mg kg^{-1} (n = 3)	%RSD	Mean mg kg^{-1} (n = 5)	%RSD	Mean mg kg^{-1}	%RSD
1,2,3-Trichlorobenzene	0.591	5.4	0.639	8.1	0.623	10.3
Pentachlorobenzene	1.420	4.9	1.588	5.0	1.326	20.5

and the intercomparison exercise. However, differences are noted with the results obtained in this work.

4 Conclusions

It has been shown that it is possible to determine the level of chlorobenzenes in contaminated soil samples using both aqueous calibration standards and the method of standard additions. Concurrent results were obtained by both approaches.

References and Bibliography

1. Y.F. Li and E.C. Vordner, *Sci. Total Environ.*, 1995, **160**, 201.
2. Hazardous Substances Data Bank, MEDLARS Online Information Retrieval System, National Library of Medicine, 1994.
3. N.I. Sax and R.J. Lewis, Sr. (Eds.), *Hawley's Condensed Chemical Dictionary*, 11th Edn., Van Nostrand Reinhold, New York, 1987.
4. C.R. Pearson, in *The Handbook of Environmental Chemistry, Anthropogenic Compounds*, ed. B. Hutzinger, Part B, Springer, Berlin, 1982, p. 1.
5. B.G. Oliver and K.D. Nicol, *Environ. Sci. Technol.*, 1982, **16**, 532.
6. International Programme of Chemical Safety, Environmental Health Criteria 128, Chlorobenzenes other than hexachlorobenzene, World Health Organization, Geneva, Switzerland, 252 pp., 1991.
7. R. Morlando and S.E. Manahan, *Environ. Sci. Technol.*, 1997, **31**, 409.
8. J.L. Graham, D.L. Hall and B. Dellinger, *Environ. Sci. Technol.*, 1986, **20**, 703.
9. D. Makay, W.Y. Shiu and K.C. Ma, *Illustrated Handbook of Physical-chemical Properties and Environmental Fate for Organic Chemicals. Vol. 1, Monomeric Hydrocarbons, Chlorobenzenes and PCBs*, Lewis Publishers, Boca Raton, FL, 1992.
10. Toxics Release Inventory, Office of Pollution Prevention and Toxics, US EPA, Washington, DC, pp. 84, 94, 1994.
11. R.A.F. Matheson, E.A. Hamilton, A. Trites and D. Whitehead, Surveillance Report EPS-5-AR-80-1, Environment Canada, Environmental Protection Service, Halifax, Nova Scotia, Canada, March 1980.
12. W.N. Beyer, *Bull. Environ. Contam. Toxicol.*, 1996, **57**, 729.
13. F.J. Santos, M.N. Sarrion and M.T. Galceran, *J. Chromatogr. A*, 1987, **771**, 181.

See also

J. Pawliszyn, *Solid Phase Microextraction. Theory and Practice*, Wiley-VCH, New York, 1997.

CHAPTER 19

Analysis of Solid Samples by Hot Water Extraction–SPME

HIROYUKI DAIMON

Solid Phase Microextraction (SPME) was applied to analyse extracts from hot water extraction. Two different SPME approaches, dynamic and static, are described in this chapter. In the dynamic extraction technique, analytes leached by hot water from a matrix were collected in a vial and simultaneously extracted from the water with SPME fibre. In the static approach, hot water extraction of non-polar semivolatile analytes from solid samples was performed using a high-pressure extraction cell with an SPME fibre inserted in the vessel. During the extraction step the cell was heated to release analytes into the water or headspace from the solid matrix. When the cell cooled down, the analytes partitioned into the fibre coating. For quantitative analysis of complex real samples, isotopically labeled internal standards were used to compensate for the partial readsorption of analytes on the solid matrix at lower temperature. Static hot water extraction–SPME method uses only simple apparatus and procedures, and extends the SPME technique to semivolatile analytes tightly bound to solid samples.

1 Introduction

Applications of Solid Phase Microextraction (SPME) include a variety of different analytes in aqueous, gaseous, and solid samples. Concerning SPME extraction from a solid sample, several methods have been used. For the extraction of polar analytes in soil, mixing the soil with water and extracting directly from the water under ambient temperature has been applied using poly(acrylate) coated fibre.[1] Water extraction followed by SPME was found to be a very useful approach for polar compounds, such as herbicides. Volatile non-polar analytes have also been quantitatively extracted using the headspace SPME technique with addition of small amount of water and heating of the matrix.[2] In order to release analytes quantitatively from the matrix into the headspace, extraction temperature was increased. However, the coating/sample distribution coefficients of analytes were decreased, resulting in the reduction of extracted amount. To prevent loss of sensitivity, the internally cooled SPME device was used simultaneously with sample heating

Recently, hot water has been demonstrated to be a potentially useful analytical extraction fluid for a range of polar and non-polar organic chemicals in environmental solids.[3–5] This is possible because the dielectric constant of water decreases with temperature increase.[6] The attractive point in the hot water extraction is that the extraction is performed under milder conditions compared to supercritical water extraction ($T > 374\,°C$, $P > 218$ atm). Polycyclic aromatic hydrocarbons (PAHs) and polychlorinated biphenyls (PCBs) have been efficiently extracted from solid matrices at $250\,°C$ and 50 atm.[3,4] It was found that extraction efficiencies for polar and non-polar organics using water depend primary on extraction temperature of the extraction as long as sufficient pressure is used to maintain the extractant water in the liquid state. In their approach, hot water extraction was performed in a procedure similar to conventional supercritical fluid extraction (SFE). Solid sample was put into an extraction cell. The extraction cell was heated in an oven and extraction was performed at constant pressure with water delivered from a pump. The analytes extracted were collected by inserting the outlet of the restrictor into a vial containing several ml of organic solvent. After extraction, the collection solvent was removed from the water layer in the collection vial and prepared for analysis. The concentrations were then determined using gas chromatography–mass spectrometry (GC–MS) by conventional syringe introduction technique.

The SPME technique has been proven to be a relatively simple and time efficient method for sample preparation. The poly(dimethylsiloxane) coating is very effective in extracting aromatic hydrocarbons and PAHs from water.[7–9] The advantages of hot water extraction and SPME have led to the more recent development of coupling hot water extraction with SPME.[10,11]

In this chapter, the use of SPME to quantify non-polar semivolatile analytes removed from a solid matrix by dynamic and static hot water extraction is described. These methods eliminate organic solvent in the collection step, which is typically used in hot water extraction technique.[3–5] The potential of hot water extraction–SPME–GC for the determination of PAHs is discussed.

In the dynamic approach, the sample placed in an extraction cell was constantly swept with fresh hot water. The analytes removed were collected in a vial and extracted by SPME fibre from water. In the static approach, the extraction was performed in a high-pressure cell with an inserted SPME fibre. Hot water acted as an extraction solvent, facilitating the release of analytes from the solid matrix. SPME fibre, on the other hand, was used to isolate and concentrate the analytes simultaneously from water into the coating. A cool-down period was applied to allow the analytes to partition into the fibre coating more efficiently. For quantitation, the SPME fibre was removed from the collection water or the high-pressure cell, and transferred to a gas chromatograph injection port to analyse extracts without any clean-up or pre-concentration. Both methods enabled determination of PAHs in solid matrices without the use of organic solvents. Static hot water extraction–SPME has an additional advantage of not requiring a high-pressure pumping system.

2 Experimental

Dynamic hot water extraction was performed in a procedure similar to conventional SFE. An ISCO 260D syringe pump (ISCO, Lincoln, NE, USA) was operated in the constant pressure mode to supply water to the system. HPLC stainless steel columns (4.6 mm i.d., 100 mm length) were used as the extraction cells. A Varian 6500 GC oven was used to control the extraction temperature. A standard stock solution containing, naphthalene, anthracene, fluoranthene, pyrene and benzo[a]pyrene was prepared (50 μg mL^{-1} of each PAHs in acetone). This stock solution was spiked to a desired concentration into sand or water. One gram of sand (purified by acid, 50–150 mesh, BDH Inc., Toronto, Canada) was loaded into the extraction cell. In the study of collection methods, the outlet of the extraction cell was connected to a flow restrictor constructed of a 30 μm i.d., 10 cm long piece of fused silica tubing (Polymicro Technologies, Phoenix, AZ, USA). The flow rate was 1.2 mL min^{-1} at 300 atm. For the extraction of urban air particulates (SRM 1649, National Institute of Standards and Technology; NIST, Gaithersbug, MD, USA), a 50 μm i.d., 10 cm long restrictor was used. This restrictor yielded a flow of 1.1 mL min^{-1} at 50 atm. The flow rate was as liquid water measured at the pump.

In order to investigate the collection efficiency after dynamic hot water extraction, 5 μL of standard solution was spiked onto sand loaded in an extraction cell. The water extraction was performed at 300 °C and 300 atm for 15 min. Figure 1 shows a schematic diagram of the collection apparatus for dynamic hot water extraction. Collection of the extracted analytes was performed by inserting the outlet of the restrictor into a 40 mL glass vial containing 10 mL of water with a stirrer bar. All collection vials were silanized prior to use. The collection vial was cooled to less than 5 °C with a cooling bath to avoid analyte evaporation during collection. At the same time, 30 μm poly(dimethylsiloxane) fibre (Supelco Canada, Mississauga, Ontario, Canada) was exposed to the collection water with rapid stirring to extract analytes. After the dynamic hot water extraction, the volume of water collected was adjusted to 30 mL by adding several mL of water. Then, the analytes in the collected water were extracted with the SPME fibre for 70 min with rapid stirring. After SPME the fibre was inserted into a GC injection port and thermally desorbed at 270 °C for 5 min. To establish an equivalent 100% recovery, an appropriate spike of the same standard mixture was spiked into 30 mL water in the 40 mL glass vial. When urban air particulates were investigated, quantitative calibration was performed on the basis of external standard solutions made in pure water. An internal standard was added to both the collected water and the water calibration solution.

For static hot water extraction and simultaneous extraction with SPME fibre, 50–100 mg of urban air particulates were placed in a 20 mL high-pressure cell. The high-pressure cell was constructed of stainless steel in the Science Machine

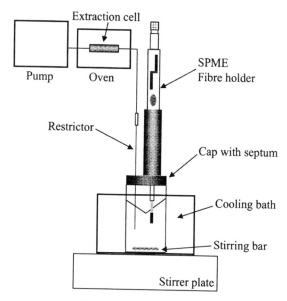

Figure 1 *Schematic diagram of the apparatus for water collection after dynamic hot water extraction*

Shop at the University of Waterloo (Figure 2). It was confirmed that a magnetic stirring bar worked well in the cell. The cell consists of a SLIPPER connector (Keystone Scientific, Bellafonte, PA, USA), a regular 0.4 mm i.d. M-2B GC ferrule (Supelco Canada) and a Viton O-ring (SPAE-NAUR Inc., Ontario, Canada). The O-ring used to seal the cell is stable up to 288 °C, according to the

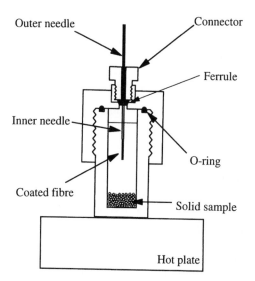

Figure 2 *Schematic diagram of high-pressure extraction cell for static hot water extraction with simultaneous SPME*

manufacturer. Isotopically labeled standards, [10-^{2}H]fluoranthene, [10-^{2}H]pyrene, [12-^{2}H]benzo[*a*]pyrene (Cambridge Isotope Laboratories, Andover, MA, USA) were spiked as internal standards into the air particulates. 18 mL water was added to the 20 mL cell. The remaining 2 mL headspace is important to keep the pressure inside of the cell at the vapor pressure of water at high temperature. So long as the cell has a headspace, the inside pressure does not exceed the steam–water equilibrium pressure. Should the pressure become too high, the ferrule sealing the SPME inner needle fails as the weakest point, thus releasing the pressure. SPME was performed with 30 μm poly(dimethyl-siloxane) fibre. A SLIPPER fitting was closed tightly to seal around the SPME inner needle. Static hot water extraction was performed for 120 min at different extraction temperatures. To heat the cell, a hot plate (Series 400HPS, VWR Scientific, Ontario, Canada) was used. The extraction temperature shown was set at the hot plate. The whole cell was not at the set temperature during the entire heating step. After static hot water extraction was completed, the cell was cooled down for 120 min by switching off the hot plate. Once the extraction was completed, the fibre was withdrawn back into the needle, removed from the cell, and analysed immediately with a GC–MS.

A Varian 6000 gas chromatograph equipped with a flame-ionization detector (FID) was used for the separation and analysis of all compounds, except for those in urban air particulates. A Supelco SPB-5 column (30 m × 0.25 mm i.d., 0.25 μm film thickness) was used for separations. The oven temperature was initially set at 50 °C for 5 min; it was programmed at 15 °C min^{-1} to 310 °C and held for 10 min. The injector and detector temperatures were maintained at 270 °C and 300 °C, respectively. Quantitation of the spiked test compounds was based on GC–FID analysis. Analysis of extracts from urban air particulates by GC–MS was performed with a Varian Saturn gas chromatograph-ion trap mass spectrometer (Mississauga, Ontario, Canada).

3   Results and Discussion

SPME Following Dynamic Hot Water Extraction

To characterize the performance of the 30 μm thick poly(dimethylsiloxane) fibre coating for five PAHs collected in water, several parameters were studied: absorption time profile, linearity and reproducibility. The absorption time profiles for PAHs from directly spiked 30 mL (8.3 ppb) water to a fibre coated with 30 μm poly(dimethylsiloxane) were determined. The absorption time profile is a graph of absorbed amount as a function of exposure time. The exposure time must be long enough for equilibrium to occur or for the rate of absorption to have slowed in order to improve precision. Four PAHs (naphtha-lene, anthracene, fluoranthene and pyrene) reached equilibrium in 60 min. Although the equilibrium was not reached at 60 min for benzo[*a*]pyrene, the absorption time was set at 70 min for the subsequent experiments for practical convenience. The RSDs for three replicates of a 70 min SPME were very good, ranging from 4% to 9%.

Table 1 *Amount of PAHs extracted by SPME from collection water under different conditions after dynamic hot water extraction of spiked sand as % of amount extracted by SPME from spiked water (%RSD)**

Compound	Direct SPME from spiked water A	No cooling, SPME after water extraction B	Cooling, SPME after water extraction C	No cooling, SPME simultaneous with water extraction D	Cooling, SPME simultaneous with water extraction E
Naphthalene	100 (4)	19 (18)	110 (8)	19 (18)	96 (7)
Anthracene	100 (5)	70 (11)	99 (10)	76 (8)	97 (5)
Fluoranthene	100 (6)	101 (6)	107 (6)	102 (6)	111 (8)
Pyrene	100 (9)	102 (6)	103 (6)	102 (10)	111 (8)
Benzo[a]pyrene	100 (4)	158 (29)	50 (8)	162 (18)	102 (7)

*All experiments were performed in triplicate. Amount spiked to sand equals amount spiked to water. A: Extraction with SPME fibre was performed on spiked 30 mL water, without hot water extraction. B: The collection device was with cooling bath and extraction with SPME fibre after the water extraction was complete. C: The collection device was with cooling bath and extraction with SPME fibre after the water extraction was complete. D: The collection device was without cooling bath and with simultaneous extraction with SPME fibre. E: The collection device was with cooling bath and simultaneous extraction with SPME fibre. For B, C, D and E, spiking was performed on sand placed in extraction cell. Extraction was performed at 300 °C, 300 atm and 1.2 mL min^{-1} for 15 min.

The main purpose of the initial investigation was to evaluate SPME after the water extraction and not necessarily to study the hot water extraction mechanism. Therefore, spiked sand was utilized to investigate the collection efficiency. The effects of water cooling and simultaneous dynamic hot water leaching/ SPME on recoveries were investigated (Table 1). The amounts extracted corresponding to four different collection conditions were compared to those obtained for water spiked directly.

Without the use of a cooling bath (Table 1 columns B, D), naphthalene and anthracene were lost during water extraction because of their volatility. In addition, the amount of benzo[a]pyrene extracted was higher compared with spiked water because the temperature of collected water rose to 40 °C, which increased the mass transfer of the analytes between water and the fibre coating. Compounds with lower molecular mass were not affected since they had already reached equilibrium. On the other hand, when water was cooled to 5 °C with a cooling bath (C, E), the amounts extracted of naphthalene and anthracene were almost 100%, but the amount of benzo[a]pyrene extracted was very low without simultaneous extraction with SPME fibre (C). The benzo[a]pyrene extracted might have immediately precipitated from the collection water or adsorbed onto the vial walls. The importance of simultaneous SPME extraction of collection water during dynamic hot water extraction, as opposed to SPME extraction following hot water extraction, was less significant for the other four compounds. When SPME was performed simultaneously with hot water extraction and the cooling bath was used, the recoveries of all the tested compounds were good, ranging from 96 to 111% of amounts extracted from spiked water at room temperature. The extraction amounts also showed excellent reproducibility, with relative standard deviations for all analytes being less than 10% for triplicate extractions of spiked sand.

In simultaneous hot water–SPME extraction (D, E), the total extraction time of SPME was actually 85 min (simultaneous with water extraction for 15 min and SPME extraction for 70 min). Therefore, the extraction with SPME on B, C should have been 85 min after water extraction to be able to compare the extraction amounts between B, C and D, E. The simultaneous extraction (D, E) increased the extraction time by SPME after the water extraction compared with B, C. The collected water was cooled to prevent evaporation of naphthalene during the water extraction, and to control the amount of benzo[a]pyrene extracted by SPME. The results clearly indicate the importance of simultaneous hot water–SPME extraction and cooling of the system.

As an application of the optimized collection method after dynamic hot water extraction, extraction of PAHs from urban air particulates was performed for 15 min at 250 °C and 50 atm (Table 2). The concentration was estimated based on an external calibration curve corresponding to spiked water, and compared with the certified value supplied by NIST. Relative standard deviations were based on triplicate dynamic hot water extractions. Under the extraction conditions, water easily carried the small particles from the extraction cell, through a 0.5 μm frit, to the restrictor. Therefore, restrictor plugging often occurred. In order to prevent this problem, the extraction cell was filled as followed: a filtering fibre (Pyrex wool), 0.2 g of sand, sample and more sand.

The water extraction and optimized collection method provided concentrations above 85% of the concentrations certified by NIST for fluoranthene and pyrene. Since the extraction conditions were not fully optimized, the estimated concentration of benzo[a]pyrene was lower, and the relative standard deviations were also higher, compared with the certified values supplied by NIST. One of the reasons for higher RSDs is associated with the presence of traces of acids

Table 2 *Quantification of PAHs from Urban Air Particulates (NIST 1649) with dynamic hot water extraction**

Compound	Cert. conc. μg g^{-1} (%RSD)	Estimated concentration as % of certified concentration (%RSD): 250 °C, 50 atm
Fluoranthene	7.1 (7)	134.0 (16)
Pyrene	7.2 (7)	87.5 (15)
Benzo[a]pyrene	2.9 (17)	72.0 (29)

* Estimated concentrations versus certified values supplied by NIST. Relative standard deviations are based on triplicate dynamic hot water extractions.

originating from acid washed sand used for prevention of restrictor plugging, which was detected in the extract and interfered in the PAH peaks. The results demonstrate that the present collection technique after dynamic hot water extraction is potentially useful since it does not use organic solvents and includes sample preparation and convenient introduction for subsequent analysis.

SPME Following Static Hot Water Extraction

Hageman *et al.* developed the coupled static hot water extraction–SPME method for extraction of semivolatile organic pollutants from environmental solids.[11] They investigated static hot water extraction in detail. A stainless steel pipe was used as the extraction cell. After sample and water were put into the cell, the cell sealed with a cap was placed in an oven. After 15–60 min extraction, the cell was cooled, the water was removed from the cell into a vial and analytes in the water were determined by SPME–GC–MS. The analytes were quantitatively removed from solids into water at 250 °C. When the water in the cell was cooled, analytes readsorbed to the solid sample. Additional losses occurred when the water was put into the vial for SPME extraction, since the analytes in an over-saturated water phase precipitated to glassware and therefore the quantity of analyte extracted by SPME was possibly reduced.

Static Hot Water Extraction with Simultaneous SPME

For extraction of analytes strongly adsorbed to the matrix, or semivolatile analytes, by headspace SPME, heating of the sample helps to release the analytes from their matrix into the headspace, enhance the mass transfer process, and increase the vapor pressure of analytes. Adding some water to a clay matrix can facilitate the desorption and vaporization of analytes at 50 °C or 100 °C sampling temperature.[12]

Usually, the presence of 5–15% of water in 2 g of a sample contained in a 4 mL vial does not cause significant pressure build up as the sample vial is heated to over 100 °C. On the other hand, a very large amount of water added into the sample (> 50%) produces a pressure increase after heating the vessel, which causes leaking of an ordinary vial design.[2] Therefore, a specially constructed high-pressure container, which allowed fibre insertion, was used in the static hot water experiments. The container keeps the pressure only to maintain the liquid state for effective hot water extraction, since hot water extraction efficiency is only slightly dependent on pressure.

The procedure consists of adding the sample, spiking isotopically labeled internal standards and adding water to the vial, then inserting the fibre and seating it. For quantitative analysis of complex real-world samples, the use of isotopically labeled internal standard is a very effective and accurate quantitation method, since target compounds and the corresponding internal standards have very similar physio-chemical properties. This approach to calibration was used for static hot water extraction combined with simultaneous SPME.

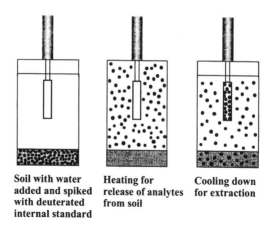

Soil with water	Heating for	Cooling down
added and spiked	release of analytes	for extraction
with deuterated	from soil	
internal standard		

Figure 3 *Mechanism of static hot water extraction with simultaneous SPME process*

Figure 3 shows the mechanism of the static hot water extraction–SPME process. The cell is heated to release the analytes from the solid matrix into the water. When the cell cools down, the analytes partition into the fibre coating and simultaneously partially re-adsorb to the solid matrix.

Urban air particulates were heated to 270 °C and 300 °C for 120 min in the presence of water. The analytes released were extracted with SPME fibre while the cell was cooling down for 120 min. Table 3 shows the effects of different experimental conditions on the relative amounts extracted from the air particulates by the static hot water–SPME technique. Concentrations were estimated based on internal standards, and compared with the certified value supplied by NIST.

For the PAHs tested, this method gave good quantitative results. The estimated concentration of benzo[*a*]pyrene increased with a decreasing amount of the sample. Increasing the extraction temperature to 300 °C did not increase

Table 3 *Quantitation of PAHs from Urban Air Particulates (NIST 1649) with static hot water extraction**

| | | Estimated concentration as % of certified concentration (%RSD) | | |
Compound	Cert. conc. μg g^{-1} (%RSD)	100 mg 270 °C	50 mg 270 °C	50 mg 300 °C
Fluoranthene	7.1 (7)	137 (6)	149 (8)	128 (2)
Pyrene	7.2 (7)	113 (9)	121 (7)	105 (7)
Benzo[*a*]pyrene	2.9 (17)	49 (14)	72 (10)	70 (4)

* Estimated concentrations versus certified values supplied by NIST. Relative standard deviations are based on triplicate static hot water extractions and simultaneous extraction with SPME fibre at each condition.

the estimated concentrations significantly. The results indicated that 270 °C and 50 mg of the sample were the best conditions for quantitation of PAHs with this method. The lower temperature is preferable for the method, because 300 °C is close to the operational temperature limit of the fibre and the O-ring seal of the cell.

Potential degradation of the analytes and the deuterated internal standard must be tested under high temperature conditions because of the reactive nature of hot water. To investigate the possibility of degradation in this study, a standard solution of 3 PAHs (fluoranthene, pyrene, and benzo[a]pyrene), and a standard solution of three deuterated PAHs [10-^{2}H]fluoranthene, [10-^{2}H]pyrene, and [12-^{2}H]benzo[a]pyrene) with the same concentration of each analyte were spiked into water in the cell without sample matrix. The static hot water extraction was carried out under the same conditions as described above. The ratio between a PAH and the corresponding deuterated PAH obtained with SPME was the same as that obtained with solvent injection method for GC–MS.

Poor recoveries were obtained after the fibre was used several times. When the fibre was immersed into the water containing urban air particulates, the particulates were attached to the coating of the fibre so that its color turned to black. In order to clean the coating, the fibre was wiped with tissue paper and rinsed with acetonitrile before conditioning in a GC injection port. Though the fibre was conditioned before each extraction, the lifetime of the fibre was quite short. With membrane protection, direct SPME was used successfully for extraction of analytes from complex aqueous.[13] This technique is useful for static hot water extraction–SPME.

In static hot water extraction, headspace SPME can also be used. Benzo[a]-pyrene was extracted at 250 °C from soil (60 min extraction step and 30 min cooling step) by headspace SPME. The amount of water added into the cell was 12 mL, which means that headspace SPME was the configuration inside the cell at the start of heating. Headspace SPME allows extraction of a variety of matrices without matrix interference.

The above results demonstrate that the hot water–SPME method can be successfully applied to the determination of non-polar semivolatile compounds present in solid samples. The advantage of the dynamic technique is associated with simple external calibration and fibre protection, since the coating is exposed to relatively pure extraction water at low temperature condition. Static hot water extraction with simultaneous SPME, on the other hand, uses only a simple apparatus and procedure, and facilitates extraction of non-polar semivolatile analytes tightly bound to sample matrices. This simple technique preserves advantages of SPME, such as on-site sampling. Additional investigations on the water amount, using water purged with carbon dioxide as modifier, salting-out effect, agitation and cooling method are expected to improve the accuracy of the method. Much work is still required to establish the method as routine.

The financial support of Supelco, Varian and the Natural Sciences and Engineering Research Council of Canada is gratefully acknowledged.

References

1. A. Boyd-Boland and J. Pawliszyn, *J. Chromatogr. A*, 1995, **704**, 163.
2. Z. Zhang and J. Pawliszyn, *Anal. Chem.*, 1995, **67**, 34.
3. S. Hawthorne, Y. Yang and D. Miller, *Anal. Chem.*, 1994, **66**, 2912.
4. Y. Yang, S. Bowadt, S. Hawthorne and D. Miller, *Anal. Chem.*, 1995, **67**, 4571.
5. Y. Yang, S. Hawthorne and D. Miller, *Environ. Sci. Technol.*, 1997, **31**, 430.
6. L. Haar, J.S. Gallagher and G.S. Kell, *National Bureau of Standards/National Research Council Steam Tables*, Hemisphere Publishing Corp., Bristol, PA, 1984.
7. Z. Zhang and J. Pawliszyn, *Anal. Chem.*, 1993, **65**, 1843.
8. D.W. Potter and J. Pawliszyn, *J. Environ. Sci. Technol.*, 1994, **28**, 298.
9. J.J. Langenfeld, S.B. Hawthorne and D.J. Miller, *Anal. Chem.*, 1996, **68**, 144.
10. H. Daimon and J. Pawliszyn, *Anal. Commun.*, 1996, **33**, 421.
11. K.J. Hageman, L. Mazeas, C.B. Grabanski, D.J. Miller and S.B. Hawthorne, *Anal. Chem.*, 1996, **68**, 3892.
12. Z. Zhang and J. Pawliszyn, *J. High Resolut. Chromatogr.*, 1993, **16**, 689.
13. Z. Zhang, J. Pörschmann and J. Pawliszyn, *Anal. Commun.*, 1996, **33**, 219.

CHAPTER 20

Field Analysis by SPME

LAURA MÜLLER

1 Introduction

Recently, the need for fast on-site analysis of field contaminants is rapidly increasing and the long term trend is to conduct analytical investigations in the field.[1-3] The development of a variety of portable instrumentation, such as portable GC coupled to different detectors, chemical sensors and immunochemical techniques, has made possible the screening and the quantitation of pollutants directly at the source.[1] The time of the analysis has been reduced moreover with the introduction of high-speed gas chromatography, with which separation and quantitation of volatile compounds can be achieved at the seconds/minutes scale.[1,4-6]

Some of the most important advantages of field analysis are that it allows real-time decisions, interactive sampling and cost-effective solutions to the problems faced at the time of the investigation. In case of emission of hazardous compounds, from fires or chemical accidents for instance, a fast on-site identification and quantification of the spillage in the environment can prevent a disaster.[7] Rapid on-site screening is also very important in monitoring process emissions, for example at industrial waste plants, where a continuous real-time measurement can warn about accidental spillage of contaminants into the environment (see Chapter 12). Moreover, analysis of environmental matrices directly in the field can avoid any losses and degradation of some components during the collection, transportation and storage of the sample, giving more representative quantitative results.

What Makes SPME Suitable for Field Applications?

Many characteristics of SPME make this technique feasible for field applications. As previously described in Chapter 1 of this book, SPME is based on an equilibrium process in which the volume of the sample does not influence the amount of the analyte extracted, as long as the volume of the sample is very large in comparison with the distribution constant between the fiber and the sample (K_{fs}) multiplied by the volume of the fiber coating (V_f) [$K_{fs}V_f \ll V_s$].

269

If this condition is respected, only the initial concentration of a sample (C_0), together with the volume of the polymer coating (V_f) and the partition coefficient between the fiber and the sample (K_{fs}), are relevant to quantify the analyte in the sample, and the relationship (equation 6), reported in Chapter 1, can be applied. This means that for field quantitations by SPME it is not necessary to know the volume of the matrix sampled and the SPME fiber can be exposed directly to the water of a lake, or the air of a room, in order to quantify the analytes of interest, without having to collect a certain volume of the sample. This is the largest difference between SPME and all other extraction techniques that require a quantitative extraction of a well defined portion of sample collected in the field (SPE, LLE, purge and trap, *etc.*).[5]

Since only a small amount of the analyte is removed from the sample by the fiber, SPME extraction does not disturb the system, so that the real composition of the sample can be evaluated. For example, flavours directly released in the air by a living flower can be extracted and quantified, without removing it from the plant.[8]

Moreover, the SPME device itself is portable (pocket-sized), inexpensive and easy to handle. The absence of solvents and the rapid extraction together with the fast desorption of the analytes from the fiber make the technique feasible to be coupled to fast GC.[4-6]

Several standard procedures are currently used in field to analyse volatiles and semivolatiles in air, water and soil, but different procedures and equipments have to be used, depending upon the matrix and the compound investigated. For example, sorbent materials are usually used for air sampling, 'purge and trap' or 'spray and trap' for volatiles in water, liquid–liquid extraction (LLE) for semivolatiles in water, headspace analysis for volatiles in soil, and so on.[7] SPME allows the same sampling procedure with the same equipment to be used for a large variety of compounds and matrices, which means easier operative conditions even for operators that are not experienced analytical chemists.

The different modes of SPME extraction, described in detail in Chapter 1 (direct immersion, headspace or membrane protected extraction), allow the application of this technique to a wide variety of matrices, even dirty and complex ones. Even heavily contaminated samples, such as dirty industrial wastes or landfill leachates, can be analysed directly in the field by exposing the fiber to the headspace if the target compounds are volatiles, or by protecting the fiber with a membrane for less volatile analytes. In this way the interferences due to high molecular weight compounds in the matrix, such as humic acids, are minimised and the damaging of the fiber is prevented. By SPME extraction, any time-consuming clean-up process for sample preparation of dirty matrices can be avoided, so that a faster characterisation of the sample is achieved.

Field gaseous samples, such as outdoor or indoor air, can be easily and rapidly equilibrated with the fiber, since the natural flow of air is frequently sufficient for fast equilibration. The analysis of aqueous samples sometimes needs improvement in sample agitation, even though a moving water stream, as a river or an industrial wastewater discharge, could provide the necessary agitation of the sample.[9] It is more difficult to analyse solid samples directly on site, since a warm up is often required to release the analytes from the matrix;[7] hot water extraction, described in detail by Daimon in Chapter 19, represents a valid application of SPME for solid analyses, and the procedure can be optimised for field investigations.

From this rapid overview, SPME seems to be ideal for field applications. This chapter will review some of the most recent developments of SPME for field analyses, in order to emphasise the promising perspectives of this technique for the future *in situ* investigations.

2 Field Applications

Many aspects have to be considered in order to ensure reliable results from field investigations by SPME. In order to understand the advantages of field SPME applications, it is important to compare quantitative results obtained in the field with results of traditional sampling methods that require storage and transportation of the sample to the laboratory for instrumental investigation. Field SPME investigations of chlorinated hydrocarbons in water samples have been compared with traditional methods.[10] In the study, 62 samples were analysed both in the field by SPME (PDMS–DVB)–portable GC and in the laboratory by standard methods, showing results with an average difference of 18.7%; more field investigations are in progress to produce more accurate results. This is a very good agreement taking into account that the analysis was carried out in the field, and it was shorter by an order of magnitude than laboratory analysis.

The time between sampling and instrumental analysis has to be as short as possible, and, even if storage after on-site sampling is required, losses of analytes from SPME fibre have to be somehow minimised. For this reason, the coupling to portable instrumentation and the optimisation of volatile storage onto the fiber are extremely important for field applications of SPME.

Development of New SPME Portable Samplers for Field Applications

If on-site sampling is carried out in a location where the transportation of a portable instrument is difficult, or the investigation requires an instruments that is not portable, such as GC–MS, even the short time to reach the nearest laboratory can be crucial to give representative results. An important aspect in these cases is to preserve the integrity of the sample in the period of time between sampling and instrumental analysis. This is the reason why an increasing importance has been given to the optimisation of the storage of volatile compounds on SPME fibers. A proper SPME device for field sampling should guarantee storage of the analytes, and should also be simple, easy to handle and to transport, inexpensive and, in some cases, disposable, in order to avoid any cross contamination due to multiple extractions of dirty matrices. Our challenge is to make SPME a tool that can always be kept in your pocket,

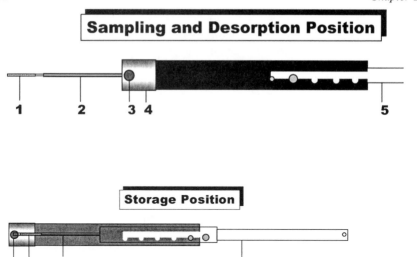

Figure 1 *Schematic description of Supelco field sampler: (1) SPME fiber, (2) needle,
(3) Thermogreen-LB2 septum (Supelco), (4) aluminium nosepiece, (5) plunger*

ready to be used whenever and wherever you are interested in knowing the
composition of a gaseous, liquid or solid sample.

To date, different approaches have been reported in the literature to prevent
analyte losses from the SPME fiber during storage.[11] A field SPME sampler,
commercialised by Supelco, is presented in Figure 1.[12] In this semi-disposable
device, where the lifetime of the device is related to the longevity of the fiber that
is not replaceable, storage is ensured by retracting the tip of the needle (2) into
the sealing septum (3) (Thermogreen-LB2 septum, Supelco) placed into the
aluminium nosepiece (4) of the device.

Two alternative devices were designed and built for field sampling (Figures 2
and 3). Figure 2 presents an aluminium body field sampler with a retractable
needle and a leaf closure. The barrel, kept in place by a thumbscrew (3), allows
the regulation of the length of the exposed part of the needle, while an internal
plunger (6), guided by a Z-slot, allows the exposure of the SPME fiber (11). The
tip of the needle, in the storage position, is squeezed by a nylon leaf closure (8),
which is tightened simply by screwing a finger-tight nut (9). The commercial
fiber SPME assembly (4) has been mounted on this prototype, just removing the
coloured hub from the internal tubing (5). The ability to use the commercial
fiber assembly is very important in order to ensure a good reproducibility of the
fiber performance. For sampling, the nut (9) is slightly unscrewed to allow the
two halves of the closure (8) to open and the SPME needle to be pulled out. To
protect the fiber during air or water sampling, a removable nylon shield (10) can
be optionally mounted on the nut (9). With this 'pen-shaped' device, no cap has
to be removed from the body of the sampler in order to expose the fiber, so that
sampling can really be as easy as taking a pen from your pocket!

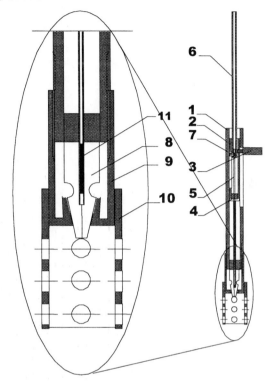

Figure 2 *SPME field sampler with nylon leaf closure: (1) aluminum body, (2) movable barrel, (3) thumb screw, (4) SPME fiber assembly, (5) inner tubing, (6) plunger, (7) set screw, (8) leaf closure (nylon), (9) finger-tight nut, (10) nylon shield, (11) SPME fiber*

Figure 3 presents a rugged prototype of a cheap and disposable field sampler, with a nylon body (1) and a tight Teflon cap (9) that seals the needle during storage. A nylon shield (7) can be mounted to protect the SPME needle (8). The barrel (1), guided by a small screw (5) that moves in a Z-slot (6), regulates the exposure of the SPME fiber during sampling. A commercial SPME fiber assembly can be mounted on this device as well.

If required, the fiber can be replaced in both of the two prototypes described above.

Storage capacity of these new field samplers was compared with that of the SPME manual holder sealed by a Thermogreen-LB2 septum (Supelco)[11] and of the Supelco SPME field sampler (Figure 1).[12] Both of them use a silicon rubber septum to seal the tip of the needle where the fiber is retracted during storage. The devices in Figures 2 and 3 were designed to avoid the use of a septum to seal the SPME fiber. It has been demonstrated that contamination of the fiber can occur from the septum rubber itself or from the outer environment: if a clean blank fiber (PDMS 100 μm) is kept at room temperature sealed with a new

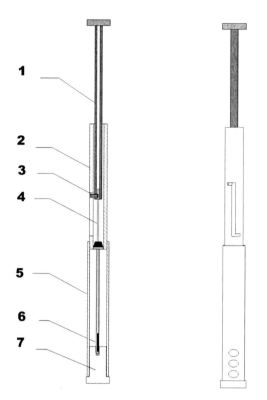

Figure 3 *Disposable SPME field sampler with Teflon: (1) plunger, (2) nylon body,*
(3) set screw, (4) SPME fiber assembly, (5) nylon shield, (6) SPME fiber,
(7) Teflon cap

Thermogreen-LB2 septum (Supelco), after 24 hours some unknown contaminants interfere with the chromatographic investigation (Figure 4).

The field samplers investigated in this study, and listed in Table 1, were tested
for storage in the laboratory with gaseous standards of compounds of a wide
range of volatility: chloroform, 1,1,1-trichloroethane, benzene, toluene, tetrachloroethylene and 1,1,2,2-tetrachloroethane. Air standards (30 μg L^{-1}),*
prepared freshly for each different experiment in a 1 L glass bulb, were extracted
by the SPME fiber for 3 minutes in static conditions. Different coatings and
several storage parameters were investigated (Table 2). For cold storage, the
device was maintained at low temperature for the entire period; only the storage
in dry ice required *ca.* 1 minute of warm up of the device body at room
temperature before injecting, to avoid any mechanical problem in the extraction
of the fiber from the needle. At the scheduled time ($\pm$ 3 s), after storage, the fiber
was desorbed in the GC injector and the analysis by GC–FID (Varian 3400)
was carried out. A SPB-5 column (30 m, 0.25 mm, 1 μm film thickness) from

---

* Standards of 60 μg L^{-1} were used for experiments with PDMS 100 μm, due to a lower response of
  this fiber toward the analytes investigated.

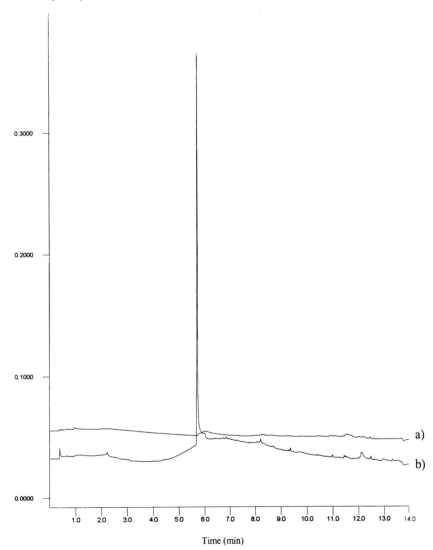

Figure 4 *Contamination of a PDMS* 100 μm *fiber due to a* 24 h *sealing with a rubber septum (Thermogreen-LB2, Supelco): (a) chromatogram of a clean fiber blank; (b) chromatogram obtained by desorption of the same SPME fiber after* 24 h *of sealing at room temperature* (24 °C) *with the septum*

Supelco was used for separation of this volatile compounds with the following GC parameters: oven temperature program 40 °C for 2 min, 15 °C min^{-1} to 150 °C hold for 1 min; injector temperature 250 °C; carrier gas pressure (H$_2$) 30 psi.

Interesting results were obtained by comparing storage capacities of the selected sealing methods. The most interesting results are presented by the

Table 1 *Details of the field samplers investigated for storage and compared in the study*

Sampler	Description	Sealing material	Available coatings	Further details
Device 1	SPME manual holder (Supelco) + septum	Silicon rubber septum Thermogreen-LB2 (Supelco)	Any of the commercial SPME fibers	Ref. 11
Device 2	Prototype of aluminium holder with leaf closure	Nylon leaf closure	Any of the commercial SPME fibers	Figure 2
Device 3	Prototype of nylon holder with Teflon cap	Teflon cap	Any of the commercial SPME fibers	Figure 3
Device 4	Supelco field sampler	Silicon rubber septum Thermogreen-LB2 (Supelco)	PDMS 100 μm CX/PDMS 75 μm	Ref. 12

Table 2 *Storage conditions tested in this study with the selected field samplers*

SPME fibers	Storage temperatures	Storage periods
PDMS 100 μm	Room temperature (24 °C)	5 min
PDMS/DVB 65 μm[a]	Fridge temperature (2–8 °C)[b]	30 min
Carboxen/PDMS 75 μm	Dry ice temperature (-70 °C)	3 h
		24 h

[a] Supelco field sampler (Device 4) was not available with PDMS/DVB 65 μm fiber.
[b] Fridge temperature was not constant during all the experiments: the monitored temperature was between 2 and 8 °C.

PDMS 100 μm fiber, which is the coating that showed the lowest capacity to retain these analytes after extraction. The best storage was achieved by sealing the needle with a Teflon cap, as can be seen in the bar graph in Figure 5, which shows the amount (%) of selected compounds retained by a PDMS 100 μm fiber after refrigerated storage (2–8 °C) for different periods of time. Even after 24 hours from the extraction, 89% of 1,1,2,2-tetrachloroethane was still on the fiber of the Teflon capped device. As expected, the most volatile analytes are retained to a lesser extent by the fiber during storage. Chloroform could not be stored at fridge temperature by PDMS 100 μm for 24 hours with any of the samplers, with the exception of the Teflon capped sampler that still retained 38.2% of the compound on the SPME fiber. The low storage capacity of the SPME holder sealed with septum and of the Supelco field sampler, especially for long periods, confirmed that silicon rubber septa are not suited to prevent volatile releases from the fiber: the silicon rubber of the septum itself could absorb these volatile compounds, since it is substantially the same polymer as

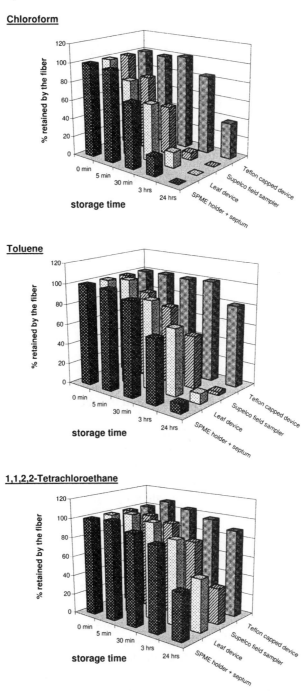

Figure 5 *Amount (%) of some selected compounds retained by a PDMS* 100 μm *SPME fiber, after refrigerated storage (2–8 °C) for different storage periods. Percentages are calculated over the* 0 min *storage area counts*

the fiber coating (PDMS) and a partition between the SPME fiber and the septum can occur.

CX/PDMS 75 μm coating showed a high storage capacity itself, no matter how the needle was sealed. For instance, even after 24 hours of storage in dry ice, over 70% of chloroform was still on the fiber with any of the devices investigated. In this case a good seal is important to prevent any contamination from the environment during storage, since in this study a significant increase of small unknown peaks in the chromatograms after storage, was observed. The good storage capacity of the CX/PDMS fiber is also reported in the literature,[12] where it has been demonstrated that a 3 day storage, even of very volatile compounds, can be achieved without considerable losses from the fiber.

Even with a PDMS/DVB 65 μm fiber, storage with Teflon capped device is better than with the other samplers. In general, storage capacity with this coating is consistently enhanced by decreasing the temperature of storage: after 24 h of storage at room temperature only 19% of chloroform was still found on the fiber, whereas in dry ice almost 90% of chloroform was retained by the SPME fiber of device 3.

In general, from the promising storage performances of the Teflon capped device in this study, we can assure that Teflon, as a sealing material, could be a valid alternative to other methods to achieve a complete isolation of the SPME fiber from the environment, avoiding any contamination problem, which occurs when using rubber septa to seal the needle. The operating conditions of the prototype with the nylon leaf closure are more rapid and practical since no cap has to be placed/removed for sampling and storage, even if the design has still to be optimized in order to avoid any loss from the cut between the two halves of the closure (see Figure 2).

The concept of storing the SPME fiber inside the barrel of a sealed gas-tight syringe, together with a small volume of the air sample, could be applied to preserve the integrity of volatile samples on the fiber during storage. In this way, even if volatiles are released by the fiber, they will be injected in the GC, together with the air in the barrel.[13] The idea will be optimised and a selected device will be developed for this purpose in the future.

These prototypes will be tested in the field for sampling of real matrices, in order to confirm the results obtained in the laboratory.

In situ Analysis of Groundwater and Soil Gas by SPME

In situ measurements of volatiles in groundwater and soil are of utmost importance since relevant losses can occur during sampling, pumping or excavation. The introduction of SPME to monitor underground pollution has been demonstrated to be an effective alternative to other methods that showed some limitations for in-field applications.[14] A device designed for this purpose (Figure 6) was developed and tested in the field for the determination of volatile organic compounds in groundwater and soil gases.[14] The SPME sampling probe described in literature was lowered into the well, to reach groundwater for direct sampling, or was placed in the headspace, close to the surface of the

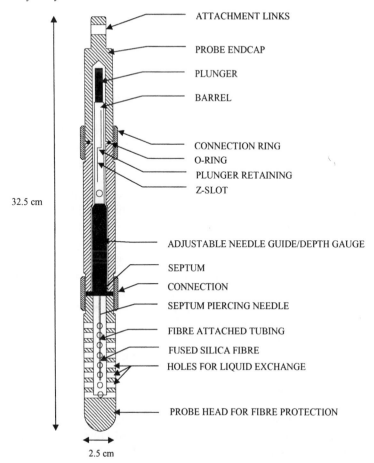

ATTACHMENT LINKS

PROBE ENDCAP

PLUNGER

BARREL

CONNECTION RING
O-RING
PLUNGER RETAINING
Z-SLOT

32.5 cm

ADJUSTABLE NEEDLE GUIDE/DEPTH GAUGE

SEPTUM

CONNECTION

SEPTUM PIERCING NEEDLE

FIBRE ATTACHED TUBING

FUSED SILICA FIBRE

HOLES FOR LIQUID EXCHANGE

PROBE HEAD FOR FIBRE PROTECTION

2.5 cm

Figure 6 *Design of a prototype SPME device for* in situ *sampling of groundwater and soil gas*
(Adapted by permission from reference 14)

water, to monitor underground soil gas. After the extraction step, the probe was retracted to the surface and the SPME fiber was immediately desorbed in the GC of a mobile laboratory. A comparison between underground SPME sampling and samples collected from the same wells with a traditional procedure and analysed by SPME in a mobile laboratory was performed. Various VOCs were detected in the groundwater of the wells investigated. Results showed that *in situ* SPME extractions of groundwater provide, in general, higher results than traditional sampling methods, where the pumping of the sample on the surface causes volatile compound losses.[14] Toluene and naphthalene were detected by underground soil gas investigation and their presence was confirmed by extracting the same samples with Tenax tubes.

A second prototype probe was also designed to fit directly in the head of a cone penetrometer, so that sampling can occur even during the excavation procedure, and different ground levels can be monitored. For further details the reader is addressed to the reference.[14]

Optimisation of SPME Fast–GC System for Field Trace Analysis

The rapid high-speed GC could be exploited to increase the advantages of field analyses for monitoring and real-time screenings.[1] The use of SPME is particularly indicated for fast-GC, since it is a solvent-free technique and also thin coatings can provide a very fast desorption of analytes at high temperatures. Some effective instrumental modifications were performed to date, allowing a successful coupling of SPME to fast-GC.[4–6]

An SPME–fast-GC portable system was optimised for field analyses.[4] A proper injector, designed for the purpose, was built and mounted on a commercial SRI portable gas chromatograph (model 9300B) (Figure 7). The portable instrument SRI model 8610C, equipped with this injector, is presently commercially available. A capacitive discharge allowed very fast heating of the desorption area of the injector (4000 °C s^{-1}), and very narrow injection bands were observed, as required by fast-GC. The system was tested in laboratory with standards of BTEX and purgeables A, and then was carried to the field for trace analysis of trichloroethylene (TCE) in soil.[4]

For field determinations a photoionization detector (PID) was used, so that fast analysis was possible without the use of a make-up gas. The detector, mounted on the portable system, was modified in order to reduce its internal volume, and an excellent improvement in resolution for separation of BTEX was achieved.[4]

Field determinations were performed by quantifying TCE in clay samples and the migration of the pollutant was investigated by collecting core samples from depths up to 5 m. Soil samples were extracted with MeOH, then an aliquot of the solvent was used to spike pure water, and the aqueous solution was then vigorously stirred to equilibrate with the headspace. A PDMS/DVB (65 μm) SPME fiber was exposed to the headspace of the static water sample for the

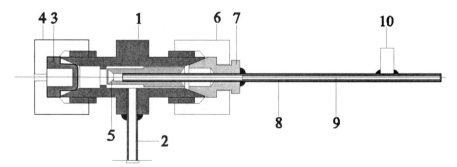

Figure 7 *Modified injector for SPME/field portable fast GC: (1) nut, (2) molded septum, (3) needle guide, (4) modified Swagelok fitting, (5) staintess steel tubing, (6) nut, (7) blind ferrule, (8) stainless steel tubing, 9–0.53 mm i.d. fused silica capillary, (10) electrical contact*

extraction. Then instrumental analysis was successfully carried out with the portable system described above. The entire process of extraction of the spiked water samples, desorption and instrumental analysis took 3 minutes to be completed, and over 500 samples could be quantified in 10 days. This study is an example of an effective application of SPME coupled to fast-GC for rapid field investigations. For further details the reader is referred to reference 4.

Field Analysis of Indoor Air

One of the most interesting applications of SPME in the field is for industrial hygiene purposes. Industrial hygienists are concerned about monitoring the levels of hazardous and undesirable exposures to which workers are subjected during their working day. A proper monitoring equipment for this purpose should be portable, and should give immediate and continuous results, alerting of excessive exposure.[15]

Air sampling by SPME can provide information both on the momentary concentration at the sampling location, if the exposure of the fiber is within second/minute range (*grab sampling*), and also on the time-weighted average indoor concentration, if the fiber is kept retracted in the assembly for a long sampling period, in the range of hours (*time-weighted average sampling*), as described in Figure 8.[16,17] For further details on this topic the reader is addressed to Chapter 11.

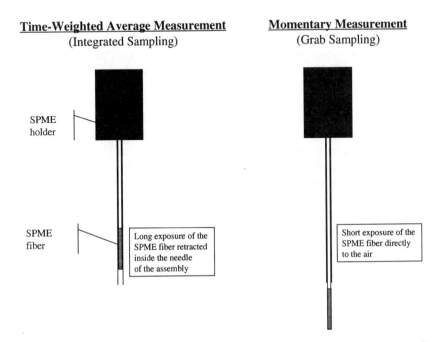

Time-Weighted Average Measurement
(Integrated Sampling)

Momentary Measurement
(Grab Sampling)

SPME holder

SPME fiber

Long exposure of the SPME fiber retracted inside the needle of the assembly

Short exposure of the SPME fiber directly to the air

Figure 8 *Time-weighted average measurement and momentary measurement procedures for indoor air sampling by SPME (for further details see Chapter 11)*

For indoor air analysis, SPME, coupled to high-speed GC, has been demonstrated to be an effective and powerful alternative to other traditional air sampling methods, both active or passive, such as sorbent material badges or sorbent traps.[16]

To date, two field investigations have been performed by SPME for indoor air analysis. Airborne styrene has been analysed in field by SPME,[16] and field analysis of airborne formaldehyde with on-fiber (PDMS/DVB) derivatization with o-(2,3,4,5,6-pentafluorobenzyl) hydroxylamine (PFBHA) was performed.[18,19]

Field sampling of airborne styrene was performed with a PDMS 100 μm SPME fiber.[16] After field sampling, the sorbent materials were stored in dry ice and transported to the laboratory where instrumental analysis was carried out, within three hours of the sampling. SPME grab and integrated sampling were compared with active sampling methods, such as charcoal tubes and a portable photoionisation detector (PID), and also with passive sampling methods, such as 3M badges. The SPME fiber was either exposed for 5 minutes to the air for momentary measurements, or was kept retracted in the needle for 30 minutes for integrated sampling of styrene in air. Results showed that SPME, either for 5 or 30 minutes sampling, can be a valid alternative to charcoal tubes or other traditional sampling methods for styrene analysis in air, with the advantages that SPME is simple, solventless, cheap, fast, sensitive, it does not require a power supply and it is easy to automate and to transport to the field.[16]

A valid approach to the analysis of airborne formaldehyde (HCHO) by SPME with on-fiber (PDMS/DVB) derivatisation with o-(2,3,4,5,6-pentafluoro-benzyl)hydroxylamine (PFBHA) has been recently described.[18] This method was also tested in the field by monitoring airborne formaldehyde in the animal pathology sample storage room of the University of Guelph (Guelph, ON, Canada).[19] The PDMS/DVB fibers utilised in this study were loaded with the derivatising agent in the laboratory, and then stored in an empty 1.8 ml Teflon capped vial for the transportation to the field. After integrated sampling (420 minutes), the fibers were stored retracted 3 cm deep into the assembly needle inserted in the same vial, and carried to the laboratory for instrumental analysis.

In order to also evaluate a momentary concentration at the same sampling locations directly in field, grab sampling with the SPME fibers, loaded with the same derivatising agent, was performed by exposing the fiber to the indoor air for 30 seconds. Fast GC separation was carried out on-site, using the same portable fast-GC/PID (SRI model 9300B), already described in this chapter. In a 3 minute procedure, including sampling, separation and detection, the concentration was evaluated in the field. Even in this investigation a comparison of both grab and integrated SPME sampling with traditional sampling methods, such as passive sampling with 3M badges or active NIOSH Method 2541 sorbent tubes sampling, was performed. Results indicate that both a momentary measurement and a time-weighted average measurement of formaldehyde in air can be successfully performed in field by SPME with on-fiber derivatisation and there is a very good agreement with values obtained with traditional sampling methods.[16,19]

Acknowledgements

Tadeusz Górecki and Perry Martos are gratefully acknowledged for their invaluable help and support during the preparation of this chapter, and for providing some of their results and material reported by the author herein. Thanks to Torben Nilsson for the precious information on his work. Darryl Basset is acknowledged for personal communications about some preliminary results of his field studies. The Gordon and Breach Publishers and the OPA (Overseas Publishers Association), together with the authors of reference 14, are acknowledged for permission to use some material in this chapter.

References

1. V. Lopez-Avila, *Anal. Chem.*, 1997, **69**, 289R.
2. E.B. Overton, H.P. Dharmasena, U. Ehrmann and K.R. Carney, *Field Anal. Chem. Technol.*, 1996, **1**, 87.
3. T. Kotiaho, *J. Mass Spectrom.*, 1996, **31**, 1.
4. T. Górecki and J. Pawliszyn, *Field Anal. Chem. Technol.*, 1997, **1**, 277.
5. T. Górecki and J. Pawliszyn, *Anal. Chem.*, 1995, **67**, 3265.
6. T. Górecki and J. Pawliszyn, *J. High Res. Chromatogr.*, 1995, **18**, 161.
7. G. Matz, W. Schroeder, A. Harder, A. Schilling and P. Rechenbach, *Field Anal. Chem. Technol.*, 1997, **1**, 181.
8. B.D. Mookherjee, personal communication to Prof. J. Pawliszyn, 1996.
9. R. Eisert and K. Levsen, *J. Chromatogr.*, 1996, **737**, 59.
10. D. Basset, personal communication, 1998.
11. M. Chai and J. Pawliszyn, *Environ. Sci. Technol.*, 1995, **29**, 693.
12. R. Shirey, V. Mani and R. Mindrup, *Am. Environ. Lab.*, 1998, **1–2**, 21.
13. Z. Zhang and J. Pawliszyn, *J. High Resolut. Chromatogr.*, 1996, **19**, 155.
14. T. Nilsson, L. Montanarella, D. Baglio, R. Tilio, G. Bidoglio and S. Facchetti, *Int. J. Environ. Anal. Chem.*, 1998, **69**, 1.
15. M. Harper, C.R. Glowacki and P.R. Michael, *Anal. Chem.*, 1997, **69**, 307R.
16. P.A. Martos, *Air Sampling with Solid Phase Microextraction*, Ph.D. Thesis, University of Waterloo, Waterloo, Ontario, Canada, 1997.
17. J. Pawliszyn, *Solid Phase Microextraction. Theory and Practice*, Wiley-VCH, New York, 1997.
18. P.A. Martos and J. Pawliszyn, *Anal. Chem.*, in press.
19. T. Górecki and P.A. Martos, personal communication, Waterloo, 1998.

CHAPTER 21

Organometallic Speciation by Combining Aqueous Phase Derivatization with SPME–GC–FPD-MS

JOSEP M. BAYONA

1 Introduction

Organometallics are the molecules which contain at least one transition metal–carbon bond. The interest in their determination in environmental samples began in the seventies following environmental accidents such as in Minimata in Japan where methylmercury bioaccumulated in fish leading to a massive intoxication in the inhabitants consuming it. There are other cases where organometallic pollution had important impacts at the socioeconomic level such as the oyster production collapse in Arcachon (France) due to the tributyltin used in antifouling paints. More recently, organometallic species have been demonstrated to play a key role in the biogeochemical cycles of metal ions[1] and their fate and toxic effects are dependent of the chemical species.[2]

Early analytical techniques used for organometallic speciation appeared in the mid-eighties but they did not become reliable until the present decade when commercially available analytical instrumentation was commercialized. Particularly important in the success of organometallic speciation is the interfacing between chromatographic and elemental selective techniques.

The objective of this chapter is to present the state of the art of the applications of SPME in the field of organometallic speciation by using commonly available analytical techniques such as GC–FPD or GC–MS and the figures of merit of these techniques will be compared with those of elemental selective detection techniques. In this work, we will focus on mercury, lead and tin since their organometallic species are relevant from the toxicological point of view. The application of elemental selective analytical techniques (AAS, AED, ICPMS) can expand the number of metal ions analyzed and it will be presented in the next chapter.

Table 1 *Sources of commonly occurring organometallics in the environment*

Source	Application	Tin	Mercury	Lead
Anthropogenic	antifouling	tributyl	–	–
	pesticides	phenyl, cyclohexyl	–	–
	antiknocking agents	–	–	tetraethyl
Bacterial	–	methyl	methyl	–
Chemical	–	methyl	methyl, ethyl	–

Organometallics in the environment can originate in industrial processes or in the biogeochemical cycles of each element (Table 1). Once in the environment, they suffer biotic and/or abiotic degradation processes which lead to the formation of metal ionic species. The behavior of metal ions and organometallic species in the environment and their toxicity is dependent on their chemical structure. Therefore, there is great deal of interest in the development of analytical techniques for their determination according to the chemical species occurring in the environment to provide a more realistic risk assessment. Moreover, organometallic species are usually toxic at very low concentration levels and, therefore, analytical techniques should provide higher sensitivity than the non-effect level.

Several organometallic compounds such as butyl- and phenyl-tins are included in the EU priority pollutant list and they are target analytes in the national monitoring programs since their occurrence is regulated in several international protocols (Rhine Basin program, Mediterranean protection against land based sources of pollution, *etc.*). Furthermore, most of the developed countries have enforced the usage of tin based antifouling paints only to vessels larger than 25 m. In order to improve the quality in their measurement, International Bodies such as the European Union through the Standards, Measurements and Testing, the National Research Council of Canada and National Institute of Environmental Studies of Japan have produced sediment and biota standard reference materials (SRMs) with certified values for butyltins, ethyllead and methylmercury (*i.e.* mussel, tuna fish, urban dust, *etc.*). Some of these reference materials have been used for the validation of the developed SPME procedures.

Since concentrations of target organometallics are below the detection limit of the analytical techniques, a preconcentration step is mandatory before the determination by GC or LC coupled to a variety of detection systems (FID, ECD, FPD, AFS, MSD) and elemental selective techniques (AAS, ICPOES, ICPMS). However, a higher sensitivity is achieved by GC techniques following a derivatization step to yield volatile species.[3]

While SPME application is getting established in the determination of volatile organic analytes, its application to organometallic speciation is still in its infancy. The main difficulty found in the application of SPME to organometallic speciation is the lack of volatility of the analytes since they occur as ions in environmental samples. Furthermore, the high toxicity of organometallic species requires the determination of extremely low concentrations, and thus large preconcentration factors are necessary in the extraction step. However, since most of the classical methods of organometallic speciation are time consuming, requiring large volumes of toxic reagents such acids and solvents, the inherent advantages of SPME make it a clear alternative in sample preparation and several authors have already evaluated its application to environmental studies.

In this work, we will present the state of the art of different approaches that have been developed for organometallic determination using SPME combined with GC techniques. Sample pretreatment, derivatization and sampling of the derivatized species will be discussed and applications to lead, mercury and tin speciation will be presented. Further developments on the application of SPME to organometallic speciation will be also addressed.

2 Sample Pretreatment Prior to SPME

Sample pretreatment is a critical issue in organometallic speciation since the stability of the species is rather limited. Degradation can occur due to biotic (*i.e.* biodegradation) or abiotic (*i.e.* photolysis, chemical reaction) transformation processes. Another important aspect is the adsorption of analytes in the walls of the sample container. Furthermore, the nonionic organometallic species are intrinsically volatile and can be lost during sample transport or storage in the laboratory before determination.

Although in the case of soil, sediments and biota most of the former problems are avoided if the sample is frozen immediately after sampling, in the case of aqueous samples this is troublesome. Several authors have investigated aqueous sample storage to minimize the degradation–adsorption processes which can lead to biased results. Most of the commonly used sample containers such as glass, polypropylene, PVC, PTFE, polycarbonate *etc.*, adsorb organotin compounds to different degrees on the wall even at low storage temperatures and at acidic pHs.[4] SPME can be a promising approach for field sampling of organometallic species since they can be adsorbed on the SPME fibers, where they should be more stable than in the aqueous matrices, and the transportation and storage expenses can be minimized. Our results show that solutions of organotin chlorides, particularly monobutyl- and monophenyl-tin, suffer of adsorption–degradation processes independently of the storage conditions (-20, -4, $25\,°C$) evaluated. Further research is needed to evaluate the possibility of field sampling with SPME for organometallic speciation but it is attractive from the easier sample handling point of view.

Aqueous Samples

The determination of organometallic species in aqueous samples usually involves surrogate addition, acidification to an appropriate pH (2–4), filtration through 0.4–0.7 μm to remove the suspended particulate matter, pH adjustment and derivatization. The latter step will be developed in the following section of this chapter. Since some organometallic species have strong affinity for particles such as OTs,[5] their determination in the particulate phase becomes necessary.

Solid Matrices

The determination of organometallic species in abiotic matrices involves an extraction step to release the contaminants adsorbed onto the matrix. Only highly volatile species (*i.e.* $HgMe_2$, SnH_4, $SnMe_4$, Et_4Pb) can be extracted from the headspace in equilibrium with the dried solid sample; otherwise a leaching step is mandatory prior to the SPME extraction. The leaching procedure depends on the matrix but usually involves acid or basic conditions followed by a pH adjustment to perform the derivatization reaction. The determination of mercury in biotic matrices has been carried out by basic digestion under sonication.[6]

3 Derivatization Reactions Used in Organometallic Speciation

Since most organometallic species lack volatility, a derivatization step is mandatory prior to GC determination techniques. Furthermore, the derivatization reaction allows one to obtain lipophilic species which can be easily preconcentrated by SPME with nonpolar polymers (see Section 4).

Alkylation Reactions

Although a variety of alkylation reactions have been used in organometallic speciation (*i.e.* methylation, ethylation, pentylation and hexylation, *etc.*), ethylation with $NaBEt_4$ is the most widely used in tin,[7] mercury[6,8] and lead[9] speciation because it can be performed in the aqueous phase whereas other alkylation reactions are performed with Grignard reagents demanding strictly anhydrous conditions in non-protic solvents. Such media are not appropriate for SPME analysis and the ethylated derivatives have significant volatility to allow headspace sampling. Furthermore, methylated derivatives should have higher volatility but this is not useful in the speciation of real samples since methylated species can already occur in the samples making speciation studies impossible. Phenylation in the aqueous phase with $NaBPh_4$ is another alternative but it has not been fully explored.

The proposed reactions for the ethylation of mercury, tin and lead with SPME are indicated here:

$$NaB(C_2H_5)_4 + RHg^+ \rightarrow (C_2H_5)HgR + (C_2H_5)_3B + Na^+$$

$$3NaB(C_2H_5)_4 + RSn^{3+} \rightarrow (C_2H_5)_3SnR + 3(C_2H_5)_3B + 3Na^+$$

$$4NaB(C_2H_5)_4 + 2Pb^{2+} \rightarrow (C_2H_5)_4Pb + 4(C_2H_5)_3B + Pb + 4Na^+$$

The ethylation reaction yield depends on the pH of the medium. Usually the acetic/acetate buffer is used to bring the pH to about 5. Cai and Bayona[7] have optimized the pH range for the ethylation reaction of organotins. They found that the derivatization reaction can be carried out over a pH range of 4–8.5. At more acidic pH, $NaEtB_4$ is hydrolyzed rapidly and the ethylation yields are poor. The amount of $NaEtB_4$ is strongly dependent on the matrix since it can react with the organic matter. In addition, alkaline ions also might depress the derivatization yield which can be particularly relevant in case of speciation studies in seawater. In the case of aqueous matrices, the amount needed is in the range of 0.8–1.0 mL of 1% (w/v) solution for a 150 mL sample. The reaction time is around 5 min under mechanical agitation.

In the case of mercury speciation in biotic samples,[6] a basic digestion is performed to digest biota tissues and then an aliquot of the digested liquor is taken and the ethylation reaction is performed in the buffered acidic media. In order to release organotin compounds from abiotic matrices, acid hydrolysis with hydrochloric or acetic acids yields the corresponding chlorides or acetates, respectively.[3,10]

Other reagents have also been evaluated for mercury methylation[11] (see Table 3). The $K_3[Co(CN)_5CH_3]$ derivatization reagent gave a high derivatization yield (80%) of the monomethylated species under highly acidic oxidative conditions. However, the main disadvantage of this reagent is the formation of methylated derivatives which can occur in several matrices such as fish since it is easily bioaccumulated. Therefore, this procedure is only useful in matrices where methylmercury is not expected.

Hydridization Reaction

The hydridization reaction can be also performed in aqueous media and leads to the formation of derivatives more volatile than the ethylated counterparts but they are less stable and they have not been evaluated yet in speciation studies using SPME. They have been used in tin, arsenic and lead speciation in on-line systems combining purge and trap and atomic spectrometry detection. However, it is reported that the matrix can affect the hydridization reaction yield and it becomes necessary to use the standard addition approach for quantitation of organometallics in real samples. The main drawback of such derivatives is their reactivity leading to adsorption and degradation reactions during GC determination, hybridization being replaced by the ethylation reaction with $NaBEt_4$. Another disadvantage of hydrides derivatives is their lower lipophilicity than ethylated species which can give lower preconcentration factors when commonly coated polysiloxanes are used as preconcentration polymers. However, the main potential of these derivatives lies in the speciation of the less volatile organometallics such butyl, phenyl and octyl compounds.

4 Speciation of Hg, Pb and Sn with SPME

Until now, speciation studies using SPME have been carried only for tin, mercury and lead by using dimethylpolysiloxane as preconcentration polymer since organometallic species are analyzed as methyl or ethyl derivatives. The species determined and the analytical conditions according to the element are reported in Tables 2–4.

The following variables have been evaluated to improve the sensitivity in the SPME speciation.

Extraction Procedure

Two different sampling procedures are possible for organometallic speciation with SPME, headspace or direct sampling from the aqueous sample. Provided that the ethylated organometallic species have a relatively high volatility, headspace sampling is the preferred procedure since carryover effect is minimized and fiber lifetime is increased. Derivatization reagents are used in the aqueous phase and they affect the fiber stability when it is exposed to high temperatures in the injector port during the desorption process.[6,11,12] The sensitivity in headspace analysis of highly volatile compounds such as the methyltins clearly depends on their partition coefficients. In this regard, the longer the alkyl chain, the higher the sensitivity on account of their higher partition coefficient. In this regard, the highest sensitivity was obtained for monomethyl- and dimethyl-tin since they were analyzed as ethylated derivatives and their lipophilicities are higher than those of tri- and tetramethyl-tin (Figure 1). The differences in the SPME response among the organometallics according

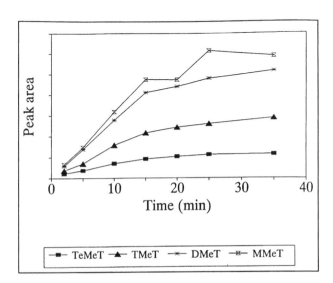

Figure 1 *Extraction profiles of ethylated methyltins sampled in the headspace*
(Reproduced by permission from *J. High Resolut. Chromatogr.*, 1995, **18**, 767)

Table 2 *SPME procedures used in tin speciation from aqueous samples*

Compound	Extraction	Sample volume (m L^{-1})	Derivatization	Film thickness (μm)	Fiber composition	Determination	LOD as tin (ng L^{-1})	Source
Methyl	headspace	5	NaBEt$_4$	100	PDMS	GC–FPD	24–125	12
Butyl	headspace	20–25	NaBEt$_4$	100	PDMS	GC–ICPMS	0.3–2.1	13
	aqueous	4	NaBEt$_4$	100	PDMS	GC–AED	ng L^{-1} range	16

Table 3 *SPME procedures used in mercury speciation*

Compound	Matrix	Extraction	Derivatization	SPME method	Determination	LOD (as Hg)	Reference
CH$_3$Hg$^+$	water (ng l^{-1})	headspace	NaBEt$_4$	100 μm, PDMS	GC–MS	7.5	6
		direct				6.7	
Hg^{2+}		headspace				3.5	
		direct				8.7	
CH$_3$Hg$^+$	fish tissue (μg g^{-1})	headspace				0.15	
		direct				0.1	
Hg^{2+}		headspace				0.07	
		direct				0.13	
CH$_3$HgPh	soil	direct	K$_3$[Co(CN)$_5$CH$_3$]	100 μm, PDMS	GC–MS	–	11

Table 4 *SPME procedures for lead speciation in aqueous samples*

Compound	Extraction	SPME method	Determination	Relative LOD (ng L^{-1} as Pb^{2+})	Reference
Et$_4$Pb	direct	100 μm, PDMS	GC–ITMS	200	9
Bu$_4$Pb	headspace	100 μm, PDMS	GC–AED	ng L^{-1} range	16

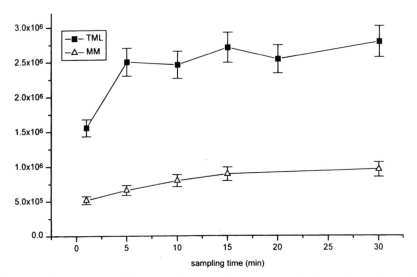

Figure 2 *Extraction profiles of ethylated methylmercury and trimethylead obtained by headspace SPME GC coupled to ICPMS*
(Reproduced by permission from *Anal. Chem.*, 1997, **69**, 1607 © 1997 American Chemical Society)

to the alkylation degree are even higher if they are underivatized and are analyzed as chlorides. In this case the monosubstituted species have a very low response because their lipophilicity is too low to be preconcentrated by non-polar polysiloxanes.[16]

Another factor which affects the response is the number of substituents because it increases the partition coefficient between the fiber and the aqueous phase since it increases with the number of alkyl substituents. This trend is apparent in the comparison between organometallics with different oxidation such as trimethyllead and methylmercury (Figure 2).

Sampling Time

Sampling time depends on the volatility of compound. The more volatile species are transferred faster into the vapor phase and *ca.* 5 min of sampling is enough. However, *ca.* 10 min gives a compromise between extraction efficiency and analysis time for most of the analytes but a high stirring rate is necessary to facilitate the equilibration between the liquid and the vapor phase. When vigorous stirring is applied to the sample, it could lead to an increase in temperature and therefore it is necessary to perform the extraction in a thermostated bath. When sampling is performed directly from the aqueous sample, the sampling time is longer because the mass transfer from the sampling media into the fiber is faster in the headspace. About 60 min has been reported as necessary to reach equilibrium in the case of tetrabutyllead,[16] which is too long on a routine basis.

Salting-out Effect

This effect has been investigated in the cases of mercury,[6] tin[12] and lead.[13] Since derivatized organometallic are nonpolar compounds, it is expected that the addition of salts into the aqueous samples will increase the ionic strength of the aqueous media and increase the equilibrium concentration of the analytes towards the apolar SPME fiber[13]. However, the addition of saturated volumes of NaCl in the aqueous media did not improve the extraction efficiency for all the organometallic species. Two reasons have been postulated to explain this fact.[13] First, derivatized organanometallics are nonpolar and the fiber–aqueous partitioning is independent of the presence of salt. The second is the existing ionic strength of the aqueous solution since a buffer (HOAc/NaAcO) is used in the derivatization reaction and the addition of salts to the solution does not lead to an increase in ionic strength of the solution.

Sampling Temperature

There are no differences for methylmercury and methyltins between 25 and 50 °C.[6,12] The equilibrium between the analyte sorbed in the SPME fiber coating and the concentration in the sample solution depends on both the solubility of the analyte in the aqueous phase and its sorption affinity onto the SPME fiber coating. Increasing the temperature will increase the partial vapor pressure of analytes in the headspace but simultaneously the sorption onto the fiber will decrease with increasing temperature, particularly for highly volatile compounds. However, in the case of less volatile compounds such as dibutyl- and tributyl-tin an increase in the response was found when the sampling temperature was increased from 20 to 60 °C.[13] At higher temperatures (80 °C), a slight decrease in the response was observed for dibutyltin because desorption from the SPME fiber is important for this analyte at such high temperatures.

Fiber Film Thickness

The film thickness of the fiber plays an important role in the sampling and desorption kinetics. The higher the film thickness the slower is the adsorption process and higher desorption temperatures are needed.[16] As expected, when thinner fibers are used (7 μm), the amount of analyte adsorbed is lower and for most of the applications the 100 μm film thickness is used.

Thermal Desorption

The injector temperature is a key parameter in organometallic speciation since the thermal stability of the species is rather limited. However, in order to minimize the organometallic carryover it is necessary to increase the injector temperature as much as possible. Butyltins and tetraethyllead are rather stable at 250 °C and it is reported that they can be completely desorbed following 1 min of desorption time. It offers a good compromise between organometallic desorption completeness and carryover effects. At lower temperatures, carry-over effects were detected for the less volatile compounds such as butyl- and phenyl-tins.[14]

Determination of Organometallic Compounds

The stationary phases used in organometallic speciation are usually nonpolar, such as dimethylphenylpolysiloxanes. However, the determination of the most volatile species such as diethylmercury (b.p. 159 °C) or trimethylethyltin (b.p. 108 °C) requires either a thicker film stationary phase or cryofocusing. From the practical point of view, the first option is preferable and only slight broadening occurs for the highly volatile tin species such as trimethylethyltin when the column temperature is kept at 55 °C (Figure 3).

The linearity range in organometallic speciation has been evaluated by several analytical techniques such as GC–FPD in the tin selective mode, GC–MS in the SIM mode for mercury (*i.e.* quadrupolar analyzer) and tetraethyllead (GC–FID). Methyltin and tetraethyllead determination exhibited three orders of magnitude in the headspace sampling (0.1–100 ppb) but it was slightly lower for methylmercury (0.03–6.7 ppb).[6] However, for the GC–ITMS (ion trap analyzer) used for tetraethyllead determination, the responses are not linear in the whole range of concentrations evaluated (0.005–100 ppb)[9]. This was attributed to the unstability of tetraethyllead molecules in the ion trap, which decompose to give very reactive free radicals. A third-order polynomical regression led to an acceptable fit.

The lowest LODs were obtained by AED and ITD–MS and are in the range of ppt (pg L^{-1}) (Table 4). The high sensitivity of these analytical techniques could allow their application to the determination of organometallics in monitoring programs where the Environmental Quality Targets for these compounds are in the low ppt level. Other analytical techniques such as GC–MS and GC–FPD possess a LOD one order of magnitude higher than the former analytical techniques and they can be useful for the determination of these compounds in biota or sediments where these compounds are concentrated.

A point of primary importance in the application of the developed analytical procedures is their validation. Standard Reference Materials are available for butyltins in sediment and mussel tissue, methylmercury in fish and trimethyllead in urban dust. Until now, speciation studies developed using SPME have been validated in the cases of butyltin sediment and methylmercury in fish.[6,13] In both

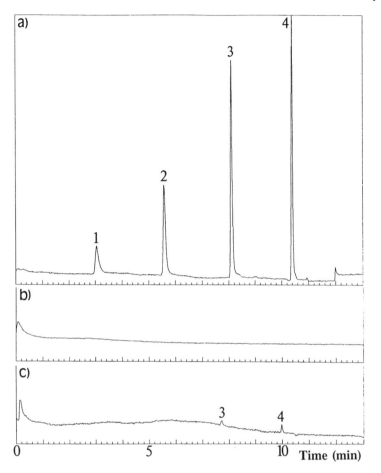

Figure 3 *GCFPD chromatograms of methyltin compounds analyzed as methylated derivatives in a DB-624 of 30 m × 0.32 mm i.d. and 1.8 μm film thickness. (a) Spiked sample followed by aqueous phase ethylation. (b) Second desorption after thermal treatment at 225 °C. (c) Blank following an extraction experiment. Compound identification: 1, TeMT; 2, TMeT; 3, DMeT; 4, MMeT*
(Reproduced by permission from *J. High Resolut. Chromatogr.*, 1995, **18**, 767)

cases, the value obtained falls within the certified range except in the case of monobutyltin where the value obtained is higher for this reference material similar to values reported by other authors.[3]

5   Future Developments in SPME Speciation

Two trends are envisaged in speciation studies in the coming years. The first is the development of more selective and effective preconcentration polymers for organometallic species. Until now, the polymers used in organometallic speciation studies are similar to those used in the determination of volatile organic compounds. However, since the derivatized organometallics have a moderate lipophilicity, the concentration factors are rather small and only highly sensitive and selective detection systems can be used in monitoring studies. Another possibility to circumvent the lack of sensitivity is the application of sol gel technology to SPME of organometallics (see Chapter 3).

The second aspect of interest in SPME speciation studies of organometallics is the development of polymers useful for the preconcentration of ionic metallic and organometallic species since some these species are not possible to be derivatized (*i.e.* Cr, Se, As). A fused silica fiber coated with poly(dimethylsiloxane) was modified with a liquid ion exchanger [di(2-ethylhexyl)phosphoric acid] to produce a micro probe with ion exchange capability and it was tested for the concentration of Bi(III). The extracted Bi(III) was desorbed into an acidic potassium iodide solution.[15] However, the full potential of the former approach should be the combination with ion chromatography or to multielemental detection systems such as ICPMS.

References

1. J.T. Byrd and M.O. Andreae, *Science*, 1982, **218**, 565.
2. K. Fent, *Crit. Rev. Toxicol.*, 1996, **26**, 1
3. M. Ábalos, J.M. Bayona, R. Compañó, M. Granados, C. Leal and M.D. Prat, *J. Chromatogr. A*, 1997, **788**, 1.
4. R.J. Carter, N.J. Turoczy and A.M. Blond, *Environ. Sci. Technol.*, 1989, **23**, 615.
5. I. Tolosa, L. Merlini, N. de Bertrand, J.M. Bayona and J. Albaigés, *Environ. Toxicol. Chem.*, 1992, **11**, 145.
6. Y. Cai and J.M. Bayona, *J. Chromatogr. A*, 1995, **696**, 113.
7. Y. Cai and J.M. Bayona, *J. Chromatogr. Sci.*, 1995, **33**, 89.
8. R. Fisher, S. Rapsomanikis and M.O. Andreae, *Anal. Chem.*, 1993, **65**, 763.
9. T. Górecki and J. Pawliszyn, *Anal. Chem.*, 1996, **68**, 3008.
10. M. Ábalos and J.M. Bayona, *Appl. Organomet.*, in press.
11. C.M. Barshick, S.-A. Barshick, M.L. Mohill, P.F. Britt and D.H. Smith, *Rapid Commun. Mass Spectrom.*, 1996, **10**, 341.
12. Y. Morcillo, Y. Cai and J.M. Bayona, *J. High Resol. Chromatogr.*, 1995, **18**, 767.
13. L. Moens, T. De Smaele, R. Dams, P. Van den Broeck and P. Sandra, *Anal. Chem.*, 1997, **69**, 1604.
14. Y. Morcillo, Y. Cai, C. Porte and J.M. Bayona, Proc. 16th International Symposium on Capillary Chromatography, ed. P. Sandra 1994, Vol. 1, Hüthig, Heidelberg (Germany), p. 864.
15. E.O. Otu and J. Pawliszyn, *Mikrochim. Acta*, 1993, **112**, 41.
16. S. Tutschku, S. Mothes and R. Wennrich, *Fresenius' J. Anal. Chem.*, 1996, **354**, 587.

CHAPTER 22

Metal Speciation by SPME–CGC–ICPMS

TOM DE SMAELE, LUC MOENS, RICHARD DAMS AND PAT SANDRA

1 Introduction

For some years, man has been aware of the toxicity of metal-containing organic compounds. The toxicity of these compounds depends highly on their chemical structure. Subtle differences in at first glance similar molecules can have severe consequences for their chemical toxicity.[1] As a consequence, speciation of organometallic compounds has gained more and more importance, and several hyphenated techniques, *i.e.* a combination of a separation technique with an element specific detection system, have been described in the literature.[2–5] Since the introduction of commercial capillary gas chromatography–microwave induced plasma atomic emission detection systems (CGC–MIP–AED), organometallic analysis can be performed routinely. In addition, the coupling of CGC to inductively coupled plasma mass spectrometry (ICPMS) has proved to yield a highly sensitive and selective method for organometallic speciation.[6–17] Since both GC and ICPMS can work independently and can easily be coupled within a few minutes by means of a transfer line, the hyphenation of these instruments is even more attactive than the CGC–MIP–AED system for environmental laboratories, where ICPMS is often the method of choice for routine total metal concentration determination.

Since most organometallic components of tin, mercury and lead occur in nature in rather unvolatile ionic species, these compounds have to be derivatized into volatile solutes prior to GC analysis. Originally, hydride generation in combination with cryogenic trapping of the volatile species[18–20] or Grignard derivatization[21–23] have been used widely. Hydride generation, however, suffers from interferences during the derivatization and the species obtained are often unstable. Grignard reaction, on the ofter hand, has the advantage of being versatile: many different alkylations such as ethylation, propylation and pentylation are possible so that nearly all alkyl-lead, -mercury and -tin species can be derivatized and determined by CGC. Grignard reactions, however, require non-aqueous, aprotic media and numerous handling steps.

The sample preparation for organometal speciation by CGC has been drastically simplified by the introduction of an aqueous *in situ* derivatization by Ashby *et al.*[24–26] Derivatization with sodium tetraethylborate, NaBEt$_4$, has been investigated for a wide range of organometallic compounds such as organo-lead, -mercury, -cadmium, -tin and -selenium. Recently, De Smaele *et al.* reported the possibilities of *in situ* propylation as a novel aqueous derivatization method with which even the ethyl derivatives of Hg and Pb can be volatilized.[27] The advantages of this derivatization technique are that alkylation takes place in the aqueous phase and extraction can be performed simultaneously. In addition, it is compatible with modern extraction methods such as solid phase microextraction[28] and purge and trap[29] or sample preparation techniques such as microwave assisted extraction.[30] Solid phase microextraction (SPME), developed by Pawliszyn and co-workers,[31–35] has been succesfully applied to metal speciation for the determination of organo-tin, -mercury and -lead compounds.[28,36–38] SPME in combination with CGC–ICPMS has been used for the first time by Moens *et al.*[28] for the simultaneous extraction and determination of organo-tin, -mercury and -lead compounds. In this chapter the SPME extraction parameters as well as the specific sample preparation for organometallic species are demonstrated. The SPME technique has been applied to real environmental samples.

2 Instrumentation, Reagents and Standards

CGC–ICPMS

The CGC–ICPMS system consisted of a Perkin Elmer Autosystem GC, equipped with a 30 m × 0.25 mm × 0.25 μm of methylsilicone capillary column, coupled to a Perkin Elmer Sciex Elan 5000 ICP mass spectrometer by means of an in-house made transfer line.[14–15] The instrumental parameters are summarized in Table 1.

ICPMS is well known for its multi-element capabilities and, coupled to a chromatographic system, this feature can be fully exploited. In the case of organometallic speciation with CGC–ICPMS, the most abundant isotopes of the different elements, leading to the most sensitive signals, *e.g.* ^{120}Sn, ^{202}Hg and ^{208}Pb for Sn, Hg and Pb, respectively can be chosen. In addition, ^{126}Xe originating from Xe, doped in a concentration of 1% (v/v) to the carrier gas, can be monitored continuously during the analyses and acts as an internal standard to correct for instrument instabilities, malfunctions or signal drifts during the GC analyses. Detection limits with ICPMS are, in comparison with CGC–MIP–AED, at least a factor of 10 superior, *i.e.* ICPMS instrumental detection limits for organo-Sn, -Hg, -Pb are in the order of 10–100 fg absolute as metal.

Table 1 *Instrumental parameters for CGC–MIP–AES and CGC–ICPMS*

ICPMS	Perkin Elmer Sciex Elan 5000
RF Power	1250 W
Sampling depth	10 mm
Carrier gas flow rate	1.10–1.25 L min^{-1}
Auxiliary gas flow rate	1.20 L min^{-1}
Plasma gas flow rate	15 L min^{-1}
Sampling cone/aperture diameter	Ni/1 mm
Skimmer cone/aperture diameter	Ni/0.75 mm
Dwell time	30–50 ms (depending on number of nuclides to be measured)
	10 ms (^{126}Xe)
Transfer line	home-made
	250 °C
Gas chromatograph	Perkin Elmer Autosystem
Column	FSOT, methylsilicone
	30 m; 0.25 i.d.; d_f = 0.25 μm
Injection technique	splitless
Injection temperature	250 °C
Temperature programme	60 °C (1 min)–20 °C/min–200 °C (0.5 min)
Carrier gas/inlet pressure	Xe/H$_2$ (1/99 mixture); 30 psi

SPME Device

A SPME fiber holder for manual injections, with a 100 μm PDMS coated fiber was used. 50 mL glass vials closed with PTFE-coated rubber septa were used for sampling. Proper mixing of the sample solutions during the SPME extractions was achieved with a magnetic stirrer.

Reagents and Standards

Monobutyltin trichloride (MBTCl$_3$, 95% purity), dibutyltin dichloride (DBTCl$_2$, 97% purity) and tributyltin chloride (TBTCl, 96% purity) were from Aldrich (Sigma-Aldrich Belgium, Bornem, Belgium). Methylmercury chloride (MMCl, 98%) and tripropyltin acetate (TPTOAc, p.a.) were purchased from Merck (Darmstadt, Germany), trimethyllead chloride (TMLCl, p.a.) from ABCR (Karlsruhe, Germany). These organometals were dissolved in ethanol (EtOH, analytical reagent grade, Merck) to obtain mono-component solutions of 1 g L^{-1} as metal. Mixed organometallic standard solutions were prepared in and further diluted with EtOH to concentrations varying between 1 and 10 μg L^{-1} and stored in the dark at 4 °C. Sodium tetraethylborate (NaBEt$_4$) is commercially available (Strem Chemicals, Bischheim, France) whereas sodium tetrapropylborate (NaBPr$_4$) has been synthesized in our laboratory. Milli-Q water (Millipore Corp.) was used to prepare all aqueous solutions.

A buffer solution of pH value of 5.3 was prepared by mixing appropriate amounts of 0.2 M sodium acetate (NaOAc, p.a., UCB, Leuven, Belgium) and concentrated acetic acid (HOAc, p.a., Merck). Methanol (MeOH, p.a., Merck) of analytical grade quality has been used for the leaching of the samples.

3 SPME in Organometal Speciation

Optimization of SPME Extraction

In situ alkylation with sodium tetraethylborate (NaBEt₄) or sodium tetrapropylborate (NaBPr₄)

All environmental relevant organometallic species of Sn, Hg and Pb occur in ionic form in nature. Therefore, they need to be derivatized into apolar volatile species by use of $NaBEt_4$ or $NaBPr_4$. Commonly, to 25 ml sample a standard solution buffered at pH 5.3 with NaOAc/HOAc, 100–1000 µL of a 1% aqueous $NaBEt_4$ or $NaBPr_4$ solution was added to the vial by means of a syringe. The vial was sealed and immediately afterwards the SPME fiber was exposed to either the sample headspace or the aqueous phase depending on the sampling method. From classical liquid/liquid extractions and CGC–ICPMS detection, it was found that the derivatization was completed in less than 10 minutes, which corresponds to the residence time of the fiber in the vial.

Relative sensitivity of SPME versus liquid/liquid extraction

Figure 1 shows the relative signals for different organo-tin, -mercury and -lead species measured with CGC–ICPMS after SPME (direct and headspace, 10

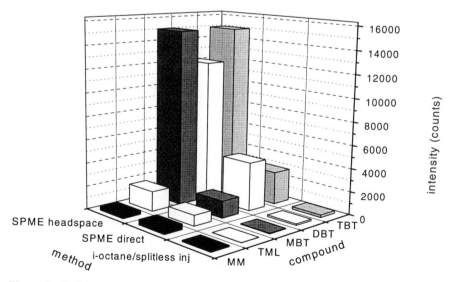

Figure 1 *Relative sensitivity for MM, TML, DBT and TBT of direct and headspace SPME and liquid/liquid extraction (1 µL splitless injection). ICPMS peak areas normalized to the peak area of the $^{126}Xe^+$ signal to correct for changes in ICPMS sensitivity*

minutes extraction time) and splitless injection of a *iso*-octane extract. The observed signal intensities were normalized so as to correct for the different concentrations in solutions used for liquid/liquid extractions (100 μg L^{-1} as metal) and SPME (2 μg L^{-1}). The sensitivity of headspace SPME is by a factor of up to 10 (MBT) higher when compared with direct SPME and by a factor of up to 324 (MBT) when compared with liquid/liquid extraction. The derivatized organometallic species are completely apolar and poorly soluble in water and tend therefore to migrate to the headspace and then to the apolar PDMS fiber. As stated in Chapter 1, this is advantageous to 'real life' sample analysis. Environmental samples are known to be very dirty and low volatile organics, coextracted in classical liquid/liquid extractions, interfere in the separation of organometals. All further analyses of the organometals mentioned here were performed in the 'headspace mode'.

Extraction time

In Figure 2, the effect of the extraction time on the relative extraction yield for organo-tin, -mercury and -lead compounds is demonstrated. After 10 minutes of extraction, 90% of the maximum extracted amount of organometals is collected. These curves are similar to those obtained for organics such as phenanthrene (Chapter 1). From this Figure, it can be deduced that the equilibrium is reached faster for more volatile MM than for TBT. The less volatile the compound is (lower K_{hs}, equation 6, Chapter 1), the lower is its concentration in the headspace and the slower the diffusion in the headspace

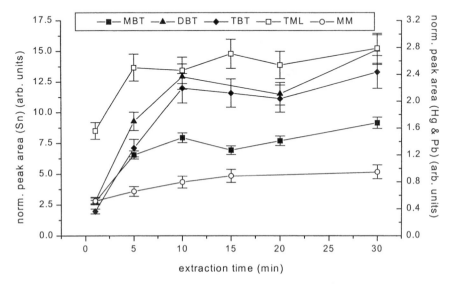

Figure 2   *Influence of sampling time on the extraction efficiency for organo-tin, -mercury and -lead compounds. Peak areas are corrected for changes in ICPMS sensitivity via the peak area of the $^{126}Xe^+$ signal*

since the concentration gradient of less volatile compounds in the headspace will be smaller. A sampling time of 10 minutes was therefore considered to yield sufficient recovery within a reasonable sampling time.

Sampling temperature

The influence of the sampling temperature was investigated by sampling standard mixtures for 10 minutes at different temperatures. The results are plotted in Figure 3 for TML, MBT, DBT and TBT. For DBT and TBT heating the sample mixture to 60 °C leads to a higher sorption efficiency. The extracted amounts of the more volatile TML and MBT, however, start to decrease from 20 °C on. As mentioned before, the equilibrium between the analyte concentration sorbed by the SPME fiber coating and the concentration of the analyte in the sample solution depends on both the solubility of an analyte in the aqueous phase (Henry's law) and its sorption affinity onto the SPME fiber coating. Increasing the temperature will increase the Henry constants of the organometallic compounds, resulting in a higher analyte partial vapour pressure in the headspace. The sorption on the other hand will decrease with increasing temperature and depends on the analyte's volatility. For DBT and TBT, the less volatile species, the temperature increase enhances the overall extraction process. For MBT and TML, however, partial desorption of the species from the SPME fiber coating occurs from relatively low temperatures on. Since the derivatization reaction is less efficient for MBT, and because performing extractions at higher temperatures is more tedious and time consuming, further extractions were carried out at 25 °C.

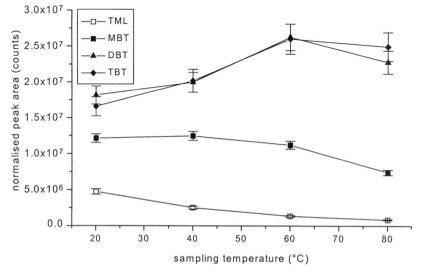

Figure 3 *Influence of sampling temperature on the extraction efficiency. Peak areas are corrected for changes in ICPMS sensitivity via the peak area of the ^{126}Xe signal*

Table 2 *Reproducibility and limits of detection of SPME–CGC–ICPMS*

Compound	RSD % (n = 10)	LOD (3 s, n = 10) (ng L^{-1} as metal)
Methylmercury	11	4.3
Butyltins	5.2–14	0.34–1.1
Methyl-, ethyl-leads	8.0–10	0.19

Analytical Performances: Figures of Merit

Reproducibility and limits of detection of SPME–CGC–ICPMS

The reproducibility of 10 subsequent SPME extractions varies between 5 and 14% (Table 2) . These values are somewhat higher than the 5% RSD mentioned in Chapter 1. SPME of organometallic compounds, however, requires derivatization prior to extraction, which of course will have an influence on the RSD values.

The limits of detection, LODs, were determined as 3 times the standard deviation of the background measured for 10 successive SPME extractions of buffer and NaBEt$_4$ only (Table 2). The LODs for MBT, DBT and TBT range between 0.34 and 2.1 ng L^{-1} and are of the same order of magnitude (0.30–0.82 ng L^{-1} as Sn) as those obtained *via* classical liquid/liquid extraction and NaBEt$_4$ derivatization. The latter method, however, requires at least 500 mL of sample, and thus a higher enrichment, when concentrations near the LOD must be determined whereas for SPME 25 mL of sample is sufficient. The LOD for TML is somewhat lower than those for organotin species. This is due to the lower background of Pb. On the contrary, the LOD of MM is higher (4.3 ng L^{-1} as Hg). This is probably due to the lower extraction efficiency of the headspace SPME technique for the relatively volatile MM in comparison with organo-tin and -lead species.

Linearity of calibration curves

For the organometallic compounds analysed, the overall analytical procedure shows a linear dynamic range (correlation coefficients between 0.9986 and 0.9994) between 10 and 1000 ng L^{-1}. At concentrations >1000 ng L^{-1}, the SPME fiber coating tends to be saturated, resulting in a lower sorbed amount of analyte. In Figure 4 a chromatogram of the simultaneous analysis of organo-tin, -mercury and -lead is shown.

Calibration and internal standardization

Environmental samples are known to be very complex. Water and sediment samples contain, besides heavy metals, a whole range of organic pollutants such as pesticides and herbicides, as well as involatile detergents, humic acids, oils

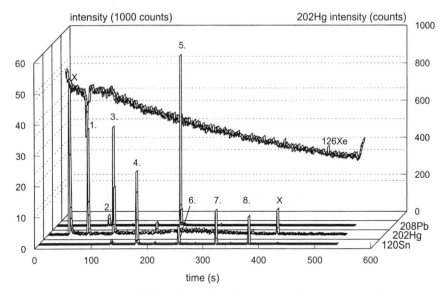

Figure 4 *Chromatogram of an organometal standard, extracted with headspace SPME.*
$^{120}Sn^+$, $^{202}Hg^+$ and $^{208}Pb^+$ were simultaneously measured. 1. MM, 2. TML,
3. inorganic mercury (diethylmercury), 4. inorganic tin (tetraethyltin),
5. inorganic lead (tetraethyllead), 6. MBT, 7. DBT, 8. TBT, X. unknown
compounds

and fats. Especially the latter hamper the sample preparation of organometallic species. They can interfere in the derivatization, making phase separations in liquid/liquid extractions difficult or even impossible, or they can act as an organic sorbent in SPME experiments. Since the derivatized organometallic compounds are completely apolar, they have indeed great affinity to sorb onto organic particles or into apolar liquids. This means that in SPME, based on equilibria, the presence of such compounds can change the equilibrium of the analytes of interest between the different phases. Therefore, single standard addition was used as a calibration method in headspace SPME instead of external calibration. TPTOAc was added to all samples, standard and blank solutions at a final concentration of 20 ng L^{-1} and functioned as an internal standard. The internal standard undergoes the complete sample preparation procedure and thus corrects for the differences in derivatization and/or sorption yields which cannot be corrected by standard addition only. It can be stated that the more organic the matrix is, the bigger the suppression of the derivatization/ extraction of the organometallic species in SPME. The organometallic species are partially extracted into the organic matrix components. This is illustrated in Figure 5 where the normalized peak area of the TPT peak is plotted *vs.* the matrix in which it was spiked. As can be seen, there is already a severe suppression in the derivatization/extraction yield in environmental water samples. Only 13.8% of the peak area compared to Milli-Q water was found. In the heavy sediment matrix, the suppression is even more pronounced: a

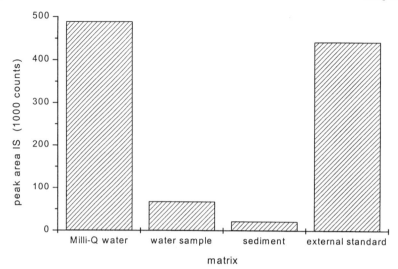

Figure 5 *Influence of matrix composition on the derivatization/extraction yield, expressed as the signal intensity of the internal standard spiked in a final concentration of 20 ng L^{-1}*

decrease of 95.6%. In addition, all peak areas are normalized to the continuous $^{126}Xe^+$ signal, originating from Xe doped into the H$_2$ carrier gas, and corrected for instrument instabilities and signal drift during the analysis.[14–16, 39]

Accuracy

The reliability of SPME–CGC–ICPMS for the determination of organo-tin and -mercury compounds was checked by the analysis of the standard reference material PACS-1 (Marine sediment) and DORM-2 (Dogfish Muscle), both from the National Research Council, Canada (NRCC). Approximately 0.2 g of sediment was weighted in a 50 mL glass vial and the internal standard, TPTOAc, was added. The organotin compounds were leached from the sediment matrix by adding 5 mL of HOAc and 5 mL of MeOH. The vial was closed and placed in an ultrasonic bath for 30 minutes. Subsequently, 250 µL of the supernatant was pipetted into another vial. 25 mL of NaOAc/HOAc buffer at pH 5.3 was added. The vial was sealed and 1000 µL 1% NaBEt$_4$ was added with a syringe. Subsequently, the SPME device was inserted and the PDMS fiber was exposed to the sample headspace for 10 minutes followed by CGC–ICPMS analysis. The fish tissue on the other hand had to be hydrolysed. The organomercury compounds are incorporated in the biological tissue matrix so that leaching only can not liberate the organometal species. To approximately 0.1 g of fish sample, 10 mL 10% KOH was added. Hydrolysis took place for 1–2 hours under ultrasonic treatment. 50 µL of the dissolved tissue was then pipetted into 25 mL of NaOAc/HOAc buffer (pH 5.3) followed by derivatiza-

Table 3 *Accuracy of SPME–CGC–ICPMS analysis certified reference materials*

Compound	Headspace SPME ($ng\ g^{-1}$ metal)	Certified values ($ng\ g^{-1}$ metal)
	NRC PACS-1 Marine Sediment	
MBT	750 ± 210*	280 ± 170
DBT	1060 ± 150	1160 ± 180
TBT	1220 ± 190	1270 ± 220
	NRC DORM-2 Dogfish Muscle Tissue	
MM	4280 ± 910	4470 ± 320

* Limit of 95% confidence ($n = 3$).

tion, SPME extraction and CGC–ICPMS analysis. The results obtained with SPME–CGC–ICPMS are summarized in Table 3. As can be seen, all concentrations are in good agreement with the certified values except for MBT. The certified value of MBT, however, is known to be too low.[27–28,40] Recoveries of 95% were found for MM, 91% and 97% for DBT and TBT, respectively. These figures prove that SPME is a reliable extraction technique for quantitative analysis.

4 Applications of SPME for Organometal Speciation in the Environment

Analysis of Organo-tin, -mercury and -lead in (Industrial) Surface Waters

The most abundantly used organometallic compounds are organo-tin, -mercury and -lead. Organotin compounds have been widely used in the shipping industry as antifouling agents in boat paints (tributyl- and triphenyl-tins), as biocides (triorganotins) and as stabilizers for poly(vinyl chloride) (mono- and dibutyl-tins).[1] Organolead compounds are well known as anti-knocking agents (tetra-ethyl- and tetramethyl-lead) in fuels whereas organomercury compounds are used as biocides and seed dressings (methylmercury).[1] Surface water samples were collected from industrial regions in the north east of Belgium (Limburg) and monitored for organo-tin, -mercury and -lead. The samples were acidified to pH 2 with concentrated HCl and stored in the dark in glass bottles at 4 °C until analysis. The sample preparation was very simple and short. 25 mL of water sample was pipetted into a sample vial, buffered with NaOAc/HOAc buffer (pH 5.3), NaBEt$_4$ was added, followed by SPME extraction during 10 minutes. The GC analysis takes about 10–12 minutes (including oven temperature re-equilibration) so that the sample preparation and analysis nearly run 'in time'. Figure 6 shows the concentrations found. The concentrations are very

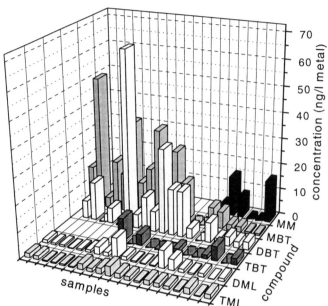

Figure 6 *Concentrations of MM, TML, DML, MBT, DBT and TBT in environmental water samples. Concentrations as* ng L^{-1} *metal*

low for all organometallic species and vary between 0.1 and 67.1 ng L^{-1} as metal. The average concentration of DBT is most abundant. The amount of TBT is low (1.4–9.0 ng L^{-1}) since TBT tend to sorb onto the sediment and solid particles. These figures illustrate the very high sensitivity of the SPME technique. 25 mL of water sample is sufficient whereas for classical liquid/liquid extractions at least 500 mL of water should be extracted and the solvent should be evaporated to a final volume of 500 μL.

Analysis of Organolead Compounds in Water and Grass

As already mentioned, organolead compounds are used in fuel as antiknocking agents. Most frequently used are tetramethyl- and tetraethyl-lead compounds, a mixture of them or mixed methylethyllead species. These compounds degrade during the combustion of the fuel to TML and TEL, which can further degrade, to DML and DEL.[1] The final degradation product is inorganic lead. These compounds enter the environment *via* exhaust fumes of leaded fuels and adsorb onto aerosol particles.

Organolead concentrations have been determined in potable and rain water, street dust and grass (rural, vicinity of highway and highway verge) by SPME–CGC–ICPMS. Instead of NaBEt$_4$ as derivatization agent, NaBPr$_4$ was used. In this way all methyl and ethyl derivatives of lead could be determined simultaneously. Water samples were treated as described previously except that ethylenediaminetetraacetate (EDTA) was added to mask and/or reduce the derivatization of inorganic lead. The grass was first frozen in liquid nitrogen and ground by brittle fracture. 0.5 g of street dust and grass sample were leached with HOAc/NaOAc and EDTA was added. After leaching, the complete sample sludge was derivatized and extracted. The results are summarized in Figure 7. As can be seen low concentrations of organolead were found in rain water (1.35–21.2 ng L^{-1}) whereas the concentrations in potable water reach the LOD. The concentrations in highway grass 2 (highway verge) are remarkably higher

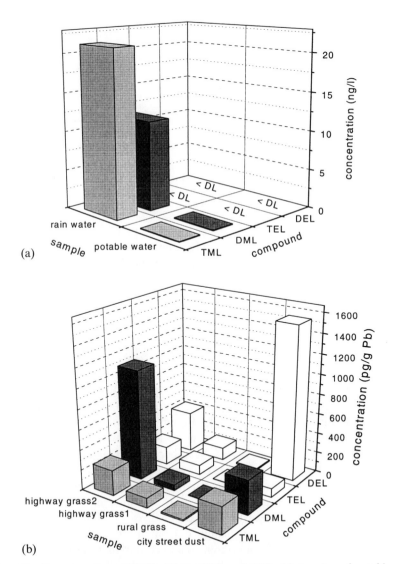

Figure 7 *Concentrations of TML, DML, TEL and DEL:* (a) *in rain and potable water* (ng L^{-1} Pb); (b) *in grass and street dust* (pg g^{-1} Pb)

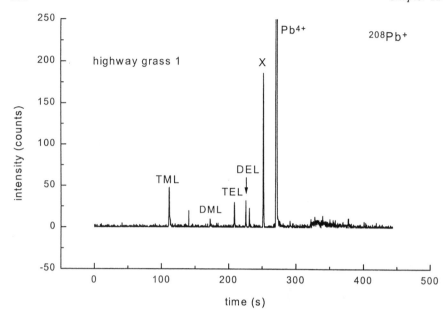

Figure 8 *Typical SPME–CGC–ICPMS chromatogram of organolead compounds in highway grass 1, taken in the vicinity (50–100 m) of a highway*

(180–1100 pg g^{-1}) than in rural grass (3.0–15.2 pg g^{-1}). Also organolead concentrations in city street dust (87–1540 pg g^{-1}) are significantly higher. Finally, in Figure 8 a chromatogram of organolead in highway grass 1 (vicinity of highway) is shown proving once again the high sensitivity of the SPME technique.

5 Conclusion

Headspace SPME is an accurate and precise alternative to classical liquid/liquid extractions of organo-tin, -mercury and lead species. Especially for environmental water samples, the sample preparation can be reduced to a strict minimum, pH adjustment, the addition of internal standard and derivatization reagent, and takes only 10 minutes. Owing to the simultaneous preconcentration of the organometallics on the PDMS fiber and the high sensitivity of the CGC–ICPMS hyphenation, only 25 mL of water sample is needed whereas with classical extraction techniques at least half a liter must be preconcentrated. Since headspace SPME was demonstrated to be more effective than direct SPME, the major advantage of headspace SPME can be fully exploited: only the organometals of interest are sampled from the complex environmental sample matrix, which reduces interferences and contaminations in the capillary column. Since SPME and CGC–ICPMS run 'in time', SPME can easily be used for routine analysis with high sample throughput.

References

1 P.J. Craig, *Organometallic compounds in the environment. Principles and reactions*, Longmans, Harlow, UK, 1986.
2. S.J. Hill, M.J. Bloxham and P.J. Worsfold, *J. Anal. At. Spectrom.*, 1993, **8**, 499.
3. L. Ebdon, S.J. Hall and W.R. Ward, *Analyst*, 1987, **112**, 1.
4. R. Smits, *LC-GC Int.*, 1994, **7**, 694.
5. R. Łobiński, *Appl. Spectrosc.*, 1997, **7**, 262A.
6. J.C. Van Loon, L.R. Alcock, W.H. Pinchin and J.B. French, *Spectrom. Lett.*, 1986, **19**, 1125.
7. N.S. Chong and R.S. Houk, *Appl. Spectrosc.*, 1987, **41**, 66.
8. G.R. Peters and D. Beauchemin, *J. Anal. At. Spectrom.*, 1992, **7**, 965.
9. A.W. Kim, M.E. Foulkes, L. Ebdon, S.J. Hill, R.L. Patience, A.G. Barwise and S.J. Rowland, *J. Anal. At. Spectrom.*, 1992, **7**, 1147.
10. E.H. Evans and J.A. Caruso, *J. Anal. At. Spectrom.*, 1993, **8**, 427.
11. W.G. Pretorius, L. Ebdon and J. Rowland, *J. Chromatogr.*, 1993, **646**, 369.
12. A. Kim, S. Hill, L. Ebdon and S. Rowland, *J. High Resolut. Chromatogr.*, 1992, **15**, 665.
13. A. Prange and E. Jantzen, *J. Anal. At. Spectrom.*, 1995, **10**, 105.
14. T. De Smaele, P. Verrept, L. Moens and R. Dams, *Spectrochim. Acta Part B*, 1995, **50**, 1409.
15. T. De Smaele, L. Moens, R. Dams and P. Sandra, *Fresenius' J. Anal. Chem.*, 1996, **354**, 778.
16. T. De Smaele, L. Moens, R. Dams and P. Sandra, *LC-GC Int.*, 1996, **9**, 138.
17. T. De Smaele, F. Vanhaecke, L. Moens, R. Dams and P. Sandra, *Proc. 18th International Symposium on Capillary Chromatography*, 1996, 20–25 May, Riva del Garda, Italy, ed. P. Sandra, Publ. Hüthig Verlag, Heidelberg, p. 52.
18. F.M. Martin and O.F.X. Donard, *Fresenius' J. Anal. Chem.*, 1995, **351**, 230.
19. O.F.X. Donard, S. Rapsomanikis and J.H. Weber, *Anal. Chem.*, 1986, **58**, 772.
20. Y. Cai, S. Rapsomanikis and O. Andreae, *Anal. Chim. Acta*, 1993, **274**, 243.
21. W.R.M. Dirkx, W.E. Van Mol, R.J.A. Van Cleuvenbergen and F.C. Adams, *Fresenius' J. Anal. Chem.*, 1989, **335**, 769.
22. M.D. Mueller, *Anal. Chem.*, 1987, **59**, 617.
23. R.J. Maguire and R.J. Tkacz, *J. Chromatogr.*, 1983, **268**, 99.
24. J. Ashby and P.J. Craig, *Appl. Organomet. Chem.*, 1991, **351**, 173.
25. J. Ashby and P.J. Craig, *Sci. Total Environ.*, 1989, **78**, 219.
26. J. Ashby, S. Clark and P.J. Craig, *J. Anal. At. Spectrom.*, 1988, **3**, 735.
27. T. De Smaele, L. Moens, R. Dams, P. Sandra, J. Van der Eycken and J. Vandyck, *J. Chromatogr. A*, 1998, **793**, 99.
28. L. Moens, T. De Smaele, R. Dams, P. Van Den Broeck and P. Sandra, *Anal. Chem.*, 1996, **15**, 1604.
29. M. Ceulemans and F.C. Adams, *J. Anal. At. Spectrom.*, 1996, **11**, 201.
30. J. Sprunar, V.O. Schmitt, R. Łobiński and J.-L. Monod, *J. Anal. At. Spectrom.*, 1996, **11**, 193.
31. D. Louch, S. Motlagh and J. Pawliszyn, *Anal. Chem.*, 1992, **64**, 1187.
32. D.W. Potter and J. Pawliszyn, *Environ. Sci. Technol.*, 1994, **28**, 298.
33. K.D. Buchholz and J. Pawliszyn, *Anal. Chem.*, 1994, **66**, 160.
34. Z. Zhang, M.J. Yang and J. Pawliszyn, *Anal. Chem.*, 1994, **66**, 844A.
35. Z. Zhang and J. Pawliszyn, *Anal. Chem.*, 1993, **65**, 1843.
36. Y. Morcillo, Y. Cai and J.M. Bayona, *J. High Resol. Chromatogr.*, 1995, **18**, 767.

37. Y. Cai and J.M. Bayona, *J. Chromatogr.*, 1995, **696**, 113.
38. S. Tutschku, S. Mothes and R. Wennrich, *Fresenius' J. Anal. Chem.*, 1996, **354**, 587.
39. M. Heistercamp, T. De Smaele, J.-P. Candelone, L. Moens, R. Dams and F.C. Adams, *J. Anal. At. Spectrom.*, 1997, **12**, 1077.
40. J. Szprunar, V.O. Schmitt, O.F.X. Donard and R. Łobiński, *Trends Anal. Chem.*, 1996, **15**, 181.

CHAPTER 23

The Application of SPME–LC–MS to the Determination of Contaminants in Complex Environmental Matrices

MONIKA MÖDER AND PETER POPP

Introduction

The SPME analysis of non- or semi-polar organic substances in samples with a simple matrix such as air, ground- and drinking water is a process which is well mastered and understood. The small amounts of accompanying matrix substances generally do not interfere with the detection of the target analytes, and only a few optimisation steps are required for routine analysis, even in trace concentration ranges of ng L^{-1}. However, the more complex the matrix, the more effort is required to prepare and verify the analysis. Moreover, the difficulties are compounded if trace amounts of polar components from highly matrix-loaded aqueous or solid samples are to be determined.

Classic sample preparation includes several steps of liquid/liquid extraction and liquid chromatographic separation for the clean-up and enrichment of the target analytes.[1] Subsequent GC analysis often requires preliminary derivatisation or special separation columns to analyse polar compounds. Laborious effort is also required to check each individual step for maximum recovery, reproducibility and precision. GC–MS has been used in combination with SPE for the trace determination of polar as well as nonpolar pesticides.[2] The high accumulation factors make SPE attractive for trace analysis but on-line operation requires a special interface and high volume injection.

The use of SPME–GC–MS (PA coating) for the detection of phenolic compounds from a lignite-derived wastewater revealed that the humic acid-like matrix only affected the extraction time but not the extraction yields of the phenols.[3] The diffusion of the analytes to the coating was delayed—as was consequently the adjustment of the partition equilibrium of the phenols. Quantitation by standard addition was recommended depending on variations in the matrix composition.

The performance of the SPME fibre was monitored using an additional internal standard. In particular, the accompanying high molecular weight material of the matrix can adversely affect the fibre coating. Changes in sampling and desorption behaviour or even shortened lifetime of the fibre can result. The blocking of the coating can be prevented by sampling from headspace or extraction with a membrane-protected fibre (see Chapter 1, Figure 2c). Both techniques require special temperature conditions to achieve extraction yields and times comparable to direct liquid-phase sampling.

Although the determination of phenols from a complex matrix is a routine SPME technique,[3] some important information was lost—for only complementary HPLC–ESI–MS analysis revealed the presence of polyhydroxylated and carboxylated aromatics. As the high polarity and partially thermal lability of these compounds prevent thermal desorption and GC separation on a medium polar column without any pre-derivatisation, HPLC is thus the practical alternative.

If the concentration of the analytes is sufficiently high, HPLC with conventional detection is the method commonly used. However, when unknown substances which may not have a chromophoric system have to be identified from a very complex mixture, mass spectrometric detection is required. Although high selectivity is available during detection, high sensitivity is not automatically guaranteed. Therefore, trace analysis often requires the enrichment or pre-separation of the target analytes. Numerous SPE techniques are at present being developed for the pre-concentration and clean-up of environmental samples using only small amounts of solvents and support material.[4–7] The on-line coupling of SPE to GC–MS[2] and HPLC–MS[8–10] has been reported for the determination of trace pollutants in aqueous samples, but considerable effort is required to optimise the column switch procedure and analyte elution, making this approach only worthwhile for large analysis series. The input required increases with the complexity and concentration of the accompanying matrix of the sample.

With this in mind, SPME was examined as a new strategy to save time and material for the pre-concentration and analysis of low-concentration polar and semi-polar organic compounds from very complex mixtures such as wastewater, sludge, soil, sediment and also biological material.[11]

Special attention was paid to the analysis of polar and semi-polar pesticides directly from soil slurries.[12] Furthermore, the determination of polar constituents of sewage sludge and sediments[13] demonstrates the advantages and problems associated with the application of SPME–HPLC–MS.

2   SPME–HPLC–MS Conditions

Mass Spectrometric Detection

HPLC coupled to MS has been favoured in an increasing number of reports dealing with the determination of environmentally hazardous substances from a variety of sample matrices.

Previously, thermospray (TSP) was one of the most widely used MS interfaces for HPLC coupling. Numerous reports have been presented on the usage of HPLC–TSP–MS for pesticide residue analysis.[14-17] However, sometimes the high temperature required for ionisation causes a few thermally labile pesticides to decompose; moreover, the information on molecular weight is limited. Another spray-ionisation technique, the particle beam interface (PBI), does not attain the high sensitivity of TSP, although it does generate electron impact spectra which are accessible to library search programs. The drawback of lower sensitivity is overcome by trace-enrichment procedures coupled on-line to LC, which also makes PBI suitable for trace analysis.[18] Nevertheless, highly polar and thermally labile analytes such as carbamate pesticides still cause problems.

Developments of atmospheric pressure ionisation interfaces (API–MS) have extended the polarity range of substances that can be analysed by MS. API comprises atmospheric pressure chemical ionisation (APCI) and electrospray ionisation (ESI) interfaces,[19-24] both of which are characterised by different ionisation efficiencies depending on the chemical nature of the analytes. Used in conjunction with HPLC, both techniques provide high sensitivity and selectivity for a broad variety of environmental pollutants. Electrospray is favoured for the analysis of semi- and highly polar components which often already exist as preformed ions in solution.

The detection limits of pesticides reported for LC–MS analysis are in the lower microgram per litre range or even less.[24] Despite this highly sensitive detection, sample enrichment procedures are often required for analyses at the 100 ng L^{-1} level as prescribed for example in the European Union's Drinking Water Guidelines.[25] In this regard SPME provides a rapid, simple method of combining sample enrichment and analysis.

The adaptation of SPME *via* HPLC to API–MS requires not only a special interface, the 'desorption chamber',[26,27] but also the careful setting of the mass spectrometric operation parameters. In addition to ion source parameters such as spray voltage and spray capillary-temperature, the set HPLC flow rate, the type of eluents and the gradient programme used all affect the mass spectrometric ionisation process—and consequently the detection sensitivity which can be achieved. Numerous reports address the optimisation of mass spectrometric operation conditions.[21,28,29] The application of a gradient elution for instance can cause undesirable variations in ion formation[28] depending on eluent composition.

The spray composition determines the type of ions produced in the gas phase. The mass spectra obtained in positive ESI mode are dominated by pseudomolecular ions (M^{+}, $[M + H]^{+}$ or $[M + Na]^{+}$) and cluster ions formed by the addition of one or more solvent molecules. The mixture of ions produced needs to be constant over a wide range of eluent composition and analyte concentration, for only then does sensitive detection using selected ion monitoring (SIM) become practicable.

If a gradient elution is used for LC-separation, the changes in ionisation behaviour have to be verified with references prior to analysis. Since the sensitivity of an API source directly depends on the concentration of the sample in the spray volume, HPLC separation must be optimised for minimum sample dilution. Most flow rates recommended for ESI are between 1 and 200 μL min^{-1}.

The ESI$^+$ and ESI$^-$ conditions used in our experiments were optimised with selected compounds as shown in Table 1. Various eluent additives such as acetic acid, trifluoroacetic acid and ammonium acetate were examined to improve the ion abundances. As no significant changes in ion intensities were observed, overall ESI-analysis was performed without any additives.

Table 1 *Compounds studied by SPME–HPLC–ESI$^\pm$–MS*

Pesticides investigated in soil from a former industrial region

	Molecular weight	ESI$^+$ target ion
Aminocarb	208	209
Asulam	230	253 (M + Na)
Chloropham	213	212 (−)
Methomyl	162	185 (M + Na)
Oxamyl	219	242 (M + Na)
Promecarb	207	208/230 (M + Na)
Propham	179	180/138 (M − 43 + H)
Carbofuran	221	222/244 (M + Na)
Bendiocarb	223	246 (M + Na)
Propoxur	209	210/232 (M + Na)
Simazine	201	202
Atrazine	230	216 (M − 15 + H)
Propazine	229	230
Prometryn	241	242

Compounds studied* and identified$^\#$ in Canadian sludge and a river sediment

	ESI target ion
Heptanedioic acid*	159 (−)
Decanedioic acid*	201 (−)
Dodecanoic acid*	199 (−)
Hexadecanoic acid*,$^\#$	255 (−)
Heptandecanoic acid*	269 (−)
Octadecanoic acid$^\#$	283 (−)
Dibutyl phthalate*,$^\#$	279 (+)
Di(2-ethylhexyl) phthalate*,$^\#$	391 (+)
Benzoic acid*	121 (−)
Dihydroxybenzenes*,$^\#$	109 (−)
Polyhydroxy ether$^\#$	e.g. 421
Alkylsulfonates$^\#$	e.g. 285 (−)
Trichlorophenols$^\#$	195 (−)
Succinate and adipic acid ester$^\#$	133, 147, 165 (+)

(−) ESI$^-$; (+) ESI$^+$

APCI–MS with flow injection was only applied for comparative purposes and operated with methanol/water (50:50, v/v, 1% acetic acid). APCI–MS was of lower sensitivity than the ESI interface. For example, the pesticide Chloropham was detected via flow injection-APCI$^+$ at about 1 ng mL^{-1}. Although the negative mode only allowed determination in the 100 ng mL^{-1} range, ESI$^+$ detection was possible up to 200 pg mL^{-1}. Furthermore, in APCI$^+$ among the desired [M + H]$^+$ ions, additional fragments with varying abundances appeared, such as [M − methyl]$^+$, or in the case of the pesticide Barban for instance [M − C$_2$H$_5$]$^+$. As a changing split in ion distribution is not acceptable for quantitation, electrospray ionisation was preferred for all experiments.

The combination of low flow rates for ESI–MS (0.05–0.2 mL min^{-1}) and narrow bore LC columns corresponds to the advantages of SPME.

SPME–HPLC–ESI–MS analysis of pesticides, polyhydroxylated benzenes, alkylated and aromatic carboxylic acids and esters from environmental compartments produced optimum results at the following conditions. The temperature of the spray capillary was maintained at 200 °C and the manifold temperature was set to 70 °C. The spray voltage was maintained at 4.5 kV. Nitrogen was used as sheath gas for drying and nebulising at a pressure of 55 psi. During all experiments an octapol potential (CID offset) of 10 V was adjusted to reduce abundant cluster ions. A Finnigan SSQ7000 mass spectrometer was used in SIM mode with a scanning rate of 3 amu s^{-1} and a span of 0.3 amu.

HPLC separation was performed on a 150 × 2.1 mm i.d. stainless steel column packed with Supelcosil LC18 (5 μm, Supelco, Deisenhofen, Germany). For pesticide analysis, gradient elution at a flow rate of 0.2 mL min^{-1} with water (A) and methanol (B) as solvents was applied using the following program: linear gradient from 100% A to 100% B within 2 min; 18 min isocratic period; linear gradient to 50% of both A and B to 22 min and returning to 100% A to 30 min. The stability of the selected SIM target ions during the entire gradient program was examined using a reference pesticide mixture. In addition to the ESI–MS data, the UV signal at 225 nm was used for detection. For the characterisation of sludge and sediment components, the following gradient elution at a flow rate of 0.1 mL min^{-1} was preferred. After an isocratic period of 5 minutes using a mixture of water/methanol (50:50, v/v), a linear gradient to 100% methanol within 20 minutes was applied. An isocratic period of 10 minutes followed and finally, the eluent composition returned to the initial mixture.

Suitability of SPME Coatings

The initial SPME–HPLC application concerned the determination of PAH's and surfactants from aqueous samples.[26,27] All commercially available fibre coatings were examined and PDMS proved to be the most suitable for the SPME–HPLC analysis of PAH's and of non-ionic model surfactants.

The selection of SPME fibre coating corresponded to the nature of the analytes, a primary feature of which is their polarity. The new generation of fibre coatings equipped with a high extraction capacity, especially for polar compounds is based on Carbowax. In principle, the rates and reversibility of the adsorption/desorption mechanisms occurring make this fibre material interesting for LC application.

Thus, in the applications presented here, it came as no surprise that of all coatings examined the solid, polymeric sorbents Carbowax and polyacrylate proved to be the most acceptable for the SPME–HPLC–MS analysis of polar and semi-polar compounds as listed in Table 1.

Fibre Comparison

Figure 1 shows the results of an inter-fibre comparison using carbamate pesticides as examples. The SPME fibres investigated for an HPLC–MS coupling were Carbowax/divinylbenzene (CW/DVB) of 65 μm thickness, Carbowax/Template Resin (CW/TPR) of 50 μm thickness and a 85 μm polyacrylate coated fibre, all commercially available from Supelco. The poly(dimethylsiloxane)/divinylbenzene (PDMS/DVB) fibre was not tested because first investigations showed poor extraction yields for some carbamates like methomyl, oxamyl, and carbofuran.[30]

The fibres examined were conditioned prior to use in methanol at room temperature for one hour. A series of experiments was carried out to examine the performance of the various fibre coatings. Both Carbowax coatings produced sufficient extraction yields for appropriate analysis at ppb levels.

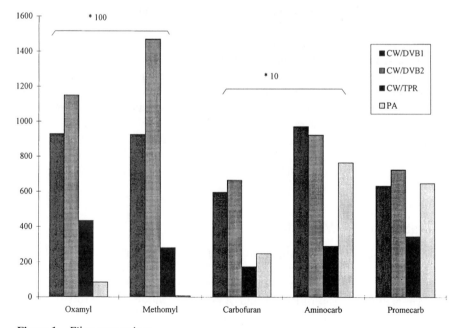

Figure 1 *Fibre comparison*

Using the CW/TPR coating, an extraction yield of about 50% compared with the CW/DVB fibre was obtained. The PA fibre proved to be unsuitable for this application owing to the low extraction yields observed for some carbamates. When comparing the performance of newly conditioned (CW/DVB2 in Figure 1) and a Carbowax fibre used for 20–30 extraction/desorption cycles (CW/DVB1 in Figure 1), a mean loss of about 9% was observed and, finally, the coating of the multiple used fibre was partially destroyed. The same experiments carried out with the CW/TPR coating showed greater stability in the extraction results, even after a certain period of usage. This coating was used for more than 50 extraction/injection cycles with a loss of only about 3% in efficiency and without any remarkable destruction features which reflects also the higher ruggedness of this coating.

The reproducibility of the extraction of different charges of newly conditioned CW/TPR fibres was examined. Sufficient reproducibilities between 1.6 and 12% (mean 6%) were determined. The mean reproducibility of one coating was about 8%. The more rugged CW/TPR fibre is especially recommended for SPME–HPLC application despite its lower extraction efficiencies as the greater durability of the CW/TPR coating compensates for this disadvantage.

Carryover

Unfortunately, a high carryover of between 11 and 20% was observed for compounds with high affinities (partition coefficients) to the coating, such as triazine pesticides or phthalates. Heating the interface during desorption to 60 °C was unable to eliminate a certain carryover. In these cases, thorough cleaning before each analysis is required. A washing step of 10 min using pure methanol can overcome carryover effects. The high-polarity compounds such as asulam, oxamyl and methomyl were desorbed completely from the fibre by rinsing with methanol during the desorption step in the SPME–HPLC interface.

In the case of a high content of accompanying impurities the carryover observed necessitates a careful analysis strategy with blank runs before each analysis. If required a thermal treatment is recommended to prevent the accumulation of matrix components on the fibre, or an additional hollow fibre must be placed around the coating to keep it away from polymeric matter (*e.g.* humic acids).

SPME Interface and Desorption of Analytes

The extracted analytes are transferred to the HPLC column *via* a specially designed SPME interface (see Chapter 1, Figure 1b). This SPME device uses a conventional Reodyne (six-port) injection valve where the original injection loop has been replaced by a SPME desorption chamber.

The separation of the static desorption from the injection into the HPLC column allows the application of different solvent mixtures for desorption. The typically less polar desorbing solvents (methanol, ethanol) show higher eluting power than the initial solvent composition used in reversed phase HPLC.

However, the application of arbitrary solvents for static desorption has its limits. During the application of pure nonpolar, organic solvents, the fibre coatings can be affected by swelling, and partial loss of coating parts may even occur. Furthermore, the peak resolution of very polar, quickly eluting compounds (C-18 column) is jeopardised when pure organic solvents such as methanol are used for desorption. Therefore, no separate desorption solvent was used and the dead-volume of the desorption chamber was reduced ($< 10\ \mu L$) by cutting the tubing as short as possible.

Various experiments with the dead-volume modified SPME chamber using different mixtures for desorption (water/methanol, methanol/ethanol) proved that the HPLC peak shapes were less affected by different solvent compositions.

The commercially available SPME/HPLC interface introduced by Supelco uses a larger volume for desorption (*ca.* 70 μL). In this case the selection of the appropriate desorption solvent is very important to avoid peak broadening in HPLC separation. In the SPME–HPLC determination of PAHs using the commercial interface and fluorescence detection, well shaped peaks were obtained when very similar solvent mixtures (water/acetonitrile) for desorption and HPLC elution were used.

For the desorption of polar pesticides using the low-volume SPME interface, pure methanol proved to be the most efficient solvent. The analytes extracted from sludge were of a more nonpolar nature and required a mixture of methanol/ethanol (80:20, v/v) for sufficient desorption.

Before the fibre was inserted into the desorption chamber, 50 μL of the corresponding desorbing solvent was injected to guarantee complete flushing of the fibre. When the fibre was sealed in the interface, desorption was started. Although 5 min were generally used for desorption, further experiments indicated that the kinetics of the fibre desorption were fast, with 80–90% of the analytes being released within the first minute of desorption.

After the fibre was removed from the interface, it was flushed five times with 100 μL methanol to avoid carryover.

Matrix Influence

Different sample preparation techniques were examined to study the SPME results with respect to changing matrices.

Pure water solutions and soil leachates were spiked with reference pesticides. The corresponding SPME sampling conditions applied were similar to those discussed in detail by Pawliszyn.[31] The best results were obtained with salt saturation, adjusting the pH of the solution to 2 and vigorous magnetic stirring (1000 rpm).

The extraction mechanism of the Carbowax coating differs from that of the polymeric liquid coatings like PDMS. The solid porous surface of the Carbowax coating mainly enables the reversible adsorption of the analytes. At constant temperature the extraction time is an important factor influencing the extraction yield. The extraction time profiles of selected triazines, phthalates and benzoic acid were determined using the original sample matrix of a soil leachate prepared as described below.

The extraction time profiles with CW/TR indicated that most of the pesticides reached the maximum extraction amount at 30 minutes—the exception being prometryn, which exhibited increasing extraction yield up to 60 minutes. The high yield and the kinetics of the extraction behaviour are probably reflected by parameters such as the organic carbon sorption coefficient K_{oc}, which characterise the sorptive properties of a compound. The corresponding results of such sorption experiments[32] using clay as a target clearly showed the highest values for prometryn with $K_{oc} = 611$ compared with, for example, simazine of $K_{oc} = 155$ or atrazine of $K_{oc} = 120$.

An extension of the extraction time to 15 hours results in a small decrease of all extraction yields, and so one hour was preferred overall in the experiments as the standard extraction time.

In the cases of dibutyl phthalate and di(2-ethylhexyl) phthalate, an extraction time of just 15 minutes provided the maximum yield (Figure 2). These short equilibrium times reflect the high affinities of these compounds to the Carbowax coating. The corresponding octanol–water partition coefficients, $\log K_{OW}$, of 4.72 and 7.6 are high compared with those of the pesticides extracted (simazine $\log K_{ow} = 2.1$, propazine $\log K_{ow} = 3.0$, atrazine $\log K_{ow} = 2.38$)[32,33] and determine their extraction behaviour.

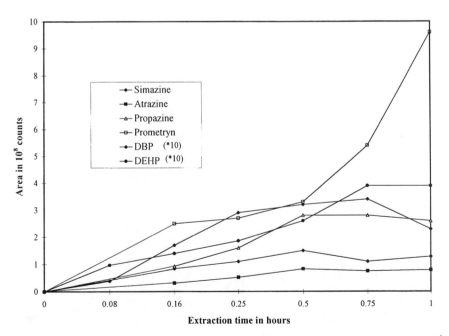

Figure 2 *Extraction time profiles obtained for selected substances in the $\mu g\,mL^{-1}$ range using CW/TPR: DBP = dibutyl phthalate; DEHP = di(2-ethylhexyl) phthalate*

SPME of a slurry containing solid particles differs from the conditions studied above. The competitive equilibria which exist between the solid matter and the water phase influence the distribution constants of the analytes. The extraction rate is largely determined by the kinetics of the analyte released from the solid matter. Consequently, the optimisation of the extraction time is an important step in slurry analysis to adjust maximum sensitivity. By way of an example, the pesticide contents of soil leachates and corresponding soil slurries were determined and the results compared (Figure 3). Air-dried soil (90 g) was leached with 900 mL water at pH 4 for 24 hours, and 4 mL was then analysed by SPME–HPLC–MS (1 hour extraction time). By comparison, a slurry of only 200 mg air-dried soil and 4 mL of NaCl saturated water was prepared and also sampled for 1 hour. Following the assumption that the analyte equilibrium between soil particles and water has to be adjusted before SPME sampling, in the next experiment the soil slurry was stirred for 110 minutes and then directly sampled with CW/TPR for 60 minutes. The results in Figure 3 show that in the case of simazine and atrazine, SPME of a slurry made from a very small initial amount of soil provided results comparable to the much more time-consuming leaching procedure. The extraction of propazine was improved by a factor of 1.3 when the slurry was preconditioned. Nevertheless, the extraction yield of the corresponding leachate was not reached. By contrast, the extraction of prometryn was most successful in combination with leaching. Probably, the high K_{oc}

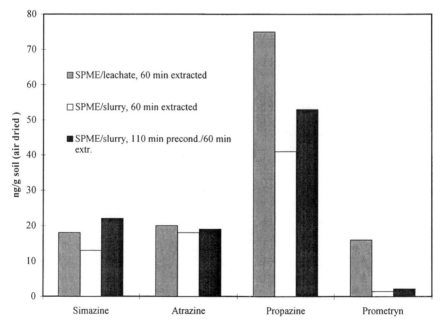

Figure 3 *Comparison of SPME–HPLC–MS results obtained from a soil leachate and the corresponding soil slurry*

value points to a strong sorption on the soil particle and the kinetics of release are correspondingly slower. An increase in the sampling temperature ought to affect the results observed as the water solubility of analytes can significantly depend on the temperature. As an example, the water solubility of simazine for instance varies between 6 mg L^{-1} at 22 °C and 84 mg L^{-1} at 85 °C, while for atrazine the differences are even larger, namely from 33 mg L^{-1} at 22 °C to 320 mg L^{-1} at 85 °C.[32,33] The corresponding SPME experiments were performed at 40 °C during the entire sample preconditioning and extraction period. This moderate temperature increase was chosen to avoid the undesired lower partitioning of analytes at higher temperatures. The resulting extraction yields were only improved by about 2% for particular pesticides. The extraction results of other constituents of the leachate matrix such as the comparison of the respective UV traces and the total ion current traces even remained constant.

3 Quantitation of Pesticides

The samples used to demonstrate the performance of a SPME–HPLC–MS coupling originated from the waste disposal of a disused chemical plant near Bitterfeld (central eastern Germany), where a great variety of waste and by-products of chemical production were deposited over a long period. Today the polar and water-soluble residues pose a threat to the groundwater quality of the surrounding region.

Detection Limits, Precision and Reproducibility

The pesticides shown in Table 1 were determined by using the target ions listed.

The LOD were calculated using a signal-to-noise ratio of 3. In general, the limit of detection achieved with ESI$^+$–MS was about 10 ng mL^{-1} or less (Figure 4). Using a 1 hour exposure time, the method's accuracy was measured for four triazine herbicides originally determined in soil leachates from Bitterfeld. The precision obtained using the 50 μm CW/TPR fibres varied between 5.7 and 15% RSD ($n = 4$) depending on the compound.

Comparison between pure water and original leachates spiked with references showed that the additional matrix components do not interfere with the analysis of the triazines, chiefly since selected ion monitoring improves the signal-to-noise ratio. The extraction results of both experiments showed close correlation, and thus the effect of the accompanying matrix on the SPME results was almost insignificant. The fibre was used for over 50 extraction/injection cycles without any significant loss in efficiency.

Linearity

Quantitation with CW/TPR was performed through standard addition to compensate for changes in matrix composition.

The calibration curves of four triazines and selected carbamates created in the original matrix by standard addition showed a small linear range between lower

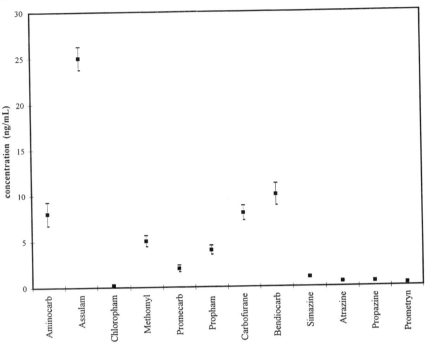

Figure 4 *Ranges of detection limits of compounds analysed by SPME–HPLC–ESI⁺–MS*

ng mL^{-1} and 100 ng mL^{-1} concentrations. The interpretation of the UV signal indicated the same effects as the ESI signal, although a higher LOD was achieved.

It is assumed that both the ionisation interface and the use of the porous polymer fibre coating are the cause of the problem. The ion generation process applied to highly concentrated analyte solutions is often not efficient enough to ionise all the analyte molecules present in the sample. As a result, too low an amount of analyte is detected. The concentration that limits the linearity of the calibration curve depends on the eluent composition, the flow rate and the nature of the analyte. The second possible reason for a narrow linear calibration range may arise from the adsorption/desorption mechanism of the CW/TPR coating, which determines the extraction success of the analytes. Thus the careful selection of the standard concentrations is necessary for quantitation with the standard addition mode.

The soil samples from the Bitterfeld region were analysed under optimised SPME–HPLC–ESI⁺–MS conditions. Four triazine pesticides were identified and determined. The contents of between 19 and 53 ng g^{-1} air-dried soil are shown above in Figure 3. Other pesticides of a more polar nature were not detected owing to their high water solubility and related mobility in wet soil. Furthermore, the polar carbamate pesticide is bound to undergo rapid partial decomposition processes under genuine conditions.[34]

4   Investigation of Sludge and Sediments

The increasing application of sewage sludge in agriculture advances the broad spread of partially toxic compounds into the environment. The analysis of even apparently harmless compounds such as phthalates, surfactants and fatty acids has to be considered under environmental parameters, since the results of toxicological investigation have indicated special hormone-like effects affecting human health.[35,36] The stock of data concerning these compound classes is still small. The highly complex mixture of bio-organic, partially polymeric material and chemical residues originating from domestic usage and industry make analysis very difficult.

The most popular methods used for the determination of organics from sludge are based on solvent extraction. They include accelerated solvent extraction (ASE) or SFE,[37,38] and involve laborious, material-consuming sample preparation. Because GC analysis of polar constituents often requires derivatisation steps, reversed phase HPLC is one of the most commonly used and most suitable techniques, especially when coupled with SPME and MS. The characterisation of sludge or sediment constituents ought to be facilitated by the mass spectrum information; however, the simplicity of the ESI or APCI mass spectra often does not assist the identification of unknown species. Additionally, the diversity of cluster ions appearing depends on both eluent composition and ionisation conditions. This deficiency can be overcome by the use of MS–MS if available or by measuring references. The suitable choice of these standards can reflect and summarise the properties of substance classes in API–MS detection. The compounds in Table 1 marked with an asterisk were selected for the optimisation and validation of SPME–HPLC–ESI$^{\pm}$–MS conditions for sludge and sediment analysis. The LOD of most compounds from sludge and sediments were found to be in the range of 10–100 ng mL^{-1} with precision between 8 and 15%.

The compositions of the Canadian sludge (Leslie Sewage Treatment Plant, Burlington, Canada) and the sediment of the River Elster (sampled near Leipzig, Germany) clearly differ both qualitatively and quantitatively, as shown by Figure 5.

The SPME of just a few milligrams of air-dried and screened sludge and sediment allowed the identification of most of the water-soluble substances within an extremely short time compared with classic sample clean-up and preparation procedures. It should be emphasised that only the interplay of negative and positive ESI–MS was able to provide full information about the very complex mixture of water soluble substances.

Owing to the unsatisfactory separation of individuals within the same substance class, only semi-quantitative conclusions result. The semi-quantitative results of the summarised substance classes listed in Table 2 were obtained by taking the different response factors into account. In both samples the class of alkyldicarboxylic acids account by far for the largest share, which is doubtless caused by their high water solubility. One remarkable difference is the presence of chlorophenols in the Elster river sediment, which clearly points to the impact

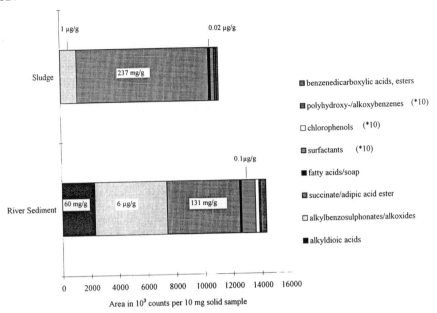

Figure 5 *Distribution of water-soluble substance classes found in a Canadian sludge and a river sediment*[13]

Table 2 *Contents of characteristic water-soluble compound classes found in a Canadian sewage sludge and a river sediment*

	River sediment $\mu g\ g^{-1}$	Sludge $\mu g\ g^{-1}$
Alkyldicarboxylic acids	191 000	238 000
Fatty acids/soaps	800	1050
Surfactants	6	1
Chlorophenols	2	0
Polyhydroxy-/alkoxy-benzenes	100	130
Benzenedicarboxylic acids, esters	400	770

of chemical industry. Otherwise the concentrations are in the lower mg g^{-1} to μg g^{-1} range. Overall the portion of water-soluble components was larger in the river sediment compared to the sludge sample.

5 Summary

SPME used in conjunction with HPLC–MS proved to be suitable for the quantitation of polar and semi-polar organic compounds from aqueous slurries of soils, sediments and sewage sludge. The following brief list of the main features of the SPME–HPLC–MS set-up should encourage further application.

1. Time-saving direct sampling from slurries is possible, although the detailed determination of equilibrium times is necessary to characterise matrix influences.
2. The extraction step is solvent-free but the desorption process requires appropriate solvents or mixtures with sufficient desorption capacity. The combination of suitable desorbing solvents and a SPME desorption chamber with a small volume boosts sensitivity.
3. Reproducible results can be obtained from milligram initial sample amounts.
4. The porous polymer coatings CW/DVB and CW/TR provided the highest extraction capacities, but the monitoring of carryover is more necessary the higher the analytes' adsorption affinities. If necessary, additional fibre-cleaning and the use of blanks before analysis prevents carryover and misinterpretation of results. At present CW/TPR is the most suitable fibre for the determination of polar and semi-polar analytes.
5. When using API–MS, the detection limits of pesticides at ng mL^{-1} level were obtained but unfortunately also had small linear ranges of the calibration curves.
6. This combination of methods is distinguished by high selectivity and identification capacity for unknown compounds, especially since MS–MS options are available. The careful choice and tuning of MS ionisation conditions define the selectivity and sensitivity of the entire method combination.
7. Last but not least, the possibility of automation[39–41] is a major advantage for industrial applications and monitoring programs.

References

1. P. Parrilla and J.L.M. Vidal, *Anal. Lett.*, 1997, **30**, 1719.
2. K.K. Verma, A.J.H. Louter, A. Jain, E. Pocurulli, J.J. Vreul and U.A.T. Brinkman, *Chromatographia*, 1997, **44**, 372.
3. M. Möder, S. Schrader, U. Franck and P. Popp, *Fresenius' J. Anal. Chem.*, 1997, **357**, 326.
4. C. Aguilar, F. Borrull and R.M. Marce, *J. Chromatogr. A*, 1997, **771**, 221.
5. D. Gorlo, J. Namiesnik and B. Zygmunt, *Chem. Anal. (Warsaw)*, 1997, **42**, 297.
6. S.J. Lehotay and A.V. Garcia, *J. Chromatogr. A*, 1997, **765**, 69.
7. K. Bester and H. Huhnerfuss, *Fresenius' J. Anal. Chem.*, 1997, **69**, 2742.
8. U.A.T. Brinkman, *Chromatographia*, 1997, **45**, 445.
9. J. Slobodnik, A.C. Hogenboom, J.J. Vreuls, J.A. Rontree, B.L.M. van Baar, W.M.A. Niessen and U.A.T. Brinkman, *J. Chromatogr. A*, 1996, **741**, 59.
10. C. Crescenzi, A. DiCorcia, E. Guerriero and R. Samperi, *Environ. Sci. Technol.*, 1997, **31**, 479.
11. M. Möder, H. Löster, R. Herzschuh and P. Popp, *J. Mass Spectrom.*, 1997, **32**, 1195.
12. M. Möder, R. Eisert, J. Pawliszyn and P. Popp, in press.

13. M. Möder, P. Popp and J. Pawliszyn, *J. Microcol. Sep.*, 1998, **10**, 225.
14. H.Y. Lin and R.D. Voyksner, *Anal. Chem.*, 1993, **65**, 451.
15. D. Volmer, K. Levsen and G. Wünsch, *J. Chromatogr. A*, 1994, **660**, 231.
16. D. Volmer, A. Preiss, K. Levsen and G. Wünsch, *J. Chromatogr.*, 1993, **647**, 235.
17. R.B. Geerdink, P.J. Berg, P.G.M. Kienhuis, W.M.A. Niessen and U.A.Th. Brinkman, *J. Int. Environ. Anal. Chem.*, 1996, **64**, 265.
18. J. Slobodnik, M.E. Jager, J.F. Hoeckstra-Oussoren, M. Honing, B.L.M. van Baar and U.A.T. Brinkman, *J. Mass Spectrom.*, 1997, **32**, 43.
19. D. Barceló, *J. Chromatogr.*, 1993, **643**, 117.
20. J. Slobodnik, B.L.M. van Baar and U.A.T. Brinkman, *J. Chromatogr. A*, 1995, **703**, 81.
21. N.H. Spliid and B. Køppen, *J. Chromatogr. A*, 1996, **736**, 105.
22. H.F. Schroeder, *Environ. Monit. Assess.*, 1997, **44**, 503.
23. D.A. Volmer, D.L. Volmer and J.G. Wilkes, *LC GC—Magazine of Separation Science*, 1996, **14**, 216.
24. C. Molina, M. Honing and D. Barceló, *Anal. Chem.*, 1994, **66**, 4444.
25. EEC Drinking Water Guidelines, 80/779/EEC, EEC No. L229/11-29, EEC, Brussels, 1980.
26. J. Chen and J. Pawliszyn, *Anal. Chem.*, 1995, **67**, 2530.
27. A.A. Boy-Boland and J.B. Pawliszyn, *Anal. Chem.*, 1996, **68**, 1521.
28. D. Barceló, G. Durand, R.J. Vreeken, G.J. De Jong, H. Lingeman and U.A.T. Brinkman, *J. Chromatogr.*, 1991, **553**, 311.
29. H. Itoh, S. Kawasaki and J. Tadano, *J. Chromatogr. A*, 1996, **754**, 61.
30. Application Note 121, SUPELCO, 1997.
31. J. Pawliszyn, *Solid Phase Microextraction. Theory and Practice*, Wiley-VCH, New York, 1997.
32. *CIBA GEIGY Corporation—Toxicology Data*, 1989, in *ARS Pesticide Properties Database*, Remote Sensing & Modeling Laboratory (RS & ML).
33. *Herbicide Handbook of the Weed Society of America*, 5th Edn., 1983, in *ARS Pesticide Properties Database*, Remote Sensing & Modeling Laboratory (RS & ML).
34. G.R. Chaudhry, *Biological Degradation and Bioremediation of Toxic Chemicals*, Chapman & Hall, London, 1994, pp. 198.
35. A.M. Soto, H. Justicia, J.W. Wray and C. Sonneschein, *Environ. Health Perspect.*, 1991, **92**, 167.
36. W. Klein, W. Kördel, G.H.M. Krause and J. Wiesner, *Kriterien zur Beurteilung organischer Bodenkontaminationen: Dioxine (PCDD/F) und Phthalate*, ed. DECHEMA e.V., Frankfurt am Main, 1995.
37. P. Popp, P. Keil, M. Möder, A. Paschke and U. Thuss, *J. Chromatogr. A*, 1997, **774**, 203.
38. T.L. Chester, J.D. Pinkston and D.E. Raynie, *Anal. Chem.*, 1994, **66**, 106R.
39. R. Eisert and J. Pawliszyn, *J. Chromatogr. A*, 1997, **776**, 293.
40. R. Eisert and J. Pawliszyn, *Anal. Chem.*, 1997, **69**, 3140.
41. R. Eisert and J. Pawliszyn, *Crit. Rev. Anal. Chem.*, 1997, **27**, 103.

CHAPTER 24

SPME–HPLC of Environmental Pollutants

ANNA A. BOYD-BOLAND

1 Introduction

The advantage of HPLC over GC for the analysis of thermally unstable or low to non-volatile analytes lies in the fact that no derivatisation steps are required. The disadvantage can lie in the fact that, owing to the lower sensitivity of many of the detection systems used with HPLC, analytes can require further concentration or larger injection volumes than GC. This can be a concern when the quantity of an analyte is limited (for example in forensic applications) or when concentration methods lead to the loss of some analytes, for example by adsorption to vials *etc.* To some extent, HPLC–MS can compensate for these disadvantages; however, HPLC–MS systems are not the most economical solution for many labs and are not always a viable alternative.

SPME combined with HPLC provides a means by which simple, rapid concentration of analytes can be achieved together with a means of introduction of the concentrated analytes to the HPLC system. This eliminates the need for larger injection volumes and avoids the derivatisation step required if the analytes were to be detected by GC. Although attempts have been made to incorporate derivatisation reagents in the SPME device, these methods are in the developmental stages and until these methods are developed, SPME–HPLC provides the simplest alternative for thermally unstable or non-volatile analytes.

This chapter provides an overview of the application of SPME–HPLC developed for analysis of chromium, comparison of the new SPME method with other methods and the application of the method to real samples.

2 Chromium Speciation and Analysis Methods

Chromium occurs naturally in trace amounts in natural waters through weathering of rocks and erosion of soils. Chromium at higher levels can be found from contamination of waters, for example by tannery waste, chrome plating industries and wood treatment plants. The two stable oxidation states in water are Cr(III) and Cr(VI) at ratios that are dependent upon the pH and E_h values of the water in question. Although Cr(III) is the more stable state in natural waters, Cr(VI) is always present to some extent and is the dominant species in sea water. Furthermore, Cr(III) can be oxidised to Cr(VI) and as Cr(VI) is by far the more toxic of the two species[1] methods that only measure total chromium concentration can seriously underestimate or over-estimate the toxicity of the water sample, depending upon which form of chromium is present. It is thus important to develop methods that are capable of distinguishing between the two oxidation states.

The instability of chromium species in solution after extraction can lead to variation in the concentration of chromium present. It is thus important to analyse the two chromium species with as little sample handling as possible and as rapidly as possible after sampling occurs. Many speciation methods involve separation of the two chromium oxidation states using different pre-treatment methods, or separation by formation of different complexes. These methods can cause serious loss of one species or another.

Cathodic stripping voltametry is a sensitive technique for measuring total chromium. To determine the individual chromium species, formation of an inert complex of the Cr(III) allows the levels of Cr(VI) to be determined directly and the Cr(III) by difference.[2]

There are several HPLC methods that enable determination of the different species of chromium; however, they are not particularly sensitive to Cr(VI).[3–9] SPME combined with HPLC provides a means by which the HPLC methods can be made more sensitive and the sample handling minimised. The method also allows for simultaneous preconcentration and extraction of the two chromium species.

3 Extraction Principle for Chromium Species

SPME relies on extraction of analytes due to their partition coefficients between the sample matrices and the coated phase. In the case of chromium, the two species present in water are the anionic Cr(VI) and the cationic Cr(III). In order to analyse the two oppositely charged species using the one coating and with one HPLC system, formation of a complex with either Cr(III) or Cr(VI) is necessary. EDTA forms stable 1:1 complexes with most cationic metal species, and Cr(III) is no exception. The EDTA forms the Cr(EDTA)⁻ complex with Cr(III) and does not complex with the Cr(VI) species.

Once the chromium species are both anionic, a cationic ion-pairing reagent can be used to elute the species from a reversed phase LC-8 column. The ion-pairing reagent selected is tetrabutylammonium hydroxide (TBAOH) which is combined with acetonitrile to form the mobile phase. The chromium species are desorbed by the mobile phase and partitioned between the stationary phase and the mobile phase, resulting in the usual chromatographic separation. Retention times of the species are obviously affected by the ratio of acetonitrile:TBAOH in the mobile phase with an optimal ratio that was determined experimentally. Increasing the TBAOH concentration tended to cause an increase in retention time, whereas increased acetonitrile caused decreased retention times with an eventual loss of resolution.

4 Optimised Conditions

Table 1 provides details of the optimum extraction and desorption conditions as well as the GC parameters used. The important factors involved in optimisation of the chromium method, apart from the usual SPME parameters (extraction time, desorption time, salt, coating selection *etc.*), were the quantity of complexation reagent added, the pH and the effect of other metal species in the solution. Typically the pH values are adjusted in SPME methods to enhance extraction levels; however, for this method the more important consideration was to select a pH that maintained the original distribution of chromium oxidation states and encouraged formation of the EDTA–Cr(III) complex.

Cr(VI) is always present as a negatively charged complex in the pH range of natural waters (either as $HCrO_4^-$ or CrO_4^{2-}). Cr(III) complexation with EDTA is favoured at pH 6, whereas at higher pHs precipitation of $Cr(OH)_3$ occurs. EDTA can be oxidised by Cr(VI) at low pHs. A study of the effect of pH on Cr(III) formation from a Cr(VI) standard solution to which EDTA was added showed that at pHs below 5 there was appreciable oxidation of EDTA and reduction of Cr(VI) to Cr(III).

The rate of complexation of Cr(III) with EDTA increases with increasing pH and temperature; however, increasing temperature also increases the oxidising power of Cr(VI). The complexation rate is also enhanced by increasing the ratio of EDTA:Cr; however peak broadening was observed for very high (30:1) ratios. For this reason, the optimal ratio of EDTA to Cr(III) was determined to be 10:1. When the temperature was increased from 25 to 40 °C, the time for complete complexation of the chromium was four times faster at both pH 4 and pH 6; however the rate at pH 6 was significantly faster than the rate at pH 4.

The optimal pH of the extraction was thus determined to be pH 6.[10] At this pH there is no evidence of any change in the oxidation states of the Cr species

Table 1 *Optimal conditions for chromium speciation*

Parameter	Conditions
Extraction Time	60 min
Desorption Time	2 min
HPLC run time	15 min
Column	RP-C8 (15 cm × 4.6 mm × 5 μm)
Mobile Phase	Isocratic: 0.021 M TBAOH, 12% acetonitrile, adjusted to pH 6 with NaOH
Detection	UV vis, 242 nm
Linear Range	1–40 ppm
LOD	17 ppb Cr(III); 5 ppb Cr(VI)
Precision	2%

and the reaction rate is sufficiently fast (complete complexation occurring within 25 min for 10:1 EDTA:Cr ratio, or 6 min for 30:1 EDTA:Cr ratio).

The stability of the species over time was also considered as an important factor in the analysis. When samples were left for five hours in the presence of the EDTA complex, it was found that the RSD in peak height was less than 6%. After 48 hours, the amount of Cr(VI) decreased, while the concentration of Cr(III) increased.[10] It is thus recommended that samples not be stored with the EDTA for longer than 5 hours.

The interference of other metal ions was considered by investigating the recovery of chromium in the presence of Co^{2+}, Cu^{2+}, Fe^{3+} and Pb^{2+}. None of these species were found to interfere with Cr recovery or chromatographic detection, as the peaks had significantly different retention times. It was noted that Co^{2+} and Cu^{2+} EDTA complexes had the same retention times.[10]

The coatings evaluated for the method were CWAX/TR, PDMS/DVB, and several prototype fibres called: SPME3/anion exchange, PDMS/anion exchange, PDMS/sol gel with octyl group. The CWAX/TR provided the best extraction efficiency for the two chromium species and the best precision.

Under the optimum conditions described above, the limit of detection, linear range and precision were determined (see Table 1). The linearity of the method for concentrations between 1 and 50 ppm was extremely good, with correlation coefficients of 0.997 and 0.999 for Cr(III) and Cr(VI), respectively. Investigation of the linearity above 50 ppm showed some decreased sensitivity, possibly as a result of insufficient EDTA for complete complexation to occur within the time selected. For high concentrations of chromium it is thus recommended that samples be diluted or further optimisation be undertaken for high level work.

The limit of detection of the method was determined to be 17 ppb for Cr(III) and 5 ppb for Cr(VI). Approximately 16% of the chromium species are extracted from the water sample by the SPME fibre.

Figure 1 shows a chromatogram of the chromium species under the optimum

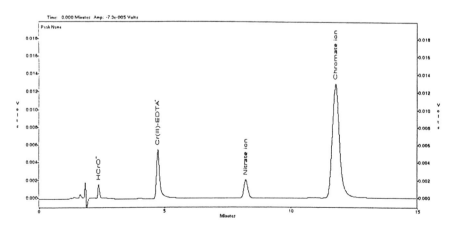

Figure 1   *Chromatogram of a standard solution of Cr(III) and Cr(VI) by SPME–HPLC under optimum conditions*

Table 2 *Comparison of HPLC methods for chromium speciation*

Cr Species	Column/detector	LOD (ppb)	Reference
Cr(VI)	RP-C8/AAS	120	4
Cr(III)		60	
Cr(VI)	RP-C18/AES	nd	5
Cr(III)		5–10	
Cr(VI)	RP-C18/UV	40	7
Cr(III)		23	

conditions. The peaks are identified in the figure and are representative of the ratios of the species found in the standard prepared from potassium dichromate and chromic nitrate. All the species are separated within 13 minutes.

5 Comparison with Other HPLC Methods

Table 2 provides a comparison of the limits of detection of various HPLC methods for Cr(VI) and Cr(III). The new SPME–HPLC method provides substantially improved detection limits for Cr(VI) despite the use of the relatively less sensitive UV detection system (compared with AAS). Cr(III) detection limits are generally better than other methods or at worst similar. Although there are more sensitive techniques available, such as ICP–AAS and ICP–AES, these methods are generally not portable and are more expensive than conventional HPLC systems.

The method developed for the extraction of chromium results in approximately 16% of the chromium in a sample being extracted by the fibre. To achieve the same transfer of analytes to the HPLC by injection of a 40 ppm solution would require 0.6 mL to be injected. This is not practical when resolution is important as large volumes can lead to peak broadening, or when sample sizes are limited.

6 Real Samples

Untreated and treated tannery waste were analysed by the new SPME–HPLC method and the results compared with those obtained by the analytical laboratory that regularly analyses the waste. Figure 2 shows the chromatogram obtained by SPME–HPLC of the untreated tannery waste. The results obtained using the two methods for untreated and treated waste were the same within experimental error. The untreated samples contained 51 ppm total chromium (50 ppm determined by the analytical laboratory) and the treated waste contained 0.18 ppm (0.17 ppm by the analytical laboratory).[10] The HPLC method alone was not able to detect Cr(III) or Cr(VI) below 1 ppm with a 10 μL injection, hence it would be unable to determine any of the chromium levels in the treated tannery waste.

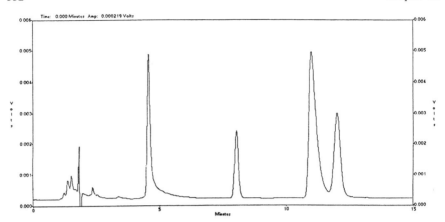

Figure 2 *Chromatogram obtained from SPME–HPLC of* 4 mL *of untreated tannery*
waste

7   Conclusion

The development of smaller volume desorption chambers to enable simple
coupling of SPME with HPLC–MS will provide simple accurate screening of
water samples for non-volatile and thermally unstable analytes. The advent of
autosampler methods and on-line SPME methods that have recently been
published should lead to a further increase in the types of applications available
as well a further reduction in the labour requirements for the methods.

References

1. G. Vos, *Fresenius' J. Anal. Chem.*, 1985, **320**, 556.
2. G. Ou-Lien and J. Jen, *Anal. Chim. Acta*, 1993, **279**, 329.
3. M. Sperling, S. Xu and B. Welz, *Anal. Chem.*, 1992, **64**, 3101.
4. A. Syty, R.G. Christiansen and T.C. Rains, *J. Anal. Atom. Spectrom.*, 1988, **3**, 193.
5. I.S. Krull, K.W. Panaro and L.L. Gersheim, *J. Chromatogr. Sci.*, 1983, **21**, 460.
6. S. Ahmed, C. Ramesh and V. Satya, *Analyst*, 1990, **115**, 287.
7. J. Jen, G. Ou-Yang and C. Chen, *Analyst*, 1993, **118**, 1281.
8. M. Sperling, S. Xu and B. Welz, *Analyst*, 1992, **117**, 629.
9. P.A. Sule and J.D. Ingle, *Anal. Chim. Acta*, 1996, **326**, 85.
10. V. McKelvie, *Chromium Speciation and Quantification of Environmental Water
    Samples by SPME/HPLC*, Honours Thesis, 1997, University of New England,
    NSW, Australia.

CHAPTER 25

Analysis of Industrial Pollutants in Environmental Samples

EMILIO BENFENATI, LAURA MÜLLER, L. PERANI
AND P. PIERUCCI

1 Introduction

Industrial processes produce thousands of chemicals that in different ways may enter the environment causing severe pollution problems.

Approximately 60 000 chemicals are in daily use[1] and 1000–1500 new chemicals are produced every year. Most of the pollutants found in the environment come from industry. Industrial pollutants belong to many different chemical classes, and the definition refers to their origin, though pollutants that are generally identified as being of a different origin may, in some cases, result from industrial release.[2]

This picture alone appears daunting, but in fact it is even more complex, because industrial pollutants undergo transformation processes which generate new progenies of pollutants. In some cases these processes start in biotreatment plants inside the factory and continue in the environment. Biotransformation and chemical and photochemical processes can produce compounds which in some cases are even more toxic then the parent compounds. The task of the analytical chemist called to investigate environmental pollution is therefore complicated.

In most cases the first step in the process leading to (eco)toxic evaluation and later to legal regulations is the analytical discovery of the chemical in an environmental compartment. It is clear that analytical environmental chemistry has an important role at this stage.

Today our knowledge of the pollution phenomenon is only partial and, in most cases, monitoring follows consolidated paths looking mainly for well-known and dangerous pollutants. This reflects the limited funds available for control, but the result is that many pollutants may escape detection.[3]

This brief overview illustrates for instance the importance of analytical methods able to monitor the levels of known components once an effluent has been characterized, and to identify new individual components in the matrix.

The industrially polluted sample may contain many different compounds from a variety of chemical classes. A challenging step in the investigation of industrial effluents, which are normally heavily contaminated, is sample preparation before instrumental analysis, during which there is always the risk of losing some important components. Additionally, analytical methods should be cheap and simple, and easily automated, to satisfy the need for monitoring compounds in a world where resources are limited, including funds for environmental control.

In this context SPME presents some interesting characteristics, since it can be used for the analysis of a variety of pollutants, so the same technique can be applied both for research studies and for routine control analysis. It is suitable for automated on-line monitoring (see Chapter 12) and can be applied directly in-field (see Chapter 20). SPME can be used even with complex and dirty matrices, by extracting the headspace or with membrane protection (see Chapter 1, Figure 2), so it is highly suitable for industrial pollutant investigations.

In this chapter we shall look closely at identification studies and quantitative analysis of industrial pollutants using SPME.

2   Qualitative Studies: Screening for Industrial Pollutants by SPME

Industrial pollutants have long been recognized as important.[4] However, screening studies of pollutants from landfills, effluents or industrial emissions often do not receive the same attention as other pollutants in scientific communications. An interesting experience, which introduces an innovative approach, is a project coordinated by Levsen, in which we are involved. It uses SPME for *in situ* quasi on-line analysis of industrial effluents. Details are provided in Chapter 20 by Levsen and Grote in this book.

In monitoring an industrial waste landfill (IWL), a parallel qualitative analysis of leachate and well water is useful to check for dispersion of pollutants in groundwater and to exclude external contamination (for example by-products from other activities or agricultural residues). Leachate is a complex mixture, usually highly concentrated, and can be considered a good material for setting up a qualitative analysis. In such cases a headspace analysis with SPME can give good results in terms of the number of compounds identified.

Even though the number of peaks obtained in SPME extraction is still high, SPME chromatograms are 'cleaner' than those obtained with solid phase extraction (SPE). SPE extraction chromatograms are often complicated by the presence of particulate matter and phthalates released by the cartridges. Table 1 reports results with SPME (PDMS 100 μm) and SPE (C-18) for two case studies. An important point in the comparison of SPE and SPME is the number of chromatographic peaks, which however is affected by many different variables, such as the sample volume. A second parameter is the number of

Table 1 *Identification of IWL leachate components, with SPE–GC–MS and SPME–GC–MS analyses: results of two case studies*

Sample	Results with SPE (C-18)	Results with SPME (PDMS 100 μm)
IWL leachate 1	71 total peaks 19 identified peaks 27% identified	106 total peaks 72 identified peaks 68% identified
IWL leachate 2	111 total peaks 61 identified peaks 55% identified	73 total peaks 52 identified peaks 71% identified

peaks identified. In these qualitative investigations SPME gives a larger percentage of peak identification, mainly because the chromatograms are cleaner and, as a consequence, backgrounds are lower; this means both better quality of the mass spectra and better match with library spectra.

A sample of leachate, called IWL leachate 1 in Table 1, was analyzed with the headspace technique using four different fibers, PDMS 100 μm, PDMS/divinylbenzene (PDMS/DVB) 65 μm, Carbowax/divinylbenzene (CW/DVB) 65 μm, and polyacrylate (PA) 85 μm. PDMS and PA fibers were also used on the same sample saturated with NaCl in order to improve extraction of polar analytes.[5]

A similar approach was used to characterize an industrial effluent, but using direct extraction (dipping the SPME fiber directly in the liquid sample). A coating of Carboxen/PDMS 75 μm (CX/PDMS) was also used for the investigation on this sample. The number of peaks found in the chromatograms, indicating the efficiency of extraction, the number of compounds identified and the range of abundances, are reported in Table 2. In both samples a large number of pollutants were extracted by all the fibers.

One cannot generalize about how the fibers react towards the different compounds present in a sample because the results vary significantly from sample to sample. However, certain groups of compounds with similar characteristics (molecules of the same class or with the same functional group) are extracted better by a specific polymer (Figure 1), in the leachate samples.

A group of mono-heterocyclic compounds containing nitrogen, and several aliphatic alcohols, were extracted distinctly better from the IWL leachate 1 with fibers containing PDMS. These results can be explained by the medium/low boiling points of these substances (semivolatiles), which can easily diffuse in a thick coating (100 μm). In contrast benzo-heterocycles, like indoles and benzothiazoles, are extracted similarly by all the fibers.

Phenols showed no particular affinity for any one coating, but generally recovery was better with the headspace (HS) technique when NaCl was added to the samples.

Table 2 *Results of SPME–GC–MS experiments with different fibers on an IWL*
leachate and the effluent from a chemical plant

Sample	Fiber	Technique	Total peaks	Ident. Peaks	Ident. %	Range of abundance
IWL leachate 1						
	PDMS 100 μm	HS	106	72	68%	11 000–332 000
	PDMS 100 μm/NaCl	HS	105	66	63%	9 000–440 000
	PDMS/DVB 65 μm	HS	105	74	70%	14 000–1 250 000
	CW/DVB 65 μm	HS	86	61	71%	8 000–760 000
	PA 85 μm	HS	51	35	68%	8 500–760 000
	PA 85 μm/NaCl	HS	57	43	77%	10 000–1 000 000
Industrial effluent						
	PDMS 100 μm	Immersion	112	62	55%	12 000–850 000
	PDMS/DVB 65 μm	Immersion	52	16	30%	5 000–72 000
	CW/DVB 65 μm	Immersion	88	59	67%	12 000–430 000
	PA 85 μm	Immersion	69	50	73%	7 000–520 000
	CX/PDMS 75 μm	Immersion	58	37	63%	8 000–1 840 000

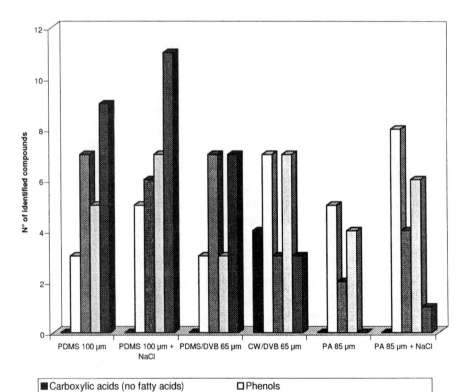

Figure 1 *Selected groups of compounds identified with different SPME coatings*

A series of carboxylic acids (acetic acid, butanoic acid, phenylacetic acid and 2-phenylpropanoic acid) was recovered only in the chromatogram related to the extraction with CW/DVB. This behavior is in good agreement with the coating structure, which is particularly suited to extract polar acidic groups, not mitigated by the presence of long alkylic chains, and volatiles which have high affinity toward the porous divinylbenzene particles.

In the industrial effluent, the homogeneity of the chemical content of the sample—mostly halogenated aromatic hydrocarbons—led to a certain uniformity in the extraction and no particular sets of compounds were extracted preferably by any single fiber.

If a qualitative investigation of a complex industrial matrix has to be performed, the best choice for the coating is the one that gives the broadest overview of the chemical content of the sample. For this reason, on the basis of the results obtained in this study, it appears that a good choice for a qualitative investigation of a complex matrix may be a fiber with PDMS in a high thickness, especially if its efficiency to extract polar analytes is increased by adding salt to the sample during the extraction. Moreover, this fiber does not present problems of displacements in complex matrices since it has a liquid coating (absorptive extraction rather than adsorptive) and it is thick enough to extract volatile compounds, even if a longer equilibration time is necessary. It has to be considered, however, that some compounds, such as carboxylic acids, will not be easily found in a qualitative screening performed with a PDMS coating. If it is necessary to analyze a specific class of compounds quantitatively, it can be useful to use a more specific coating, having the best sensitivity toward the chemicals of interest.

3 Quantitative Analysis of Industrial Pollutants

SPME is also useful for quantitative analysis of industrial pollutants. Most applications for the quantitation of industrial pollutants have used aqueous matrices. For quantitative studies of strongly contaminated industrial wastewaters by SPME, many important aspects have to be considered. Firstly, industrial wastes are usually very complex matrices, with a high content of organic matter and inorganic pollutants, that can interact with the SPME fiber and change its performance and properties toward the compounds of interest. If heterogeneous matrices are analyzed, such as samples containing oily suspensions or particles of hydrophobic humic material, the distribution constant (K_{fs}) and the volume of the coating can change during the extraction. Strong matrix effects can also impact linearity and precision, so appropriate investigations are needed before quantitation. Quantitation of complex samples must be performed using an internal standard, isotopically labeled spikes, or standard addition, so that the standard interacts with the matrix similarly to the target compound.[6] As previously reported a high content of humic acids and surfactants, for example, extends the equilibration time of phenols on a PA coating.[7] This is related to an increase of the density and the viscosity of the sample, causing slower diffusion of the analytes through the matrix.

Competition can also occur between the compounds of a complex matrix[6,7] if an adsorption process is involved in the extraction. Solid porous polymer coatings, such as PDMS/DVB and CW/DVB, extract the analytes through adsorption. This means that only a limited number of molecules can be 'accommodated' on the surface of the coating, and displacement can occur when compounds with high affinity for the fiber are present at high concentration in a complex sample. These fibers are consequently more affected by interferences than liquid polymeric sorbents (PDMS and PA).[6]

Industrial effluents may also differ in their salt content, pH and temperature, that can vary over time and space.[8] All these parameters have to be carefully considered when quantitatively evaluating industrial pollutants in complex matrices.

SPME has been used for many different classes of industrial pollutants. This section gives a rapid overview of the quantitative analysis of different classes of industrial pollutants. It is beyond the scope of this chapter to list the numerous analytes studied so far. Readers may refer to the web address *http:// www.cm.utexas.edu/groups/brodbelt/spme_refs.htm* which is regularly updated with papers on SPME.

Some of the most representative cases are listed in Table 3 (p. 340), where the sample matrix, instrumental technique, LOD achieved, CAS identification numbers and the reference for more detailed information are reported. The list includes VOC, PAH, heterocyclic compounds, chlorinated alkenes, aromatic amines, carboxylic acids, phenols and other industrial contaminants.

SPME has been mainly used with fibers (including optical fibers),[9] but new SPME techniques, interesting for industrial applications, have been developed using for instance Parafilm M,[10] PDMS chips[11] or pencil lead[12] as sorbents.

Volatile Organic Compounds (VOC)

The presence of VOC in industrial wastewaters is well known.[13,14] VOCs discharged directly with wastewater effluents have highly toxic effects on the ecosystem and are considered harmful for human beings.

Rapid screening for VOC, in chemical, pharmaceutical, petrochemical industrial waste waters, and other contaminated samples, has been successful with SPME coupled to GC–MS using both PA (85 μm) and PDMS (100 μm) coatings, depending on the effluent. For PDMS coating, the 100 μm fiber was preferred. For petrochemical industry wastewaters, the HS extraction was 1.6–7.8 times more sensitive than direct immersion extraction, due to the volatility of BTEX.[13]

There have also been recent reports on extensions of the technology. Chips of absorbing material (PDMS) were used in a recent study to extract BTEX, analyzed by UV.[11] Special devices have been developed for *in situ* analysis of chlorinated compounds and alkylbenzenes in groundwater.[15]

Nonpolar Compounds

Many quantitative investigations of nonpolar compounds in industrially polluted matrices have been carried using SPME. PAHs have been widely analyzed with different instrumental approaches (Table 3), and a class of very interesting industrial pollutants, the polychlorinated 1,3-butadienes (PCBD), was quantified in the groundwaters of an industrially contaminated area.[16]

Polar Compounds

Quantitation of polar compounds by SPME in complex matrices is a challenge, since many matrix effects (such as pH, salt content *etc.*) influence their extraction.[5] The introduction of polar coatings such as PA and CW/DVB, which have high affinity for polar compounds, has meant that more applications have been developed for polar analytes, both with and without derivatization.[17]

Phenols have been quantitated in wastewater matrices by SPME using a PA coating, in an evaluation of the matrix effect in strongly contaminated waters.[7]

A method for quantitation of aromatic amines in water matrices by SPME without derivatization was recently developed.[18] A CW/DVB (65 μm) coating was used to quantify aniline chlorinated derivatives at the ppt level in groundwater contaminated by industrial wastes.

For short-chain fatty acids, derivatization in the SPME fiber coating lowered the limits of detection by 1–4 orders of magnitude.[17]

Non-volatile Compounds

For non-volatile or thermally unstable compounds, SPME can be combined with HPLC. This technique could be useful for industrial applications. SPME coupled to HPLC has been employed to determine alkylphenol-ethoxylate surfactants in water, often found in the effluent streams of sewage treatment plants of textile and pulp and paper operations.[19]

The latest innovation for SPME/HPLC applications is 'in-tube SPME' in which the analytes are extracted while flowing through a capillary coated internally with polymeric sorbent, and then analyzed directly by HPLC. The advantage of this procedure is that it can be fully automated, and with further optimization it could be useful for on-line monitoring of compounds that are usually treated by HPLC.[20]

Organometallic Compounds and Metal Ions

Metal ions on their own are not suitable for GC analysis, but they can be derivatized and then extracted by SPME and analyzed by GC. This method has been applied for determination of tetraethyllead (TEL) and inorganic lead.[21,22]

Table 3 *A selection of SPME methods using different analytical techniques, for various environmental pollutants*

Substances	Matrix	Instrumentation	LOD	CAS No.	Ref.
1,1,1-Trichloroethane	Air	GC–MS	2 μg L^{-1}	71–55–6	36
1,1,2,2-Tetrachloroethane	Air	GC–MS	0.06 μg L^{-1}	79–34–5	36
1,1,2,4,4-PCBD [a]	Water	GC–MS	0.025 μg L^{-1}	–	16
2-Hexanone	Water	GC–FID	40.6 n L^{-1}	591–78–6	30
Acetaldehyde	Water	GC–FID	5.6 μg L^{-1}	75–07–0	30
Acetic acid	Air	GC–FID	760 μg L^{-1}	64–19–7	37
Aniline, 2,4-dichloro-	Water	GC–MS	0.02 μg L^{-1}	554–00–7	18
Aniline, 2,5-dichloro-	Water	GC–MS	0.02 μg L^{-1}	95–82–9	18
Aniline, 3,4-dichloro-	Water	GC–MS	0.007 μg L^{-1}	95–76–1	18
Aniline, 3,5-dichloro-	Water	GC–MS	0.005 μg L^{-1}	626–43–7	18
Aniline, *p*-chloro-	Water	GC–MS	0.010 μg L^{-1}	106–47–8	18
Benzene	Water	GC–MS (Scan)	60 μg L^{-1}	71–43–2	13
Benzene	Water	UV	97 μg L^{-1}	71–43–2	11
Benzene	Water	IR	9.63 mg L^{-1}	71–43–2	10
Benzene	Water	GC–FID	3.6 μg L^{-1}	71–43–2	29
Benzene	Water	GC–MS (SIM)	15 μg L^{-1}	71–43–2	40
Benzene	Air	GC–MS	0.6 μg L^{-1}	71–43–2	36
Benzene	Air	GC–FID	1.2 μg L^{-1}	71–43–2	38
Benzene, 1,2,4-trimethyl-	Water	UV	8.1 μg L^{-1}	95–63–6	11
Benzene, 1,3,5-trimethyl-	Air	GC–FID	0.06 μg L^{-1}	108–67–8	38
Benzene, ethyl-	Water	GC–MS (Scan)	70 μg L^{-1}	100–41–4	13
Benzene, ethyl-	Water	UV	12 μg L^{-1}	100–41–4	11
Benzene, ethyl-	Water	IR	4.74 mg L^{-1}	100–41–4	10
Benzene, ethyl-	Water	GC–FID	0.31 μg L^{-1}	100–41–4	29
Benzene, ethyl-	Water	GC–MS (SIM)	2 μg L^{-1}	100–41–4	40
Benzene, ethyl-	Water	GC–MS (Scan)	10 μg L^{-1}	100–41–4	13
Benzofuran	Water	GC–FID	1.5 ng L^{-1}	271–89–6	13
Benzyl acetate	Water	CGC–FPD[b]	0.4 ng L^{-1}	140–11–4	30
Butyl tin	Water	CGC–FPD[b]	0.4 ng L^{-1}	–	34
Butyl tin	Water	CGC–ICPMS[c]	0.3–0.8 ng L^{-1}	–	21

Compound	Sample	Method	Detection limit	CAS number	Ref.
Butyl tin	Water	CGC–ICPMS[c]	0.3–2.1 ng L^{-1}	—	21
Butyl, phenyl tin	Water	CGC–FPD[b]	1.0–10 ng L^{-1}	—	32
Butyl, phenyl tin	Water	CGC–MIP–AES[d]	0.1 ng L^{-1}	—	33
Butyric acid	Air	GC–FID	122 μg L^{-1}	107–96–2	37
Carbon tetrachloride	Air	GC–MS	1.7 μg L^{-1}	56–23–5	36
Chloro-anilines	Soil	GC–ECD	μg L^{-1} range	—	39
Chloro-benzenes	Soil	GC–ECD	μg L^{-1} range	—	39
Chloroform	Water	GC–MS (Scan)	68 μg L^{-1}	67–66–3	13
Chloroform	Air	GC–MS	2 μg L^{-1}	67–66–3	37
cis-1,1,2,3,4-PCBD[a]	Water	GC–MS	0.025 μg L^{-1}	—	16
Decane	Water	GC–FID	5.6 ng L^{-1}	124–18–5	30
Decanoic acid	Air	GC–FID	0.02 μg L^{-1}	334–48–5	37
D-Limonene	Air	GC–FID	0.04 μg L^{-1}	138–86–3	38
EPA VOC Mix 7[e]	Water	GC–MS	~1.25 μg L^{-1}		25
Ethanol	Water	GC–FID	0.76 μg L^{-1}	64–17–5	30
Ethyl decanoate	Water	GC–FID	0.366 ng L^{-1}	110–38–3	30
Ethylbenzene	Air	GC–FID	0.16 μg L^{-1}	100–41–4	38
HCBD[f]	Water	GC–MS	0.05 μg L^{-1}	87–68–3	16
Heptanoic acid	Air	GC–FID	0.11 μg L^{-1}	111–14–8	37
Hexanoic acid	Air	GC–FID	0.5 μg L^{-1}	142–62–1	37
Inorganic lead	Water	GC–FID	200 ppt	—	22
m/p-Xylene	Water	GC–FID	0.36 μg L^{-1}	108–38–3/106–42–3	29
m/p-Xylene	Water	GC–MS (SIM)	1 μg L^{-1}	108–38–3/106–42–3	40
Methanol	Water	GC–FID	1.2 mg L^{-1}	67–56–1	26
Methyl lead	Water	CGC–MIP–AES[d]	0.20 ng L^{-1}	—	31
Methyl mercury	Water	CGC–MIP–AES[d]	0.60 ng L^{-1}	—	31
Methyl tin	Water	CGC–FPD[b]	4.5–27 ng L^{-1}	—	35
Methyl tin	Water	CGC–MIP–AES[d]	0.15 ng L^{-1}	—	31
Monocyclic hetero-aromatic compounds	Water	GC–MS	1.0–10 μg L^{-1}	—	23
Naphthalene	Water	UV	0.40 μg L^{-1}	91–20–3	11
Naphthalene, 1-methyl-	Water	UV	0.41 μg L^{-1}	90–12–0	11

(continued)

Table 3 *(continued)*

Substances	Matrix	Instrumentation	LOD	CAS No.	Ref.
n-Hexane	Air	GC–FID	2.6 μg L^{-1}	110–54–3	38
Nitro-anilines	Soil	GC–ECD	μg L^{-1} range	–	39
Nitro-benzenes	Soil	GC–ECD	μg L^{-1} range	–	39
Nonanoic acid	Air	GC–FID	0.03 μg L^{-1}	112–05–0	37
n-Pentane	Air	GC–FID	5.5 μg L^{-1}	109–66–0	38
n-Undecane	Air	GC–FID	0.02 μg L^{-1}	112–02–14	38
Octanoic acid	Air	GC–FID	0.04 μg L^{-1}	124–07–2	37
o-Toluidine	Water	GC–MS	0.025 μg L^{-1}	95–53–4	18
o/m/p-Xylene	Water	UV	7.8/5.5/5.5 μg L^{-1}	95–47–6/108–38–3/ 106–42–3	11
o-Xylene	Water	IR	1.08 mg L^{-1}	95–47–6	10
o-Xylene	Water	GC–FID	0.36 μg L^{-1}	95–47–6	29
o-Xylene	Water	GC–MS (SIM)	1.5 μg L^{-1}	95–47–6	40
o-Xylene	Air	GC–FID	0.12 μg L^{-1}	95–47–6	38
PAH	Water	GC–MS	1–20 ng L^{-1}	–	24
PAH	Water	CD-modified-CE^g	8–75 μg L^{-1}	–	9
PAH	Water	GC–FID	0.03–0.42 μg L^{-1}	–	29
Phenol	Water	GC–MS	0.8 μg L^{-1}	108–95–2	27
Phenol, 2,4,6-trichloro-	Water	GC–MS	0.08 μg L^{-1}	88–06–2	27
Phenol, 2,4-dichloro-	Water	GC–MS	0.02 μg L^{-1}	120–83–2	27
Phenol, 2,4-dimethyl-	Water	GC–MS	0.02 μg L^{-1}	105–67–9	27
Phenol, 2,4-dinitro-	Water	GC–MS	1.6 μg L^{-1}	51–28–5	27
Phenol, 2-chloro-	Water	GC–MS	0.024 μg L^{-1}	95–57–8	27
Phenol, 2-methyl-4,6-dinitro-	Water	GC–MS	0.44 μg L^{-1}	534–52–1	27
Phenol, 2-nitro-	Water	GC–MS	0.38 μg L^{-1}	88–75–5	27
Phenol, 4-chloro-3-methyl-	Water	GC–MS	0.01 μg L^{-1}	59–50–7	27
Phenol, 4-nitro-	Water	GC–MS	0.75 μg L^{-1}	100–02–7	27
Phenol, pentachloro-	Water	GC–MS	0.11 μg L^{-1}	87–86–5	27

Compound	Sample	Method	Detection limit	CAS	Ref.
Polycyclic hetero-aromatic compounds	Water	GC–MS	0.02–0.3 μg L^{-1}	–	23
Propionic acid	Air	GC–FID	280 μg L^{-1}	79–09–4	37
p-Xylene	Water	GC–MS (Scan)	45 μg L^{-1}	106–42–3	13
p-Xylene	Air	GC–FID	0.15 μg L^{-1}	106–42–3	38
Sarin[h]	Water	GC–MS (SIM)	0.05 μg L^{-1}	107–44–8	28
Sarin[h]	Water	GC–NPD[i]	0.05 μg L^{-1}	107–44–8	28
Soman[h]	Water	GC–MS (SIM)	0.05 μg L^{-1}	96–64–0	28
Soman[h]	Water	GC–NPD[i]	0.05 μg L^{-1}	96–64–0	28
Tabun[h]	Water	GC–MS (SIM)	0.05 μg L^{-1}	77–81–6	28
Tabun[h]	Water	GC–NPD[i]	0.05 μg L^{-1}	77–81–6	28
TCBD[l]	Water	GC–MS	0.05 μg L^{-1}	–	16
Tetrachloroethane	Air	GC–MS	0.05 μg L^{-1}	630–20–6	36
Tetraethyl lead	Water	GC–FID	100 ppt	78–00–2	22
Tetraethyl lead	Water	GC–ITMS	5 ppt	78–00–2	22
Toluene	Water	GC–MS (Scan)	170 μg L^{-1}	108–88–0	13
Toluene	Water	UV	10 μg L^{-1}	108–88–3	11
Toluene	Water	GC–FID	0.93 μg L^{-1}	108–88–3	40
Toluene	Water	GC–MS (SIM)	5 μg L^{-1}	108–88–3	36
Toluene	Air	GC–MS	0.2 μg L^{-1}	108–88–3	38
Toluene	Air	GC–FID	0.41 μg L^{-1}	108–88–3	19
Triton X-100 (each single ethoxamer)[m]	Water	HPLC–UV	1.57 μg L^{-1} (average)	–	
Valeric acid	Air	GC–FID	3.1 μg L^{-1}	109–52–4	37
VX[h]	Water	GC–MS (SIM)	1 μg L^{-1}	50782–69–9	28
VX[h]	Water	GC–NPD[i]	0.5 μg L^{-1}	50782–69–9	28
α-Pinene	Air	GC–FID	0.09 μg L^{-1}	80–56–8	38

[a] Polychlorobutadiene; [b] Capillary gas chromatography flame photometric detector; [c] Capillary gas chromatography inductively coupled plasma mass spectrometer; [d] Capillary gas chromatography microwave induced plasma atomic emission spectrometer; [e] Chloroform, benzene, bromodichloromethane, dibromochloromethane, bromoform, 1,4-dichlorobenzene; [f] Hexachlorobutadiene; [g] Cyclodextrine modified capillary electrophoresis; [h] Chemical warfare agents; [i] Nitrogen phosphorous detector; [l] Tetrachlorobutadiene; [m] Octylphenol, polyethoxylated.

4 Conclusions

From this brief overview of the many applications of SPME in industrial studies it is clear that this technique is useful for screening investigations and quantitative analysis of industrial pollutants. It is cheap, simple, clean, easily automated and versatile, and can be used for industrial monitoring as a valid alternative to SPE and liquid/liquid extraction (LLE). The wide range of fibers commercially available means SPME can be applied for all kinds of industrial pollutants, with a certain selectivity toward some classes, depending on the nature of the polymeric coating. From our investigations, fibers with PDMS have been demonstrated to give a qualitative characterization of an industrially contaminated sample with the widest range of compounds identified.

Quantitative analysis of industrial pollutants in complex matrices is also possible. Even though the matrix may influence extraction (salt content, pH, organic matter content, temperature *etc.*), careful investigation of the operating conditions can permit reliable quantitation of a variety of classes of chemicals.

Acknowledgements

We thank the EC project ENV4 CT95-0021 for financial support. P. Pierucci is recipient of a fellowship from 'Fondazione Lombardia per l'Ambiente'.

References

1. J.L. Schnoor, *Environmental Modeling: Fate and Transport of Pollutants in Water, Air and Soil*, John Wiley & Sons Inc., New York, 1996.
2. J.B. Marr, R.M. Facey, B. Tansel, W. Ying, L.J. Weathers, J.L. Walsh, Jr., C.C. Ross, G.E. Valentine, Jr., T.S. Reeves, D. Srinivasan, J.P. Unwin, M.S. Bahorsky, D.H. Bryant, K.S. Ro and J.A. Libra, *Water Environ. Res.*, 1995, **67**, 503.
3. C.M. Joy, K.P. Balakrishnan and A. Joseph, *Wat. Res.*, 1990, **24**, 787.
4. J.W. Grisham, *Health Aspects of the Disposal of Waste Chemicals*, Pergamon Press, New York, 1986.
5. K.D. Buchholz and J. Pawliszyn, *Anal. Chem.*, 1994, **66**, 160.
6. J. Pawliszyn, *Solid Phase Microextraction: Theory and Practice*, Wiley-VCH, New York, 1997.
7. M. Möder, S. Schrader, U. Franck and P. Popp, *Fresenius' J. Anal. Chem.*, 1997, **357**, 326.
8. I. Valor, J.C. Moltó, D. Apraiz and G. Font, *J. Chromatogr. A*, 1997, **767**, 195.
9. A. Nguyen and J.H.T. Luong, *Anal. Chem.*, 1997, **69**, 1726.
10. D.L. Heglund and D.C. Tilotta, *Environ. Sci. Technol.*, 1996, **30**, 1212.
11. B.L. Wittcamp, S.B. Hawthorne and D.C. Tilotta, *Anal. Chem.*, 1997, **69**, 1197.
12. H.B. Wan, H. Chi, M.K. Wong and C.Y. Mok, *Anal. Chim. Acta*, 1994, **288**, 218.
13. K.J. James and M.A. Stack, *Fresenius' J. Anal. Chem.*, 1997, **358**, 833.
14. S. Al-Muzaini, H. Khordagui and M.F. Hamouda, *Wat. Sci. Tech.*, 1994, **30**, 79.
15. T. Nilsson, L. Montanarella, D. Baglio, R. Tilio, G. Bidoglio and S. Facchetti, *Int. J. Environ. Anal. Chem.*, 1998, **69**, 1.
16. E. Fattore, E. Benfenati and R. Fanelli, *J. Chromatogr. A*, 1996, **737**, 85.
17. L. Pan and J. Pawliszyn, *Anal. Chem.*, 1997, **69**, 196.

18. L. Müller, E. Fattore and E. Benfenati, *J. Chromatogr. A*, 1997, **791**, 221.
19. A.A. Boyd-Boland and J. Pawliszyn, *Anal. Chem.*, 1996, **68**, 1521.
20. R. Eisert and J. Pawliszyn, *Anal. Chem.*, 1997, **69**, 3140.
21. L. Moens, T. De Smaele, R. Dams, P. Van Den Broeck and P. Sandra, *Anal. Chem.*, 1997, **69**, 1604.
22. T. Górecki and J. Pawliszyn, *Anal. Chem.*, 1996, **68**, 3008.
23. S. Johansen and J. Pawliszyn, *J. High Resol. Chromatogr.*, 1996, **19**, 627.
24. D.W. Potter and J. Pawliszyn, *Environ. Sci. Technol.*, 1994, **28**, 298.
25 A.A.M. Hassan, E. Benfenati, G. Facchini and R. Fanelli, *Toxic. Environ. Chem.*, 1996, **55**, 73.
26. Z. Penton, *www.varian.com/inst/csb/gcnotes/spme08.html*.
27. K.D. Buchholz and J. Pawliszyn, *Environ. Sci. Technol.*, 1993, **27**, 2844.
28. H.A. Lakso and W.F. Ng, *Anal. Chem.*, 1997, **69**, 1866.
29. J.J. Langenfeld, S.B. Hawthorne and D.J. Miller, *Anal. Chem.*, 1996, **68**, 144.
30. R.J. Bartelt, *Anal. Chem.*, 1997, **69**, 364.
31. M. Ceulemans and F.C. Adams, *J. Anal. At. Spectrom.*, 1996, **11**, 201.
32. J. Szpunar, V.O. Schmitt, O.F.X. Donard and R. Lobinski, *Trends in Anal. Chem.*, 1996, **15**, 181.
33. M. Ceulemans, R. Lobinski, W.M.R. Dirkx and F.C. Adams, *Fresenius' J. Anal. Chem.*, 1993, **347**, 256.
34. P. Michel and B. Averty, *Appl. Organomet. Chem.*, 1991, **5**, 393.
35. Y. Morcillo, Y. Cai and J.M. Baiona, *J. High Resolut. Chromatogr.*, 1995, **18**, 767.
36. M. Chai and J. Pawliszyn, *Environ. Sci. Technol.*, 1995, **29**, 693.
37. L. Pan, M. Adams and J. Pawliszyn, *Anal. Chem.*, 1995, **67**, 4396.
38. P.A. Martos and J. Pawliszyn, *Anal. Chem.*, 1997, **69**, 206.
39. A. Fromberg, T. Nilsson, B.R. Larsen, L. Montanarella, S. Facchetti and J.O. Madsen, *J. Chromatogr. A*, 1996, **746**, 71.
40. D.W. Potter and J. Pawliszyn, *J. Chromatogr.*, 1992, **625**, 247.

Food, Flavour, Fragrance and Pheromone Applications

CHAPTER 26

Analysis of Food and Plant Volatiles

ADAM J. MATICH

1 Introduction

SPME is a solvent-free sampling technique developed by Pawliszyn and co-workers[1-3] to measure organic compounds in groundwater. The technique, as originally proposed, involved immersion of a bare silica fibre into a liquid sample, allowing adsorption of compounds onto the fibre, followed by de-sorption in a GC injection port. This technique was extended to measuring caffeine levels in beverages using a bare silica fibre and by spiking the beverages with isotopically labelled caffeine (^{13}C), and preparing a calibration curve of ratio of ^{13}C-labelled caffeine to unlabelled caffeine.[4] However, immersion sampling can cause fouling of the SPME fibre, thus shortening its life. Contaminating the GC injection port may eventually introduce artefactual errors in quantification. Only a small amount of work describing immersion sampling in 'dirty' samples has been reported. Headspace sampling volatile compounds removes many effects of fibre contamination by dirty samples (*e.g.* greasy water, blood samples and beverages which contain sugars, fats and proteins) and is often the only practical method of sampling levels of volatiles in foodstuffs or plant matter.

Quantification of volatiles from liquid foods can be done simply by standard addition of analytes to the sample, but because of the heterogeneous nature of solid foodstuffs and plants, quantification of their volatiles by headspace sampling can be challenging. Some solid samples can be ground or powdered, thus increasing the sample surface area. This enables realistic times for equilibration of standard solutions of analyte volatiles with the sample matrix, and for equilibration between the sample, the air, and the fibre. However, other samples such as fruits and flowers are living systems and their aroma profiles are altered by tissue and cell damage resulting from such treatments. Similarly, optimising sampling conditions (temperature for example) may alter the aroma profiles of living samples by metabolic perturbation or damage to the sample.

Despite the problems associated with sampling trace volatiles from foods and plants, there is a rapidly expanding literature on the application of SPME to analysis of trace volatiles present both in the headspace and in the samples themselves. This chapter reviews literature up to February 1998 and details progress made in using SPME for the quantitative and semi-quantitative analysis of volatiles in complex food and plant systems. Many of the problems encountered in the use of the SPME sampling method are discussed, along with how various workers have overcome them. Key parameters involved in SPME sampling are discussed, along with the possibilities for, limitations to and problems associated with their optimization. This chapter is organized in order of increasing sampling difficulty. It commences with measurement above liquid foodstuffs by headspace SPME and discusses the factors affecting quantification. This is followed by discussion of examples of the analysis of volatiles from solid food and plant samples, and the chapter concludes with examples of measuring headspace concentrations of volatiles to determine their levels within solid samples.

2   Headspace Sampling above Liquid Food Samples

Factors Affecting SPME Quantification

Sample/air/fibre partitioning of volatiles has been shown to depend upon many factors.[5-12] These include:

- the sample matrix
- the lipid content in the sample
- salt concentration (ionic strength)
- pH
- presence of interfering compounds
- choice of standards
- sample and headspace volumes
- agitation
- temperature
- fibre selectivity.

Matrix Effects

It is not always clear how the sample matrix will effect partitioning of analytes between the sample, the air, and the fibre. In the analysis of volatiles in wine, while the calibration curves for the levels of terpene alcohols were linear, the slopes of these curves were dependent upon the particular type of wine and terpenol being analysed.[13] For example, the geraniol calibration curve slope for an Australian Muscat (18% ethanol) was 50% of that for the 12% ethanol–water reference solution, whereas for linalool the slope was 72%. Thus the amount of terpenol extracted decreases with increasing ethanol content, but the decrease is terpenol dependent. This analyte and matrix dependency of efficiency of extraction of analytes highlights the importance of determining calibration curves for each analyte of interest. With terpenols, adjusting the ethanol concentration by dilution with water would overcome the problem of variable ethanol content. This should also increase sensitivity by decreasing terpenol solubilities in the sample, thus increasing air/wine partition coefficients (unpublished results from this laboratory).

Lipid Content

In a model system with halogenated volatiles dissolved in water, equilibration onto the fibre occurred within 30 minutes with sub-ppb (w/w) limits of detection (LODs).[6] Addition of vegetable oils to the solution reduced recovery of volatiles from the solution by up to an order of magnitude, the largest reductions being for the nonpolar, low-volatility compounds which in clean water had the lowest LODs. Similarly, efficiencies for extracting halogenated volatiles from milk samples were between 20 and 96% lower for milk containing 3.4% *versus* 0.1% butterfat. Extraction of airborne analytes from milk samples was significantly lower than from fruit and aerated drinks. Therefore there is a reduced sensitivity for measurement of nonpolar compounds in the headspace above samples with substantial fat contents, and a considerable variability in recoveries from samples with different fat contents.

Ionic Strength and pH

Saturation of aqueous solutions, containing halogenated volatiles, with salt increased volatile recovery by between 2 and 20 fold.[6] Similar enhancements were observed for in-fibre/derivatization extraction of C_2–C_{10} free fatty acids,[8] with further 6–55 fold increases by reducing the pH from 7 to 1.5 to produce the non ionized, volatile form of the organic acids. When common flavour volatiles in water, ground coffee, fruit juice and a butter flavour in vegetable oil were analysed, most compounds (in water) displayed increasing sampling sensitivity with increasing NaCl concentration.[7] Limonene, anethole and β-ionone showed steady decreases. Improved extraction efficiencies of fruit juice flavour volatiles were also obtained by addition of salt (36% w/v was optimal), the exception being α-pinene for which extraction efficiencies were also reduced.[10] Limonene in orange juice is not suitable for quantitation by static headspace.[14] It is present as an emulsion at concentrations exceeding its solubility limit in the liquid phase, and thus its vapour concentration is not proportional to its liquid concentration. Divergence of limonene and the above mentioned compounds from the expected 'salting out' effect may relate to the formation of emulsions.

Interfering Compounds

Direct injection of a fruit punch flavour mix into a GC gave three large solvent peaks, *viz.* ethanol, propylene glycol and glycerin, the latter masking a number of aroma compounds.[9] The only flavour compound detected was ethyl hexanoate sitting on the glycerin peak. A Likens–Nickerson extraction resulted in numerous artefacts from thermal decomposition of sugars during the steam distillation. When headspace SPME sampling was used, thirteen compounds (in addition to ethyl hexanoate) were detected. Glycerin was not detected because of its low volatility. Suspended solids in liquid samples can also cause problems. It was necessary to centrifuge solids from fruit juice samples before standard addition of analytes would produce linear calibration plots.[10]

Choice of Standards

An alternative to the standard addition approach is addition of a compound chemically similar to the analyte of interest (an internal standard). The ethanol/propanol response calibration curve is linear for ethanol concentrations between 0.1 and 20%,[9] which enables propanol to be used as an internal standard for ethanol. This approach avoids the need for sampling and then re-sampling (of the same sample) after addition of the internal standard. Other internal standards are isotopically labelled analytes which would be expected to behave indistinguishably from the non-labelled analyte. An example is quantification of 2,4,6-trichloroanisole (TCA), a very common cork taint compound in wines, using a deuterated analogue ($[^2H_5]$-TCA) as an internal standard.[15]

Sample and Headspace Volume

Higher masses of analytes on the fibre have been achieved with larger sample volumes, while keeping the sample:headspace volume ratio the same. Increasing this ratio will also increase loading on the fibre and reduce equilibration times.[7] However, this general behaviour is not observed for highly polar compounds which favour partitioning into an aqueous phase. No difference in fibre loading was observed when the sample volume of 100 ppm methanol in a 2 mL vial was tripled from 200 μL to 600 μL.[11]

Stirring

The utility of stirring liquid samples when headspace sampling appears to depend upon the volatility of the analyte. Stirring had minimal effect on recovery and equilibration times for headspace SPME of amphetamines in urine samples.[12] This was attributed to very rapid solution/headspace equilibration so that the bulk of the sample was in the headspace prior to introducing the fibre. However, for less volatile compounds (polynuclear aromatic hydrocarbons) in water, equilibration times were substantially reduced by stirring.[5] Generally stirring is used for liquid samples.[10]

Temperature

Reproducibility is generally poorer if sampling is not allowed to reach equilibrium. Sampling at elevated temperatures can reduce times required to reach equilibrium without serious reductions in yields of analyte on the fibre.[8,12]

Table 1 *Yield of amphetamines on the fibre* (ng) *at different sampling times and temperatures. The values in bold are those obtained for the sampling conditions favoured by these workers. Taken from Figure 1, ref. 12*

	Amphetamine		Methamphetamine	
Sampling time (min)	15	90	15	90
Temperature (°C)				
20	35[a]	125[a]	125[a]	375[a]
40	90[a]	140[a]	275[a]	450[a]
60	**95**	105	**325**	375
73	75	80	300	305

[a] Sampling was not at equilibrium.

The fibre/air partition coefficient (K_{fa}) decreases with temperature, with a reduced mass of analyte on the fibre. This reduced adsorption is counteracted by a concomitant increase in the air/solution partition coefficient (K_{as}) which increases the concentration of analyte in the headspace, thus favouring increased adsorption of analyte by the fibre. Headspace sampling for amphetamines in urine at *ca.* 60 °C, enabled equilibration within 15 min with acceptable yields of analyte on the fibre.[12] At 40 °C, equilibration took longer than 90 min, and at above 60 °C reduced partitioning from the headspace to the fibre produced significant reductions in analyte yield without appreciably reducing equilibration times. Table 1 shows how good yields and short equilibrium times can be obtained by optimising the sampling temperature.

An SPME study of the humulene/caryophyllene (H/C) ratio in the headspace above hops revealed that sampling at 50 °C resulted in H/C ratios closely matching those obtained by the traditional distillation and solvent extraction methods.[16] However, sampling at 90 °C (to improve sensitivity) produced higher H/C ratios, which was attributed to oxidative loss of caryophyllene.

Headspace SPME of nineteen freshly grated cheeses (Swiss, Cheddar and Romano) was performed at 60 °C to increase the headspace sampling sensitivity.[17] However, a number of dairy flavour compounds are thermal degradation products. Lactones are produced from hydroxy acids or triglycerides in dairy products under relatively mild conditions.[18,19] and so their concentrations would be overestimated at elevated sampling temperatures. Other thermal degradation products include β-ionone from β-carotene,[20] which is present in milk from pasture fed cattle,[21] aldehydes such as nonanal and hexanal from oxidation of lipids[18,22] and methyl ketones from decarboxylation of β-keto acids formed by oxidation of free fatty acids.[18,23] Whilst sampling at elevated temperatures can improve the sensitivity of SPME, caution needs to be exercised to avoid creating experimental artefacts.

If the SPME sampling regime itself is not harsh, other parts of the analytical system, such as the GC injection port, may well be. A comparison was made between cryotrapping/direct injection, cryotrapping/SPME, and solid phase

(Tenax-GC) extraction for sampling sulfur compounds in the odour of cut garlic, leek and onions.[24] Following SPME and tenax trapping, GC–MS analysis revealed a number of disulfides. These are degradation products of the thiosulfinate compounds which were identified, by cold trapping followed by HPLC–UV and HPLC–MS, as true components of the aroma of freshly cut garlic. Thiosulfinates are thermally labile[25] and would be expected to break down to the disulfides in the GC injection port/column/transfer line. Instead of SPME–GC, the recently developed SPME–HPLC[26] might be more applicable to analysis of such thermally unstable compounds.

Fibre Selectivity

Polyacrylate (PA) fibres are more selective towards polar compounds such as acids, alcohols, phenols, and aldehydes[9,10,27] compared with nonpolar compounds such as hydrocarbons. Thus a PA fibre was used to measure slightly polar flavour additives in tobacco,[27] with the advantage that it would also discriminate against the numerous natural nonpolar hydrocarbons present in tobacco. If these compounds had similar retention times to the flavour compounds, they would have interfered with quantification. The selectivity of nonpolar (PDMS) fibres is shown in a study of orange juice flavour volatiles[10] in which the PDMS fibre extracted more of only the terpenes (γ-terpinene, limonene, α-pinene, and β-myrcene) than did the PA fibre. The PA extracted more of all the other flavour compounds, which included alcohols, esters, and aldehydes.

PDMS and polyacrylate coated fibres were compared for sampling the headspace above freshly grated cheeses,[17] with convincingly higher levels of volatiles being adsorbed by the polyacrylate fibre (3–20 × depending upon the class of compound). Volatiles extracted from the headspace were predominantly polar and included mostly free fatty acids (FFAs), some alcohols and δ-lactones, which are flavour compounds with creamy or milky notes.[28] Distinct differences were detected between the aroma profiles both within varieties and between different varieties of cheese. Static headspace (1 mL) sampling favoured the more volatile components and with this method 2,3-butanediol was the largest component detected.

Another feature of SPME fibre selectivity is discrimination towards high MW volatiles.[9,29] Extraction efficiencies of low MW volatiles are low, because of their volatility and higher solubility in the sample with respect to the fibre. This problem can been addressed by using a Carboxen™–PDMS fibre which is a carbon-coated PDMS fibre with a high affinity for gaseous and low MW compounds. This fibre will adsorb small hydrocarbons such as ethane and ethene, volatile organic compounds such as chloromethane and vinyl chloride, sulfur compounds including carbon disulfide, hydrogen sulfide, and sulfur dioxide in white wine, and carbon dioxide in beer.[30] In ripening studies (unpublished results from this laboratory) of apricots and peaches, ethyl acetate and ethanol were not detected by a PA fibre, but were readily measured with a Carboxen fibre. When comparing fibre sensitivity to atmospheres of

Table 2 *LOD for in fibre derivatization SPME and direct SPME on a PA fibre. Taken from ref. 31*

	LOD (ng mL^{-1})	
Compound	Derivatization SPME[a]	Direct SPME
Propanoic acid	2.5	290
Butanoic acid	1.4	122

[a] Derivatization with pyrenyldiazomethane.

ethanol and ethyl acetate (6.7 μL L^{-1}), a 75 μm Carboxen fibre adsorbed 11 fold more ethyl acetate and 1.5 fold more ethanol than did a 85 μm PA fibre.

Altered Selectivity by Derivatization/SPME

Further gains in sensitivity to analytes with low air/water (K_{aw}) partition coefficients, without heating the sample, can be made by derivatization SPME. This technique was introduced to increase the sensitivity to polar analytes in aqueous solution.[31] Derivatization was examined in the sample matrix, in the fibre itself, and in the GC injection port to both increase the volatility and decrease the polarity of analytes, thus increasing K_{aw} and K_{fw}. Derivatization in the fibre has the advantage that, given the amount of derivatizing agent in the fibre is greater than the amount of analyte in the sample, exhaustive extraction of analyte from the fibre can be achieved instead of equilibrium extraction.

In-fibre derivatization headspace SPME of fatty acids produced up to 100 fold increases in sensitivity over direct SPME (Table 2). Immersion SPME (PDMS fibre) phenol extraction was improved by derivatizing the phenols with acetic anhydride.[32] In-fibre derivatization headspace SPME sampling for phenols has not been reported, but increases in sensitivity to phenols might well be expected with appropriate derivatization reagents.

3 Headspace Sampling above Solid Samples

SPME has been found ideal for headspace sampling above liquid samples because of the ease of quantification by standard addition. Of increasing interest is measuring aroma and flavour volatiles in foods and plants by headspace SPME above the heterogeneous or intact solid samples. This is useful for determining how storage and processing effects flavour/aroma levels in foodstuffs,[33,34] the quality of and contaminants in produce,[6,35,36] the botanical origin of produce from its aroma profile,[37] and direct identification of volatiles.[9]

Quantification by standard addition of analytes of interest to solid samples is not necessarily the simple matter it is for liquid samples and in many cases only qualitative and semi-quantitative (or comparative) sampling are practical.

Qualitative and Semi-Quantitative Headspace Sampling

SPME of fruit aroma volatiles as an alternative to vacuum or steam distillation has been discussed.[9] Sampling above the flesh cut from cantaloupes revealed a number of esters, ranging in MW from ethyl acetate to 2-ethylhexyl acetate. SPME extracted both reported and previously unreported aroma compounds from bananas (3-hexenyl caproate) and Bartlett pears (isoamyl acetate, cis-3-hexenyl acetate and carveol propionate). However, some previously reported Bartlett pear aroma compounds (butyl acetate, hexyl acetate, methyl trans-2-cis-4-decadienoate and α-farnesene) were not detected by SPME. These compounds are normally not problematical for SPME, and so they may not have been present in that particular sample or were at below the threshold level for the fibre used. A fibre more sensitive (the new Carboxen–PDMS fibre for example) than the PDMS fibre used, or another method of volatile sampling may be necessary to measure these compounds if they are present at very low concentrations. Alternatively, their detection by other analytical methods may be an artefact of those extraction techniques.

Favourable results were obtained when comparing headspace SPME (PDMS fibre) with solvent assisted supercritical fluid extraction (SFE) for differentiating between true cinnamon and *cassia*.[37] SPME gave a higher recovery of the more volatile components and the fibre adsorbed a number of terpenes that either were not detected or were at trace levels in the SFE extracts. Reasonably linear correlations were obtained between compounds extracted by SPME and those by SFE. The authors pointed out that without calibration SPME is only a representative sampling method whereas SFE is truly quantitative. Despite this limitation, differentiation between true cinnamon and *cassia* only required monitoring the relative levels of four compounds (eugenol, benzyl benzoate, coumarin and δ-cadinene) and that this was easily achievable by SPME without the need for 'true' quantitation.

Quantitative Headspace Sampling

Theoretical and experimental work[5] enables quantification of airborne volatile levels by use of partition coefficients (K). K values for partitioning of analytes between the fibre and the air (K_{fa}) have been determined by two methods. The first method described herein was used to determine K_{fa} values for apple aroma volatiles[34] and for humulene and caryophyllene (sesquiterpenes) in the headspace above female hop cones and male hop lupulin.[16] They used[5] equation 1:

$$K_1^i = \frac{(A_f V_g)}{(A_g V_f)} \tag{1}$$

where A_f is the area counts for analyte i on the fibre, V_g is the vapour phase volume, A_g is the total area counts for analyte i in the headspace, and V_f is the volume of the fibre. Experimentally, this involves equilibrating a mixture of the analytes of interest in a sealed glass jar followed by equilibrium SPME sampling, from which A_f is determined. The headspace is then sampled with a syringe to obtain A_g.

Hop cones favoured for adding flavour and aroma to beer have high humulene to caryophyllene ratios, as do male lupulin favoured for breeding purposes. The female hop cones are traditionally extracted by steam distillation, whilst for the male lupulin, in which the concentrations of these compounds are very low, solvent extracts are obtained. These extracts contain non volatile compounds which build up in GC columns thus reducing resolution and column life. SPME (with a PDMS fibre) was used, with K_{fa} determinations, to quantify these compounds in the headspace above a number of different samples of female hop cones and male lupulin.[16] Good correlations were obtained with the steam distillation ($r^2 = 0.962$) and solvent extraction ($r^2 = 0.969$) techniques for determining the humulene to caryophyllene ratios. Sampling times of 4 h at 50 °C produced satisfactory results, a considerable saving over the investment of materials and time required for the steam distillation and solvent extraction methods.

Apple aroma volatiles[34] were sampled by SPME for 6–8 minutes followed by 4 minute GC runs (the sesquiterpene α-farnesene being the last peak) on a specialized time-compressed chromatography system (rapid GC thermal gradient with resolution of overlapping peaks by time of flight mass spectrometry), compared with 100–120 min for a traditional purge and trap thermal desorption GC system. Partition coefficients obtained for apple aroma[34] and hop[16] volatiles are given in Table 3. One note of caution was expressed for the apple aroma analysis[34] when sampling with a syringe to determine A_g. Both hexyl acetate and 6-methyl-5-hepten-2-one adsorbed onto the syringe needle requiring the plunger to be pumped 10–15 times in order to obtain a consistent GC response. It was suggested that this procedure enabled saturation of the adsorption sites within the needle. Significant adsorption of volatiles onto the glass of the syringe itself have been observed by other workers[29] who reported that up to 45% of the

Table 3 *Partition coefficients for a* 100 μm *PDMS fibre determined by comparison between SPME and direct headspace sampling with a syringe of a standard atmosphere*

Compound	Partition coefficient
Butyl acetate[a]	2.7×10^4
Ethyl 2-methylbutanoate[a]	3.35×10^5
Hexyl acetate[a]	2.3×10^5
Butanol[a]	1.7×10^4
Hexanol[a]	2.1×10^5
6-Methyl-5-hepten-2-one[a]	5.65×10^5
α-Humulene[b]	1.41×10^4
β-Caryophyllene[b]	9.44×10^3

[a] Taken from ref. 34; [b] Taken from ref. 16.

gaseous α-farnesene drawn into a 5 mL gas-tight syringe, from a headspace, was adsorbed onto the glass walls.

Monoterpene levels within conifer needles and in the headspace above them[38] were determined by SPME (PDMS fibre) using distribution constants (K_{fa}), between the gas phase and the fibre, determined from Kovats retention indices. This was performed using a polydimethylsiloxane GC column (which was necessarily the same phase as the fibre). The net retention time (I) in isothermal GC should be directly proportional to the distribution constant (K_{fa}) *viz.*

$$\log K \propto I \tag{2}$$

The distribution coefficient (K_{ref}) for only one of the analytes (by direct headspace injection versus SPME[16,34]) was determined. From this, Ks for all the other analytes in the standard mixture could be determined from retention times on the GC column: thus

$$\log \frac{K_{fa}}{K_{ref}} \propto I \tag{3}$$

Monoterpene levels above the *Pinus peuce* needles were measured by SPME[38], using the K values determined by direct headspace sampling using a gas-tight syringe. The K value for tricyclene (the lowest b.p. monoterpene studied) was used, along with the Kovat's indices for the other monoterpenes, to predict their K values using equation 3. The calculated relative distribution factors were in reasonable agreement with the experimentally determined values. The dependence of distribution factor on the Kovat's retention index (I) was exponential, *i.e.*

$$K_{fa} = 0.00009e^{0.00766I} \tag{4}$$

where $r = 0.975$. These authors proposed using a homologous series of *n*-alkanes as standard substances to determine the dependence of K values on Kovat's indices over a wide range of boiling points.

K values can also be obtained from linear temperature programming retention indices (LTPRI). This approach has the flexibility of being applicable to temperature-programmed GC and a linear calibration plot of measured K_{fa} *versus* LTPRI, for a series of *n*-alkanes, was obtained.[39] The theoretical relationship between K_{fa} and the LTPRI is

$$\log K_{fa} = a + b(\text{LTPRI}) \tag{5}$$

and the experimentally obtained relationship using a PDMS fibre for alkanes (C_5–C_{14}) was

$$\log K_{fa} = 0.0042(\text{LTPRI}) - 0.188 \tag{6}$$

with $r^2 = 0.99989$. This relation provided good results for isoparaffinic and aromatic compounds. A modified version of this equality[40] and Henry's Law coefficients were used to determine the fibre/water partition coefficients for isoparaffins, cyloalkanes and aromatics in aqueous solution, when sampled by headspace SPME with a PDMS fibre. However, air/water and therefore fibre/water partition coefficients will be dependent upon the sample matrix and so this approach may be applicable only to relatively clean samples. Furthermore, partitioning between solid samples and the air may not be predictable because of matrix variability (particle size, porosity, and heterogeneous composition). In solid/air systems K_{as} needs to be determined, by standard addition for example, otherwise the above relation can only be used to quantify headspace concentrations and not those in the solid sample.

Quantification of Analytes within Solid Samples by Headspace SPME

There has been a relatively small amount of work using headspace SPME to determine the concentrations of analytes present within solid samples based upon airborne concentrations.

Flavour volatiles (benzaldehyde, tetramethylpyrazine, menthol and anethole in ethanol) and additives (mandarin orange oil, nutmeg oil and sweet fennel oil) in tobacco were measured[27] by their standard addition to 1 g tobacco allowing 2 h for equilibration. LODs for flavour additives ranged from ppm to ppb and sampling times of 15 min were found satisfactory. Better sensitivity was obtained for benzaldehyde and menthol by adding 1 mL 3M KCl, while tetramethylpyrazine and anethole recoveries were reduced. Larger volumes of KCl solution reduced recoveries below levels obtained without KCl addition. This may merely reflect dilution of volatiles by the larger volumes of solution used. Similarly, benzaldehyde, menthol and anethole recoveries were improved at higher temperatures (95 °C), whereas that of tetramethylpyrazine was reduced by *ca.* 40%. A broad glycerol peak, which masked the menthol peak, was detected as the temperature was raised above 95 °C. There was also evidence of Maillard reactions between amino acids and sugars at 145 °C, which also limited the attainable sampling temperature.

As discussed previously, levels of monoterpenes were measured above conifer needles using partition coefficients.[38] These workers also measured monoterpene levels in the needles by spiking the samples, to determine the fibre/pine-needle partitioning. SPME quantification was compared with SDE (Likens–Nickerson liquid–liquid co-distillation). Good correlations were obtained between the two methods for the lower boiling point compounds (Table 4). However, recoveries of the higher boiling point compounds were lower with SPME, reflecting incomplete vaporization of these compounds. These workers reported that different essential oil contents of various needles and difficulties with reliability of the spiking experiments made quantification of levels within the needles problematical.

There is scope for measuring volatile contents within solid biological matrices

Table 4 *Comparison between concentrations* ($\mu g\ g^{-1}$ *fresh weight) of analytes measured in Pinus peuce needles by SPME and by SDE. Taken from ref. 38*

Analyte	SPME (SD)[a]	SDE
Trycyclene	48 (2)	62
α-Pinene	2280 (63)	2960
Camphene	594 (17)	853
β-Pinene	562 (16)	968
Sabinene	4.0 (0.1)	9.2
Myrecene	41 (2)	111
α-Phellandrene	16 (0.5)	33
β-Terpinene	4.8 (0.1)	3.8
Limonene	52 (2)	187
β-Phellandrene	52 (2)	169
Terpinolene	9.6 (0.5)	66

[a] $n = 8$.

by headspace SPME, although spiking the matrices with standards is not always without problems. Variability from sample to sample[13,38] would suggest that in some cases the efficiencies of extraction of volatiles need to be determined for each individual sample and not necessarily just for a 'representative' sample from which concentrations in the solid matrix can then be determined for the remainder of the samples. This may be manageable by addition of chemically similar or isotopically labelled internal standards.

However, quantification above and within solid samples can be hindered by excessive equilibration times. Lower recoveries reported[38] for higher boiling point compounds from pine needles were attributed to non-equilibrium sampling. Similarly, equilibrium was not achieved when sampling higher MW apple aroma compounds.[34,29] Song et al.[34] reported that both α-farnesene and hexyl-2-methyl butanoate had not equilibrated with the fibre after 24 min of sampling. Matich et al.[29] sampled for 90 min and whilst hexyl-2-methyl butanoate did appear to be approaching equilibrium by this time, the increase in area of the α-farnesene peak area with sampling time remained linear (Figure 1). The low volatility, of α-farnesene, coupled with its relatively high concentrations (15%) in the apple wax of some cultivars,[41] would cause both long equilibration times between the fruit and the headspace, and also between the fibre the fruit and the headspace once the fibre is inserted into the headspace container. Unpublished work in the authors' laboratory revealed that headspace levels of α-farnesene did not reach equilibrium until five days after a 'Granny Smith' apple was sealed in a headspace jar at room temperature.

4 Conclusions

Since its inception for immersion sampling of aromatic compounds in ground water, solid phase microextraction has been used to measure volatiles from an

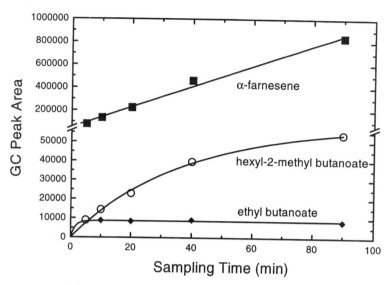

Figure 1 *Equilibration times for SPME sampling (100 μm PDMS fibre), at 20 °C, of aroma volatiles of different boiling points above two 'Granny Smith' apples in a 1.5 L headspace jar. Taken from ref. 29*

increasingly diverse range of sample types. SPME of volatiles from and within foodstuffs and plant matter has proven to be an interesting and challenging goal. Measuring levels of compounds within liquid samples by SPME now appears to be a straightforward and practical technique for the majority of samples by using internal standards or standard addition techniques. Sampling conditions (temperature, pH, ionic strength, sample and headspace volume) need optimization and workers should be aware of possible pitfalls. With the development of a theoretical basis for relating air/fibre partition coefficients to GC column retention times, rapid measurement of headspace volatiles without calibration against standard atmospheres, measured by static headspace sampling, can develop into a routine procedure. Progress has been made into measuring volatiles within solid samples by headspace SPME, through standard addition of volatiles to the headspace. However, long equilibration times between some solid sample matrices and standards added to the headspace, and substantial sample to sample matrix variability dictates that quantification of analytes within such samples is still problematical. It is clear that further development of approaches to quantitative SPME of volatiles within solid samples is desirable.

References

1. D.W. Potter and J. Pawliszyn, *J. Chromatogr.*, 1992, **625**, 247.
2. C.L. Arthur and J. Pawliszyn, *Anal. Chem.*, 1990, **19**, 2145.

3. C.L. Arthur, L.M. Killam, S. Motlagh, D.W. Potter and J. Pawliszyn, *Environ. Sci. Technol.*, 1992, **26**, 979.
4. S.B. Hawthorne, D.J. Miller, J. Pawliszyn and C.L. Arthur, *J. Chromatogr.*, 1992, **603**, 186.
5. Z. Zhang and J. Pawliszyn, *Anal. Chem.*, 1993, **65**, 1843.
6. B.D. Page and G. Lacroix, *J. Chromatogr.*, 1993, **648**, 199.
7. X. Yang and T. Peppard, *J. Agric. Food Chem.*, 1994, **42**, 1925.
8. L. Pan, M. Adams and J. Pawliszyn, *Anal. Chem.*, 1995, **67**, 4396.
9. A.D. Harmon, in *Techniques for Analysis of Food Aroma*, Food Science and Technology, ed. R. Marsili, Marcel Dekker, 1997, p. 81.
10. A. Steffen and J. Pawliszyn, *J. Agric. Food Chem.*, 1996, **44**. 2187.
11. Z.E. Penton, *Adv. Chromatogr.*, 1997, **37**, 205.
12. H.L. Lord and J. Pawliszyn, *Anal. Chem.*, 1997, **69**, 3899.
13. Z.E. Penton, *SPME Application Note No. 6*, 1997, Varian Chromatography Systems, Walnut Creek, CA.
14. R. Marsili, in *Techniques for Analysis of Food Aroma*, Food Science and Technology, ed. R. Marsili, Marcel Dekker, 1997, p. 237.
15. T.J Evans, C.E. Butzke and S.E. Ebeler, *J. Chromatogr. A*, 1997, **786**, 293.
16. J.A. Field, G. Nickerson, D.D. James and C. Heider, *J. Agric. Food Chem.*, 1996, **44**, 1768.
17. H.W. Chin, R.A. Bernhard and M. Rosenberg, *J. Food Sci.*, 1996, **61**, 1118.
18. D.A. Forss, *J. Dairy Res.*, 1979, **46**, 691.
19. L. Moio, P. Etievant, D. Langlois, J. Dekimpe and F. Addeo, *J. Dairy Res.*, 1994, **61**, 385.
20. T. Yamanishi, M. Kawakami, A. Kobayashi, T. Hamada and Y. Musalam, in *Thermal Generation of Aromas*, ACS Symposium Series 409, American Chemical Society, Washington, 1989, p. 310.
21. F. Visser, I. Gray and M. Williams, *Composition of New Zealand Foods, 3. Dairy Products*, Design Print, Auckland, 1991.
22. P.S.W. Park and R.E. Goins, *J. Agric. Food Chem.*, 1992, **40**, 1581.
23. O.W. Parks, M. Keeney, I. Katz and D.P. Schwartz, *J. Lipid Res.*, 1964, **5**, 232.
24. S. Ferary and J. Auger, *J. Chromatogr. A*, 1996, **750**, 63.
25. E. Block and E.M. Calvey in *Sulfur Compounds in Foods*, ed. C.J. Mussinan and M.E. Keelan, ACS Symposium Series 564, American Chemical Society, Washington, 1994, p. 63.
26. R. Eisert and J. Pawliszyn, *Anal. Chem.*, 1997, **69**, 3140.
27. T.J. Clark and J.E. Bunch, *J. Agric. Food Chem.*, 1997, **45**, 844.
28. N. Osawa, *Koryo*, 1987, **153**, 37.
29. A.J. Matich, D.D. Rowan and N.H. Banks, *Anal. Chem.*, 1996, **68**, 4114.
30. R.E. Shirey and V. Mani, Supelco Inc., presentation 497015 at the 1997 Pittcon Conference, Atlanta, Georgia, 1997.
31. L. Pan and J. Pawliszyn, *Anal. Chem.*, 1997, **69**, 196.
32. K.D. Buchholz and J. Pawliszyn, *Anal. Chem.*, 1994, **66**, 160.
33. R.J. Stevenson and X.D. Chen, *Food Res. Int.*, 1996, **29**, 495.
34. J. Song, B.D. Gardner. J.F. Holland and R.M. Beaudry, *J. Agric. Food. Chem.*, 1997, **45**, 1801.
35. D. Garcia, S. Magnaghi, M. Reichenbacher and K. Danzer, *J. High Resol. Chromatogr.*, 1996, **19**, 257.
36. L. Urruty, M. Montury, M. Braci J. Fournier and J.-M. Dournel, *J. Agric. Food Chem. Rapid Commun.*, 1997, **45**, 1997.

37. K.J. Miller, C.F. Poole and T.M.P. Pawlowski, *Chromatographia*, 1996, **42**, 639.
38. B. Schäfer, P. Hennig and W. Engewald, *J. High Resol. Chromatogr.*, 1995, **18**, 587.
39. P.A. Martos, A. Saraullo and J. Pawliszyn, *Anal. Chem.*, 1997, **69**, 402.
40. A. Saraullo, P.A. Martos and J. Pawliszyn, *Anal. Chem.*, 1997, **69**, 1992.
41. F.E. Huelin and I.M. Coggiola, *J. Sci. Food Agric.*, 1968, **19**, 297.

CHAPTER 27

Application of SPME to Measure Volatile Metabolites Produced by Staphylococcus carnosus and Staphylococcus xylosus

REGINE TALON AND M.C. MONTEL

1 Introduction

In the volatile fraction of dry sausages, different classes of chemical compounds can be recognised: alkanes, alkenes, alcohols, carboxylic acids, esters, ketones, aldehydes, terpenes, sulfur compounds, furans and pyrazines.[1-3] Even though there are increasing numbers of publications about volatiles in meat products, little research has been undertaken on the effect of microorganisms on flavour development. The most demonstrative results have been those of Berdagué et al.,[1] Stahnke[3,4] and Montel et al.[5] They showed that strains of Staphylococcus added as starter culture in fermented sausage modulated the levels of volatile compounds. Sausages inoculated with Staphylococcus carnosus or Staphylococcus xylosus were characterized by high contents of branched-chain aldehydes and acids, ethyl esters, ketones and low amounts of aldehydes (hexanal, nonanal). These sausages developed a strong dry cured aroma. Hinrichsen and Andersen[6] also mentioned that inoculation of Vibrio sp. in cured pork increased the level of 3-methylbutanal and 2-methylbutanal, improving the flavour of the product. Branched chain compounds such as 3-methylbutanol, 3-methylbutanal and 3-methylbutanoic acid arise from leucine catabolism during sausage manufacture. These compounds have a strong odour and a low detection threshold, so they could contribute to the dry sausage aroma. Ethyl esters with fruity aroma were detected in higher amounts in sausages inoculated with S. carnosus[5] or S. xylosus.[4] These two species have intracellular and extracellular esterases that hydrolyse esters with different acid chain lengths.[7] The oxidation of unsaturated fatty acids is well recognized as a source of volatile products which are perceived as off-flavors,[8] so it is crucial to limit this oxidation. In dry sausage, it is often mentioned that staphylococci are able to limit lipid oxidation but no clear evidence has been published.

We are still a long way from understanding the different pathways involved in flavour production in dry sausages. The mechanisms of the production of volatile molecules by microorganisms have not yet been established. In complex media such as sausage or cured pork it is difficult to elucidate the pathway. Our laboratory has focused its research on the catabolism of leucine, on the production of esters and on the oxidation of unsaturated free fatty acids by *S. carnosus* and *S. xylosus* in laboratory media, in order to highlight the metabolic pathways that could be involved in sausages. To assess the antioxidant activity of *S. carnosus* we can measure its interaction on the oxidation of unsaturated fatty acids. To measure this oxidation we proposed to assay hexanal, which is recognised as a good marker of oxidation of unsaturated free fatty acids.[9] To accomplish these goals, we had to select an appropriate analytical method.

2   Why SPME?

The analysis of a mixture of volatile compounds having different chemical natures (acids, alcohols, aldehydes, esters) in liquid aqueous media is very difficult because of the different polarities of the compounds and their interactions with the matrix. The difficulty is even greater because the production of volatiles by bacteria is presumed to be low, so a sensitive method is needed. Also, to screen a lot of bacteria for their capacity to produce aromatic compounds a rapid method is required. Many different analytical methods have been developed to determine flavor compounds. Present headspace methods include steam distillation/solvent extraction/gas chromatography, static headspace/gas chromatography and dynamic purge and trap/gas chromatography.

These methods are very time-consuming, require exhaustive concentration steps and/or require dedicated gas chromatographs equipped with headspace sampling devices. Therefore we chose a new, quicker and simpler technique, solid-phase microextraction (SPME). SPME has been shown to be a very sensitive method for the analysis of flavor and fragrance compounds.[10] This method compared favorably to the commonly used purge and trap type analysis.[11] The use of a fiber for extraction can also enhance the selectivity of the analysis because one may choose the stationary phase that best suits the analytes.

To use SPME–GC, it was necessary to optimize the extraction conditions for the different metabolites.[12] For this purpose, the effects of the sampling mode (headspace or direct) and coating phases (PDMS, PA) were characterized. Esters were better extracted by direct sampling with the PDMS phase, whereas all the other compounds were better extracted by headspace sampling at 80 °C with the PA phase. To quantify the metabolites by SPME–GC, we used an external calibration. For esters, hexanal and the three metabolites from leucine, standard curves were generated. The detection limit and the linear range of the compounds are mentioned in Table 1.

Table 1 *Linear ranges and detection limits of compounds studied*

Compounds	Linear range (ppm)
3-Methylbutanal[b]	0.5[a]–10
3-Methylbutanol[b]	0.2–10
3-Methylbutanoic acid[b]	0.2–10
Ethyl acetate[c]	0.2–2
Ethyl butanoate[c]	0.06–200
Ethyl 2-methylbutanoate[c]	0.01–2
Ethyl 3-methylbutanoate[c]	0.01–2
Ethyl valerate[c]	0.01–2
Ethyl hexanoate[c]	0.01–2
Ethyl heptanoate[c]	0.01–2
Ethyl decanaote[c]	0.01–2
Hexanal[b]	0.2–50

[a] The lowest value is the limit of detection.
[b] Extraction of all these compounds was done by SPME with PA coating phase, sampling was done on the headspace of 0.7 mL of liquid sample in a 1.5 mL vial, samples were equilibrated for 15 min prior to 15 min of sampling at 80 °C under magnetic stirring, samples were acidified to pH = 3.0 by adding HCl, samples were saturated with NaCl and extraction time lasted 15 min (see reference 12 for more details).
[c] Extraction of esters was done by SPME with PDMS coating phase, direct sampling was on 1.3 mL of liquid sample in 1.5 mL vial, samples were saturated with NaCl and extraction time lasted 15 min at ambient temperature under magnetic stirring (see reference 12 for more details).
All the samples were analysed by gas chromatography as described by Vergnais *et al.*[12]

3 Application of SPME–GC to Staphylococcal Metabolites

Catabolism of Leucine by *Staphylococcus carnosus* and *Staphylococcus xylosus*

The catabolism of leucine by the two staphylococci was studied by Masson.[13] This study showed that these species catabolized leucine into 3-methylbutanal, 3-methylbutanol and 3-methylbutanoic acid (Table 2). 3-methylbutanoic acid was found to be the dominant metabolite, representing more than 95% of the total.

The production of 3-methylbutanoic acid was influenced by the preculture conditions for *S. carnosus*, with higher production recorded in a complex medium (Table 3). The presence of nitrate in the medium sharply inhibited the production of 3-methylbutanoic by both species. The effect of agitation during incubation (static or shaking) was only important for the production of the acid by *S. xylosus* (Table 3).

Table 2 *Leucine catabolism by S. xylosus (16) and S. carnosus (833)*

Compounds*	3-Methylbutanal	3-Methylbutanol	3-Methylbutanoic
S. xylosus	0.86	1.39	52.14
S. carnosus	2.46	0.45	43.57

Reprinted with permission (Vergnais *et al.*[12])
* Quantities are expressed in nmol ml^{-1} of bacterial cell suspension with an optical density of 1.0.

Table 3 *Production of 3-methylbutanoic acid by S. carnosus (833) and S. xylosus (16) under different conditions*

	S. carnosus	S. xylosus
Preculture conditions		
Synthetic medium[a]	19.49	39.47
Complex medium[a]	38.38	36.50
Nitrate content		
0%	55.23	60.72
0.03%	2.63	15.25
Stirring conditions		
Static	26.68	55.57
Shaking	31.19	20.40

Results adapted from Masson.[13]
* Quantities are expressed in nmol mL^{-1} of bacterial cell suspension with an optical density of 1.0.
[a] Complex media was PYS described by Lechner *et al.*[20] with 1% of glucose, synthetic media was described by Hussain *et al.*[21] without glucose.

Different pathways can lead to 3-methylbutanoic acid from leucine.[14] However, there is very little literature data on amino acid catabolism and, in our laboratory, research is in progress to characterize the enzymatic pathways involved.

Production of Esters by *Staphylococcus carnosus* and *Staphylococcus xylosus*

Resting cells and extracellular enzymes of *S. xylosus* and *S. carnosus* were prepared as described by Talon and Montel.[7] These preparations, incubated separately with ethanol and different acids, were able to synthesize esters. *S. xylosus* produced a higher level of esters with either resting cells or extracellular enzymes than *S. carnosus* (Table 4). The two species esterified ethanol with acids varying from acetic to decanoic (Table 4); however, the resting cells of *S. xylosus* preferentially esterified butanoic acid and then valeric and hexanoic acids (Table 4). The extracellular enzymes of *S. xylosus* esterified these three acids at approximately the same rate. The resting cells of *S. carnosus* did not exhibit a

Table 4 *Ethyl esters synthesis with different acid chain lengths, by resting cells and extracellular enzymes of S. carnosus (833) and S. xylosus (16)*

Ethyl esters	EC2	EC4	E2C4	E3C4	EC5	EC6	EC10	Total
Resting cells*								
S. xylosus	18.27	270.06	7.81	26.06	120.29	78.56	9.10	530.15
S. carnosus	0.00	3.98	2.97	0.86	2.08	2.54	0.00	12.43
Extracellular enzymes*								
S. xylosus	3.53	16.03	10.44	2.91	16.56	19.05	9.10	77.62
S. carnosus	0.00	6.87	2.76	1.58	15.03	7.25	0.00	33.49

EC2 ethyl acetate, EC4 ethyl butanoate, E2C4 ethyl 2-methylbutanoate, E3C4 ethyl 3-methyl-butanoate, EC5 ethyl valerate, EC6 ethyl hexanoate, EC10 ethyl decanoate, Total = sum of the esters EC2 to EC10.

* Quantities are expressed in nmol g^{-1} of wet cells for the resting cells (conversion factor $\times 9$ from nmol mL^{-1} of media with OD = 1.0) and in nmol mg^{-1} of proteins for the extracellular enzymes (proteins were assayed by the method of Bradford).

pronounced specificity; they esterified butanoic, 2-methylbutanoic, valeric and hexanoic acids at similar rates (Table 4). The extracellular enzymes of this strain synthesized ethyl valerate at the highest rate, and then ethyl hexanoate and ethyl butanoate.

Short (acetic) and long (decanoic) chain acids were poorly esterified by the two species. Isoacids were esterified even less, primarily by *S. xylosus* esterases.

Temperature affected the synthesis of ethyl butanoate (Table 5). The activity of the resting cells and, to a lesser extent, that of the extracellular enzymes of *S. xylosus*, was higher at 14 °C than at 24 °C. For *S. carnosus*, the activity of both enzymatic extracts was similar at the two temperatures (Table 5).

Table 5 *Ethyl butanoate synthesis by resting cells and extracellular enzymes of S. carnosus (833) and S. xylosus (16) at two temperatures*

	14 °C	24 °C
Resting cells*		
S. xylosus	183.84	101.92
S. carnosus	1.79	1.88
Extracellular enzymes*		
S. xylosus	33.13	26.64
S. carnosus	20.20	31.32

Results adapted from Talon et al.[19]

* Quantities are expressed in nmol g^{-1} of wet cells for the resting cells (conversion factor $\times 9$ from nmol mL^{-1} of media with OD = 1.0) and in nmol mg^{-1} of proteins for the extracellular enzymes (proteins were assayed by the method of Bradford).

Table 6 *Ethyl butanoate synthesis by resting cells and extracellular enzymes of S. carnosus (833) and S. xylosus (16) at two pH values*

	pH 5.5	pH 7.0
Resting cells*		
S. xylosus	48.28	237.48
S. carnosus	1.23	3.74
Extracellular enzymes*		
S. xylosus	23.01	36.75
S. carnosus	21.59	29.93

Results adapted from Talon *et al.*[19]
* Quantities are expressed in nmol g^{-1} of wet cells for the resting cells (conversion factor $\times 9$ from nmol mL^{-1} of media with OD = 1.0) and in nmol mg^{-1} of proteins for the extracellular enzymes (proteins were assayed by the method of Bradford).
Experiments were carried out in phosphate buffer, 0.1 M, pH 7.0 or 5.5.

Acid pH strongly inhibited the activity of the resting cells of both staphylococci (Table 6). Ethyl butanoate production at pH 5.5 was 20% and 32% of the activity measured at pH 7.0 for *S. xylosus* and *S. carnosus* respectively. The activities of the extracellular enzymes were equivalent at the two pH values (Table 6).

Esterifications by the two species corresponded to those for the hydrolysis of *p*-nitrophenyl esters.[7] *S. xylosus* showed high hydrolysis and esterification activities, whereas *S. carnosus* had low esterase activies both in hydrolysis and esterification. Concerning the substrate specificity, very close profiles of synthesis and hydrolysis of esters were noticed for the different enzymatic extracts of the staphylococci. It appears that the same esterases are involved in both hydrolysis and esterification.

Our results for ethyl esters synthesised by staphylococci and assayed by SPME–GC were comparable to those of thioesters produced by coryneform bacteria, *Micrococcaceae* and lactococci, and measured by dynamic headspace GC.[15] SPME therefore appears to be a good method to extract volatile compounds from aqueous media.

Effect of *Staphylococcus carnosus* on the Oxidation of Unsaturated Free Fatty Acids

To characterize antioxidant potential, *S. carnosus* was grown in a medium supplemented with a mixture of unsaturated fatty acids (C18:1, C18:2, C18:3). The comparison of the level of hexanal in the control and in the inoculated sample indicated clearly that *S. carnosus* inhibited the oxidation of fatty acids (Table 7). The strain inhibited oxidation during the entire incubation period (9 days). In the sterile control, oxidation of unsaturated fatty acids reached a maximum after 3 days.

Table 7 *Effect of S. carnosus (833) on the level of hexanal*

Time (days)	0	3	9
Control	1.86	16.6	15.9
S. carnosus	1.86	1.62	1.10

Reprinted with permission (Vergnais *et al.*[12]).
Quantities of hexanal are expressed in nmol mL^{-1} of media.

It is difficult to compare these data to literature results because there is very little existing data on the antioxidant properties of *Staphylococcus aureus*.[16,17] Those of other staphylococci are still unknown. The measurement of hexanal by SPME seems to be a good method to estimate the oxidation of unsaturated fatty acids. It could replace the thiobarbituric acid test (TBARS) used by numerous authors, but not always correlated to the oxidation of lipids alone.[18]

5 Conclusion

SPME–GC proved to be a good tool to study the production of aromatic compounds by bacteria in aqueous media. As it is easy to handle, it should be useful to screen new starter cultures. With this method we showed that *S. xylosus* and *S. carnosus* had high enzymatic potential. These strains catalysed leucine in 3-methylbutanol, 3-methylbutanal and 3-methylbutanoic acid, compounds having strong odors. They esterified acids from diverse origins with ethanol to yield esters with fruity notes. Esterification was achieved at low or high temperatures or pH depending on the enzymatic extracts of the staphylococci (resting cells and extracellular enzymes). Furthermore, *S. carnosus* inhibited oxidation of unsaturated acids and so could inhibit the production of rancid flavours. These two species could produce compounds which are involved in dry sausage aroma.

Acknowledgments

This work was supported by EU program (No. AIR2 CT94 1517) entitled 'Optimisation of endogenous and bacterial metabolism for improvement of safety and quality of fermented meat products', coordinator D. Demeyer.

References

1. J.L. Berdagué, P. Monteil, M.C. Montel and R. Talon, *Meat Sci.*, 1993, **35**, 229.
2. G. Johansson, J.L. Berdagué, M. Larsson, N. Tran and E. Borch, *Meat Sci.*, 1994, **38**, 203.
3. L.H. Stahnke, *Meat Sci.*, 1995, **41**, 179.
4. L.H. Stahnke, *Meat Sci.*, 1994, **38**, 39.

5. M.C. Montel, J. Reitz, R. Talon, J.L. Berdagué and S. Rousset-Akrim, *Food Microbiol.*, 1996, **13**, 489.

6. L. Hinrichsen and H.J. Andersen, *J. Agric. Food Chem.*, 1994, **42**, 1537.

7. R. Talon and M.C. Montel, *Int. J. Food Microbiol.*, 1997, **36**, 207.

8. J.R. Vercellotti, A.J.S. Angelo and A.M. Spanier, in *Lipid Oxidation in Food*, ed. A.J.S. Angelo, American Chemical Society, Washington, DC, 1992, p. 1.

9. C.W. Fritsch and J.A Gale, *J. Am. Oil Chem. Soc.*, 1976, **54**, 225.

10. Z. Zhang and J. Pawliszyn, *Anal. Chem.*, 1993, **65**, 1843.

11. B. MacGillivray and J. Pawliszyn, *J. Chromatogr. Sci.*, 1994, **32**, 317.

12. L. Vergnais, F. Masson, M.C. Montel, J.L. Berdagué and R. Talon, *J. Agric. Food Chem.*, 1998, **46**, 228.

13. F. Masson, Thesis, University of Blaise Pascal, Clermont-Ferrand, 1998.

14. R. Martin, V.D. Marshall, J.R. Sokatch and L. Unger, *J. Bacteriol.*, 1973, **115**, 198.

15. G. Lamberet, B. Auberger and J.L. Bergère, *Appl. Microbiol. Biotechnol.*, 1997, **48**, 393.

16. H.D. Lilly, J.L. Smith and J.A. Alford, *Can. J. Microbiol.*, 1970, **16**, 855.

17. J.A. Alford, J.L. Smith and H.D. Lilly, *J. Appl. Bact.*, 1971, **34**, 133.

18. J.R. Vercellotti, O.E. Mills, K.L. Bett and D.L. Sullen, in *Lipid Oxidation in Food*, ed. A.J.S. Angelo, American Chemical Society, Washington, DC, 1992, p. 232.

19. R. Talon, C. Chastagnac, L. Vergnais, M.C. Montel and J.L. Berdagué, in *2nd International Congress of Meat Sciences and Technology*, ed. Matfork, 1997, p. 536.

20. M. Lechner, H. Markl and F. Götz, *Appl. Microbiol. Biotechnol.*, 1988, **28**, 345.

21. M. Hussain, J.G.M. Hasting and P.J. White, *J. Med. Microbiol.*, 1991, **34**, 143.

CHAPTER 28

Application of SPME Methods for the Determination of Volatile Wine Aroma Compounds in View of the Varietal Characterization

DEMETRIO DE LA CALLE GARCÍA AND
MANFRED REICHENBÄCHER

1 Introduction

There are more than a thousand different organic compounds registered in the bibliography of wines,[1] a large number of which have been analysed using gas chromatography.

In the last 20 years, sample preparation for the GC analysis of wines has been mainly carried out by means of liquid/liquid extraction,[2] but solid phase extraction,[3] dynamic headspace,[4] supercritical fluid extraction,[5] direct injection[6] and simultaneous distillation-solvent extraction[7] have also been used. Other well-known sampling methods, such as static headspace, are not sensitive enough for the extraction of the most volatile aroma compounds in wines.[8]

Recently, SPME was used as sampling method for the GC analysis of volatile compounds in wine.[9,10] The effectiveness of this sampling method is similar to liquid\liquid extraction with 1,1,2-trichlorotrifluoroethane (Kaltron),[8] and it is also useful for headspace sampling with GC analysis. However, SPME does not require the use of halogenated hydrocarbon solvents, which are required for liquid/liquid extraction, and there is a good potential for method automation using SPME.

Of all the many volatile wine compounds, the terpenes form a characteristic part of the grape bouquet. According to the work of Rapp,[1] terpenes are not changed by yeast metabolism during fermentation, a fact which allows for the characterization of wines made from different grape varieties. Therefore, for the instrumental-analytical characterization of wine varieties, an effective extraction of terpenes is required to produce the so-called aromagrams, which represent the so-called 'finger print' of wines. The aromagrams for some

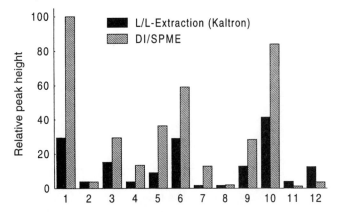

Figure 1 *Terpene pattern of a Morio-Muskat wine obtained with DI-SPME(PA)–GC and with a gas chromatogram from a Kaltron extraction 1 geranoic acid, 2 terpendiol II, 3 geraniol, 4 nerol, 5 citronellol, 6 α-terpineol, 7 terpendiol I, 8 trans-rosen oxide, 9 hotrienol, 10 linalool, 11 cis-linalool oxide furanoside, 12 trans-linalool oxide furanoside*

important terpenes obtained by means of SPME sampling are similar to those obtained by means of liquid/liquid extraction, as shown by Figure 1.

Because the analysis of terpenes alone does not always supply a correct classification of an unknown wine, it is necessary to consider not only terpenes but also other aroma compounds which significantly constitute the particular varieties of a certain vintage. Furthermore, the statistical classification of wine varieties requires data of the aromagrams obtained by many wines. This requires an automated sampling technique for the gas chromatographic analysis, which may be only realized by SPME.

Wine bouquet components other than the terpenes belong to classes of compounds with very different polarities; for example hydrocarbons, carbonyl compounds, alcohols, and acids. Therefore, the question of which fiber was appropriate for the highly efficient extraction of a certain group arose, and the extraction of different wine aroma compounds using all commercially available SPME fibers was studied.

Furthermore, SPME–GC is an appropriate method for the quantitative analysis of volatile aroma compounds of wines, and in the multiple headspace (MHE) sampling mode SPME is also applicable for the quantitative analysis of such compounds in the skin tissue of berries and the seeds.

2 Optimization of the SPME–GC Analysis of Wine Bouquet Components

The experimental conditions required for the high sensitivity and good precision of SPME–GC analysis are described in Chapter 1. A wine sample is character-

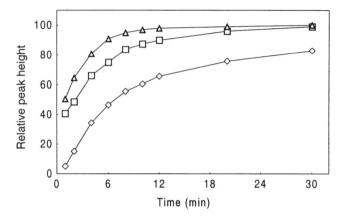

Figure 2 *Influence of the agitation speed upon peak height of terpenes:* ◇ *no agitation,*
□ *speed 550 rpm,* Δ *speed 1500 rpm. Standard 2 mL vial for autosamplers, stir*
bar: 8 × 3 mm
(Adapted with permission from ref. 9)

ized by a very high number of volatile components from various classes of compounds, which have different polarities, by its relatively high ethanol content and by the existence of thermal equilibria which must be considered in direct (DI) as well as headspace (HS) SPME.

The following parameters influence the adsorption of wine bouquet components on the SPME fiber.

Agitation Technique and Sampling Mode

When magnetic stirring is used, the required equilibrium time is about 25–30 minutes in the direct extraction mode, as shown in Figure 2 for the DI–SPME of linalool from an aqueous solution, to which 10% ethanol was added in order to simulate the wine matrix.[9]

Owing to the lower volatility of most of the wine bouquet components, the equilibrium times are longer in the headspace mode.[11] In general, 60–90 minutes is required for the equilibrium.

Ultrasonic waves (35 kHz, 1 h, 25 °C) do not increase extraction effectivity, but change the terpene aromagrams.

Sample Volume and Phase Ratio β

According to equation 1 derived from equation (4) in Chapter 1, the number of moles of analyte n is determined not only by the constants K_{fs}, K_{hs} and the sample volume V_s, but also by the phase ratio β, *i.e.* the ratio of the headspace

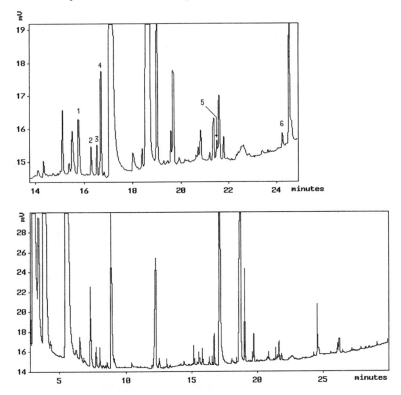

Figure 3 *HS-SPME(PA) gas chromatogram of a Riesling wine with 75 µL sample volume: 1* trans-*linalool oxide furanoside, 2* cis-*linalool oxide furanoside, 3-linalool, 4 hotrienol, 5 geraniol, 6 geranois acid*

volume V_h to the sample volume V_s.

$$n = \frac{K_{fs} \cdot c_o \cdot V_s}{\dfrac{V_h}{V_s} + \dfrac{1}{K_{hs}}} = \frac{K_{fs} \cdot c_o \cdot V_s}{\beta + \dfrac{1}{K_{hs}}} \qquad (1)$$

Using a 2 mL vial, a sample of about 700 µL ($\beta \approx 1.85$) is enough for an effective extraction.[11]

Owing to the high sensitivity of headspace SPME–GC, only minimal sample volumes are needed, which is demonstrated by the HS–SPME capillary gas chromatogram with 75 µL Riesling wine, as shown in Figure 3. This fact could be of interest if only a small sample amount is available; for instance, in forensic analysis.

Effect of Extraction Parameters

Owing to the complexities of the wine matrices mentioned above, the components of sample matrices and extraction conditions have an important influence

Table 1 *Effect of temperature on the extraction of 3-decanol (10 ppm) from water which has a content of 10% ethanol*

Temperature (°C)	25.0	50.0	60.0	70.0	80.0
Relative peak area	100.0	28.0	23.0	12.5	12.6

on SPME sampling. Whereas pH in the range 2–12 does not have a marked influence on the extraction of terpenes and other wine bouquet components, the mass extracted by the SPME fiber is essentially determined by the sampling temperature, salting methods and ethanol concentration.[9]

The important influence of the sampling **temperature** is shown by the SPME extraction of 10 ppb 3-decanol added to a Riesling wine as shown in Table 1. According to van't Hoff's equations for the constants K_{fh} and K_{hs}, increasing the temperature results in lower peak intensities.[12] Furthermore, in the case of terpene sampling, the existence of acid-catalyzed thermal equilibria[1] results in unusual temperature profiles.[11] Therefore, room temperature is necessary to avoid changes in the terpene finger print during the sampling step.

The content of **ethanol** is about 10^5–10^6 times higher than the wine bouquet components. It is also extracted, and its high concentration diminishes the extraction of the other aroma compounds (see Figure 4). In spite of the considerable decrease of the extractable mass for wines with high ethanol concentrations, SPME is an appropriate extraction method to produce aroma-grams for chemometric studies.[13] The influence of different alcohol contents on the extracted aroma compounds in wine samples is eliminated by using an internal standard.

The addition of salt to samples (**salting**) can increase or decrease the amount of analyte extracted, depending on the polarity of the analytes and the salt

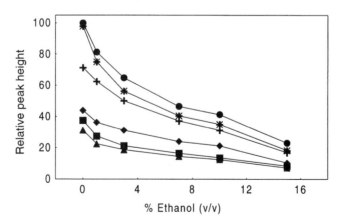

Figure 4 *Influence of the ethanol content upon peak height of terpenes:* ● *nerol,* * *nerol oxide,* + *α-terpineol,* ◆ *citronellol,* ■ *geraniol,* ▲ *linalool*
(Adapted with permission from ref. 9)

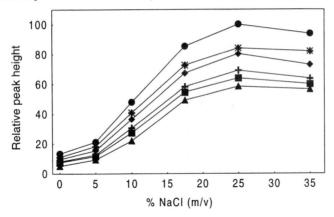

Figure 5 *Influence of NaCl concentration on peak height of terpenes:* ● *nerol,* ∗ *nerol oxide,* + *α-terpineol,* ◆ *citronellol,* ■ *geraniol,* ▲ *linalool* (Adapted with permission from ref. 9)

Table 2 *Salting effect on the extraction of linalool and hotrienol from a Riesling wine shown by the peak areas (arbitrary units) from HS-SPME(PA)– GC chromatograms. All salts were used in saturated concentration*

	NaCl	*KCl*	*KBr*	*NH₄Cl*
Linalool	2.56	2.23	1.97	1.95
Hotrienol	1.56	1.18	1.15	1.00

concentration. Since the salting effect is directly related to compound polarity, salting should increase the extracted amount of polar terpenes (see Figure 5). NaCl shows the best salting effect as shown for the extraction of linalool and hotrienol from a Riesling wine. As shown in Table 2, the highest terpene amount extracted is achieved with a NaCl-saturated wine solution. However, such high salt concentrations accelerate the ageing of the SPME fiber in the direct sampling mode. Therefore, after each extraction step, it is necessary to clean the fiber with water and to equilibrate it in the GC-injector for several minutes; otherwise, the headspace sampling mode should be used.[11]

Selection of Fiber Coating

According to the general rule 'like dissolves like' the polar polyacrylate fiber (PA) should be considered for the extraction of the middle polar terpenes. Figure 6 shows that the terpenes are best extracted using a PA fiber of thickness 85 μm both in the direct and in the headspace sampling mode. This SPME fiber is very rugged and allows high injector desorption temperatures, up to 300 °C.

To investigate the extraction yields of the other wine aroma compounds, the

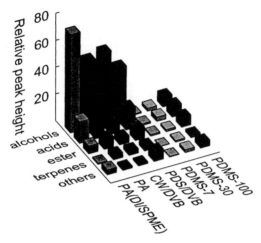

Figure 6 *Sum of the peak areas of five groups of wine aroma compounds after HS-SPME
of a Morio-Muskat wine*

following SPME fibers were used:[14] poly(dimethylsiloxane) 100 μm (PDMS-
100), 30 μm (PDMS-30) and 7 μm (PDMS-7), PA, CW/DVB and PDMS/DVB.
For reasons of comparability all tests were carried out with the same wine
sample. The aroma compounds listed in Table 3 were summarized in five
groups: alcohols, acids, esters, terpenes and other substances. Ethanol was not
considered.

The appropriate fiber for the extraction of certain substance classes by DI-
SPME was earlier described.[14] The fiber PA should be considered in direct
SPME–GC wine analysis of alcohols, terpenes and organic acids. For the acids
the CW/DVB-65 is appropriate as well, and the terpenes may also be effectively
extracted with the PDMS-100. This coating is also applicable for the extraction
of esters. The SPME fibers PDMS-7, PDMS-30 and PDMS/DVB generally
produce poor extraction yields for volatile wine aroma compounds.

The results of headspace SPME–GC of a Morio-Muskat wine with different
fibers are given in Figure 6. In order to compare HS-SPME and DI-SPME the
results using the PA fiber in the direct sampling mode are considered. The
85 μm PA is the most appropriate fiber for the extraction of terpenes by DI-
SPME.[14]

In general, HS-SPME using the fibers PA, CW/DVB and PDMS/DVB is a
more effective sampling method than the DI-SPME mode.[14] The superiority of
HS-SPME or DI-SPME in the method development for the SMPE–GC analysis
of such wine aroma compounds is determined by the volatilities of the
individual compounds.

The extractions of terpenes in the headspace sampling mode, with the fibers
PA, CW/DVB and PDMS/DVB-65 produce similar sensitivities, but the PA
fiber is recommended for the HS-SPME GC analysis of terpenes in wines, owing
to its greater ruggedness.

Thermodesorption

Desorption conditions for the SPME extracted wine aroma compounds are similar to those described by Pawliszyn,[12] but because of the relatively high concentrations of the volatile aroma compounds, a split injection should be used to avoid overloading of the GC column (see Figure 7).

Optimized Experimental Conditions for SPME–GC Analysis of Wines

The experimental conditions optimized for SPME sampling of wines with GC analysis are summarized in Table 4. For SPME–GC analysis, the headspace sampling mode should be used. In addition to the well-known advantages of the headspace sampling mode in comparison to the DI-SPME, terpenes and other volatile wine components show a higher sensitivity with HS-SPME.[8,11] Only compounds which have low volatility are extracted more effectively with the direct sampling mode.

In Figure 8 a typical capillary GC–MS chromatogram of a Morio-Muskat wine using DI-SPME(PA) sampling is shown. It clearly demonstrates the applicability of SPME as a sampling method for capillary gas chromatography in the analysis of volatile wine components.

The sums of peak areas of the most important classes of volatile wine bouquet components from a DI-SPME(PA) gas chromatogram are listed in Table 5. The concentration of terpenes in wine is small.[1] Therefore, the sum of peak areas of terpenes is only about 4% of the overall peak sum. With optimized conditions however, the SPME–GC produces chromatograms appropriate for the quantitative analysis of terpenes.

Cryofocusing ($0 \,^{\circ}C$) increases the precision of retention times, as shown by Figure 9. This is important for the peak identification of the high number of volatile wine components in the autosampling mode.

3 SPME–GC Analysis of Volatile Wine Components

Peak Identification

The adjusted retention times t' ($t' = t_r - t_o$) of eighty-nine compounds identified by GC–MSD, or with reference substances in the gas chromatogram of wines using the SPME fiber PA and the column DB/WAX,[14] are listed in Table 3. Many of these compounds are 'key substances' for the chemometric classification of wines.

An interesting method for the analysis of halogen hydrocarbon compounds in wines is the SPME–GC using the atomic emission detector (AED), as shown in Figure 10. The combination of the effective SPME sampling method with the highly specific AED allows the identification of compounds of very low concentrations in component-rich gas chromatograms. This is important in the residue analysis of wines.

Table 3 Wine aroma compounds identified in the SPME(PA-85)–GC(DB/WAX)–MS chromatograms (Figure 8) of a Morio-Muskat wine. t' adjusted retention time (t' = t_r − t_0). Class of compounds for Figure 6: A alcohols, S acids, E esters, T terpenes, O other compounds, IS internal standard

No.		t'	Name	No.		t'	Name
1	E	9.97	ethyl 2-methylbutanoate	46	IS	72.29	3-decanol
2	E	11.14	ethyl 3-methylbutanoate	47	T	73.43	hotrienol
3	O	13.59	2-ethenyltetrahydro-2H-pyran	48	A	73.98	2-(2-ethoxyethoxy)ethanol
4	A	14.90	2-methyl-1-propanol	49	E	76.06	ethyl decanoate
5	E	15.15	3-methylbutyl acetate	50	A	78.63	2-furanmethanol
6	A	21.08	1-butanol	51	S	79.53	butanoic acid
7	E	25.85	ethyl hexanoate	52	E	80.47	diethyl succinate
8	A	27.52	2-methyl-1-butanol	53	T	82.47	α-terpineol
9	A	28.06	3-methyl-1-butanol	54	A	84.31	3-methylthio-1-propanol
10	E	30.74	hexyl acetate	55	T	85.07	2,6-dimethyl-3,7-octadiene-2,6-diol
11	A	31.49	1-pentanol	56	O	85.41	2(5H)-furanone
12	E	32.90	ethyl hydroxyacetate	57	O	89.25	3-methylcyclopentanone
13	O	35.32	1-hydroxy-2-propanone	58	E	90.26	2-phenylethyl formate
14	O	36.82	hydroxyacetaldehyde	59	E	90.89	ethyl benzoate
15	A	38.57	4-methyl-1-pentanol	60	T	91.32	citronellol
16	A	39.98	3-methyl-1-pentanol	61	A	93.46	2-(2-butoxyethoxy)ethanol
17	E	41.21	2-hydroxyethyl propanoate	62	E	93.93	2-phenylethyl acetate
18	A	43.35	1-hexanol	63	T	94.40	nerol
19	A	44.10	cis-3-hexen-1-ol	64	S	98.87	hexanoic acid
20	A	46.32	trans-3-hexen-1-ol	65	T	100.20	geraniol
21	E	46.87	methyl 2-propenoate	66	A	101.00	benzyl alcohol
22	A	48.34	3-octanol	67	A	102.02	4,5-dimethyl-2-hepten-3-ol

23	A	49.20	2-hexen-1-ol
24	E	51.39	ethyl octanoate
25	T	52.53	*trans*-linalool oxide
26	S	53.50	acetic acid
27	O	53.73	2-furancarboxaldehyde
28	T	54.47	nerol oxide
29	A	54.90	7-octen-4-ol
30	A	55.50	1-heptanol
31	T	55.97	*cis*-linalool oxide
32	O	56.31	2,3-butanedione
33	A	56.68	6-methyl-5-hepten-2-ol
34	S	59.08	formic acid
35	A	59.66	2-ethyl-1-hexanol
36	T	60.88	terpene (unknown)
37	S	64.35	propanoic acid
38	E	65.09	ethyl 2-hydroxy-4-ethylpentanoate
39	A	65.60	2,3-butanediol (*threo*)
40	T	66.57	linalool
41	A	67.76	1-octanol
42	S	68.00	2-methylpropanoic acid
43	O	69.92	2-hydroxy-3-hexanone
44	A	70.03	2,3-butanediol (*erythro*)
45	O	71.30	butyrolactone
68	O	102.67	*N*-butylacetamide
69	A	105.21	phenylethyl alcohol
70	A	107.85	1,4-butanediol
71	S	110.08	2-ethylhexanoic acid
72	S	111.06	2-hexenoic acid
73	E	111.39	3-hexenyl butanoate
74	A	114.18	phenol
75	E	118.87	diethyl hydroxybutanoate
76	S	121.16	octanoic acid
77	E	130.93	ethyl 2-hydroxypropanoate
78	O	134.52	5-ethyldihydro-2(3*H*)-furanone
79	O	138.27	2,3-dihydro-3,5-dihydroxy-4*H*-pyran-4-one
80	S	141.00	decanoic acid
81	A	144.20	1,2,3-propanetriol
82	T	146.21	*trans*-geranoic acid
83	E	149.76	hexylethyl succinate
84	S	152.42	benzoic acid
85	O	158.50	5-hydroxymethyl-2-furancarboxaldehyde
86	S	159.24	dodecanoic acid
87	S	161,82	tetradecanoic acid
88	O	165.42	*N*-(2-phenylethyl)acetamide
89	E	171.54	2-phenylethyl octanoate

Adapted with permission from ref. 14.

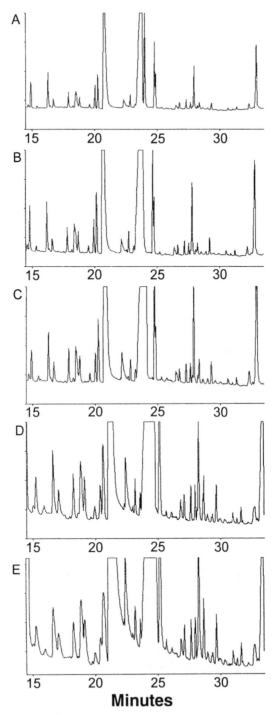

Figure 7 *Influence of the split ratio on the gas chromatogram using SPME sampling:*
A 12:1; B 7.5:1; C 3:1; D 1:1; E splitless

Table 4 *Optimized experimental conditions for the SPME–GC analysis of terpenoids and other volatile compounds in wines*

SPME	*DI-SPME*	*HS-SPME*
Volume of the sample	1300 μL	600 μL
NaCl content	200 g L^{-1}	300 g L^{-1}
Temperature of the sample	room temperature	
SPME-fiber	PA^a	
Extraction time	60 minutes	
Desorption temperature	300 °C	
Desorption time	3 minutes	
GC		
Column	**DB-WAX**	**DB-5**
Column length	25 m × 0.25 mm;	60 m × 0.25 mm;
	d_f = 0.25 μm	d_f = 0.35 μm
Temperature program	30 °C (5 min),	30 °C (5 min),
	1 °C min^{-1} → 200 °C,	5 °C min^{-1} → 200 °C
	20 °C min^{-1} → 220 °C	
Internal standard	3-decanol	2,6-dimethyl-5-hepten-2-ol

^{a} For terpenes. For recommended fibers for volatile compounds other than terpenoids see Figure 6.

HS-SPME–GC Aromagrams

In the case of the PA fiber, peak intensities of a HS-SPME capillary gas chromatogram depend upon the number of extraction cycles in which the fiber was used; see Figure 11 for 3-decanol. Therefore, the use of an internal standard is necessary for the quantitative evaluation of HS-SPME capillary gas chromatograms. 3-Decanol or 3-octanol are appropriate internal standards for HS-SPME–GC of wine aroma components. Through the use of these compounds the influence of the aging of the fiber can be corrected, as is schematically shown in Figure 12.

With regard to the classification of wines, the relative peak area Q_i is calculated for each component i according to $Q_i = A_i/A_{IS}$. A_i and A_{IS} are the peak areas or peak heights of the analyte i and the internal standard, respectively. The graphical representation of the Q_i values gives the so-called **aromagram** of a certain wine. Although the aromagrams are not based on absolute concentrations of aroma compounds, they may be used as basis for the classification of wines because all aromagrams are produced under the same conditions.

Using the aromagrams of terpenes as a 'finger print' for a certain wine, a good amount of precision of Q_i values is required for the respective analytes. In Table 6 the RSD are shown for the Q_i values of ten terpenes, which are comparable with those obtained by other extraction methods.[2]

There are optimal concentration ranges for the IS depending on the concentration of the aroma compounds in wines because % RSD increases with lower Q_{IS}. Therefore, it is recommended to use lower IS concentrations for wines which have only small concentrations of aroma compounds (*e.g.* Silvaner).

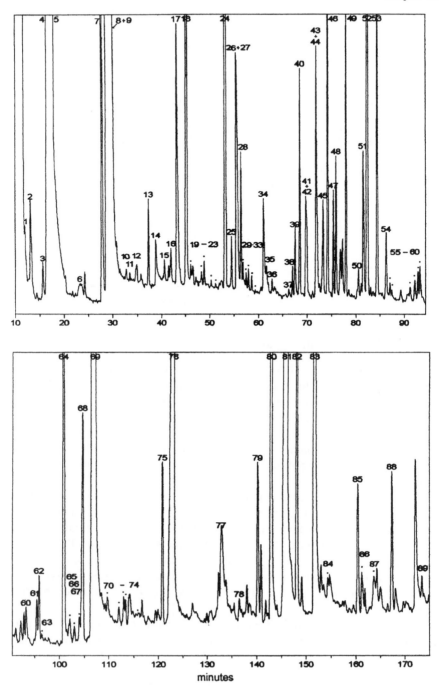

Figure 8   *A representive gas chromatogram of a Morio-Muskat wine with DI-SPME sampling using a PA fiber, and a DBWAX column. The names of peak numbers are shown in Table 3*
(Adapted with permission from ref. 14)

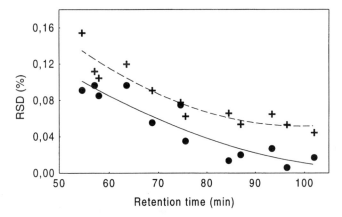

Figure 9   *Influence of cryofocusing on the relative standard deviation (RSD) of adjusted retention times t' from a SPME gas chromatogram of a Morio-Muskat wine;*
$t' = t_r - t_o: +30°C,$ ● $0°C$

Table 5   *Sum of peak areas of the most important classes of volatile wine bouquet components from a DI-SPME(PA) gas chromatogram*

Class of compound	% Sum of peak area
Alcohols	60
Acids	22
Ester	9
Terpenoids	4
Others	5

Table 6   *Relative standard deviation (% RSD) for the adjusted peak areas Q_{IS} of terpenoids obtained with HS-SPME(PA)–GC(DB/WAX)*

Terpenes	$Q_{IS} = A_i/A_{IS}$	% RSD
trans-Linalool oxide furanoside	0.0532	19.33
Nerol oxide	0.3914	13.05
cis-Linalool oxide furanoside	0.0676	15.60
Linalool	12.1538	6.73
Hotrienol	4.0275	7.77
α-Terpineol	2.9786	9.34
Terpendiol I	0.6507	13.52
Citronellol	0.7010	13.00
Geraniol	2.9336	9.86

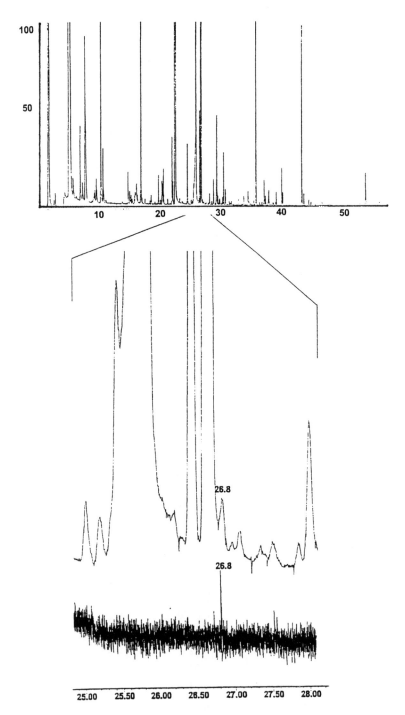

Figure 10   *HS-SPME–CGC–MSD chromatogram of a Morio-Muskat wine (upper) and the HS-SPME–CGC–AED detection of 1,3,5-trichlorobenzene (lower)*

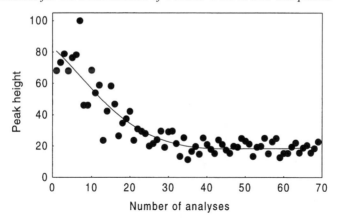

Figure 11 *The ageing effect of a PA fiber verified by the SPME–GC of 3-decanol in wines. 10 ppm 3-decanol was added to all wine samples*
(Adapted with permission from ref. 9)

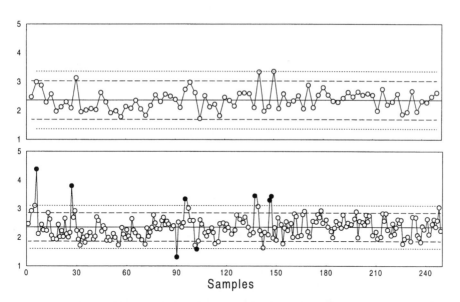

Figure 12 *Mean value control chart (MVCC) for the quotient Q_{IS} in 250 wine samples: single determination (lower), threefold determination (upper). $Q_{IS} = A_{3\text{-}decanol}/A_{3\text{-}octanol}$ with the internal standards (IS) 3-decanol and 3-octanol*

HS-SPME Methods for Quantitative Analysis of Volatile Wine Aroma Compounds

The quantitative analysis of volatile wine aroma compounds may be carried out by the well-known **standard addition** method using 3-decanol as internal standard. To demonstrate the HS-SPME–GC as an appropriate method for

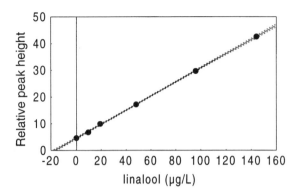

Figure 13 *Quantitative analysis of linalool in a Riesling wine by means of the standard addition method*

the quantitative analysis of volatile aroma compounds, linalool was determined in a Riesling wine to (17.3 ± 2.7) μg L^{-1} (see Figure 13).

Multiple headspace SPME (MHE-SPME)[8] is an alternative method for the quantitative analysis of volatile aroma compounds not only in wines but also in solid matrices, as well as skin tissue of the berries and grape seeds. As compared with conventional MHE, MHE-SPME demonstrates a significantly higher sensitivity.[8,11]

After i extraction steps, the peak area A_i is given by equation 2, and for the sum of peak areas of all extraction steps ΣA_i equations (3) and (4) were derived. The constant K^* considers all SPME equilibria.

$$A_i = A_1 \cdot (1 - K^*)^{i-1} \tag{2}$$

$$\Sigma A_i = \Sigma A_1 \cdot (1 - K^*)^{i-1} = \frac{A_1}{K^*} \tag{3}$$

$$\Sigma A_i = \frac{A_1^2}{A_1 - A_2} \tag{4}$$

Equation 4 is similar to that of the well-known MHE method.[15] The analyte concentration c_o is proportional to the sum of peak areas ΣA_i. According to equation 4, c_o may be calculated with only two SPME extraction steps in liquid or solid matrices. The applicability of MHE-SPME analysis is shown in Figure 14 with the example of linalool in a Morio-Muskat wine. The content of linalool determined by MHE-SPME was 17.4 ± 2.4 μg L^{-1}. This value was verified by the standard addition method.

If the constant K^* for certain analytes in wines as well as their detector response have been determined, the sum of peak areas ΣA_i may be calculated with only *one* extraction step according to equation 3. The constants K^* determined for the volatile aroma compounds α-terpineol, ethyl decanoate, 1-octanol and hexanoic acid in a Silvaner, a Portugieser, and a Mueller-Thurgau wine are listed in Table 7. With the detector response obtained by the injection

$$Ai=A1 \ (1-K)^{\wedge}(i-1)$$

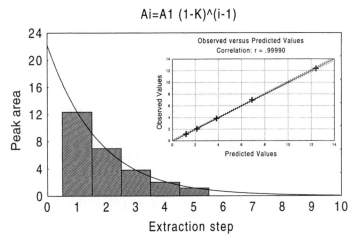

Figure 14 *MHE-SPME extraction of linalool from a Morio-Muskat wine with fitted function to equation 2*
(Adapted with permission from ref. 8)

Table 7 *Adjusted peak area for the first extraction step (A_1) and the constant K^* which considers all SPME equilibria for the MHE-SPME(PA)–GC analysis of α-terpineol, ethyl decanoate, n-octanol, and hexanoic acid in wines. (% RSD refers to the K^* values)*

		Silvaner	Portugieser	Mueller-Thurgau
α-Terpineol	A_1	5.6	2.7	4.2
	K^*	0.368	0.402	0.35
	% RSD	–	6.53	8.74
	r	0.995	0.991	0.990
Ethyl decanoate	A_1	21	15	8
	K^*	0.642	0.619	0.679
	% RSD	12.34	3.72	14.02
	r	0.998	0.980	0.992
n-Octanol	A_1	15	7	4
	K^*	0.471	0.520	0.471
	% RSD	5.33	5.55	–
	r	0.999	0.997	0.996
Hexanoic acid	A_1	37	88	19
	K^*	0.099	0.101	0.108
	% RSD	–	15.892	4.99
	r	0.963	0.995	0.995

of standard solutions of these compounds, their concentrations may be calculated by the peak areas from only one SPME extraction step.

SPME–GC Method in Routine Analysis of Wines

In the SPME–GC routine analysis of wines ageing effects of the fiber used should be considered. This effect is shown in Figure 11 for the HS-SPME–GC analysis of wine samples using a PA fiber. The efficiency of this fiber decreases with the number of extractions. After sampling twenty times, the peak intensity of the analyte extracted is nearly constant. This investigation shows that it is necessary to use an internal standard for the elimination of the ageing effect of the fiber in the course of measuring series. Using this procedure 'true' aromagrams of volatile aroma compounds can be obtained. This is important for the chemometric study of wines, for instance.

The mean value control chart (MVCC) is recommended for the internal quality control of the analytical method in the routine practice. The MVCC for the quotient Q_{IS} of the peak areas of 3-decanol and 3-octanol ($Q_{IS} = A_{3\text{-decanol}}/A_{3\text{-octanol}}$) added to the wine samples is shown in Figure 12. This quality control chart is helpful to recognize outlyers but it is difficult to declare out of control situations. The latter are best detected by the CUSUM chart which is demonstrated in Figure 15.

Owing to the fact that the use of a new SPME fiber has only a small influence on the precision of the results, CUSUM charts make it possible to recognize faulty operation of the gas chromatograph (*e.g.* leaking injector or defective detector filament) during autosampling.

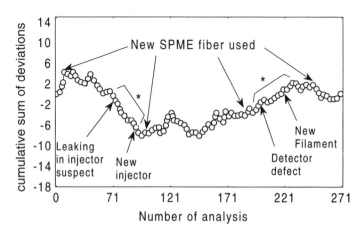

Figure 15 *CUSUM chart for the quotient Q_{IS} ($Q_{IS} = A_{3\text{-decanol}}/A_{3\text{-octanol}}$) obtained by single determinations in more than 270 wine samples. The use of a new SPME fiber does not influence the precision. An out of control situation (*) is caused only by instrument failures*

4 Classification of Wine Varieties on the Basis of HS-SPME–GC Data Using Chemometrical Methods

From the HS-SPME–GC chromatograms of wines, it is possible to get aromagrams which can be evaluated by multivariate data analysis (discriminant analysis) with the aim of recognizing the grape variety of a certain wine as shown by Figure 16.

The varietal characterization for 90 German wines from five growing areas (Blankenhornsberg, Alzey, Durbach, Freyburg-Unstrut, Oppenheim) and nine different years (1988–1996) was studied.[13] The use of the terpene pattern and

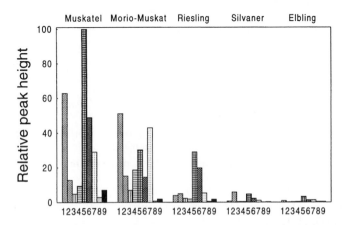

Figure 16 *Terpenoid patterns in various wine samples obtained with DI-SPME(PA): 1 trans-geranoic acid, 2 geraniol, 3 nerol, 4 citronellol, 5 terpineol, 6 hotrienol, 7 linalool, 8 cis-linalool oxide furanoside, 9 trans-linalool oxide furanoside* (Adapted with permission from ref. 14)

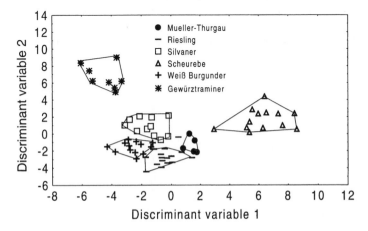

Figure 17 *Variety classification of various German wines from Blankenhornsberg by means of discriminant analysis* (Adapted with permission from ref. 8)

some vintage depending compounds (1-hexanol, 2-furancarboxaldehyde, diethyl succinate, phenylethyl alcohol, benzyl alcohol, hexanoic acid) allow a correct classification of varieties of more than 97% for Mueller-Thurgau, Riesling, Silvaner, Weißer Burgunder and Gewürztraminer wines from Blankenhornsberg (Figure 17).[13] This demonstrates the particular applicability of SPME–GC analysis for the varietal characterization of wines.

References

1. A. Rapp in *Wine Analysis*, ed. H.F. Linskens and J.F. Jackson, Springer-Verlag, Berlin and Heidelberg, 1988, p. 29.
2. A. Rapp, H. Hastrich and L. Engel, *Vitis*, 1976, **15**, 29.
3. C.G. Edwards and R.B. Beelman, *J. Agric. Food Chem.*, 1990, **38**, 216.
4. A.C. Noble, R.A. Flath and R.R. Forrey, *J. Agric. Food Chem.*, 1980, **28**, 346.
5. G.P. Blanch, G. Reglero and M. Herráiz, *J. Agric. Food Chem.*, 1995, **43**, 1251.
6. J. Villén, F.J. Senoráns, G. Reglero and M. Herráiz, *J. Agric. Food Chem.*, 1995, **43**, 717.
7. G.P. Blanch, G. Reglero, M. Herráiz and J. Tabera, *J. Chromatogr. Sci.*, 1991, **29**, 11.
8. D. De la Calle García, M. Reichenbächer, K. Danzer, C. Hurlbeck, C. Bartzsch and K.-H. Feller, *Fresenius' J. Anal. Chem.*, 1998, **360**, 788.
9. D. De la Calle García, S. Magnaghi, M. Reichenbächer and K. Danzer, *J. High Resolut. Chromatogr.*, 1996, **19**, 257.
10. (a) L. Urruty and M. Montury, *J. Agric. Food Chem.*, 1996, **44**, 3871;
    (b) L. Urruty, M. Montury, M. Braci, J. Fournier and J.-M. Dournel, *J. Agric. Food Chem.*, 1997, **45**, 1519.
11. D. De la Calle García, M. Reichenbächer, K. Danzer, C. Hurlbeck, C. Bartzsch and K.-H. Feller, *J. High Resolut. Chromatogr.*, 1998, **21**, 373.
12. J. Pawliszyn, *Solid Phase Microextraction. Theory and Practice*, Wiley-VCH, New York, 1997.
13. D. De la Calle García, M. Reichenbächer and K. Danzer, *Vitis*, 1998, **37**, 181.
14. D. De la Calle García, M. Reichenbächer, K. Danzer, C. Hurlbeck, C. Bartzsch and K.-H. Feller, *J. High Resolut. Chromatogr.*, 1997, **20**, 665.
15. B. Kolb, *Headspace-Gaschromatographie mit Kapillarsäulen*, Labor-Praxis, 1992, **5**.

CHAPTER 29

Analysis of Vodkas and White Rums by SPME–GC–MS

LAY-KEOW NG

1 Introduction

Distilled spirits are composed mainly of water and ethanol. In addition, each spirit contains a range of components (or congeners) characteristic of the raw materials used in fermentation, the methods of distilling, ageing and any other production processes. Many of the congeners are common to different spirit types. However, these spirits differ analytically in terms of the relative concentrations of the congeners. Thus quantitative analysis provides valuable congener profiles for differentiating types and in some cases, brands, of spirits.[1,2] One of the major groups of congeners consists of volatile and semi-volatile esters of fatty acids. While these components are present at mg L^{-1} level in flavorful spirits such as whiskies, brandies, cognacs, amber and dark rums, they are found only at μg L^{-1} level in the neutral spirit product, vodkas, and the light-flavor spirit, white rums. Determination of these congeners requires the use of a preconcentration technique. Liquid/liquid extractions, often applied to heavy-flavor spirits, are not efficient in extracting the trace components from vodkas and white rums, and multiple extractions using a large volume of the spirit sample must be performed to increase the enrichment factor. This procedure is labor-intensive and suffers from the formation of artefacts. SPME, known for its simplicity and high concentration power, is an ideal technique for the trace analysis of congeners in spirits.

SPME can be used for both direct and headspace sampling. For higher fatty esters which are semi-volatiles and present in trace amounts, direct sampling is considered superior to headspace sampling. In direct sampling, the analytes partition between the sample matrix and the stationary phase on the fibre. Direct sampling has been demonstrated to be very efficient for trace analysis of contaminants at μg L^{-1} levels in environmental water samples, because organic compounds, in general, have a higher affinity for the organic fibre coating than water. Applications of SPME to similar analysis in organic matrices have rarely been published, mainly owing to undesirable extracting-phase/sample matrix distribution constants, and incompatibility between the fibre coating and some organic solvents. However, concentration of analytes from aqueous matrices containing a significant portion of ethanol such as wines[3] and spirits[4] has been reported. SPME analysis of spirits is particularly notable because of the high ethanol content, typically 40% v/v, in the aqueous matrix.

This chapter describes the extraction of fatty acid esters from vodkas and white rums by SPME in direct sampling mode, followed by quantitative GC–MS analysis.[4,5] The study was carried out based on a 40% v/v ethanol matrix, which is the alcohol strength of most of the commercially available distilled spirits. The fifteen congeners investigated include five isoamyl esters of C_2, C_8, C_{10}, C_{12} and C_{14} acids, and ten ethyl esters of C_6, C_8, C_{10}, C_{12}, C_{14}, $C_{16:1}$, C_{16}, $C_{18:2}$, $C_{18:1}$ and C_{18} acids. These compounds vary widely in polarity and volatility which consequently affects their extraction and chromatography characteristics. Therefore, various SPME and instrumental variables were adjusted in order to achieve satisfactory analytical character-istics for all the analytes, while maintaining the simplicity and practicality of the method. The quantitative features of the method, which is based on extraction before partition equilibria of all analytes are reached, are pre-sented.

2 Selection of Experimental Parameters

Instrumental Variables

The analytical instrument is a HP GC–MSD system. It is composed of a 5890 series II GC, a 5970 mass selective detector equipped with a Chaneltron 5772 electron multiplier, and G1701 MS software. To increase the sensitivity, the voltage of the electron multiplier was raised to a value which was 200 V in excess of the autotune setting. Extracted samples were injected into the GC–MS system in a splitless mode. The congeners were analyzed on a 5% phenyl polydimethylsiloxane column.

Injector Insert

The GC oven temperature program was designed to elute all the congeners at a fast rate while maintaining good resolution, peak shape and sensitivity. Isoamyl acetate and C_6 ethyl ester, the low boilers, appeared as broad peaks when a 4 mm i.d. injector insert was used. The peak shapes of these components were remarkably improved when the insert was replaced by a 0.75 mm i.d. splitless inlet sleeve (Figure 1). The use of this insert also allowed the column temperature to start at a higher temperature (60 °C *vs.* 40 °C) without adversely affecting the peak shape and resolution of the early eluters and other analytes, thereby shortening the analysis time significantly.

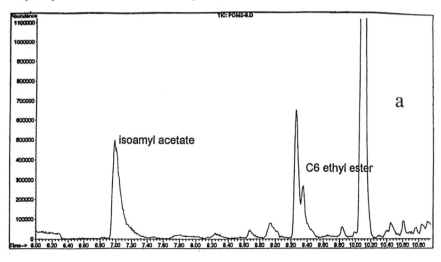

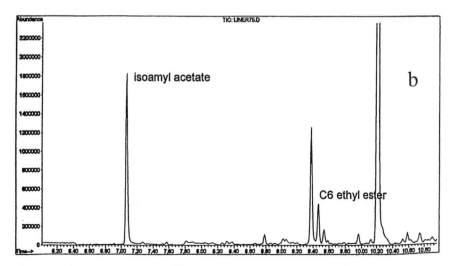

Figure 1 *Effect of internal diameter of injector liner on peak shapes:* (a) 4 mm *i.d.* (b) 0.75 mm *i.d.*

Injector Temperature

Another parameter which controls the desorption process in the GC injector is the injector temperature. While low boilers can be efficiently desorbed at relatively low temperatures, less volatile analytes require higher temperatures. It was found that the responses of the high boilers, ethyl esters of C_{16} to C_{18} acids and isoamyl tetradecanoate, were highly dependent on the injector temperature, while the remaining analytes tested were not so demanding. For example, the response of ethyl oleate was increased 4-fold while that of the ethyl octanoate remained unchanged as the temperature was raised from 220 to 250 °C. The injector temperature was set at 250 °C, since a further increase in temperature did not show any significant improvement in sensitivity.

To ensure that the extracted analytes were completely removed from the fibre, the fibre was desorbed for a 2 min period followed by a 3 min purging of the injector port. No carryover of analyte species was detected when the same fibre was desorbed for a second time after the initial desorption.

Fibre Position in the Injector

Because of the temperature gradient within the injector, the position of the fibre in the injector of the gas chromatograph has a significant impact on the desorption process and might affect the precision and the sensitivity. The fibre was inserted into the GC injector until the bottom of the depth gauge touched the septum retainer nut of the injector. The fibre position in the injector, indicated by the scale on the barrel of the SPME assembly, can be adjusted by turning the depth gauge of the device.[6] The responses of individual analytes corresponding to the fibre positions at settings 3.5 and 4.2 were found to be comparable, while at setting 2.8, the sensitivity was reduced by 10%. No significant difference in the precision of measurement was observed at the three fibre positions. All subsequent experiments were carried out using the fibre position corresponding to setting 4.

Extraction Variables

The most common fibre coating, PDMS, was employed to extract the fifteen relatively non-polar analytes from the spirits in direct sampling mode. A thick coating of 100 μm was used to increase the sensitivity. In all cases, the sample vial was screw-capped with a laminated Teflon-rubber disk to minimize change of the matrix caused by evaporation during sampling, and the fibre was introduced into the sample through the disk. Sampling was carried out using high speed magnetic stirring to achieve a vortex of about 0.5 inch deep. To avoid contamination from the preceeding analysis, a new sampling vial was used for each analysis, and the Teflon-coated stirring bar was thoroughly cleaned by soaking in acetone for at least 15 min and then rinsing with 40% ethanol before being reused.

Sampling Temperature

The operational principle of SPME is based on partitioning of analytes between the coating and the sample matrix, and the partition coefficient is affected by temperature variation. For ease of operation, the extraction was carried out at room temperature. The small temperature change (1–2 °C) associated with most controlled laboratory environments has been found to have little effect on the precision of the method. However, in manual sampling, the stirrer tends to be heated up after prolonged stirring. Therefore, it is advisable to lift the sample vial a small distance ($\sim$0.5 cm) above the base plate of the stirrer to minimize transfer of heat generated from the stirrer.

Modification of the Matrix

The effect of salt was investigated using standard solutions containing NaCl of concentrations ranging from 0 to 10% w/v. As shown in Figure 2, the responses of the ethyl esters of C_{12}, C_{16} and $C_{18:1}$ acids decreased as the salt concentration increased. The 'salting out' effect normally found in water was not observed here. This phenomenon can be explained by the fact that water, which has a higher dielectric constant than ethanol, is more efficient in solvating the added salt. In effect, the analytes were 'salted out' into an environment containing a higher percentage of ethanol. This resulted in higher solubility of the analytes in the salt-containing matrix, and consequently smaller amounts of analytes were extracted since absorption increases with decreasing analyte solubility in the matrix given the same coating.

In principle, addition of water to the 40% ethanol matrix will reduce the solubility of analytes . At the same time, however, smaller amounts of analytes will be absorbed due to dilution. These two factors work against each other and the net effect may not be the same for all analytes. The profile of a 40% v/v ethanol solution of C_8, C_{12}, C_{16} and $C_{18:1}$ ethyl esters was changed dramatically

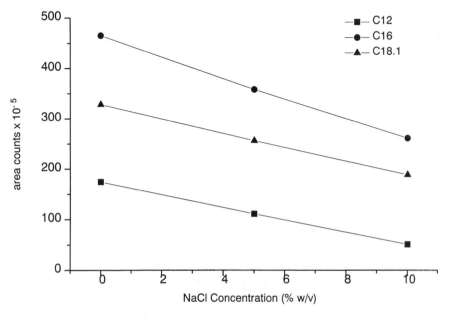

Figure 2 *Effect of salt on the responses of analytes*

after the addition of two volumes of water. Overall, the response of C_8 ethyl ester was increased by 3.5 folds while those of C_{12}, C_{16} and $C_{18:1}$ ethyl esters were reduced considerably (Figure 3). The results of these experiments indicated that there was no advantage to modifying the matrix by addition of salt or water.

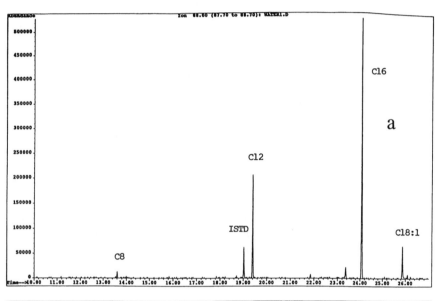

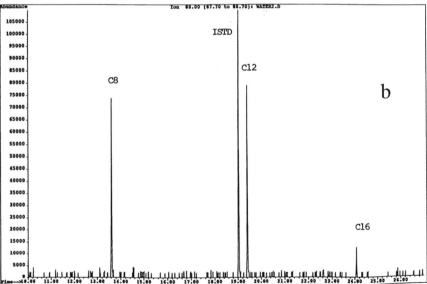

Figure 3   *Effect of addition of water:* (a) *40% v/v ethanol solution;* (b) *after addition of two volumes of water to one volume of 40% v/v ethanol solution*

Determination of K_{fs}s and the Extraction Time

Table 1 lists the equilibration time and the distribution constant of each analyte determined from the corresponding absorption time profile.[6] The results indicate that the equilibration time and the K_{fs} value increase with the carbon chain length of the homologs, and saturated compounds have higher K_{fs} values and longer equilibration times than the unsaturated homologs of the same carbon number. These observations are expected since the longer the carbon chain of an ester, the less polar it is. Consequently it has a higher affinity for the nonpolar PDMS fibre and lower solubility in the polar sample matrix, resulting in a higher K_{fs} value. The presence of unsaturation renders the compound more polar, and consequently a lower K_{fs} value is obtained for the unsaturated counterpart(s). The high MW compounds are also expected to have longer equilibration times than the low MW analytes. Because larger distribution coefficients are associated with heavier compounds, more analyte molecules have to diffuse across the thin static liquid layer surrounding the fibre and into the polymeric coating before equilibrium is reached.[7] This results in longer equilibration times for the heavier compounds.

Table 1 shows that some compounds have not reached equilibrium even after 4.5 h of absorption. Although maximum sensitivity is obtained by allowing the analytes to reach equilibrium, it is not practical to do so when equilibration times are excessively long. It has been demonstrated that SPME quantification using fibres coated with a polymeric liquid such as PDMS is feasible before an

Table 1 *SPME characteristics**

Compound	Equilibration time (h)	Distribution constant, K_{fs}
Isoamyl ester of fatty acid		
C_1	0.25	6
C_8	0.25	610
C_{10}	1	1740
C_{12}	4.5	4950
C_{14}	>4.5	>12 320
Ethyl ester of fatty acid		
C_6	0.25	18
C_8	0.25	70
C_{10}	0.5	240
C_{12}	0.5	1060
C_{14}	1.5	2740
$C_{16:1}$	2.5	3490
C_{16}	4.5	6750
$C_{18:2}$	4.5	4820
$C_{18:1}$	4.5	8230
C_{18}	>4.5	>18 960

*The equilibration times and the distribution constants were determined from the corresponding absorption time profiles established over a period of 18 h (volume of the test solution = 200 mL).

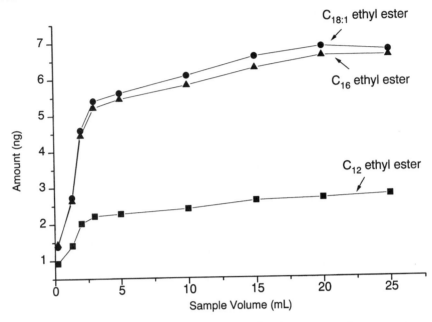

Figure 4 *Effect of sample volume on the amounts of analytes extracted*

absorption equilibrium is reached, and that the amount of analyte absorbed is proportional to the initial concentration in the sample.[8] To maximize sample throughput, 30 min was chosen to be the extraction time, which was also the instrumental time required for a GC–MS analysis to complete.

Sample Volume

Figure 4 illustrates the effect of sample volume on the amounts of analytes extracted from a test solution containing 6 μg L^{-1} of each of the ethyl esters of C_{12}, C_{16} and $C_{18:1}$ acids. Given the extraction time of 30 min, the amounts of analyte extracted increased sharply as sample volume changed from 1.3 to 3 mL, and then rapidly levelled off thereafter. It is preferable to select a sample size in the plateau area where the response is least sensitive to change in volume, and therefore, 10 mL sample was used for all the studies described below.

3 Performance Characteristics of the Method

Quantitation of congeners is based on an internal standard method using external calibration. The internal standards (ISTDs) were added to the calibration solutions and samples before SPME sampling. Methyl octanoate served as the ISTD for the low boilers: isoamyl esters of C_1, C_8 and C_{10} acids, and ethyl esters of C_6, C_8, C_{10} and C_{12} acids, methyl pentadecanoate was the ISTD for the ethyl esters of C_{14}, $C_{16:1}$ and $C_{18:2}$ acids and isoamyl dodecanoate, and methyl stearate was the ISTD for the remaining esters. Congener-free 95% ethanol was used to prepare the 40% ethanol calibration solutions.

Linearity of Standard Curves

The standard curves were generally linear over the selected concentration ranges shown in Table 2, with correlation coefficients of linear regression in the range of 0.996–0.999. Their 95% confidence bands also included the origin. The only exception was that of ethyl ester of C_{16} acid, which had a correlation of 0.986. It yielded a standard curve which was linear up to 60 μg L^{-1} and then showed a tendency to level off at higher concentrations. This is typical of MS detection which has a smaller linear range than FID. This phenomenon was not observed with other analytes because their K_{fs}s and/or upper working concentration limits are relatively small, such that only small amounts of analytes were extracted, and the responses fell in the linear ranges. Although the standard curve of the C_{16} ethyl ester could be made linear by reducing the sampling time, it was decided not to do so because the limits of quantitation of other analytes would be adversely affected. Quantitation of the congeners was carried out based only on one-point calibration. This was justified since the standard curves of most of the analytes passed through the origin and were linear over the corresponding working concentration ranges. A standard solution containing the fifteen analytes, of concentrations intermediate of their respective working ranges, was used to calibrate the procedure, bearing in mind that the predicted

Table 2 *Parameters of standard curves*

Compound	*Working concentration range* (ppb)	*Slope*	*Intercept*	*Corr. coeff.*
Isoamyl ester of fatty acid				
C_1	113–4530	0.00038	0.00964	0.9996
C_8	0.22–8.7	0.08005	−0.00046	0.9984
C_{10}	0.21–8.5	0.1614	0.01473	0.9992
C_{12}	0.13–5.1	0.28145	0.00569	0.9993
C_{14}	0.13–5.2	0.37227	−0.01233	0.9998
Ethyl ester of fatty acid				
C_6	7.8–310	0.00109	0.00204	0.9998
C_8	26–1040	0.00586	0.08484	0.9997
C_{10}	8–320	0.02939	0.11605	0.9989
C_{12}	3.6–144	0.06548	0.34470	0.9968
C_{14}	1.2–46	0.13978	0.07074	0.9993
C_{16}	2.7–107	0.02494	0.22140	0.9859
$C_{16:1}$	2.8–110	0.02615	0.11864	0.9963
$C_{18:2}$	1.5–60	0.02346	0.02591	0.9980
$C_{18:1}$	1.7–68	0.03901	0.09958	0.9966
C_{18}	0.95–38	0.12461	0.08779	0.9984

concentration of C_{16} ethyl ester at levels above 60 μg L^{-1} will be slightly lower than the true value.

Repeatability and Reproducibility

The within-run repeatability of the response ratios, analyte/ISTD, at various concentrations of analytes was determined using standard solutions. The measurements ($n = 3$) were found to be repeatable with 2–6% RSD, although for some analytes, the precisions deteriorated at low concentrations. This reasonably good short-term precision suggests that one analysis per sample would give a reliable profile.

The long-term reproducibility of the congener profiles of spirits was determined based on one-point calibration and using several white rum samples which contain low and high levels of congeners. The samples were analyzed in triplicates, one analysis per week for three consecutive weeks. For each week, the calibration standard was analyzed once with the samples. As shown in Table 3, the determination of the analyte concentration was generally reproducible within 1.5–10% RSD, although higher variations were observed as the concentrations of the analytes approached the respective limits of quantitation. These results show that the quantitative profiles can be reproducibly generated any time the sample is analyzed, and therefore the method is suitable for use in

Table 3 *Reproducibility of congener profiles**

	Sample 1 mean ppb (%RSD)	Sample 2 mean ppb (%RSD)	Sample 3 mean ppb (%RSD)	Limit of quantitation (ppb)
Isoamyl ester of fatty acid				
C_1	79.82 (7.34)	2149.67 (8.51)	71.91 (28.26)	30.7
C_8	0.58 (30.30)	1.37 (17.25)	4.39 (9.37)	0.13
C_{10}	4.68 (4.09)	0.40 (8.10)	7.49 (4.36)	0.07
C_{12}	2.92 (1.54)	0.10 (17.32)	0.43 (2.70)	0.05
C_{14}	0.18 (9.62)	0.00 (173.20)	0.02 (65.46)	0.04
Ethyl ester of fatty acid				
C_6	19.36 (9.37)	219.85 (6.27)	24.58 (13.48)	4.19
C_8	21.73 (8.78)	297.97 (6.43)	142.68 (8.13)	2.83
C_{10}	74.93 (5.55)	147.68 (8.85)	289.54 (9.81)	0.64
C_{12}	115.60 (1.35)	15.09 (8.59)	91.01 (2.85)	0.15
C_{14}	45.13 (4.99)	0.95 (7.03)	4.85 (3.82)	0.06
C_{16}	137.69 (2.95)	3.60 (5.97)	13.85 (7.99)	0.16
$C_{16:1}$	57.14 (7.57)	0.12 (91.13)	6.43 (5.87)	0.51
$C_{18:2}$	117.75 (6.59)	0.47 (29.55)	3.59 (3.89)	0.64
$C_{18:1}$	108.80 (9.70)	1.17 (6.16)	19.02 (11.90)	0.27
C_{18}	10.20 (9.12)	0.14 (4.22)	1.69 (5.82)	0.03

*Mean concentrations are averages of triplicate analyses, 1 analysis/week for three consecutive weeks.

establishing a reference databank which is essential for the authentication of the spirits.

Sensitivity of Congener Profiles to Slight Changes in Ethanol Content

In some cases, the spirits might have the ethanol content differ by $< \pm 2\%$ v/v from the alleged 40% v/v. It is, therefore, necessary to determine how these changes would affect the congener profile. It was found that in general, higher responses were obtained in solutions of lower ethanol content. These results were expected because of the reduced solubility of the analytes in the 'less organic' matrix. This effect was more pronounced for esters of small carbon number. As shown in Table 4, a difference of 2% v/v from the normal ethanol strength (40% v/v) of spirits changed the responses significantly by 10–37%, and the hydrophobic compounds, ethyl esters of C_{16}, C_{18} and $C_{18:1}$ acids and the isoamyl ester of C_{14} acid, were much less sensitive to the variation in the ethanol content of the matrix. Unlike the absolute responses, the response ratios of all analytes were not significantly affected by the ethanol content, and the percentage deviations shown in the last two columns of Table 4 were largely due to experimental errors. These results suggest that the ISTDs used were reasonably effective in correcting for the matrix effects caused by small variation in ethanol concentration.

4 SPME–GCMS Profiles of Commercial Vodkas and White Rums

Typical SPME–GCMS profiles of vodkas and white rums are shown in Figure 5. In general, the congener content in vodkas is much lower than in white rums, and isoamyl esters, though often detected in white rums, are usually absent in vodkas. Other chemicals which are not constituents of the spirits are also present, the most common ones being 2,6-di-*t*-butyl-4-methyl-phenol (BHT) and bis(2-ethylhexyl) phthalate (DEHP). They are probably contaminants originating from the hoses or storage tanks used in the manu-facturing processes. 5-hydroxymethyl-2-furaldehyde (5-HMF) and triethyl citrate (TEC) were also detected in American vodkas.[2] The presence of these two compounds is believed to be related to the addition of citric acid and sugar syrup which is allowed by the US Federal Regulations. 5-HMF is known as a decomposition product of the syrup and TEC is the reaction product between ethanol and citric acid formed during storage.

5 Conclusions

SPME in direct sampling mode has been demonstrated to be a feasible technique for extracting various ester congeners at trace level (μg L^{-1}) from aqueous media containing 40% v/v ethanol. Quantitative congener profiles can

Table 4 *Effects of slight changes in the alcohol strength of the matrix*

	Peak area counts*			% Deviation**		Response ratio*, analyte/ISTD			% Deviation†	
	42% v/v	40% v/v	38% v/v	42% v/v	38% v/v	42% v/v	40% v/v	38% v/v	42% v/v	38% v/v
Methyl octanoate (ISTD)	2 101 417	2 834 974	3 641 638	−25.9	28.5	1	1	1		
Isoamyl acetate	1 481 547	1 815 545	2 169 856	−18.4	19.5	0.71	0.64	0.60	10.1	−7.0
Ethyl hexanoate	553 315	738 050	915 638	−25.0	24.1	0.26	0.26	0.25	1.1	−3.4
Ethyl octanoate	2 607 679	3 637 917	4 781 948	−28.3	31.4	1.24	1.28	1.31	−3.3	2.3
Ethyl decanoate	5 031 142	7 199 241	9 784 000	−30.1	35.9	2.39	2.54	2.69	−5.7	5.8
Isoamyl octanoate	231 973	339 239	463 570	−31.6	36.7	0.11	0.12	0.13	−7.7	6.4
Ethyl dodecanoate	1 487 377	2 295 124	2 981 166	−35.2	29.9	0.71	0.81	0.82	−12.6	1.1
Isoamyl decanoate	478 351	653 316	839 664	−26.8	28.5	0.23	0.23	0.23	−1.2	0.1
Methyl pentadecanoate (ISTD)	1 400 979	1 843 181	2 154 541	−24.0	16.9	1	1	1		
Ethyl tetradecanoate	3 647 722	4 847 040	5 706 898	−24.7	17.7	2.60	2.63	2.65	−1.0	0.7
Isoamyl dodecanoate	669 657	803 390	901 020	−16.6	12.2	0.48	0.44	0.42	9.7	−4.1
Ethyl palmitoleate	833 109	1 074 788	1 240 368	−22.5	15.4	0.59	0.58	0.58	2.0	−1.3
Ethyl linoleate	692 256	850 457	926 888	−18.6	9.0	0.49	0.46	0.43	7.1	−6.8
Methyl stearate (ISTD)	10 706 806	11 950 996	11 918 732	−10.4	−0.3	1	1	1		
Ethyl hexadecanoate	1 042 739	1 216 768	1 294 512	−14.3	6.4	0.10	0.10	0.11	−4.3	6.7
Isoamyl tetradecanoate	1 013 534	1 146 164	1 212 572	−11.6	5.8	0.09	0.10	0.10	−1.3	6.1
Ethyl oleate	1 381 510	1 505 207	1 644 047	−8.2	9.2	0.13	0.13	0.14	2.4	9.5
Ethyl stearate	3 520 840	3 832 392	3 938 531	−8.1	2.8	0.33	0.32	0.33	2.5	3.0

* The numbers are averages of four replicates.
† These are % deviations from the results obtained in 40% v/v alcohol matrix.

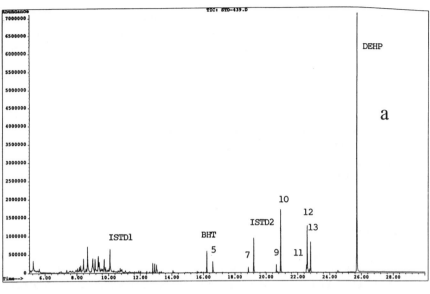

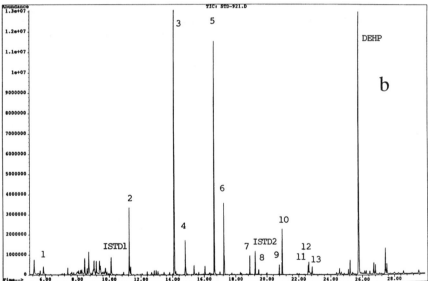

Figure 5 *Typical SPME–GCMS profiles of* (a) *vodka and* (b) *white rum: 1 = isoamyl acetate, 2 = ethyl octanoate, 3 = ethyl decanoate, 4 = isoamyl octanoate, 5 = ethyl dodecanoate, 6 = isoamyl decanoate, 7 = ethyl tetradecanoate, 8 = isoamyl dodecanoate, 9 = ethyl hexadecenoate, 10 = ethyl hexadecanoate, 11 = ethyl oleate, 12 = ethyl linoleate, 13 = ethyl stearate, ISTD1 = methyl octanoate, ISTD2 = methyl pentadecanoate, BHT = 2,6-di-t-butyl-4-methyl-phenol, DEHP = bis(2-ethylhexyl) phthalate*

be reproducibly generated by SPME–GCMS for characterizing and authenticating neutral and light spirits such as vodkas and light rums. With slight modifications to the extraction parameters and conditions, the method described herein can be easily applied to the analysis of the flavorful spirits, whiskies, brandies, rums and gins.

Acknowledgments

Thanks are due to Jean Harnois and Dennis Moccia of the Excise Laboratory Division of LSSD for initiating these studies and for their continuous participation and support. Thanks are also due to Pierre Lafontaine for his extensive contribution during the duration of the project, and Michel Hupé for his initial involvement.

References

1. W.A. Simpkins, *J. Sci. Food Agric.*, 1985, **59**, 367.
2. D.B. Lisle, C.P. Richards and D.F. Wardleworth, *J. Inst. Brew.*, 1978, **84**, 93.
3. D. De la Calle García, S. Magnaghi and M. Reichenbächer, *J. High Resolut. Chromatogr.*, 1996, **19**, 257.
4. L.-K. Ng, M. Hupé, J. Harnois and D. Moccia, *J. Sci. Food Agric.*, 1996, **70**, 380.
5. L.-K. Ng, P. LaFontaine, J. Harnois and D. Moccia, in preparation.
6. J. Pawliszyn, Chapter 1 of this book.
7. D. Louch, S. Motlagh and J. Pawliszyn, *Anal. Chem.*, 1992, **64**, 1187.
8. Jiu Ai, *Anal. Chem.*, 1997, **69**, 1230.

CHAPTER 30

Analysis of Food Volatiles Using SPME

TERRY J. BRAGGINS, CASEY C. GRIMM AND
FRANK R. VISSER

1 Introduction

To better understand the complex reactions responsible for flavour and odour development in foods, flavour chemists are constantly investigating new analytical methods that are more sensitive, faster, require less expensive equipment and are less prone to contamination or interference, yet are reproducible and give a good representation of volatile compounds in the sample or its headspace.

More traditional methods used to analyse volatile compounds from foods included steam distillation, simultaneous distillation and extraction,[1] solvent extraction, static headspace sampling[2] or dynamic headspace sampling with porous polymer (*e.g.* Tenax, carbopack, *etc.*) trapping.[3] Identification and quantitation usually involve separation by gas chromatography and subsequent detection (*e.g.* mass spectrometry, flame ionization, thermal conductivity, *etc.*) and/or olfactometry analysis.

Unfortunately, none of the separation methods mentioned offers the perfect solution for the flavour chemist. For example, during simultaneous distillation and extraction (SDE) and solvent extraction, contaminants of the solvent might be concentrated to unacceptable levels when the extract is reduced in volume before GC analysis. Furthermore, highly volatile compounds could be lost during the concentration step, when nitrogen gas is blown over the solvent surface. Steam distillation requires high temperatures for distillation that might produce compounds not associated with the original odour or flavour of the food under study. Solvent extraction can produce many peaks (*e.g.* fatty acids and lipids) in a GC chromatogram, which can confuse the chromatographer, making identification difficult. The results may bear little relevance to the odour and flavour profile of the original material. With SDE, a combination of steam distillation and solvent extraction, there may be solvent impurities[4] and thermally induced changes.[5] To get as close as possible to the true odour profile of a food product, most researchers rely on either direct headspace sampling or dynamic purge and trap headspace analysis. A popular trapping material is a porous polymer based on 2,6-diphenyl-paraphenylene oxide (Tenax-GC or -TA). Tenax-TA has been the porous polymer of choice because of its high thermal stability (up to 450 °C), relatively low water retention and reduced bleed.[3]

Concern has been noted about possible thermal conversion of labile compounds during thermal desorption of food volatiles from Tenax traps.[6] Although volatile compounds are trapped at relatively low temperatures, thermal desorption takes place at temperatures up to 260 °C for periods as long as 20 minutes.[7] Some researchers prefer to elute volatile compounds from adsorbent traps with a solvent such as ethyl ether,[8,9] or supercritical carbon dioxide.[10–12] Solvent elution, however, could introduce solvent-borne contaminants and lose very volatile compounds during the volume reduction step before GC analysis. Supercritical carbon dioxide elution has an advantage of eluting volatile compounds at relatively low temperatures but requires costly equipment.

More recently, solid phase microextraction (SPME) has been used as an alternative analytical tool to measure volatile and semi-volatile compounds in foods.[13,14] Because most food matrices are highly complex, flavour analysis using SPME is usually restricted to headspace analysis. SPME headspace analysis has the advantage of being a relative quick, one-step method of extracting volatile compounds from the headspace above foods that does not require solvents or costly equipment. Furthermore, the SPME fibre can be inserted directly into the injector of a gas chromatograph or HPLC to facilitate separation and quantitation of the extracted volatile compounds.

This chapter will cover three examples where SPME has been used effectively to analyze volatile compounds in foods: the volatile profiling of maturing cheeses using static-headspace SPME–GC, analysis of specific aroma compounds in rendered sheep fat by dynamic-headspace SPME–GC, and quantitative analysis of 2-methylisoborneol (2-MIB) and geosmin taint in pond water and catfish by SPME–direct MS–MS.

2 Volatile Profiling of Maturing Cheeses Using Static-headspace SPME–Gas Chromatography

The aroma of cheese can be analyzed in various ways. On the one hand, there is the purely chemical method of aroma isolation, separation, and identification of the many different organic compounds that generate the aroma. On the other hand, there is the option of aroma comparison by measuring a pattern of responses using a so-called electronic nose. The chemical method is time consuming and not very convenient because of the typical composition of the cheese matrix. However, the results give an absolute description of the compounds that are present in the cheese aroma. In contrast, an electronic nose gives results much faster, but only as an aroma pattern or fingerprint without any indication about the nature of the aroma components. Such results are good for comparative work, but give no information about the chemistry. These two techniques should be seen as complementing each other, not as competing.

Depending on the nature of the problem at hand, another approach, in between the two extreme options mentioned above, is to do an aroma profile based on peak size and distribution in the GC traces of the aromas. Decisions about more in-depth aroma analysis can then be made based on the results of this profiling. Such aroma profiling by SPME–GC shortens the whole process of sample preparation, aroma isolation, and aroma concentration to no more than 30 min, and requires no organic solvents or costly equipment. With 1 h of GC time and 10 min cooling added to this, six profiling runs can be made in one day by staggering extractions and GC runs. In contrast, the classical process of vacuum steam distillation, liquid/liquid extraction, drying the solution, and evaporation of the solvent takes from 2 to 2.5 days for one sample. Consequently, a massive saving in time is achieved.

The term 'aroma profiling' is not totally correct: not all peaks that are observed in a GC trace represent compounds that contribute to the aroma and, on the other hand, highly potent aroma compounds may be present in such small amounts that they are not detected by GC, but still contribute to the aroma. Nevertheless, the method gives a chromatographically detectable 'fingerprint' of the headspace, just as the aroma is a physiologically detectable characterization of the headspace. Although it is not totally correct to state that the headspace composition completely describes the aroma and *vice versa*, there is much analogy between the two.

Experimental

For SPME–GC, residual volatiles from whatever source must not be present on the fibre prior to exposure to a sample headspace. Desorption of the fibre for 2 min at 220 °C in the injector of the GC with the splitter valve wide open is sufficient to 'clean' the fibre completely, at least for cheese aroma work. It has also been shown that a clean fibre can be kept free of contaminants for at least 40 min when the needle is tightly wrapped in several layers of aluminium foil. Consequently, successive runs can be staggered, with the exposure of one sample beginning 40 min after injection of the previous sample, without contamination of the fibre between desorption and exposure.

For a typical SPME–GC profiling run, a piece of cheese 8 cm high and at least 10 cm long and wide is cut. The under-side of the cheese is made perfectly flat to ensure immobility, once secured in place. Cheese of a soft or crumbly texture is wrapped in packing tape for mechanical support. A hole is then drilled completely through the sample at right angles to the flat under-side, using a 25 mm cork borer. The plug is removed and the piece of cheese placed, flattened side down, on a sheet of aluminium foil on the base of a laboratory stand and immobilized using masking tape, if necessary. The SPME unit is supplied with a suitably wide, 1 cm thick rubber ring and covered with one layer of aluminium

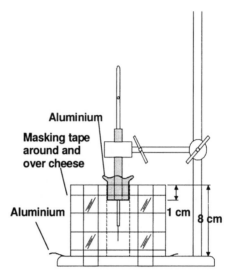

Figure 1 *Static-headspace SPME extraction setup for the analysis of volatile compounds from cheese samples*

foil. The unit is gently lowered into the hole until the top of the rubber ring is flush with the sample surface. This gives sufficient seal between the cheese and the SPME unit to avoid interference with the analysis by exchange of 'headspace' in the sample with ambient air. The unit is loosely clamped in position and checked for vertical. Then the fibre is extended into the hole and exposed to the cheese aroma for 30 min (see Figure 1). After exposure, the fibre is desorbed in the injector of the GC for 2 min.

The idea of extracting cheese aroma from a hole in the cheese is not entirely new. Manning *et al.* (1976),[15] using a syringe, withdrew 5 mL of air from a 30 mL large, capped hole drilled in a piece of cheese, and injected it into a gas chromatograph. However, the combination of SPME and direct sampling from a piece of cheese has not yet been published.

For our experiments a commercially available SPME unit and three different types of fibre were used: 100 μm polydimethylsiloxane (PDMS), 85 μm polyacrylate (PA) and 65 μm polydimethylsiloxane–divinylbenzene (PDMS–DVB), all from Supelco, Bellefonte, PA, USA.

GC analyses were done on an HRGC5300 Megaseries gas chromatograph (Carlo Erba, Milan, Italy), equipped with a 30 m carbowax capillary column (EconoCap, cat. no. 19655, Alltech, Auckland, New Zealand), 0.25 mm i.d., film thickness 0.25 μm. The carrier gas was helium at 80 kPa inlet pressure and a flow rate of 1.2 mL min^{-1}. The injector was equipped with a narrow liner (Carlo Erba, part no. 45300300) and a low bleed septum (Thermogreen LB-2, Supelco, Bellefonte, PA, USA), and was operated splitless at 220 °C and a septum purge rate of 2.4 mL min^{-1}. The fibre was desorbed for 2 min. The GC temperature programme was 2 min at 35 °C, at 5 °C min^{-1} to 220 °C and hold at 220 °C for 21 min. The flame ionization detector was operated at 240 °C with

50 kPa hydrogen and 150 kPa air for the flame and 100 kPa nitrogen make-up. The signal, obtained at range = 0 and attenuation = 4 (unless stated otherwise), was recorded by a Linseis L6512 chart recorder at 1 cm min^{-1}.

Heights of relevant GC peaks were measured as a percentage of full-scale deflection (% FS) at range = 0 and attenuation = 4. For large peaks, recorded at higher attenuation, the measured heights were converted to what they would have been at attenuation = 4. Basic statistical calculations, two-sample *t*-tests, analysis of variance and principal components analysis (PCA) were all done with Minitab®.

Results and Discussion

The decision to use certain GC peaks for profiling and reject others is governed by a number of criteria:

- Small peaks, consistently showing in blank runs, should be rejected as contaminants. For certain types of fibre even a large contaminant may be present, possibly originating from the fibre coating itself. A well-founded selection can be made only after a number of analyses and blanks have been run, to establish any consistent pattern.
- Peak size repeatability is essential; peaks with poor repeatability must be rejected for profiling. We measured repeatability as the coefficient of variation (CV). For the study reported here, CV values ranged from <1% to 75% for all three types of fibre, with the high values mostly for small peaks. Of the three types of fibre we used, the PDMS–DVB fibre gave the highest proportion of low CV values (73% of all peaks had a CV <10%).
- Day-to-day reproducibility should be good; peaks with poor reproducibility should be excluded from profiling. We established this by doing four replicate analyses on each of three different days for the same cheese. A *P* value of 0.2 was taken as the lower limit for the *F*-test to be justified in stating that there was no day effect.
- Peaks that are 1% FS or less and peaks that show outliers also should be rejected.

The peaks that remain after this process of elimination are reliable indicators and can be used for profiling, even if the data are acquired over a period of several days.

The procedure described above was followed for the profiling of Cheddar cheese (24 months old) and Colby cheese (5 months old) using three different types of fibre. Three analyses were done of each of the two cheeses for each fibre type. The results for the PDMS–DVB fibre were the most promising and were subjected to the peak elimination process just described. This left 17 reliable peaks whose sizes are shown in Table 1. They were used for PCA to look for a difference in pattern. The first three eigenvalues explained 99.0% of the variation, indicating a very successful analysis. A plot of the first two principal

Table 1 *Height of selected peaks in the GC traces for Colby and Cheddar cheese, using a PDMS–DVB fibre and three runs for each cheese*

Peak number	Colby			Cheddar		
	Run 1	Run 2	Run 3	Run 1	Run 2	Run 3
2	1.0	0.9	0.9	1.6	1.6	1.5
3	6.1	5.2	5.4	3.3	3.4	3.1
6	4.7	4.4	4.7	4.5	5.0	4.8
8	18.8	15.8	16.0	12.8	12.7	12.2
12	18.5	17.1	18.3	20.2	22.2	21.8
13	0.4	0.4	0.5	1.5	1.7	1.6
14	1.4	1.4	1.5	2.5	2.6	2.5
15	6.1	5.3	5.8	7.2	7.9	7.5
16	11.2	7.9	7.8	13.2	13.8	12.9
17	2.5	2.4	2.1	1.1	1.2	1.2
18	2.1	2.1	2.0	2.4	2.5	2.5
19	69.3	65.3	65.5	33.8	34.1	33.4
21	189	194	200	305	315	334
23	2.8	2.8	3.0	0.7	0.5	0.7
24	44.1	42.8	47.2	67.2	71.7	73.3
25	9.5	9.2	10.7	14.5	15.4	16.7
26	4.8	5.4	5.7	5.4	5.5	6.2

Note: Heights are in % FS at range = 0 and attenuation = 4.

components clearly shows good distinction between the two types of cheese (Figure 2).

This static-headspace SPME method for profiling cheese headspace has been proven to be acceptable for two types of cheese. It is a credible assumption that it can also be applied successfully to other cheese types or for grading purposes

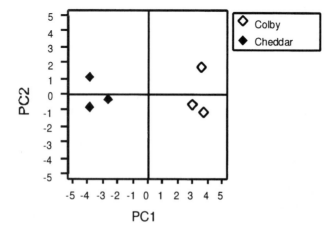

Figure 2 *Principal component analysis of selected peaks from chromatograms of volatile compounds of two different cheeses*

within one type of cheese. The headspace volatiles of different cheeses have many compounds in common, in addition to those that are typical for the type, but their relative abundances can vary enormously. It is, therefore, likely that many of the same peaks that were involved in this study will also be useful indicators for comparative work with other cheeses. The above results have also demonstrated that, for this type of work, PCA is an effective statistical technique to reveal differences between samples.

3 Analysis of Specific Aroma Compounds in Rendered Sheep Fat by Dynamic-headspace SPME–Gas Chromatography

With static-headspace SPME the volume of the headspace influences the amount of analyte adsorbed on the SPME fibre. The amount adsorbed decreases with increasing headspace volume, so to obtain higher sensitivity the sample headspace should be kept as small as possible. Vigorously stirring the liquid phase also enhances sensitivity, as it generates a continuously fresh liquid surface.[16] Analyte concentration in the headspace can also be enhanced by heating the sample matrix, because partition coefficients are temperature dependent. However, there is an optimum temperature for SPME extraction. At higher temperatures the coating–headspace partition coefficient decreases, resulting in the fibre losing its ability to adsorb analytes. This potential limitation is overcome by simultaneously cooling the fibre while heating the sample.[17]

Initial attempts to use static-headspace SPME to extract semivolatile branched chain fatty acids from rendered sheep fat were disappointing, even when a sample was both agitated and heated to 100 °C. Therefore, dynamic-headspace SPME was investigated as a method of measuring the branched chain fatty acids 4-methyloctanoic and 4-methylnonanoic, thought responsible for the characteristic flavour and odour of sheepmeat.

In more traditional purge and trap analysis the sample is continually purged by an inert gas and the volatile compounds are trapped on a porous polymer such as Tenax-TA. Similarly, in dynamic-headspace SPME, the headspace is continually swept from the sample by an inert purge gas and passed across the exposed SPME fibre.

Experimental

For this experiment an SPME 100 μm polydimethylsiloxane (PDMS) fibre (Supelco, Bellefonte, PA, USA) was used. A 2 g sample of rendered sheep fat was melted and pipetted into the bottom of a 50 mL Quickfit test tube. The sample was spiked with 3 ppm each of 4-methyloctanoic acid, 4-methylnona-noic acid, and 2-methyloctanoic acid (as internal standard). The extraction flask was fitted with a purge gas inlet tube and outlet tube to accommodate the SPME fibre or, when appropriate, a glass tube trap containing 200 mg of Tenax-TA

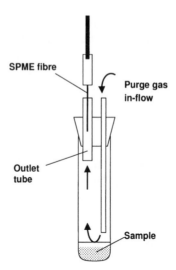

SPME fibre

Purge gas
in-flow

Outlet
tube

Sample

Figure 3   *Dynamic-headspace SPME extraction setup for the analysis of volatile com-*
pounds from rendered sheep fat

(Figure 3). The extraction vessel was heated to 100 °C and a preconditioned
SPME fibre positioned in the nitrogen-purge gas outlet tube.

Optimum purge conditions were investigated by varying the nitrogen flow
from 6 to 330 mL min^{-1} and the exposure time from 1 to 30 minutes. At the end
of the sampling period the SPME fibre was desorbed for 2 minutes into the split/
splitless injector set at 200 °C, operated in the splitless mode, of a Hewlett
Packard 5890 Series II GC/FID fitted with a Free Fatty Acid Phase (FFAP)
(30 m × 0.53 mm i.d., 1.0 μm film thickness) column. The initial linear velocity
of the column was 63 cm s^{-1} (8.5 mL min^{-1} helium, P = 50.3 kPa at 100 °C).
The temperature programme was 60 °C for 1 min, then raised to 175 °C at a rate
of 50 °C min^{-1}, held for 5 min, then raised to 220 °C at 5 °C min^{-1} and held for
10 minutes. Peak areas were integrated using Maxima integration software
(Waters Inc.).

After optimizing purge and exposure conditions for the extraction of the
branched chain fatty acids, the chromatographic profile of the dynamic-headspace SPME technique was compared with that of purge and trap using Tenax-
TA. The same purge and exposure conditions were used for both methods. For
this comparison the SPME fibres and the Tenax-TA traps were desorbed for 5
minutes at 200 °C and 250 °C, respectively. The volatile compounds were
cryogenically cooled (− 10 °C) onto the head of an FFAP (30 m × 0.25 mm
i.d., 1.0 μm film thickness) capillary column housed in a Fisons 8000 GC. The
chromatography conditions were as follows: column head pressure, 103.4 kPa;
split flow (for Tenax desorption only), 26 mL min^{-1}; split ratio of 32:1 (splitless
for SPME) and column flow of 2.0 mL min^{-1} (measured at − 10 °C); temperature programme was − 10 °C for 5 min, then raised to 60 °C at a rate of
50 °C min^{-1}, then raised to 240 °C at 5 °C min^{-1} with a final hold time at
240 °C of 5 minutes.

The capillary column was connected to a Fisons MD 800 mass spectrometer with a transfer line temperature of 280 °C and source temperature of 200 °C. Mass spectra were generated at 70 eV and a detector setting of 350 V. Data were recorded from 40 to 350 mass range by MASSLAB integration software (Fisons) in the total ion monitoring mode. Spectra were compared with an NIST mass spectral data base supplied with MASSLAB.

Results and Discussion

The recovery efficiency of dynamic-headspace SPME is strongly dependent upon the purging flow rate and the sampling time. Figure 4 compares the relative adsorption efficiency between two example compounds, 2,3-octane-dione and 2-methyloctanoic acid (similar optima to 2-methyloctanoic acid were recorded for 4-methyloctanoic and 4-methylnonanoic acids) for flow rates between 6 and 330 mL min^{-1} and exposure times between 1 and 30 minutes.

A   2,3-octanedione

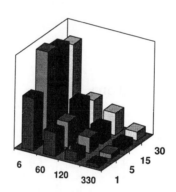

B   2-methyl octanoic acid

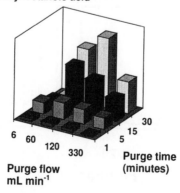

Figure 4 *Relative adsorption efficiencies between 2,3-octanedione and 2-methyloctanoic acid as the purge flow is varied from 6 to 330 mL min^{-1} and the exposure time is increased from 1 to 30 min*

For the less polar and more volatile 2,3-octanedione, the highest recovery rate was observed at a low flow rate and a relatively short adsorption period (6 mL min^{-1} for 5 min), while for the more polar semi-volatile branched chain fatty acids (*e.g.* 2-methyloctanoic acid), the highest recovery rate was observed at a relatively high flow rate and long exposure time (120 mL min^{-1} for 30 minutes).

For successful analysis of different chemical species by dynamic-headspace SPME, both the purge flow and exposure time must be optimised for each species. To ensure good repeatability, a stable isotope or other standard should be added. Dynamic-headspace SPME is superior to static-headspace SPME for detecting polar compounds that have a low partitioning coefficient between the headspace and the sample matrix, as the dynamic-headspace technique helps remove such compounds from the sample.

Direct comparison between dynamic-headspace SPME and more traditional purge and trap using Tenax-TA polymer showed similar chromatographic profiles for the rendered sheep fat sample under similar extraction conditions (Figure 5). Dynamic-headspace SPME has the advantage of rapid desorption (2 min) of trapped volatiles that do not require cryofocusing onto the capillary column, as is the case with longer (>5 min) thermal desorption of Tenax traps.

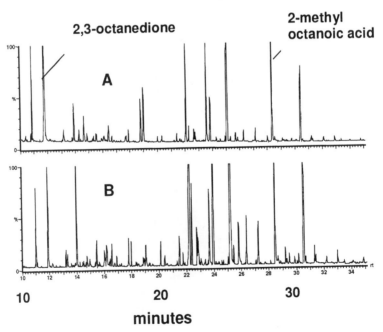

Figure 5 *Total ion count (TIC) chromatogram of volatile compounds purged from the headspace above 2 g of rendered sheep fat and trapped on and subsequently desorbed from (A) 200 mg Tenax-TA and (B) PDMS–SPME fibre*

4 Analysis of 2-Methylisoborneol and Geosmin in Drinking Water and Catfish Using Two SPME Techniques

The major contributors to off-flavour in drinking water systems are 2-methylisoborneol and geosmin.[18] Blue green algae often produce these compounds and release them into the water column, but bacteria and fungi also produce them. Associated with algal blooms in late summer, the compounds are perceptible to the human nose at the low parts per trillion range.[19] They plague drinking water systems and are particularly problematic in the warm water aquaculture production of farm-raised catfish and shrimp. Current methods of analysis are closed loop stripping[20] and purge and trap[21] with GC–MS detection. However, a technique employing SPME–GC–MS has recently been reported for the analysis of these compounds at the parts per trillion level.[22] The SPME technique has proven to be an excellent method for the concentration of volatile compounds. Combined with gas chromatography for separation, and with mass spectrometry for detection, SPME provides a state-of-the-art analytical tool.

The rate-limiting step in the process is the time required to perform a GC separation. Elimination of the gas chromatograph would permit maximum sample throughput. In certain cases, the gas chromatographic or separation step can be performed using tandem mass spectrometry. With this technique, all compounds are introduced directly into the source of the mass spectrometer, but only a specific ion indicative of the targeted compound is collected; normally this is the molecular ion. The selected or parent ion is then fragmented to produce progeny ions. The abundance of selected progeny ions is then measured, to provide quantitative information. Hence, the mass spectrometer is used both as a filter, to allow only the parent ion for subsequent analysis, and as the separation step. The tandem mass spectrometric determination serves as the detector. Problems arise when isotopic compounds, similar in structure to the compound of interest, are present and produce similar progeny ions.

Experimental

Geosmin and 2-methylisoborneol (2-MIB) were obtained from Waco Chemical (Osaka, Japan). Stock solutions of 1 part per thousand were made up in ethanol, with subsequent dilutions in Milli-Q water. Fish, obtained commercially, were trimmed and 20 g of the fillet was used for analysis. Volatile compounds were concentrated using microwave desorption.[23] SPME fibres were obtained from Supelco (Ringoes, NJ). For the fish and detection limit determinations, a 100 μm film of polydimethylsiloxane was used. Analyte volumes (8 mL) were placed in 12 mL vials and sufficient NaCl (3 g) was added to saturate the solution. The sample was stirred using a magnetic stir bar and the sample placed in a water bath held at 40 °C. The SPME fibre was exposed for 10 minutes to the headspace of the solution, then was then immediately placed in the GC injection port and desorbed.

A Finnigan GCQ ion trap (Palo Alto, CA) was used as the tandem mass spectrometer. A 25 cm length of fused silica capillary column coated with 95% polydimethylsiloxane and 5% phenyl, possessing a 0.05 mm i.d., was used between the injection port of the GC and the mass spectrometer. The GC injection port contained a 0.7 mm i.d. injection liner and was operated in splitless mode at 250 °C. Helium was used as the carrier gas and held at an initial pressure of 140 kPa. A surge pressure was used from 0.1 to 1.0 minutes. The GC oven was held isothermally at 200 °C during each analysis, then heated to 250 °C between analyses. The MS transfer line was held at 275 °C. The source of the ion trap was held at 150 °C and the offset between the trap and the source was 10 V. Methane was used as the reagent gas for chemical ionization and only positive ions were monitored. The ion trap was operated at a $q = 0.225$, with a trap offset of 10 V, a collision energy of 0.5 eV, a parent collection time of 2 ms, and a collision time of 30 ms.

Results and Discussion

Injections of a 1 ppm solution standard employing GC–MS and electron ionization were initially used for method development. The molecular ion for 2-MIB was observed at m/z 168 at an abundance of 2% relative to the base peak at m/z 95. The molecular ion for geosmin was less than 2% at m/z 182, relative to the base peak at m/z 112. To enhance the relative abundance of the molecular ions, chemical ionization was employed using methane as the reagent gas. For 2-MIB, the abundance of the $[M + H]^+$ ion remained small, as the addition of a proton resulted in the loss of water. The pseudo molecular ion of $[M + H - H_2O]^+$ at m/z 151 gave a relative abundance of 40% relative to the base peak of m/z 95.

The 30 m capillary column was removed from the GC and replaced with a 25 cm piece of a DB-5, 0.05 i.d. column (J & W Scientific, Walnut Grove, CA). The injection liner was replaced with a 0.7 cm i.d. liner. The SPME fibre was thermally desorbed in the GC injection port. The desorbed compounds were swept into the external source of the ion trap within a few seconds, where they underwent chemical ionization. The ions were then pulsed into the ion trap. Potentials were applied to the trap to eject all ions with the exception of those falling in a 2 Dalton range centred on m/z 151. Potentials were then applied on the end caps to produce collisionally induced disassociation of the parent ion. The progeny ions were then determined by sequentially ejecting the ions, with subsequent detection at the electron multiplier. A 2 Dalton window centred on the progeny ions of m/z 81, 95, and 109 was monitored for 2-MIB. For geosmin, chemical ionization gave a pseudo molecular ion at m/z 163, $[M + H - H_2O - H_2]^+$, believed to result from the addition of a proton and the subsequent loss of a water molecule and molecular hydrogen. Progeny ions were not observed at m/z 112, but were observed and monitored at m/z 95, 109, and 135.

Efficiency of the CID process for 2-MIB was analyzed by monitoring two progeny ions and the intact m/z 151 ions. In this manner the parameters of the ion trap, collection time, ionization energy, q value and reaction time were optimized to produce the greatest abundance of progeny ions. A maximum signal was obtained by setting the collection time of the parent ion at the minimum value allowed by the software, 2 ms. The maximum allowable setting of 30 ms for the reaction time produced the highest number of progeny ions relative to the parent. The optimal ionization energy was 0.5 eV. The allowable q values of 0.225, 0.300, and 0.450 were investigated with various ionization energies, collection times, and reaction times. In all cases a q value of 0.225 gave the maximum signal.

In an earlier study[23] maximum sensitivity for the GC–MS method was obtained using the 100 μm PDMS fibre at 40 °C. This was based on the adsorption capabilities of the fibre, as the GC column served to focus the analytes. No consideration was made for the relative desorption properties of these fibres. A rapid desorption during injection could deliver the analytes in a sharper burst into the instrument, thus increasing the sensitivity. A 0.1% stock solution of 2-MIB in ethanol was used to determine whether a particular coating thermally desorbed faster, thus resulting in a sharper peak. However, no significant difference was observed between the 100 μm PDMS, 30 μm PDMS, CW–DVB, polyacrylate, or PDMS–DVB fibres.

Using the pressure surge of the GC injector increased the detection limits by producing a sharper peak at the front (Figure 6). An aqueous solution containing concentrations of 2-MIB of 0, 1, 5, 10 and 20 ppb was analyzed by SPME–MS–MS. Little or no difference was observed between the blank and the 1 ppb solution. The 5 ppb solution produced a peak maxim of 442 counts at about

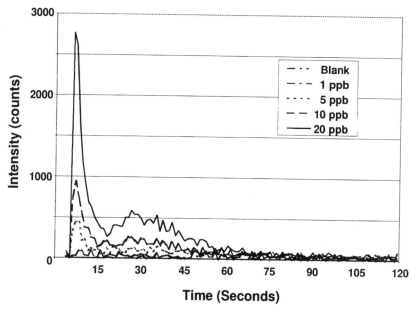

Figure 6 *Total ion count chromatograms of the m/z 81, 95 and 109 progeny ions from m/z 151 of 2-MIB at concentration of 0, 5, 10, 20 ppb*

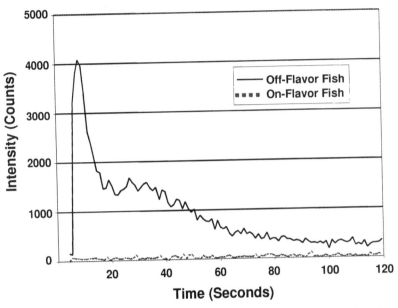

Figure 7. *Total ion count chromatograms of the m/z 81, 95 and 109 progeny ions from m/z 151 of an 'on-flavor' and 'off-flavor' catfish sample*

0.1 minute (the beginning of the surge). As the baseline is at 40 counts, this gives a $S:N$ ratio of 10:1. The limit of detection thus lies between 1 ppb and 5 ppb for the technique as described.

An acceptable 'on-flavour' catfish and an unacceptable 'off-flavour' catfish were analyzed by the SPME–MS–MS method. The fish were skinned, and 20 g of the chopped fillets were microwave desorbed while purging with N_2. The resultant steam distillate was trapped in a cold bath and yielded about 8 mL. The resultant SPME–MS–MS total ion counts of the progeny ions are shown in Figure 7. Counts for the on-flavour fish were comparable with those for the blank and the 1 ppb standard, shown in Figure 6. Counts for the off-flavour fish clearly showed that 2-MIB was present in the fillet. Comparison of the maximum height with Figure 6 puts the concentration at greater than 20 ppb. The SPME–MS–MS showed no indication of geosmin in the off-flavour fish, although a SPME–GC–MS method determined the presence of a 2 ppb concentration of geosmin. A slight rise beginning at 20 s can be observed in both Figures 6 and 7. The progeny ion ratios are consistent with 2-MIB, indicating an inefficiency in desorbing the analytes. Improved focusing would allow more molecules to arrive at the source at the same time, possibly resulting in an improved limit of detection.

This experiment demonstrates how two of the unique capabilities of the SPME technique, concentration without solvent, and rapid desorption can be exploited to provide rapid and sensitive analytical methods. SPME–MS–MS involves the optimization of a large number of variables that interact to affect the total sensitivity of the analysis. Additional improvements should yield at least an order of magnitude improvement in the limit of detection. The method as outlined is adequate to detect moderately off-flavour fish but lacks the ability to detect mildly off-flavour fish. Use of this method for water analysis would require an improvement of about three orders of magnitude. Although this is highly unlikely, it is not impossible.

5 Conclusions

The results of these studies show that SPME has a significant advantage over more traditional extraction methods for analysis of volatile compounds from three diverse food types: cheese, rendered sheep fat and catfish.

The simplicity of adsorbing volatile compounds from a hole bored into a piece of cheese onto a SPME fibre coupled with CG–PCA analysis allows rapid grading of cheese quality. Analysis times are reduced from 2.5 days for one sample by classical vacuum steam distillation, liquid/liquid extraction, drying the solution, and evaporation of the solvent, to six samples per day using the SPME method.

Similar GC chromatograms of the complex volatile compound profile found in the headspace above rendered sheep fat were found after dynamic purge and extraction with SPME or trapping with Tenax-TA. Control of purge flow rates and exposure times is required to optimise recovery of compounds of different volatility and polarity. Further targeting of compounds of interest can be achieved using SPME fibres of different polarity. SPME has an advantage over the purge and trap techniques by not requiring expensive thermal desorption equipment.

GC separation times of 75 minutes are often required for targeting specific compounds found amongst the hundreds of compounds often found in the headspace above foods. Direct SPME–MS–MS, omitting a long GC separation step, was used to measure the content of two compounds responsible for off-flavour in catfish. Combined with a rapid microwave distillation step, direct SPME–MS–MS significantly reduces analysis time and offers the potential for this technique to be used as a near-line control measure of catfish quality.

SPME has proved to be a sensitive, reliable, cost effective, and rapid extraction method for the analysis of volatile compound in foods.

References

1. S.T. Likens and G.B. Nickerson, *Amer. Soc. Brew. Chem. Proc.*, 1964, **5**.
2. C. Macku, H. Kallii, G. Takaoka and R. Flath, *J. Chromatogr. Sci.*, 1988, **26**, 557.
3. G. MacLeod and J. M. Ames, *J. Chromatogr.*, 1986, **355**, 393.
4. I.J. Jeon, G.A. Reineccius and E.L. Thomas, *J. Agric. Food Chem.*, 1976, **24**, 433.
5. H. Sugisawa, in *Analysis of Volatiles: Methods and Applications*, ed. P. Schreier, Walter de Gruyter, New York, 1984, 3.
6. M.J. Lewis and A.A Williams, *J. Sci. Food Agric.*, 1980, **31**, 1017.
7. D.S. Mottram, R.A. Edwards and H.J. MacFie, *J. Sci. Food Agric.*, 1982, **33**, 934.

8. G. Olafsdottir, J.A. Steinke and R.C. Lindsay, *J. Food. Sci.*, 1985, **50**, 1431.
9. R.B. Buttery, R. Teranishi and L.C. Ling, *J. Agric. Food Chem.*, 1987, **35**, 540.
10. S.B. Hawthorne and D.J. Miller, *J. Chromatogr. Sci.*, 1986, **24**, 258.
11. J.M. Wong, N.Y. Kado, P.A. Kuzmicky, H.S. Ning, J.E. Woodrow, D.P.H. Hsieh and J.N. Seiber, *Anal. Chem.*, 1991, **63**, 1644.
12. J.M. Snyder and J.W. King, *JAOCS*, 1994, **71**, 261.
13. X. Yang and T. Peppard, *J. Agric. Food Chem.*, 1994, **42**, 1925.
14. S.B. Hawthorne, D.J. Miller, J. Pawliszyn and C.L. Arthur, *J. Chromatogr.*, 1992, **603**, 185.
15. D.J. Manning, H.R. Chapman and Z.D. Hosking, *J. Dairy Res.*, 1976, **43**, 313.
16. Z. Zhang and J. Pawliszyn, *Anal. Chem.*, 1993, **65**, 1843.
17. Z. Zang, M.J. Yang and J. Pawliszyn, *Anal. Chem.*, 1994, **66**, 844A.
18. R.T. Lovell, *Water Sci. Tech.*, 1983, **15**, 67.
19. P.E. Persson, *Water Res.*, 1980, **14**, 1113.
20. M.J. McGuire, S.W. Krasner, C.J. Hwang and G. Izaguirre, *J. Am. Water Works Assoc.*, 1981, **73**, 530.
21. P.B. Johnsen and S.W. Lloyd, *Can. J. Fish. Aq. Sci.*, 1992, **49**, 240.
22. S.W. Lloyd, J.M. Lea, P.V. Zimba and C.C. Grimm, *Water Res.*, 1998, in press.
23. S.W. Lloyd and C.C. Grimm, *J. Agric. Food Chem.*, 1999, **47**, 164.

CHAPTER 31

Analysis of Volatile Contaminants in Foods

B. DENIS PAGE

1 Introduction

The majority of the analytical applications using SPME have focussed on the determination of nonpolar trace contaminants in environmental matrices at low or sub-μg kg^{-1} levels. Although SPME has also been applied to food products, most of these applications involve only a qualitative evaluation of the volatile profiles of naturally occurring chemicals, particularly those related to aromas. Only a few reported methods describe the quantitative determination of volatile contaminants in foods, the topic of this chapter. Quantitative methods are essential for data gathering purposes or, when suitably validated, to enforce guidelines or regulations. When required, this chapter will refer to many of the theoretical and practical aspects of SPME, particularly headspace (HS) SPME, which have been described and referenced in a recent book[1] and in Chapter 1.[2] Thus, calibration procedures, magnetic stirring, matrix modification, optimization of extraction conditions and other practical aspects of SPME will not be described in detail. In this chapter, the term 'volatile' will apply to an analyte whose isolation from the sample matrix is dependant on its volatility, usually at room temperature. Analyte volatility, however, can be increased by raising the SPME temperature, thus enlarging the scope of the definition to include 'semi-volatiles'.

Comparing the SPME of volatiles from foods to that from environmental matrices, there are two important differences that affect the complexity of analysis. One difference influences the sample preparation step; the other affects the SPME efficiency, that fraction of the analyte present in the sealed HS vial which, at equilibrium, is extracted by the fibre coating and can be transferred to the GC injector for analysis. These will be discussed in turn.

2 Sample Preparation

Volatile contaminants in foods may be inhomogeneously distributed in the bulk of the food. Blending of the food itself or blending by dispersion in water is required to obtain a representative aliquot for analysis. Such blending or homogenization, as well as any other sample manipulations such as dilutions or transfers, must be conducted with minimal volatile loss and contamination from the laboratory environment. Analyte loss by volatilization can be reducing by sample storage and method execution at reduced temperatures. Development of procedures that minimize sample exposure time and exposed sample area will also reduce volatile loss, as well as reducing the possibility of contamination from the laboratory environment. Nonpolar contaminants in a food sample will be associated mainly with any nonpolar lipid material, if present. Such lipid-containing foods are less prone to volatile analyte loss during exposure or manipulation compared to those with a lesser proportion of lipid.

It can be assumed that nonpolar volatile contaminants in packaged liquid foods are homogeneously distributed throughout the bulk of the food. The unopened cold package can be mixed by inversion before a representative test portion is taken for analysis. Nonpolar contaminants in solid or semi-solid foods, however, may be distributed inhomogeneously. Volatile contaminants in foods with large discrete particles such as nuts or whole grains, may have a higher analyte concentration towards the interior of an aged food particle or a lower concentration towards that of a recently contaminated sample. Foods contaminated in the package either from or through the packaging material would have higher contaminant concentrations nearer the food-package inter-face. These foods require that the entire contents of the package be homo-genized, usually in water, so a test sample representative of the entire package can be obtained. Such homogenization also facilitates the release of volatiles trapped within the food particle.

Dry foods can also be subdivided by blending the cold sample ($-20\,°C$) in a cold, sealed container.[3] The test aliquot can then be obtained with minimal volatile loss or contamination, transferred to the HS vial, sealed in the vial, and dispersed in water to facilitate release of volatiles for SPME.

The viscosity of the test portion for HS-SPME must permit efficient magnetic stirring. With liquid foods 1–5 g added to give 15 mL liquid is a satisfactory dilution. For most solid or semi-solid foods requiring homogenization or dispersion in water, a 10% (w/w) content of food in water is satisfactory. With some dry foods, however, a lower food content may be required for efficient mixing. Experience and convenience in our laboratory has shown that a 15–20 g test portion, containing 1–2 g of solid or semi-solid food dispersed in water with sufficient sodium chloride for saturation gives repeatable sampling and low or sub-$\mu g\,kg^{-1}$ sensitivity by HS-SPME, providing lipid food components are not present.[4] With the axial HS insertion of the SPME fibre, a 30 mL vial is required.

3 SPME of Foods

Food Lipid Matrix Effects

Another important difference between food and environmental analysis using SPME impacts on the SPME step itself. Foods are usually heterogeneous and may contain water, carbohydrates, proteinaceous material, fibre, polar and nonpolar lipids, volatile oils or flavours, minerals, vitamins, and other components from trace to high percent levels. When foods or aqueous food dispersions are extracted by SPME the fibre is normally suspended in the HS over the sample matrix in a sealed HS vial as described in Chapter 1. A direct extraction, with the coated fibre contacting the aqueous food dispersion, could transfer particulates or dissolved non-volatiles (*e.g.* sugars or salts) to the GC injector or contaminate the fibre coating with nonpolar lipid material. The SPME efficiency, that portion of the analyte initially present which is extracted by the fibre, is strongly affected by the nonpolar food components, mainly the triglycerides and, to a lesser extent, any volatile oils or flavours.[4]

The theoretical aspects of HS-SPME for the three phase system, as described in Chapter 1, can be expanded to include four or more phases.[2] With lipid-containing foods, the water immiscible triglyceride constitutes a fourth phase. The dielectric constants of the common triglycerides,[5] such as tristearin (2.7), triolein (3.1), tripalmitin (2.9), and tributrin (5.7) are similar to those of both the PDMS (2.6–2.8) and PA (2.6–3.6) fibre coatings as well as readily extracted organic compounds.[1] In a rough approximation, the food lipid could be considered as a remote part of the fibre coating in a three-phase system, located in the sealed vial, with an equal or similar equilibrated analyte concentration, but with variable volume and resulting capacity. Assuming such equal equilibrated analyte concentrations, the respective capacities of the coating and food lipid will be proportional to their volumes: a fixed 0.6 μL for the 100 μm thick PDMS fibre and a variable volume of about 11 μL for every 1% lipid in a 1 g food sample. Thus, a 1 g food sample with 1% lipid would contain about 18 times more analyte than the SPME coating because of its greater volume. When considering the effect of added lipid on the magnitude of the reduction in the amount of analyte extracted by the fibre coating, both the amount of lipid and the volatility of the particular analyte are important.[4] Equation 4 from Chapter 1[2] describes the mass of analyte adsorbed by the fibre coating, *n*, for a three phase HS-SPME system:

$$n = \frac{K_{fs}V_fC_oV_s}{K_{fs}V_f + K_{hs}V_h + V_s} \tag{1}$$

where K_{fs} and K_{hs} are the fibre/sample and headspace gas/sample distribution constants, respectively; V_f, V_s, and V_h are the volumes of the fibre coating, the sample matrix, and the headspace gas, respectively; and C_o is the initial concentration of the analyte in the sample matrix. About 1% of the nonpolar volatile analytes were found to be extracted by HS-SPME from a 15 mL aqueous system in a 30 mL vial, with the major fraction demonstrated to be in the HS.[4] In this instance, the $K_{hs}V_h$ term in the denominator is relatively large compared to that for semi-volatile analytes, and n is small. With the amount of analyte in the HS vial constant, and considering an added 10 mg of food lipid plus the SPME coating as a single phase, the V_f term and the numerator is increased about 18 fold, but the denominator increases only marginally. The increase in n, however, is proportionately distributed between the volumes of the lipid and the fibre coating, so there is only a slight reduction on the amount of analyte adsorbed by the fibre coating and transferred to the GC. With larger amounts of lipid material, however, the amount of volatile analyte adsorbed by the fibre will be further decreased as the V_f contribution to the denominator increases. For example, dichloromethane, chloroform, and trichloroethylene are extracted at 0.16, 1.2, and 1.9% from water. When 12 mg of lipid (vegetable oil) are added, 97, 64, and 59% of the initial amounts are still extracted; with 240 mg added lipid the amounts are reduced to 37, 18, and 6%, respectively.[4] With much less volatile analytes which are extracted at > 50%,[4] the K_{fs} term is larger, K_{hs} term is smaller and the resultant n is larger. When lipid is added in this instance, both the numerator and denominator are increased and the amount extracted does not significantly increase, but the amount of analyte adsorbed by the fibre will be diluted in proportion to the volume of the added lipid. For 1,4-dichlorobenzene and 1,2,3-trichlorobenzene, extracted from water at 22 and 57%, 8 and 4% of the original amount are be extracted when 12 mg of lipid are added and only 0.4 and 0.1%, respectively, when 240 mg of lipid are added.[4]

In summary, with aqueous systems, the more volatile nonpolar analytes are extracted at about 1% whereas the less volatile are extracted at > 50%. When a small amount (12 mg) of lipid is added there is a small decrease in the extraction of the more volatile analytes but a greater decrease in the extraction of the less volatile compounds. When larger amounts of lipid (240 mg) are added, similar reductions occur so that the actual percentage of analyte extracted and transferred to the GC range from 0.06 to 0.1% for both the volatile and the less volatile analytes. Thus, the sensitivity of the HS-SPME procedure when applied to a 1 g sample containing 24% lipid can be reduced by factors of 20–500 compared to aqueous solutions, depending on the analyte volatility. If sensitive methods are needed, the HS-SPME approach may lack the required sensitivity for lipid-containing foods.

Quantitation

With lipid-containing foods, the dependence of extraction efficiency on both the sample lipid content as well as volatility of the particular analyte makes accurate quantitation of volatiles complex. Several approaches have been used or suggested. Matrix matching of a pair of particular food samples, one found to contain contaminant(s), the other not containing the target residue(s) but spiked at approximately the same level, can be used. This procedure requires an identical matrix match, the amount of lipid being critical, and also requires that the incurred and spiked analytes are both properly equilibrated with the sample matrix and the SPME fibre coating. Without sufficient time to establish a true equilibration the SPME may be biased, as the spike, added to the aqueous sample as a whole may be more rapidly extracted than the incurred analyte, initially associated with the lipid phase. Matrix matching for quantitation in the presence of lipid could be used when the particular food is of an invariable lipid composition such as commercially produced milk or cream. The procedure of standard addition can be employed when the sample matrix is expected to affect the HS-SPME extraction efficiency.[4] Several aliquots of a sample may be taken and sealed in HS vials. One vial can be analyzed to determine the analytes present, the other vials can be spiked at appropriate levels, as required for quantitation. Single point or multipoint calibrations can be conducted.[1] Again the equilibration of added and incurred analytes and the SPME fibre coating must be assured. The use of isotopically labelled internal standards can be effectively employed when the mass spectrometer is used for detection. Internal standards of similar HS-SPME characteristics, *i.e.* similar volatility and polarity, and close GC elution to the target analytes could also be used to facilitate quantitation. In all cases linearity of response must be demonstrated.

Increasing Method Sensitivity

Volatile contaminants in foods are typically required to be determined at low or sub-μg kg^{-1} levels. Such levels are readily attainable using HS-SPME when a 1 g food sample is dispersed in 15 mL of water. As noted above, however, the attainable method sensitivity of HS-SPME can be dramatically reduced by food lipid. Therefore, analytical procedures which can improve method sensitivity for foods with significant lipid content are required. The enzymatic hydrolysis of the food triglyceride to the component fatty acids (or salts) and glycerol could reduce or eliminate the loss in sensitivity by eliminating the lipid phase. Such studies are currently underway in our laboratory. The isolation of the volatile contaminant from the food lipid prior to the SPME step could also provide a viable approach. Alternatively, the development and use of an SPME fibre with improved extraction efficiency in the presence of lipid would also improve the method sensitivity in such situations. The latter two approaches are discussed in the following sections.

Distillation

Steam distillation (SD) is a proven classical procedure used to isolate volatile and semi-volatile compounds from their non-volatile matrices. Analytically, a modified Garman SD apparatus has been successfully used to isolate nonpolar volatile contaminants from food matrices, the analytes being recovered by a liquid/liquid extraction.[6] In a solventless procedure, the Garman apparatus has been directly coupled to the purge vessel of a commercial purge and trap (P&T) concentrator.[7] The steam distillate and distillable volatiles are condensed directly in the purge vessel, then the volatiles are purged, trapped, desorbed, and analyzed using conventional procedures. The method was successfully applied to representative EPA Method 624[8] volatiles as well as the 1,2,4,5- and 1,2,3,4-tetrachlorobenzenes in a variety of foods. With this on-line system, losses of the volatile analytes and contamination from the laboratory environment are minimized. In our laboratory, this SD technique has recently been adapted as a cleanup procedure for HS-SPME[9] as shown in Figures 1 and 2. In this procedure the rheostat and the 4-port valve are set so steam passes through and heats the sample chamber, the injection head, and the 4-port valve, exiting through the needle and displacing all the air in the system. The valve is then rotated so the steam continues to exit through the vent. A 30 mL HS vial containing 7.5 g of sodium chloride and a magnetic stirring bar is sealed with a prepierced septum and placed in a cooling jacket (screw cap jar and modified lid for ice water circulation). The vial and cooling jar assembly is then raised to insert the needle through the septum. The valve is rotated counterclockwise to position 3 connecting the HS vial to the vacuum source. The ice water circulation is started, the sample is injected through the liquid injection head using a 12" 18 gauge (14 or 16 gauge for food slurries) needle directly into the sample chamber, the valve is turned clockwise 180°, and 15 mL of distillate are collected with stirring in the HS vial in 6–7 min. The HS vial-cooling jacket assembly is lowered to stop the distillate collection and the vial is removed for HS-SPME. Applications of this SD-HS-SPME technique to food analysis are described below. Apart from SD, other solventless procedures which are used to

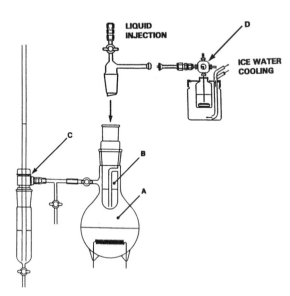

Figure 1 *Garman steam distillation apparatus modified for HS-SPME cleanup: A, boiler; B, sample chamber; C, pressure indicator; D, four-port valve, HS-SPME vial, and cooling jacket for distillate collection; and head for liquid injection.*
(Reprinted from *J. Chromatogr. A*, **788**, 131–140, 1997, with kind permission from Elsevier Science – NL, Amsterdam, The Netherlands)

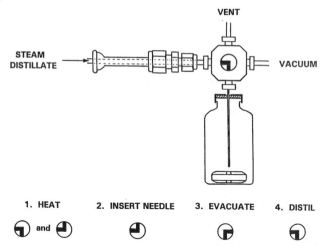

Figure 2 *Valve positions and steam path for steam preheating, vial connection, evacuation, and distillation for sample cleanup prior to HS-SPME (cooling jacket not shown)*
(Reprinted from *J. Chromatogr. A*, **788**, 131–140, 1997, with kind permission from Elsevier Science – NL, Amsterdam, The Netherlands)

isolate volatile or semi-volatile contaminants from complex matrices could also be adapted or modified to use SPME as the final sample preparation step.

Improved Fibre Extraction

The commercially availabile $80\,\mu m$ Carboxen-PDMS fibre (Supelco Inc., Bellefonte, PA) has been shown to have greater extraction efficiencies compared to the PDMS fibre for the BETX compounds and for a number of C_1- and C_2-halocarbons.[10] It was thought that the increased extraction could be effectively employed to provide sensitive methodology for volatiles in lipid-containing foods. Initial studies in our laboratories on aqueous systems and milk are presented and discussed below.

An internally cooled SPME device has been developed by Zhang and Pawliszyn.[11] The authors report that the temperature difference between the cold fibre coating and the heated HS and sample significantly increases the partition into the fibre coating. The device has been successfully used to analyze for BETX volatiles in clay. The device is not available commercially, but could provide an improved approach for the analysis of volatiles in lipid-containing foods.

4 Applications

Beverages

With the exception of milk, most widely consumed beverages contain little food lipid and the analytical problem of severely reduced sensitivity for such foods is minimal. Compared with water, however, the extraction efficiency is still somewhat reduced.[4] With most beverages, standard addition and the use of isotopically labelled standards have been successfully employed. Several examples follow.

The application of the HS-SPME technique to the determination of 2,4,6-trichloroanisole (TCA) in wine has been studied by two groups. Both Fischer and Fischer[12] and Evans and co-workers[13] have used an autosampler and the 100 μm PDMS fibre to extract TCA from wine followed by capillary GC separation and MS detection. Fischer and Fischer extracted a 5 mL sample for 30 min at 20 °C, used standard addition for quantitation and reported a LOD of 2.9 ng L^{-1}, whereas Evans *et al.* used a 10 mL sample, a 25 min extraction at 45 °C, and the fully deuterated TCA analog as an internal standard. They reported a limit of quantitation of 5 ng L^{-1}.

Gandini and Riguzzi[14] applied HS-SPME to the analysis of methyl isothiocyanate (MITC), an illegal antifermentative agent, in wine. The MITC was extracted in 30 min by HS-SPME from a 5 mL wine sample with 1.25 g of added sodium chloride using a 65 μm Carbowax-divinylbenzene fibre. Standard addition was used for quantitation. Capillary GC with nitrogen-phosphorus detection was interference free and permitted detection as low as 1 ppb. The analytical results compared favourably to an extablished liquid/liquid extraction procedure.

Page and Lacroix[4] applied HS-SPME with the PDMS coating to the analysis of a number of fruit juices, soft drinks, fruit drinks, and milks for volatile and semi-volatile analytes ranging from vinyl chloride to hexachlorobenzene, using standard addition for quantitation. Standard addition also permitted a comparison of the HS-SPME efficiency from 5 g of beverage dispersed in 10 mL of water to that of 15 mL of water. Sodium chloride (6 g) was added to all HS vials. Apple juice, cranberry raspberry drink, cola, and ginger ale apparently contain little nonpolar material as all but the least volatile penta- and hexa-chlorobenzenes were extracted at > 80%. With an orange soft drink, citrus juices, and citrus-based drinks, the SPME efficiencies for the tetrachlorobenzenes ranged from about 3–19%, those for the trichlorobenzenes, 7–32%, and those for the dichlorobenzenes, 18 to 58%, with the more volatile analytes extracted with increasing efficiency. These results demonstrate the effects of the nonpolar citrus oils on the SPME efficiency. With milk, the extraction efficiency increases with analyte volatility and decreases with increasing butter fat (BF). For example, compared to water, the dichlorobenzenes are extracted from a 5 g milk sample at about 9, 2, and 0.5% for the 0.1, 2.0, and 3.4% BF milks, respectively. With the more volatile chloroform, respective values are 86, 28, and 20%.

Quantitation using spiked matrix-matched samples were used by Forsyth and Dusseault[15] in model studies investigating the exposure of beverages to methylcyclopentadienyl manganese tricarbonyl (MMT), an antiknock gasoline additive. Capillary GC with atomic adsorption spectrometric element specific detection was used to detect MMT in spring water, apple juice, cola, coffee, and milk. Method detection limits of less than 1 pg mL^{-1} were obtained for water, cola, and apple juice and less than 4 pg mL^{-1} for coffee. For milk, however, the detection limit was 260 pg mL^{-1}, again emphasizing the dramatic effect of food lipid on the SPME efficiency.

Solid or Semi-solid Foods

Few applications of SPME to the analysis of volatiles in non-liquid foods have been reported. Page and Lacroix[4] have applied HS-SPME to 1 g dispersions of finely divided flour, flour-based products, decaffeinated teas and coffees, and selected spices in water with added sodium chloride. Standard additions of volatile and semi-volatile analytes ranging from vinyl chloride to hexachlorobenzene were employed for quantitation and permitted an evaluation of the effects of food lipid on the HS-SPME efficiency compared to a water matrix. In general, the magnitude of the reduction in SMPE efficiency was greater for the less volatile compounds, and related to the fat or lipid content of the food. In most foods, HS-SPME afforded detection at low μg kg^{-1} for all target analytes except penta- and hexachlorobenzene.[4]

Lipid-containing Foods

Steam Distillation-HS-SPME

The HS-SD-SPME technique described above and in Figures 1 and 2 has been applied in several analytical situations. Sen *et al.*[9] have reported the determination of the semivolatile *N*-nitrosodibutylamine (NDBA) and *N*-nitrosodibenzylamine (NDBzA) in smoked ham using SD-HS-SPME and a thermal energy analysis detector (nitrosamine mode). Distillates of homogenized, diluted aqueous ham aliquots (0.1 or 1.0 g) with added nitrosation inhibitor and internal standard (*N*-nitrosodioctylamine or *N*-nitrosodipropylamine) were collected in 30 mL HS vials containing acetic acid, salt, and ascorbic acid. The vial was then uncapped, and solid KOH was added. The HS-SPME was conducted using a PA fibre with stirring in an 80 °C oven for 1 h. The 0.1 g sample size gave detection limits ($S/N > 3$) of about 3 and 1 μg kg^{-1} for NDBA and NDBzA, respectively. Confirmation of residues was performed by GC–MS. The HS-SPME of other nitrosamines was also studied.

In our laboratory SD-HS-SPME has also been applied in a survey for over 30 halogenated and aromatic volatile contaminants in human milk. The dead volume of a 5 mL glass luer tip syringe, fitted with a luer stopcock and a 2" 20 gauge needle was filled with water and tared. About 1 mL of milk was drawn into the syringe and the weight of milk determined. A further 1 mL of water and 0.5 mL of air (tip up) were drawn into the syringe to rinse and displace the needle and stopcock contents. The needle was removed and an internal standard solution of [^{2}H$_6$]benzene, fluorobenzene, and six [^{13}C$_6$]chlorobenzenes was added through the stopcock, the stopcock closed, and the contents of the syringe mixed by shaking. The apparatus was set up to distil the milk for HS-SPME as shown in Figures 1 and 2. A 10" 18 gauge luer tip needle was inserted through the septum of the sample injection head into the sample chamber and heated by exiting steam. After the HS vial was evacuated, the 5 mL syringe containing the milk was attached to the 10" needle, the 4-port valve turned to connect to the SD apparatus, and the sample injected. Detection limits by GC–MS were generally less than $0.1\,\mu g\,kg^{-1}$. Volatiles were found in most samples and included toluene, ethylbenzene, xylenes, styrene, 1,4-dichlorobenzene, and hexachlorobenzene. These results demonstrate the transfer to milk of nonpolar volatiles previously reported to be present in human blood. The SD-HS-SPME procedure was demonstrated to be effective for analytes with volatility ranging from chloroform to that of hexachlorobenzene.

In the applications of SD-HS-SPME to nitrosamines in ham and to volatiles in human milk, the isolation of the analytes from the lipid-containing matrices before the SPME step permitted the analysis at low or sub-$\mu g\,kg^{-1}$ levels. Co-distillation of steam volatile food components into the HS vial, however, was shown to reduce the SPME efficiency as internal standards spiked into the HS vial containing a blank food distillate were reduced by about 30% compared with a water blank. The use of internal standards effectively compensated for this matrix effect.

Carboxen-PDMS fibre

The Carboxen-PDMS fibre was found to give a much more effective extraction of the volatile halocarbons normally determined in foods in our laboratory when compared to the PDMS fibre. Table 1 shows increases of up to 270-fold from water for extractions using the Carboxen-PDMS fibre compared to the PDMS fibre. Only hexachlorobenzene was less efficiently extracted. This efficient extraction, especially for the more volatile analytes, was shown to be applicable to matrices containing lipid material. When spiked milk or cream, containing 20 and 100 mg of BF, respectively, were extracted, low levels of most analytes were efficiently extracted. The low levels achieved would permit the routine determination of volatile contaminants, such as the trihalomethanes, in milk or cream at low $\mu g\,kg^{-1}$ levels.

5   Summary

HS-SPME has been successfully applied to the determination of volatile contaminants in a wide variety of foods. The HS-SPME is matrix dependent and internal standards or standard additions are required for accurate quantitation. The HS-SPME from lipid-containing foods can be dramatically reduced and the analysis of such foods, using the traditional PDMS fibre coating, requires separation of the volatiles from the lipid before HS-SPME. An on-line SD cleanup with HS-SPME has proven to be effective for this purpose. Initial studies show that the recently introduced Carboxen-PDMS fibre can provide a more efficient HS-SPME and improved method sensitivity, especially for the more volatile analytes from both aqueous and lipid-containing matrices. The use of HS-SPME as a quantitative tool for food contaminant analysis is

Table 1 *HS-SPME of selected chlorinated volatile contaminants from aqueous and lipid-containing food matrices: comparison of PDMS and Carboxen-PDMS (C-PDMS) fibre coatings and limits of detection*

Analyte	C-PDMS/PDMS	LOD[a] (μg kg^{-1}) C-PDMS Water	Milk[b]	Cream[c]
trans-1,2-Dichloroethylene	270	0.12	0.1	0.1
Chloroform	270	nd[d]	nd	nd
1,1,1-Trichloroethane	92	0.08	0.25	sh[e]
Carbon tetrachloride	95	0.06	0.5	0.25
Trichloroethylene	73	0.05	1.2	0.5
Bromodichloromethane	170	0.17	0.5	0.3
1,1,2-Trichloroethane	110	0.15	0.5	0.5
Tetrachloroethylene	90	0.36	2.0	1.5
1,3-Dichloropropane	110	0.1	0.2	0.2
Chlorobenzene	110	0.1	0.1	0.4
Bromoform	20	0.25	2.0	0.2
Bromobenzene	17	0.01	0.03	0.05
1,4-Dichlorobenzene	4.8	0.04	1	3
1,2,4-Trichlorobenzene	2.1	0.07	2.5	5
1,2,4,5-Tetrachlorobenzene	1.4	0.12	> 5	> 5
Pentachlorobenzene	1	0.12	> 5	> 5
Hexachlorobenzene	0.48	0.2	> 5	> 5

[a] LOD, limit of detection defined as 5 × baseline noise using Hall electrolytic conductivity detection and 1 g sample of water, milk, or cream dispersed in 14 g salt saturated water in a 30 mL (nominal volume) HS vial.
[b] Milk labelled to contain 2% butter fat.
[c] Cream labelled to contain 10% butter fat.
[d] nd, not determined. Low levels of chloroform in sample did not permit determination of value.
[e] sh, occurs as shoulder on chloroform contaminant: value about 1 μg kg^{-1}.

expected to find increased acceptance and wider application as improved techniques and newer fibre technology emerge.

Acknowledgement

The technical assistance of Mrs. Gladys Lacroix is gratefully acknowledged.

References

1. J. Pawliszyn, *Solid Phase Microextraction: Theory and Practice*, Wiley-VCH, New York, 1997.
2. J. Pawliszyn, Chapter 1 of this book.
3. B.D. Page and C.F. Charbonneau, *J. Assoc. Off. Anal. Chem.*, 1984, **66**, 757.
4. B.D. Page and G. Lacroix, *J. Chromatogr.*, 1993, **648**, 199.

5. *CRC Handbook of Chemistry and Physics*, 77th Edn., ed. D.R. Lide, CRC Press, Boca Raton, FL, 1996.
6. B.D. Page, W.H. Newsome and S.B. MacDonald, *J. Assoc. Off. Anal. Chem.*, 1987, **70**, 446.
7. B.D. Page and G. Lacroix, *J. AOAC Int.*, 1995, **648**, 199.
8. Environmental Monitoring Systems Laboratory, *Methods for the Determination of Organic Compounds in Drinking Water*, US Department of Commerce, National Technical Information Service, Springfield, VA, 1988.
9. N.P. Sen, S.W. Seaman and B.D. Page, *J. Chromatogr. A*, 1997, **788**, 131.
10. P. Popp and A. Paschke, *Chromatographia*, 1997, **46**, 419.
11. Z. Zhang and J. Pawliszyn, *Anal. Chem.*, 1995, **67**, 34.
12. C. Fischer and U. Fischer, *J. Agric. Food Chem.*, 1997, **45**, 1995.
13. T.J. Evans, C.E. Butzke and S.E. Ebeler, *J. Chromatogr. A*, 1997, **786**, 293.
14. N. Gandini and R. Riguzzi, *J. Agric. Food Chem.*, 1997, **45**, 3092.
15. D. Forsyth and L. Dusseault, *Food Addit. Contam.*, 1997, **14**, 301.

CHAPTER 32

Determination of Pesticides in Foods by Automated SPME–GC–MS

KE-WU YANG, RALF EISERT, HEATHER LORD AND JANUSZ PAWLISZYN

The application of solid phase microextraction (SPME) to different areas in analytical chemistry has been steadily increasing. It is proving beneficial for analysts to automate this microsampling technique. This chapter presents the application of a fully automated SPME coupled with gas chromatography/mass spectrometry (GC–MS) for the analysis of pesticides in fruit juice. The procedure of automated SPME–GC–MS with both static absorption and fibre vibration technique is described in some detail. The linearity and detection limit of the method was determined with the following selected pesticides: atrazine, propazine, simetryn, ametryn, prometryn, terbutryn, sulfotep, diazinon, methyl parathion, malathion, and parathion. The precision for most target analytes was below 6% relative standard derivation (%RSD). The complex fruit juice matrix decreased SPME extraction efficiency.

1   Introduction

The analysis of nutrients, flavor, and contaminated chemicals in food has been a challenge to many researchers for a long time. Separation of the target analytes from the matrix is the primary and often the most difficult step to an analytical chemist. Liquid/liquid extraction (LLE) and solid phase extraction (SPE) have been used to accomplish this task with liquid samples.[1,2] The most popular technique used for non-volatile compounds is LLE, which requires large quantities of high purity solvent, which results in contamination of the work place and environment. Further chromatographic techniques (cleanup procedures) are very often required to separate complex mixtures after extraction, owing to the high non-selectivity of LLE, which is often accompanied by loss of the analytes. The procedure itself is time-consuming, tedious, often requires preconcentration of the extracts prior to instrumental analysis, and is difficult to automate. SPE overcomes several the disadvantages encountered with LLE, but it still has some shortcomings, such as significant background interference and memory effect problems, as well as plugging and poor reproducibility between cartridges. An alternative approach is to use solid-phase microextraction (SPME) which can reduce or eliminate these problems.

The SPME method has been applied to volatile and non-volatile compounds, used for liquid or gaseous samples, can be used with any gas chromatograph (GC) or gas chromatograph/mass spectrometer (GC–MS) and can be fully automated. SPME was first introduced by Pawliszyn in 1990.[3] Detailed accounts of the theory of SPME have already been published and a summary of this technique has also been presented in Chapter 1 of this book.[4] Therefore no further account will be presented in this chapter.

Food is one of the main sources of intake of chemical contaminants, particularly pesticides.[5] However, pesticides are still economically essential to agriculture in most countries of the world. Consequently, the possible hazards associated with their use require careful study. Additionally, attention has been given to the analysis of pesticides in environmental samples. Headspace SPME has been used successfully in the analysis of flavor volatiles in some fruit juices.[6] The number of applications for SPME has been steadily increasing; however, most of these are associated with the manual SPME method.

It is important, for environmental laboratories, where there are many routine analyses, that the sample preparation be fully automated. The extraction and desorption processes of SPME–GC can be fully automated using a conventional autosampler. To date, SPME has been automated only for the static absorption mode using a commercial autosampler.[7,8] Recently, the new 8200 autosampler for SPME (Varian, Palo Alto, CA) was commercialized. One improvement is a system that incorporates an agitation mechanism consisting of a small motor and a cam to vibrate the needle (fibre vibration technique). The vibration causes the fibre to move with respect to the solution, resulting in a substantial decrease of equilibrium time due to a reduced thickness of the boundary layer compared to the static mode. This mode of agitation simplifies SPME extraction because it does not require the introduction of foreign objects to the sample prior to extraction. Thus the great potential of agitation during the absorption process will become available for automated SPME analysis. In the automated SPME–GC system, no operator is necessary for extraction and desorption steps; moreover, the precision of extraction and efficiency of the instrument are substantially improved.[9]

2   Experimental

Automated SPME

SPME has been successfully coupled with both GC and high performance liquid chromatography (HPLC). Standard GC injectors, such as split/splitless can be applied to SPME as long as a narrow insert with an inside diameter close to the outside diameter of the needle is used. The split should be turned off during SPME injection. Reference 4 describes some interfaces between automated SPME and analytical instrumentations, with particular focus on GC and HPLC.

All extractions and injections with the SPME unit were performed automatically using a Varian 8200 autosampler, which is specially designed for SPME (Varian, Palo Alto, CA) and controlled by the Varian Star Version 4.5 software through a PC. The software allows the operator to select absorption and desorption times, headspace or direct sampling, and the number of vials to be automatically analysed. Two types of vial carrousels are available from Varian for the SPME autosampler. One is the 48-vial carrousel which holds 2 mL vials. The other is 12-vial carrousel which holds 16 mL vials. The 16 mL vials were employed for this work. The SPME fibre holder for automatic use (Supelco, Bellefonte, PA) was used for automated SPME investigation.

In preliminary experiments, the following four fibre coatings were selected to investigate the effect of fibre coating on extraction efficiency for selected pesticides: 100 μm poly(dimethylsiloxane) (PDMS), 85 μm polyacrylate (PA), 65 μm Carbowax/divinylbenzene (Carbowax/DVB), 65 μm poly(dimethylsiloxane)/divinylbenzene (PDMS/DVB). All are available from Supelco (Bellefonte, PA). Static absorption and fibre vibration modes were used for these experiments. The system and fibre coating blanks were performed to ensure that no pesticides were present on the fibres before SPME extraction. In the application steps, only the PDMS fibre coating was used.

Materials

The following pesticide standards (in methanol) were added to NANOpure water and fresh fruit juice: atrazine, propazine, simetryn, ametryn, prometryn, terbutryn, sulfotep, diazinon, methylparathion, malathion, and parathion. The group includes both organonitrogen pesticides (ONP) and organophosphorus pesticides (OPP). The stock standard mixture solution was prepared using methanol as solvent. The content of methanol in the samples was controlled at constant of 1% (v/v)

Fresh oranges and carrots were purchased from the local grocery store. Home made fresh orange and carrot juices were obtained using a juice extractor (Proctor Silex®). The raw juice was centrifuged and the remaining liquid portion used for analysis. The pHs of the orange and carrot juices were 3.73 and 4.75, respectively.

Instrumentation

Gas chromatographic analysis was performed using a Varian (Palo Alto, CA) Model 3400CX gas chromatograph equipped with a septum programmable injector (SPI) and a flame ionization detector (FID). The analytes were separated on a Supelco SPB-5 column, 30 m × 0.25 mm with a film thickness of 0.25 μm. The column temperature program was as follows: 50 °C held for 5 min, ramped to 170 °C at 20 °C min^{-1}, held 1.5 min, ramped to 190 °C at 10 °C min^{-1}, held 2 min and ramped to 290 °C at 10 °C min^{-1}, with a final hold for 1 min. The carrier gas was UHP helium, maintained at a flow-rate of 1.57 mL min^{-1}. The injector was operated at 260 °C. The detector (FID) was held at 300 °C, with a nitrogen make up flow of 29.8 mL min^{-1}, hydrogen at 31.5 mL min^{-1}, and air at 303 mL min^{-1}.

Subsequent analyses were performed with a Varian Saturn II ion trap mass spectrometric detector. The column oven temperature profile was the same as described above but a DB-5 column (J & W Scientific, Folsom, CA) was used for this work. The ion-trap manifold was held at 290 °C, as was the transfer line. The detector was turned off for the first 400 s of the analysis. The mass range scanned was 40–400 amu. The mass spectrometer was operated in the electron-ionization (EI) mode and tuned to TFTBA using FC-43 (perfluorotributyl-amine).

3   Results and Discussions

In an automated direct SPME system, there are two options for absorption mode: static absorption and fibre vibration. In the static case, transport of the analyte is limited by diffusion in both the aqueous phase and the aqueous boundary layer surrounding the fibre coating surface. During the absorption process, the concentration in the boundary layer steadily decreases, thus reducing the flux into the fibre coating. However, in the dynamic case of fibre vibration, the boundary layer around fibre coating will be significantly reduced, thus reducing this effect. It can be expected that the final equilibrium time will be much shorter than in the static mode. For example, in the current work, most target analytes reached the equilibrium within 30 min in the case of fibre vibration, whereas in the static mode, equilibrium required about 60 min, using the PDMS fibre coating.

The target analytes studied here were not volatile and therefore only direct SPME was used. The fibre coating was directly exposed to the aqueous phase at room temperature (*ca.* 25 °C). In this mode, the outer needle of the syringe may contact the aqueous sample, causing water to flush inside the needle. However, the distance between the end of the outer needle and the sample surface can be adjusted by changing parameters in the software, and so the needle end is easily positioned to a set distance above the aqueous sample. By doing so, droplets of water are kept out of the needle. Peak broadening and poor precision are caused when droplets of water are injected into the GC.

When multiple samples are analysed, a selection in the software allows the fibre coating to begin absorption in the next sample, before the GC temperature program has finished for the current sample. This increases the instrument efficiency when long absorptions are being employed.

The sample vial in the autosampler carrousel is located above the hot GC injector, sometimes for an extended period of time. The sample temperature was measured at 25 °C ± 2 °C. The fibre vibration frequency was measured using a stroboscope and the average frequency determined was $v = 1100$ s^{-1}. In order to successfully desorb all of the analytes from the fibre coating and thus eliminate carryover in subsequent analysis, a 260 °C desorption temperature was selected.

It is conceivable that a high content of organics (*e.g.* solvents) in the sample precludes an efficient extraction. Eisert[10] reported that reasonable extraction may still be carried out at a methanol concentration of 10 vol.%. A methanol concentration of 1% (v/v) was maintained in this study and was found to be very satisfactory.

Equilibration Time

The absorption time profiles for all selected pesticides were studied using absorption times between 0 and 120 min, with two different modes (static absorption and fibre vibration) and with four fibre coatings. Figure 1 shows results obtained when static *vs.* fibre vibration modes are compared. The time to reach equilibrium using the fibre vibration technique is significantly shorter compared to the extraction under static conditions. The fibre vibration technique not only shortened the equilibration time, but also produced very smooth absorption-time profiles.

In the case of fibre vibration, most target analytes reached equilibrium extraction between 20 and 30 min for the PDMS fibre coating, and 60 min for the PA fibre coating; and when PDMS/DVB fibre coating was used, most ONP reached equilibrium between 60 and 120 min. No equilibration was observed within 120 min for the Carbowax/DVB fibre coating. However, equilibrium extraction was observed at a much longer time (*ca.* 60 min) for the PDMS fibre coating when static absorption was selected. For routine analysis, it is not necessary to reach a complete equilibrium, as long as satisfactory precision is

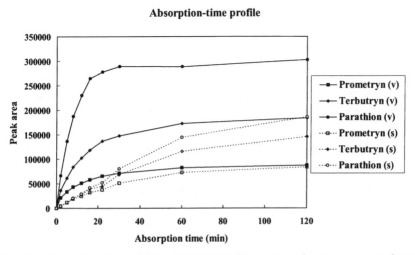

Figure 1 *Absorption–time profiles with PDMS fibre coating for prometryn, terbutryn, and parathion. Absorption: (v) fibre vibration mode and (s) static mode*

obtained or the exposure time of the fibre coating can be controlled precisely. An automated SPME system can successfully fulfill this task.

Among the four fibres tested, Carbowax/DVB and PDMS/DVB are more sensitive than the other two according to extraction efficiency under the same conditions. However, the FID flame was always extinguished when Carbowax/DVB was used. This is attributed to the affinity of water for the fibre coating. We also found that the PDMS/DVB fibre coating was easily broken. It can be expected that the selection of Carbowax/DVB and PDMS/DVB will improve the sensitivity of the method when these shortcomings are overcome. The PDMS coating is nonpolar and has been known to work effectively on a wide range of analytes, both polar and nonpolar.[13,14] Therefore the PDMS fibre coating was finally selected for the application of the method.

Precision

Method precision using different fibre coatings was determined under neutral standard conditions. In automated SPME with the 12-vial carrousel, twelve extractions from aqueous samples having a concentration of $0.1 \, \mu g \, mL^{-1}$ for single analytes were performed and analysed by GC–FID. The results are summarized in Table 1. The precision obtained with fibre vibration was optimal in comparison to the static mode. The %RSD values obtained for all four fibre coatings were less than 5% for most compounds. Precision was adequate when using fibre vibration compared with standard analytical methods, whether extraction reached the equilibrium or not. The precision obtained for static absorption was still considered to be satisfactory, although it was not as good as that for fibre vibration. Generally, the %RSD was below 10% in this case. In an automated system, the absorption time can be precisely controlled. It can be seen that precision is improved with the use of an autosampler compared to the precision obtained by manual SPME,[14] even in the case of static mode automated SPME.

The temperature of the samples may be increased if they are positioned for a long time over the hot GC injector, which is in turn dependent on the absorption time used. The temperature was measured to be $25 \, ^\circ C \pm 2 \, ^\circ C$. The extraction yield of target analytes is not significantly affected by this phenomenon. When taking equilibrium time, precision, and GC running time into consideration, a 40 min absorption time was selected for the remaining experiments.

Effect of Salt Concentration and pH, Individually and in Combination

Buffer solutions from pH 2 to pH 10 were prepared according to literature.[15] $2 \, mol \, L^{-1}$ of salt was used for extractions in combination with buffer solution. The addition of a salt can often improve recovery when conventional extraction methods are used. Sodium chloride (NaCl) and sodium sulfate (Na_2SO_4) were initially selected to investigate the effects of salt on extraction efficiency, in concentrations from $0 \, mol \, L^{-1}$ to near saturation. The results indicated that

Table 1 *Precision determinations for extraction with four fiber coatings (%RSD), fiber vibration employed, n = 12*

Fiber	PDMS			PA			Carbowax/DVB			PDMS/DVB		
Pesticide	30 min	40 min	60 min	30 min	40 min	60 min	30 min	40 min	60 min	30 min	40 min	60 min
Atrazine	6.0	4.1	5.0	4.5	5.3	4.8	3.4	4.3	4.0	8.4	3.9	1.9
Propazine	6.4	3.8	3.9	5.1	7.3	9.4	4.0	4.4	3.2	6.6	3.9	2.6
Diazinon	7.3	3.7	2.3	4.6	8.0	5.9	7.9	8.7	4.4	7.5	4.1	2.8
Simetryn	7.9	5.3	1.9	8.9	8.9	7.7	4.3	4.7	3.6	5.0	2.5	3.0
Ametryn	2.4	2.8	0.6	8.7	3.0	3.8	4.2	4.0	3.4	4.4	2.8	2.6
Prometryn	5.1	2.6	1.5	5.3	5.6	2.6	4.6	3.6	4.5	5.8	3.5	2.3
Terbutryn	4.6	2.9	1.9	5.9	5.7	2.6	3.7	2.4	2.8	6.0	2.8	1.8
Malathion	4.2	7.1	1.4	5.4	9.8	9.2	5.7	4.6	4.7	9.2	3.1	6.5
Parathion	4.9	2.9	2.2	2.6	5.0	3.4	6.9	6.3	6.0	8.8	2.8	2.2

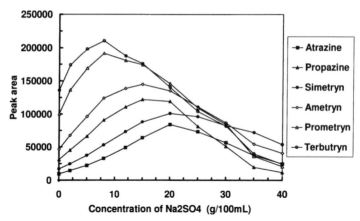

Figure 2   *Effect of concentration of Na₂SO₄ on SPME extraction efficiency*

initially the extraction yield increased with an increase in salt concentration. However, extraction was observed to reach a maximum value, and then to decrease with further increase in salt concentration. The maximum extraction yields for individual analytes were observed at different salt concentrations (see Figure 2). The same phenomenon was reported previously.[10]

This behavior can be explained by considering two simultaneously occurring processes. Initially analyte recovery is enhanced due to 'salting out', whereby water molecules form hydration spheres around the ionic salt molecules. These hydration spheres reduce the concentration of water available to dissolve analyte molecules; thus it is expected this will drive additional analytes into the fibre coating.[11] In competition with this process, however, is the fact that polar molecules may participate in electrostatic interactions with the salt ions in solution,[12] thereby reducing their ability to move into the fibre coating. Initially, it would be the interaction of the salt molecules with water that is the predominant process. As salt concentration increases further, salt molecules will begin to interact with analyte molecules. Thus it is reasonable that there should be an initial increase in analyte extracted with increasing salt concentration, followed by a decrease as the salt interaction with the analytes in solution predominates. Of further interest, it appears that the salt concentration producing maximum extraction is correlated to partition coefficient. In Figure 2, the compound with the highest initial level of extraction also has the highest partition coefficient, as all extractions were carried out under identical conditions with all analytes in equal concentration. Thus we see that the salt concentration producing maximum effect increases as analyte partition coefficient decreases.

In subsequent experiments, seven additional salts (KCl, NH₄Cl, NaBr, NaNO₃, Na₂CO₃, CaCl₂, MgCl₂) were selected to study the effect of different salts on extraction efficiency. In this work only ONP pesticides and PDMS fibre coating were used. The effect of salt concentration on extraction yield of target analytes was the same as that observed for NaCl and Na₂SO₄. The results also demonstrated that the valence state of salt ion had an effect on extraction. For example, bivalent ions had more influence than univalent ions for atrazine, propazine, simetryn and ametryn. In contrast, univalent ions had more influence than bivalent ions for prometryn and terbutryn. The commonly used salt NaCl showed a relatively lower efficiency as measured by extraction yield increase.

The effect of pH on SPME extraction yield was observed to be independent of fibre coatings. For ONP, which are weak bases, it was expected that the solution pH values would have significant influence on extraction yields. With pH 2, the extraction yields were the lowest, and increased with increased pH up to pH 8, after which it decreased. Thus, it can be concluded that the extraction yields are highest with a sample pH of about 8. In the case of OPP, sample pH didn't have significant influence on the extraction yields of diazinon, malathion and parathion, with the exception of malathion at pH 10 and diazinon at pH 2. However, pH 8 was still adequate in general. OPPs are easily decomposed in alkaline solution.

The study on the effect of the combination of pH and salt on extraction efficiency demonstrated the extraction yield mainly depends on addition of salt. The addition of salt had a significant influence compared to pH adjustment with the combination of salt and pH.

Linearity

After completing the method development, method linearity was analysed using PDMS fibre with detection by FID and MS. Aqueous standards were spiked with increasing concentrations, over a range typically from 10 to $10000\,\mathrm{ng\,mL^{-1}}$ with GC–FID and 0.01 to $500\,\mathrm{ng\,mL^{-1}}$ with GC–MS. Regression analysis was performed to generate the linear equations of the calibration curves and the correlation coefficients. All of the analytes analysed were extracted linearly over a minimum of three orders of magnitude and had correlation coefficients (R^2) from 0.993 to 1.000.

Limits of Detection

The limits of detection (LOD) were estimated as the concentration of analyte that produced a signal to noise ratio of 3. The linear range experiment provided the necessary information to calculate detection limits, by extrapolating from the lowest concentration point on the linear range calibration curve. The limit of quantitation (LOQ) can also be estimated similarly, LOQ is defined as a concentration of analyte producing a signal 10 times that of the noise. The calculated limits of detection were found to be $200\,\mathrm{ng\,L^{-1}}$ to $5000\,\mathrm{ng\,L^{-1}}$ for FID with %RSD of 1.6%–6.9%, and $1\,\mathrm{ng\,L^{-1}}$ to $20\,\mathrm{ng\,L^{-1}}$ for MS with %RSD of 0.7%–7.4%. The LODs were much higher than the permissible level for pesticide contamination set in the European Drinking Water Regulation $(0.1\,\mu\mathrm{g\,L^{-1}})$ when FID was used.[16] They were also higher than LODs of standard EPA method. The detection limits were however much lower when automated SPME–GC–MS was employed.

Fresh Juice Analysis

In order to investigate the application of automated SPME–GC–MS for the analysis of pesticides in the complex matrix of fruit juice, orange and carrot juices were selected. pH values were 3.73 and 4.75 for homemade fresh orange and carrot juice, respectively. At first, only solid sodium chloride was added to the fresh juices at a concentration of $2 \, mol \, L^{-1}$, with direct extraction with PDMS fibre coating. No background levels of target analytes were detected in the juices, according to the selected ion chromatogram (SIC). Later, the orange juice was adjusted to a pH of 6.5 with the addition of solid sodium hydroxide. Again no target analytes were detected. It was also found that no interferences were detected when the SPME fibre coating was directly exposed to the juice matrix. It can be expected that acidic substances will be generated during the storage of pH adjusted orange juice. For example, the pH of orange juice dropped to pH 5.6 from pH 6.5 after two days of storage in a refrigerator. Thus for fresh juice, analysis should be conducted as soon as possible after pH adjustment. Also, if sample pH doesn't have a significant influence on analyte extraction, it is not recommended to adjust the juice pH.

In order to investigate the effect of matrix on SPME yield, spiked samples of orange juice and aqueous buffer solution were analysed either at pH 3.73, or pH 6.5. In both cases $2 \, mol \, L^{-1}$ NaCl was used. It was found that the juice matrix had a significant influence on amount extracted. Figure 3 shows results for diazinon, parathion, simetryn, and ametryn, and demonstrates that extraction yields from aqueous solution were higher than those from orange juice samples.

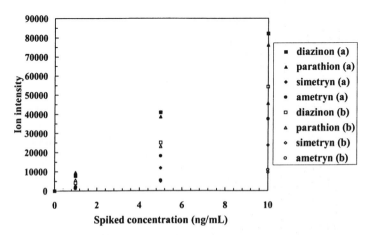

Figure 3   *Effect of matrix on SPME extraction yield with PDMS fibre coating, extraction* (a) *from pure water and* (b) *from pure orange juice; pH (3.73) and NaCl concentration* ($2 \, mol \, L^{-1}$) *were constant*

The more 'pure water' in juice, the higher is the SPME efficiency. It is apparent that we can not quantify real juice samples using the calibration curves obtained from aqueous samples. It was also found that the suspensions formed in orange juice during storage influenced the SPME efficiency. Steffen[6] reported that the particles in juice produced poor relative standard deviations (%RSD). In this case concentrated matrix prevented analytes from partitioning into the fibre coating. For the quantification of pesticide residues, we selected a reference juice sample of the same type as the sample to be analysed, but which had been shown to contain no detectable residues of the pesticides in question. In the method a matrix standard reference sample was prepared first. Pesticide standards were then spiked into the sample and quantitation was made by comparison of peak heights or areas with the unknown sample.

SPME is not an exhaustive extraction method, but rather an equilibrium method. Thus the calculation of the percentage of total analyte extraction from the matrix is not useful in determining method performance. Of greater significance is the accuracy with which the method reports analyte concentration, relative to a known concentration initially added. The concentrations of analytes spiked into orange juice, and the resulting concentrations reported by the method, are shown in the Table 2. The fourth column reports the concentrations detected as a percentage of the analyte concentrations initially added. These percentages varied from 94% to 112%. The corresponding method precision (%RSD) varied from 0.3% to 8.9%.

The pH of carrot juice was 4.74; thus only solid NaCl was also added prior to analysis. No additional modifications were made. The results were similar to those found for orange juice. The carrot matrix had a negative impact on extraction yield; *i.e.* the carrot matrix also inhibited the target analytes from partitioning into the fibre coating. From the calibration curves, correlation

Table 2 *Quantitative results for freshly prepared orange juice spiked with pesticides, n = 3*

Target analytes	Concentration spiked into sample (ng mL^{-1})	Detected (ng mL^{-1}) (%RSD)	% of spike detected
Sulfotep	0.50	0.47 (3.5)	94
Atrazine	5.0	4.8 (1.8)	96
Propazine	5.0	5.6 (2.5)	112
Diazinon	0.50	0.52 (8.9)	102
Methylparathion	5.0	5.1 (4.7)	102
Simetryn	5.0	4.7 (1.7)	94
Ametryn	5.0	4.6 (5.0)	92
Prometryn	0.50	0.56 (0.3)	112
Terbutryn	0.50	0.47 (8.2)	94
Malathion	0.50	0.47 (5.8)	94
Parathion	0.50	0.54 (4.8)	108

coefficients (R^2) from 0.998 to 1.000 were obtained. The recoveries relative to amount spiked from fortified carrot juice were from 91% to 106% with the concentrations of 0.5 ppb and 5.0 ppb.

4 Conclusions

The SPME absorption and desorption steps are fully automated with the Varian gas chromatography autosampler. The high precisions achieved when using the fibre vibration technique and lack of labor requirement in absorption and desorption are the main advantages of automated SPME–GC. Fibre vibration mode precludes the introduction of foreign objects into the vials and therefore simplifies the fibre handling and reduces potential for contamination. Linearity is obtained over a wide range of concentrations, and detection limits are in agreement with regulation levels when MS is used as detector. The precision is satisfactory whether equilibrium extraction is reached or not due to precisely controlled extraction. Thus the method is very attractive for analytes with long equilibration times. The automated SPME–GC–MS system can be applied for routine analysis of environmental aqueous sample.

The method of automated SPME with fibre vibration has been applied for the direct analysis of pesticides in fruit juice samples. Experiments with analytes spiked into orange and carrot juices indicated that the juice matrix inhibits the targets from partitioning into the fibre coating. It has been shown also that no interferences were detected when the SPME fibre coating was directly exposed to the juice matrix.

Acknowledgments

This work was supported financially by Supelco Canada Inc., Varian Canada Inc., and the National Sciences and Engineering Research Council of Canada (NSERC).

References

1. T.E. Fielding and W.A. Telliard, United States Environmental Protection Agency, Methods for the determination of nonconventional pesticides in municipal and industrial wastewater, Washington, DC, 1992.
2. W.E. Johnson, N.J. Fendinger and J.R. Plimmer, *Anal. Chem.*, 1991, **63**, 1510.
3. C. Arthur and J. Pawliszyn, *Anal. Chem.*, 1990, **62**, 2145.
4. J. Pawliszyn, *Solid Phase Microextraction: Theory and Practice*, Wiley-VCH, New York, 1997
5. WHO, *Pesticide Residues in Food*, Geneva, 1972.
6. A. Steffen and J. Pawliszyn, *J. Agric. Food. Chem.*, 1996, **44**, 2187.
7. C.L. Arthur, L.M. Killam, K.D. Buchholz, J. Pawliszyn and J.R. Berg, *Anal. Chem.*, 1992, **64**, 1960.
8. I.J. Barnabas, J.R. Dean, I.A. Fowlis and S.P. Owen, *J. Chromatogr. A*, 1995, **705**, 305
9. R. Eisert and J. Pawliszyn, *Crit. Rev. Anal. Chem.*, 1997, **27**, 103.

10. R. Eisert and K. Levsen, *Fresenius' J. Anal. Chem.*, 1995, **351**, 555.
11. Y. Marcus, *Ion Solvation*, Wiley, New York, 1985, p. 306.
12. J.E. Gordon and R.L. Thorne, *J. Phys. Chem.*, 1967, **71**, 4390.
13. A.A. Boyd-Boland, M. Chai, Y.Z. Luo, Z. Zhang, M.J. Yang, J. Pawliszyn and T. Górecki, *Environ. Sci. Technol.*, 1994, **28**, 596A.
14. A.A. Boyd-Boland and J. Pawliszyn, *J. Chromatogr. A*, 1995, **704**, 163.
15. D. D. Perrin and B. Dempsey, *Buffers for pH and Metal Ion Control*, Chapman and Hall, London, 1974.
16. EEC Drinking Water Guideline, 80/779/EEC No.L229/11–29, EEC, Brussels, August 30, 1980.

CHAPTER 33

SPME in the Study of Chemical Communication in Social Wasps

GLORIANO MONETI, G. PIERACCINI, M. SLEDGE AND S. TURILLAZZI

SPME has been used to investigate chemical communication in social wasps. Using the technique to analyse cuticular hydrocarbons (which several studies suggest to be used by wasps to recognize nestmates) and exocrine gland secretions, we demonstrate that the results are comparable with those obtained with the more classical methods that use solvents. By comparing the efficiency of extractions of hydrocarbons and glandular secretions at different temperatures, sampling times, with various fibre types and wasp body parts, we have optimized SPME in analysis of these compounds. As a result of its simplicity this technique is very suitable for research on the chemical ecology of social wasps, and on insect communication in general.

1 Introduction

Insect communication relies on signals that are transmitted in four main channels: acoustic, visual, tactile and chemical. The latter is the most widely used by insects.[1] Although a single chemical signal cannot convey information comparable to that of a visual or acoustic one, it has, generally, several distinct advantages: it has a very low cost of emission, it can be received by a great number of individuals, it can be transmitted in the dark and, in some cases, can last for a considerable length of time. Chemical communication in insects relies on substances produced in numerous exocrine glands and often released by correspondingly well defined structures of the cuticle. These substances defined as semiochemicals (and as pheromones when used intraspecifically) are used by insects for a wide range of purposes, of which one of the most studied is to facilitate encounters between the sexes for mating and reproduction (*e.g.* moths). Chemical signals are also extremely important in social insects (ants, bees, wasps, termites and other minor groups) where they are used for controlling social organization, in the co-ordination of mass activities and for the defence of the colonies (allomones).[2,3]

Research on chemical communication in social insects has increased widely recently, and the secretions of exocrine glands of several species have been characterized both functionally and chemically. Furthermore, it has been demonstrated that various social insects use compounds found on the epicuticle (the outer part of the insect cuticle). These compounds, a mixture of waxes that reduce water loss and form a barrier against pathogens, are used as contact pheromones to discriminate nestmates from non-nestmates.[4,5] This discrimination is vital to avoid invasion of the colony by unrelated individuals.[6] The main constituents of these waxes are usually many long-chained linear or branched hydrocarbons that may be saturated or unsaturated (Figure 1, Table 1).[7-10] Previous research has shown that colonies are well defined closed units and individuals foreign to a particular colony are reacted to aggressively.[2,3] The basis for this differentiation is differences in cuticular signatures between colonies. These signatures have greater similarity among individuals belonging to the same colony. In social wasps this has been demonstrated for various species of the subfamilies Vespinae and Polistinae.[6,11] We have used the SPME technique to address nestmate recognition mechanisms in *Polistes* and wasps of the subfamily Stenogastrinae, as well as the chemistry of Dufour's gland secretion.

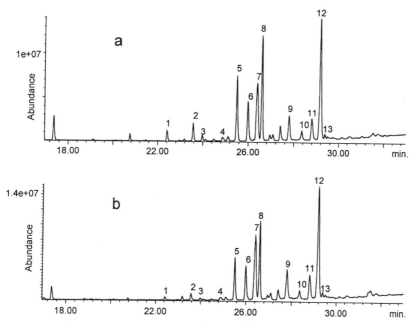

Figure 1   *Comparison of cuticular hydrocarbons from three legs of the social wasp Vespa orientalis extracted in hexane (a) and by SPME (b). Legs were extracted in hexane in an ultrasonic bath for five minutes, while SPME analysis was performed for 10 minutes in the headspace of a sample heated at 170 °C. Numbered peaks correspond to those in Table 1*

Table 1 *Cuticular hydrocarbons in leg cuticule samples of the social wasp Vespa orientalis. Peak numbers refer to identified peaks in Figure 1(a) and (b)*

Peak no.	Compound
1	n-C_{25}
2	3-MeC_{25}
3	n-C_{26}
4	4-MeC_{26}
5	n-C_{27}
6	13- and 11-MeC_{27}
7	11,15-DimeC_{27}
8	3-MeC_{27}
9	12,16-DimeC_{28}
10	n-C_{29}
11	15-, 13- and 11-MeC_{29}
12	11,17-DimeC_{29}
13	3-MeC_{29}

2 Sampling Methods

Classical Methods

Classical methods of analysing both glandular contents and the compounds present on the cuticle of insects involve extraction by solvents.[9,12] Whole glands or body parts are removed and placed in solvents such as hexane, pentane, heptane or dichloromethane. Extracts are then injected into the GC or GC–MS for analysis. This method, though, can be time consuming and results can vary depending on factors such as time of extraction and type of solvent used.[12] Unwanted contaminants, such as internal body lipids, may also be extracted in this procedure. Alternatively, insect glands and body parts can be mounted directly in the GC injector.[13] Despite this method being solventless and thus preferable to the classical method, it is cumbersome to use and involves a modification of the GC inlet. Furthermore, signals are often large as whole glands are injected. This often overloads the GC phase and results in loss of resolution (and may also decrease the life of the column). As there may also be memory effects, the injector port must be cleaned after each injection making the method time consuming.

SPME

Using SPME we have developed methods that solve, in many cases, the shortcomings of these classical methods in insect pheromone analysis. The first involves sampling from the headspace of a heated insect gland or body part with an SPME fibre.[14] This method is both solventless and very simple to perform and compares favourably with the more standard methods (Figure 1). Furthermore, SPME has also allowed for the analysis of chemical compounds released by live insects.[15] The SPME fibre can be rubbed gently on the body of individual wasps, or projected directly into glandular openings. This method proves to be extremely useful as it allows for the tracking of changes in chemical signatures in live (and the same) individuals over time. Alternatively, we have also used SPME in extracting compounds from glands already removed from the insect body by piercing the gland with the fibre.

3 Optimizing SPME Analysis of Insect Chemicals

Fibre Selection

The type of phase and thickness of the fibre play an important role in optimizing the extraction and analysis of insect semiochemicals. The influence of these two parameters is displayed in Figure 2. Using headspace sampling we show that a PDMS fibre phase of thickness 7 μm is more efficient in extracting low volatile hydrocarbons (the compounds of interest). Artefacts (such as long-chain fatty acids) are extracted in lower quantities, while those hydrocarbons of high carbon number are desorbed more efficiently in the GC injector (see inserts in Figure 2). A further advantage of this fibre type (with a bonded phase) is that it can be cleaned in a solvent removing the possibility of memory effects.

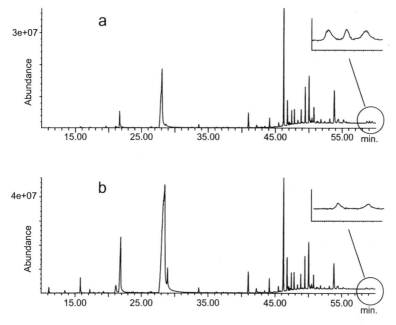

Figure 2 *Headspace extraction of cuticular hydrocarbons of Polistes dominulus by two SPME fibre types. The PDMS 7 μm (bonded) phase (a) is more efficient at extracting long chain hydrocarbons (see inserts), while lower quantities of long-chain fatty acid artefacts (those peaks present up to 35 minutes) are present than in the analysis with a PA 85 μm fibre (b)*

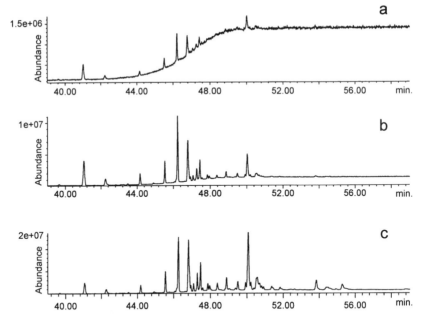

Figure 3 *SPME headspace analysis of cuticular hydrocarbons at three different temperatures:* (a) 110 °C, (b) 130 °C *and* (c) 170 °C. *Extractions were made on two legs of the wasp Polistes dominulus*

Temperature and Time of Sampling

Both the temperature and time of headspace sampling influence compound extraction by SPME. Samples heated at higher temperatures prove to be extracted more successfully (Figure 3). The length of time in which the sample is heated also plays an important role in this regard, especially for the extraction of long chain hydrocarbons (30–35 carbon atoms) (Figure 4, peaks eluting after 50 minutes). For our analyses of cuticlar hydrocarbons we selected a sampling regime, for all wasp species, of 170 °C for 10 minutes. This ensures suitable extraction of all compounds present in samples.

Qualitative and Quantitative Reproducibility

We have demonstrated that use of SPME represents a faithfully reproducible method (both qualitatively and quantitatively) of extraction by comparing extractions of wasp body parts and entire bodies (Figure 5). Solvent extractions often require the use of entire insect bodies to ensure suitable analyses. In contrast, we show that SPME analysis of single legs is as efficient as the entire body (Figure 5), as is using different parts of the same individual.

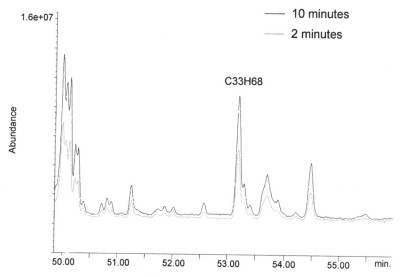

Figure 4   *The influence of extraction time (in the headspace) on the analysis of long-chain (C_{30} to C_{35}) cuticular hydrocarbons: 10 minutes (solid trace) and two minutes (interrupted trace). Extractions were made on two legs of the wasp Polistes dominulus*

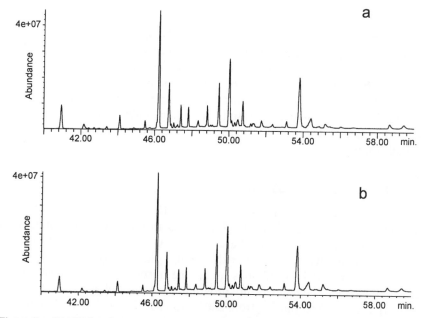

Figure 5   *SPME headspace extractions of (a) entire wasp body, and (b) two legs of an individual of Polistes dominulus. Both samples were extracted for 10 minutes at 170 °C. Peak areas are similar for both analyses because fibre saturation occurs*

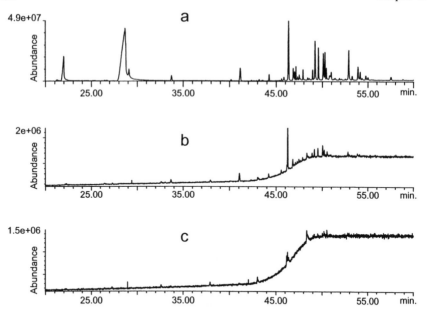

Figure 6   *Removal of residual compounds from SPME fibres:* (a) *SPME headspace extraction of cuticular hydrocarbons from three legs of Polistes dominulus,* (b) *the same fibre re-analysed directly afterwards and* (c) *after cleaning in hexane in an ultrasonic bath for 10 minutes*

Problems

Few problems are encountered in the use of SPME in headspace and live animal sampling. One of the more apparent is a memory effect. This occurs especially in the extraction of long-chained molecules that are not completely desorbed during analysis in the GC. These remaining compounds are easily removed, however, placing the fibre in hexane (in an ultrasonic bath) for 10 or more minutes. Figure 6 compares the results of a cleaned fibre with those of one re-analysed immediately after an actual sample. Cleaning of the fibre removes any residual compounds.

4   Instrument Conditions

Sample analyses were performed using a Hewlett Packard 5890A gas chromatograph coupled to an HP 5971A mass selective detector (using 70 eV electron ionization) and a Varian Saturn 2000 GC–MS. A fused silica capillary column coated with cross-linked methyl-silicone (Restek Rtx-5MS, 30 m × 0.25 mm × 0.5 μm df) was used. The injector port and transfer line were set at 280 °C. The temperature programme was as follows: from 150 °C to 200 °C at a rate of 5 °C min^{-1}, from 200 °C to 260 °C at 2 °C min^{-1}, and from 260 °C to 310 °C at 10 °C min^{-1}. The carrier gas was helium at 1 mL min^{-1}. Analyses were performed in splitless mode. For SPME analyses a narrow-bore Supelco 0.75 mm i.d. GC inlet liner was used.

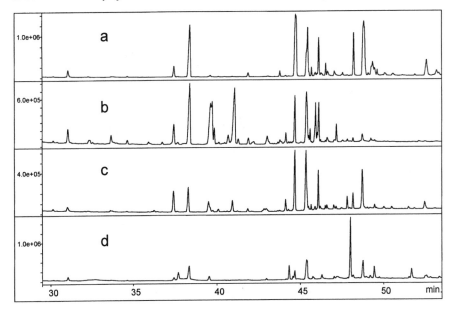

Figure 7 *Cuticular hydrocarbons of four species of social wasp:* (a) *Polistes gallicus,* (b) *Vespula germanica,* (c) *Polybia occidentalis, and* (d) *Liostenogaster vechti. Headspace extractions were performed for 10 minutes at 170 °C on legs of each species*

5 SPME in Social Wasp Communication

Cuticular Compounds

Using SPME headspace sampling, rather than the standard methods of solvent extraction, we have studied the composition of cuticular hydrocarbon mixtures in various species of social wasps. Figure 7 illustrates the differences in cuticular signature between different species. We have also been able to identify differences between nests within a single species using SPME.

The feasibility of SPME use in research on live insects has also been demonstrated,[15,16] allowing for the analysis of dynamics of recognition within insect colonies. We have used this method to assess the role of cuticular compounds in nest usurpation of *Polistes dominulus* colonies by the obligate social parasite *Polistes sulcifer.* Obligate social parasites are not able to found their own nests and do not produce workers. They thus have to usurp nests of a host species and use the resident workers to raise their own offspring. The way in which they do this appears to be based on both behavioural and chemical mechanisms.[17] Using SPME rubbed gently on the thorax of these individuals (for 1.5 min) we have been able to show that the cuticular signature of the parasite changes immediately after nest usurpation. Further analyses (up to 60 days after usurpation) show that the signature remains constant for the rest of the season (Figure 8). By analysing the cuticular signature of the same

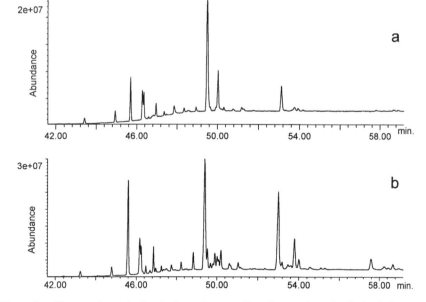

Figure 8 *Changes in cuticular hydrocarbon profile of a single individual of the social parasite wasp Polistes sulcifer: (a) before usurpation of a host nest of Polistes dominulus and (b) 40 days after usurpation. Sampling was performed by rubbing an SPME fibre on the cuticle of the wasp*

individuals over a 3 month period it was possible, using SPME, to track changes in such a dynamic system, confirming the process described by Bagnères *et al.*[18] in another species of *Polistes* social parasite.

Glandular Extraction and Analysis

The Dufour's gland is connected to the reproductive apparatus in female aculeate (with stings) Hymenoptera. In various ants it has been demonstrated to serve as a trail pheromone, where it is used to mark odorous trails by foraging ants to recruit other workers to food sources.[3] In *Polistes* wasps the secretion is very similar to the mixture of hydrocarbons found on the epicuticle[10] and thus may play a role in nestmate recognition.[11] Using SPME headspace analysis we have confirmed the presence of hydrocarbons in the Dufour's glands of *Polistes dominulus*. In addition, we have furthered the use of SPME. The size of the fibre allows for the direct piercing of individual glands dissected out of the wasp body. This procedure produces comparable results to headspace analysis.

6   Conclusions

SPME has proved to be a very useful and practical technique in research on the chemical ecology of social wasps. The results are comparable with the standard solvent methods, and more recent solid injection techniques. The new direct sampling of the cuticle and glands permits a more accurate analysis (in some cases on living insects) avoiding possible contamination due to degeneration of living tissues which is often found in solvent, solid or SPME headspace techniques.

Acknowledgements

We are grateful to Paolo Zanetti, Giancarlo Bucelli, and Francesca R. Dani for their kind assistance. Funding was obtained through the research network 'Social Evolution' of the Universities of Aarhus, Firenze, Keele, Sheffield, Uppsala, Würzburg and the ETH Zürich, financed by the European Commission *via* the Training and Mobility of Researchers (TMR) programme.

References

1. E.O. Wilson, *Sociobiology: the New Synthesis*, Belknap Press of Harvard University Press, Cambridge, MA, 1975.
2. E.O. Wilson, *The Insect Societies*, Belknap Press of Harvard University Press, Cambridge, MA, 1971.
3. B. Hölldobler and E. O. Wilson, *The Ants*, Springer-Verlag, Berlin, MA, 1990.
4. G.J. Gamboa, in *Natural History and the Evolution of Paper Wasps*, ed. S. Turillazzi and M.J. West-Eberhard, Oxford University Press, Oxford, 1996, p. 161.
5. R.W. Howard in *Insect Lipids: Chemistry, Biochemistry and Biology*, ed. D.W. Stanley-Samuelson and D.R. Nelson, University of Nebraska Press, London, 1993, p. 179.
6. M.C. Lorenzi, A.G. Bagnères and J.-L. Clément in *Natural History and Evolution of Paper Wasps*, ed. S. Turillazzi and M.J. West-Eberhard, Oxford University Press, Oxford, 1996.
7. K.H. Lockey, *Comp. Biochem. Physiol.*, 1988, **89B**, 595.
8. M.D. de Renobales, D.R. Nelson and G.J. Blomquist in *Physiology of the Insect Epidermis*, ed. K. Binnington and A. Retnakaran, CSIRO Publications, Australia, 1991, p. 240.
9. M.C. Lorenzi, A.G. Bagnères, J.-L. Clément and S. Turillazzi, *Insectes Soc.*, 1997, **44**, 123.
10. F.R. Dani, E.D. Morgan and S. Turillazzi, *J. Insect Physiol.*, 1996, **42**, 541.
11. F.R. Dani, S. Fratini and S. Turillazzi, *Behav. Ecol. Sociobiol.*, 1996, **38**, 311.
12. M.I. Haverty, B.L. Thorne and L.J. Nelson, *J. Chem. Eco.*, 1996, **22**, 2081.
13. A.G. Bagnères and E.D. Morgan, *J. Chem. Eco.*, 1990, **16**, 3263.
14. G. Moneti, F.R. Dani, G. Pieraccini and S. Turillazzi, *Rapid Commun. Mass Spectrom.*, 1997, **11**, 857.
15. T. Monnin, C. Malosse and C. Peeters, *J. Chem. Ecol.*, 1998, in press.
16. M.F. Sledge and S. Turillazzi, unpublished data.
17. S. Turillazzi, R. Cervo and I. Cavallari, *Ethology*, 1990, **84**, 47.
18. A.G. Bagnères, M.C. Lorenzi, J.-L. Clément, G. Dusticier and S. Turillazzi, *Science*, 1996, **272**, 889.

Pharmaceutical, Clinical and Forensic Applications

CHAPTER 34

Propyl Chloroformate Derivatisation and SPME–GC for Screening of Amines in Urine

METTE KROGH, STIG PEDERSEN-BJERGAARD AND KNUT E. RASMUSSEN

1 Introduction

In forensic toxicology and doping control, urine is routinely screened for drugs containing an amine function. Stimulants such as the amphetamines and the methylenedioxylated derivatives are of particular importance as analytical targets. These drugs are increasingly abused by drug addicts and for recreation by occasional users. The amphetamines are taken in relatively high doses and are to a large extent excreted unchanged into urine.

The chemical structures (Figure 1) show the presence of either a primary or a secondary amino group. Gas chromatography is the preferred screening method owing to the volatility of the compounds. The nitrogen specific detector provides sensitive and selective detection for screening and quantitative analysis. Mass spectrometry is routinely used for final identification. Most laboratories prefer liquid/liquid extraction for sample preparation as these drugs are effectively extracted into an organic solvent from a urine sample which has been made alkaline by addition of potassium hydroxide. An ether such as t-butyl methyl ether is often preferred as solvent as it provides sufficiently high recoveries of the amphetamines and a broad range of other stimulants.[1] Owing to extraction at a high pH and the use of a nitrogen specific detector, the chromatograms are normally very clean with endogenous compounds eluting in the early part of the chromatograms. There are, however, two major drawbacks with the traditional sample preparation methods: when analysing a large number of samples, the high amounts of organic solvent residues generated in the extraction process are a health and safety issue with regard to the workers and the environment, and in addition methods based on liquid/liquid extraction are difficult to automate.

Solid-phase microextraction (SPME) integrates sampling, extraction, concentration, and sample introduction into a single step and offers a solvent-free

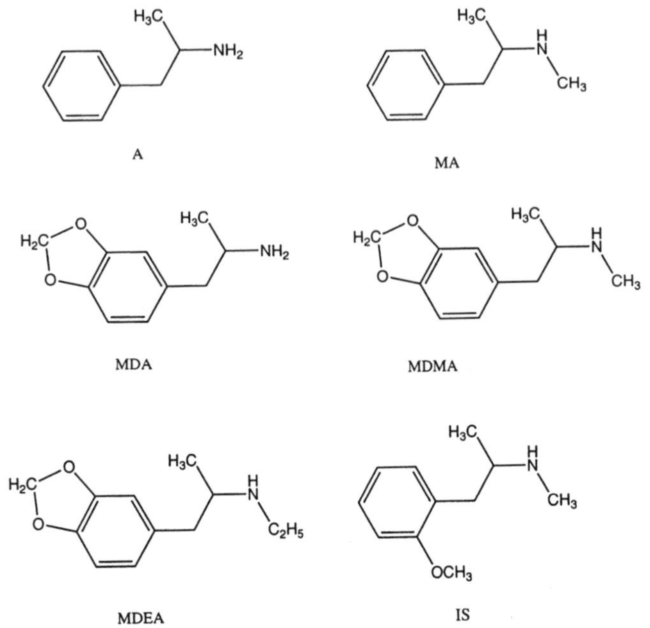

Figure 1 *The chemical structures of the amphetamines, their methylenedioxylated analogues and the internal standard: amphetamine (A), methamphetamine (MA), methylenedioxyamphetamine (MDA), methylenedioxymethamphetamine (MDMA), methylenedioxyethamphetamine (MDEA), methoxyphenamine (IS)*

alternative to the traditional methods. In addition SPME is easy to automate with the commercial autosamplers from Varian.

SPME is an equilibrium extraction method based on partitioning of organic compounds between an aqueous sample and an organic polymer coated on a slender fibre. The amount of an analyte extracted is dependent on its coating/water partition coefficient. Monitoring polar volatiles such as amphetamines by immersing the SPME fibre into complex matrixes such as urine is, however, difficult owing to small partition coefficients, long equilibrium times and interferences from the sample matrix. Headspace SPME methods may minimise these difficulties and method optimisation for the analysis of amphetamine and methamphetamine in urine has recently been published.[2] The present paper presents a validated alternative to the traditional methods for amphetamine analysis and to headspace SPME analysis. The work is based on direct derivatisation of amphetamines and their methylenedioxylated analoges in urine with propyl chloroformate prior to automated SPME–GC.

2 Derivatisation

Many GC and GC–MS methods for amphetamines analysis include derivatisation of the amino-groups after extraction in order to reduce analyte volatility and to improve the chromatography of the compounds. The reactions are carried out in organic media with reagents such as trichloroacetic anhydride, trifluoroacetic anhydride, pentafluoropropionic anhydride and heptafluorobutyric anhydride.[3–7]

In direct immersion SPME of the amphetamines and their methylenedioxylated analogues in urine, low coating/water partition coefficients are observed owing to the polar nature of the amino groups. Derivatisation of these groups into less polar analogues prior to SPME will thus increase the coating/water partition coefficient, improve the SPME efficiency and improve method sensitivity. Few papers have, however, been published on the derivatisation of amines in aqueous matrices.[8–12] These papers focus on chloroformates as derivatisation reagents. The chloroformates react with primary and secondary amines in aqueous alkaline matrices and convert the amines into their corresponding carbamates. Derivatisation of amphetamine and methamphetamine with propyl chloroformate was first reported in organic media.[13] In this work the urine was made alkaline (pH 10.8) and an extraction–derivatisation solution was added, containing hexane, chloroform and propyl chloroformate. After extraction–derivatisation and phase separation, the volume of the organic phase was reduced and an aliquot was analysed by GC–MS.

The derivatisation of amphetamines and their methylenedioxylated analogues directly in urine with propyl and butyl chloroformate prior to SPME–GC has recently been investigated.[14,15] These publications showed that amphetamine (A), metamphetamine (MA), methylenedioxyamphetamine (MDA), methylenedioxymethamphetamine (MDMA) and methylenedioxyethylamphetamine (MDEA) were completely derivatised with propyl chloroformate as no traces of underivatised analytes could be detected by either GC–NPD or GC–MS. Butyl chloroformate completly derivatised A, MA and MDMA but the derivatisation of MDA and MDEA was found to be incomplete. Therefore, propyl chloroformate should be the preferred derivatisation reagent for the amphetamines and their methylenedioxylated analogues. After derivatisation excess chloroformate is hydrolysed, with no traces seen in the chromatograms.

Based on these studies the following derivatisation procedure is recommended:[15]

An aliquot of urine (1200 μL) is placed in a 2 mL GC autosampler vial. Methoxyphenamine is added as internal standard at a final concentration of 5 μg mL^{-1} in urine. Thereafter 300 μL of a solution consisting of 2.5 M K_2CO_3–$KHCO_3$ buffer (pH 10.8) and 0.5 g NaCl is added to effectively control the pH and to enhance the partitioning onto the SPME coating. The mixture is agitated, 8 μL of propyl chloroformate reagent is added and the sample is vortexed for 10 s. The vial is then loaded into the GC autosampler for automated SPME and injection.

Figure 2 shows the reaction of MDMA with propyl chloroformate and the hydrolysis of the derivatisation reagent. The reaction of the analytes with propyl chloroformate is complete within 1 min and the sample vials can be placed in the GC autosampler immediately after the reagent has been mixed with the urine sample. The propyl chloroformate derivatives are stable in the matrix as no

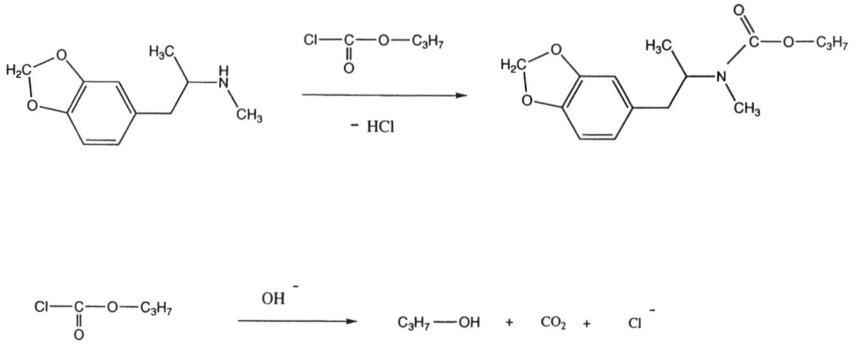

Figure 2 *Reaction of MDMA with propyl chloroformate and hydrolysis of propyl chloroformate*

change in quantitative results have been detected between samples analysed immediately and samples analysed after 24 h of storage in the GC autosampler. Excess reagent is hydrolysed into propanol. The small concentrations of propanol formed when 8 μL of propyl chloroformate was used for derivatisation did not affect the SPME recovery of derivatised amphetamines.

3 SPME and GC

Propyl chloroformate derivatisation converts the analytes into less polar compounds and the nonpolar PDMS (polydimethylsiloxane) fibre coating was found to be most efficient as well as robust in the extraction of the derivatised drugs. The more polar of the commercially available SPME fibre coatings provided lower extraction efficiencies than PDMS. The addition of sodium chloride to the sample matrix was found to further enhance the partitioning of the analytes onto the PDMS coating.[14] When the fibre was automatically agitated during sorption with the Varian autosampler the partitioning reached equilibrium after 16 min. With a sorption time of 16 min, 85 samples can be automatically analysed in 24 h.

A decrease in reproducibility and enrichment due to contamination of the fibres has been observed when the fibres were used to extract more than 150 urine samples. In order to maintain reproducibility and enrichment, it is recommended that the SPME fibres be replaced after 100 urine samples. The life-time of the fibres can be enhanced by immersion of the fibre in pure water when not in use and by placing water samples in between the urine samples to be analysed.

Figure 3 shows typical chromatograms of a urine blank and a urine spiked with A, MA, MDA, MDMA, MDEA and the internal standard methoxy-phenamine after propyl chloroformate derivatisation and SPME. The analytes are desorbed for 1 min at an injector temperature of 300 °C, separated on a SPB-1 (Supelco) polydimethylsiloxane capillary column (30 m × 0.25 min i.d.,

A

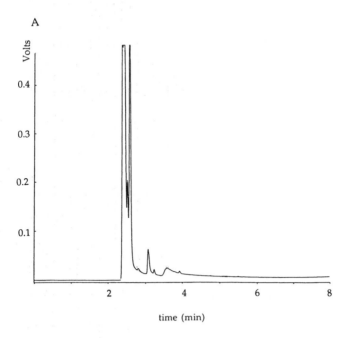

B

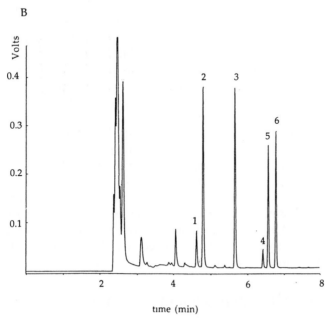

Figure 3 *Chromatogram of* (A) *a drug-free urine sample and* (B) *a urine sample spiked with 4 μg mL⁻¹ of A, MA, MDA, MDMA, and MDEA after derivatisation with propyl chloroformate: peaks: 1 = A, 2 = MA, 3 = IS, 4 = MDA, 5 = MDMA and 6 = MDEA*

0.25 μm film thickness), and detected with a nitrogen-specific detector. Chromatographic separation was achieved by temperature programming. The column temperature was held for 1 min at 180 °C during desorption and then increased at 20 °C min^{-1} to 300 °C and held for 1 min. The analytes are separated within 8 minutes. The matrix components are eluted early in the chromatogram and the formation of the less volatile propyl chloroformate derivatives increased analyte retention sufficiently to avoid coelution with matrix components. Owing to the high volatility of amphetamine and methamphetamine, interferences from volatile matrix components can be encountered if these drugs are analysed without derivatisation.

4   Confirmation by Mass Spectrometry

For the unequivocal identification of the alkyl chloroformate derivatives, capillary gas-chromatography coupled with mass spectrometry (GC–MS) is recommended. Thus, with alkyl chloroformate derivatives, GC–MS has been used in combination with SPME[14] and with liquid/liquid extraction[13,16] for the determination of both A, MA and MDMA present in human urine. The fragmentation patterns of the propyl chloroformate derivatives of A, MA, MDA, MDMA, MDEA and the internal standard are illustrated in Figure 4. Cleavage at the C_1–C_2 bond is a dominating process, with both the C_1- and C_2-containing fragments as structurally important and high abundant ions. Gain of a hydrogen atom and further fragmentation of the C_2-containing ions at the N–COO bond is also a dominating process resulting in signals at m/z 44, 58 or 72 for amphetamine, methylamphetamine and ethylamphetamine-like structures respectively. In addition, gain of a hydrogen atom and fragmentation of the C_2-containing ions at the COO–CH_2 bond is an important process for the tertiary derivatives, providing fragments at m/z 102 or 116 for the methyl- and ethyl-amphetamine-like structures. Owing to the low abundance of molecular ions for the propyl chloroformate derivatives discussed above, misidentification may in some cases occur from structurally related compounds. Thus, propyl chloroformate derivatives of the sympathomimetic amines ephedrine and pseudoephedrine may be identified as the propyl chloroformate derivative of methamphetamine owing to similar fragmentation under electron ionisation conditions.[16] However, during chemical ionisation utilising methane as the reagent gas, distinct protonated molecular ions arise from the propyl chloroformate derivatives of amphetamine-like compounds, which may eliminate the possibility of misinterpretation.[16]

5   Quantitative Analysis and Validation

Only 1.4–9.8% of the initial amount of the amphetamines and their methylenedioxylated analogues were extracted by the PDMS coated fibre after derivatisation. The carbamate derivatives of the primary amines (A and MDA) showed lower extraction recoveries than the carbamate derivatives of the secondary amines (MA, MDMA and MDEA). The highest extraction recovery was found

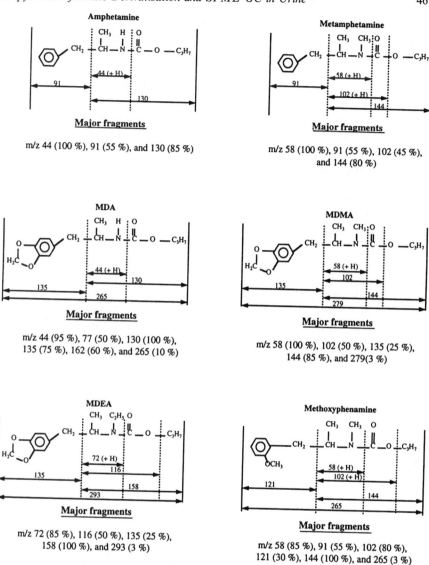

Figure 4 *Fragmentation patterns of propyl chloroformate derivatives of A, MA, MDA, MDMA, MDEA, and IS*

for MDEA. In order to correct for the variations in extraction recovery quantitative analyses were based on peak area ratios relative to an internal standard added to the urine sample prior to derivatisation and extraction. Methoxyphenamine is a suitable internal standard as it has similar chemical properties as the analytes. It is derivatised by propyl chloroformate and has the same sorption profile as the analytes. In addition it has a retention time between those of MA and MDA.

SPME is sensitive towards matrix variations and changes in salt concentration. pH may also affect analyte recovery and reproducibility.[17] Urine is a highly variable matrix as diet and liquid intake cause variations in urine ionic strength and pH. In order to overcome the difficulty with a variable urine matrix, the urine was mixed with a high ionic strength buffer and salt prior to SPME, to eliminate matrix differences.

The calibration graphs are linear in the concentration range 0.1–10 μg mL^{-1} of A, MA, MDA, MDMA, and MDEA, with correlation coefficients $r = 0.9998$ or better after derivatisation with propyl chloroformate.[15] The limit of detection in urine at a signal-to-noise ratio of 3 ($S/N = 3$) is 5 ng mL^{-1} for MA, MDMA, and MDEA, and 15 ng mL^{-1} for A and MDA, respectively. The detection limits are equivalent or lower than existing methods based on GC and GC–MS[13,16] and are sufficient for use in forensic toxicology.

In the validation of the method, testing of accuracy and precision in different urine matrices was included. In the validation procedure 15 urine samples were collected from eight individuals. These represented urine collected in the morning and throughout the day. The pH of these urine samples was in the range 5.6–7.1 and the concentrations of creatinine were in the range 1.6–18.0 mM, thus representing very low and very high urine salt concentrations. The urine samples were spiked with the target analytes and Figure 5 shows the results after spiking MDMA at a concentration of 1 μg mL^{-1}. A mean value of 0.99 μg mL^{-1} was found, and no correlation between the quantitative results, and the creatinine values, and the pH of the urine samples was observed. The intra-assay relative standard deviations in the concentration range 0.1–10 μg mL^{-1} of A, MA, MDA, MDMA and MDEA in urine were between 1.9 and 9.9% ($n = 15$). Based on these facts the method is found to be highly reproducible and robust towards natural variations in the sample matrix.

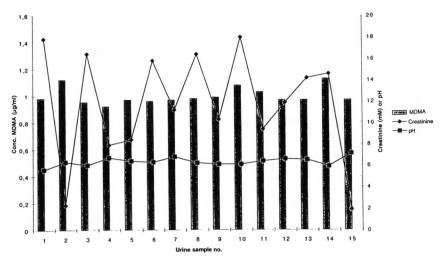

Figure 5   *Concentration of MDMA (μg mL^{-1}), concentrations of creatinine (mM) and pH in 15 urine samples spiked with 1 μg mL^{-1} MDMA*

6 Conclusion

Polar analytes such as the amphetamines and their methylenedioxylated analogues can be successfully derivatised in an alkaline urine by propyl chloroformate. The derivatisation converts the analytes into less volatile and more hydrophobic carbamates which are effectively concentrated on an immersed polydimethylsiloxane SPME fibre. Matrix variations are eliminated by adding a high ionic strength buffer and salt to the urine sample prior to SPME. The procedure can be automated by the Varian GC autosampler and 85 samples can be automatically analysed in 24 h. The method has sufficient sensitivity to be used in forensic toxicology and is a solvent free alternative to traditional methods.

References

1. P. Hemmersbach and R. de la Torre, *J. Chromatogr. B*, 1996, **687**, 221.
2. H.L. Lord and J. Pawliszyn, *Anal. Chem.*, 1997, **69**, 3899.
3. C.L. Hornbeck and R.J. Czarny, *J. Anal. Toxicol.*, 1989, **13**, 114.
4. R.J. Czarny and C.L. Hornbeck, *J. Anal. Toxicol.*, 1989, **13**, 257.
5. J.B. Jones and L.D. Mell, *J. Anal. Toxicol.*, 1993, **17**, 447.
6. R.W. Taylor, S.D. Le, S. Philip and N.C. Jain, *J. Anal. Toxicol.*, 1989, **13**, 293.
7. H. Gjerde, I. Hasvold, C. Pettersen and A.S. Christophersen, *J. Anal. Toxicol.*, 1993, **17**, 65.
8. N.O. Ahnfelt and P. Hartvig, *Acta Pharm. Suec.*, 1980, **17**, 307.
9. O. Gyllenhaal, L. Johansson and J. Vessman, *J. Chromatogr.*, 1980, **190**, 347.
10. A.P.J.M. De Jong and C.A. Cramers, *J. Chromatogr.*, 1983, **276**, 267.
11. M. Ahnoff, S. Chen, A. Green and I. Grundevik, *J. Chromatogr.*, 1990, **506**, 593.
12. P. Husek, *J. Chromatogr. B*, 1998, **717**, 57
13. R. Meatherall, *J. Anal. Toxicol.*, 1995, **19**, 316.
14. H. Grefslie Ugland, M. Krogh and K.E. Rasmussen, *J. Chromatogr. B*, 1997, **701**, 29.
15. H.G. Ugland, M. Krogh and K.E. Rasmussen, *J. Biomed. Pharm. Anal.*, accepted.
16. A. Dasgupta and A.P. Hart, *J. Forensic Sci.*, 1997, **42**, 106.
17. L. Pan and J. Pawliszyn, *Anal. Chem.*, 1997, **69**, 196.

CHAPTER 35

Isolation of Drugs and Poisons in Biological Fluids by SPME

TAKESHI KUMAZAWA, XIAO-PEN LEE, KEIZO SATO
AND OSAMU SUZUKI

1 Introduction

Interest in the field of clinical and forensic toxicology is being focused on improving methodologies, with regard to how rapidly, accurately and sensitively the chemicals can be detected. This field is highly dependent on the development of new analytical instruments or techniques. In early work SPME was considered suitable primarily for extraction of volatiles.[1-4] In 1995, however, we reported the headspace SPME for forensic analysis of tricyclic antidepressants[5] in human urine and local anaesthetics[6] in human whole blood. Our studies opened the applicability of SPME to a number of other drugs and xenobiotics of intermediate molecular weights. This chapter demonstrates that drugs and other chemicals of clinical and forensic interest can be extracted from human body fluid samples by SPME.

2 Strategies for SPME Conditions

A detailed procedure should be optimized for each compound to be analyzed, and for each matrix to be sampled (plasma, serum, urine or whole blood) experimentally. The choice of headspace or direct immersion (DI) for SPME is the first step in this method. The extent of sorption to an SPME fiber of a compound to be analyzed is always an indicator for such choice. Preliminary experiments in aqueous matrix are convenient to address this question.

The selection of an SPME fiber can be also made initially with aqueous spikes. For more polar compounds, the polydimethylsiloxane (PDMS)/divinylbenzene (DVB) or polyacrylate fiber should be tested for sorption.

At the next step, the composition of the contents in the vial should be optimized. The pH and salt additives are important parameters for achieving better efficiency of extraction. For plasma, serum and urine, such optimization experiments are not difficult; the experiments include dilution of each sample with water, addition of salt(s) [NaCl, $(NH_4)_2SO_4$, $NaHCO_3$ or K_2CO_3] and pH adjustment. The extraction temperature and the time of exposure of the fiber should be optimized with plasma, serum or urine samples.

For whole blood, the conditions can be quite different, because clot formation usually takes place during heating and stops the stirring. With addition of salts, the clot formation becomes easier for extraction or easier clot formation which would be worse for extraction. To solve these problems, deproteinization pretreatment should be evaluated. This is more suitable for basic compounds. To a blood sample (after dilution with water if necessary), a strong acid such as perchloric, trichloroacetic, hydrochloric or sulfuric acid, is added, followed by centrifugation; the resulting clear supernatant is neutralized or alkalinized, and placed in a vial for SPME. If the deproteinization process is not beneficial for recovery, the addition of NaOH solution only should be tested, in order to avoid tedious and unnecessary sample preparation steps. The strong alkalinization results in hemolysis and prevents the formation of a blood clot. After this procedure, headspace SPME is more suitable, because strong alkaline solution may cause damage to fiber coatings the DI-SPME mode.

In our laboratories, after SPME, the detection is by capillary GC. We use FID, nitrogen-phosphorus detection (NPD), electron capture detection (ECD) and surface ionization detection (SID),[7,8] according to the compounds to be tested. For GC analysis, the temperature of the GC injection port for analyte desorption and desorption time should be optimized.

3 Applications of SPME

Drugs

Tricyclic Antidepressants

As an initial study of the application of SPME to biological samples, we tested the headspace SPME for the tricyclic antidepressants amitriptyline, imipramine, trimipramine and chlorimipramine in human urine.[5] The four drugs were added to a 7.5 mL vial containing 1 mL of urine and 50 μL of 5 M NaOH solution. After heating the vial at 100 °C for 30 min, a PDMS fiber (100 μm film thickness) was exposed to the headspace of the vial at 100 °C for 15 min prior to capillary GC-FID. The method produced satisfactory linearity in the range of 50–2000 ng mL^{-1}. The detection limits for these drugs were in the range of 24–38 ng mL^{-1}.

In the case of human whole blood, the SPME procedure was essentially the same as that for urine except that distilled water was added to the sample.[9] The calibration curves for four drugs showed excellent linearity in the range of 31–1000 ng 0.5 mL^{-1}, with r-values of 0.9992–0.9998. The detection limits of amitriptyline, imipramine and trimipramine were 16 ng 0.5 mL^{-1}; that of chlorimipramine was 25 ng 0.5 mL^{-1}.

Therapeutic plasma concentrations of tricyclic antidepressnats were reported to be less than 0.24 μg mL^{-1}, and toxic effects occur with blood concentrations greater than approximately 0.3 μg mL^{-1}.[10] At toxic levels (500 mg/kg body weight), imipramine could actually be detected from rat blood by GC–FID after headspace SPME.[9] We attempted to measure therapeutic concentrations of imipramine in blood obtained 4 h after oral administration of 50 mg of imipramine-HCl to a 31-year-old male volunteer weighing 74 kg, using the headspace SPME method and GC–FID. As a result, it was difficult to detect the drug by GC–FID, but a sharp peak of imipramine could be detected by GC–SID; with a resulting concentration of 23.9 ng mL^{-1} in whole blood.[9]

Ulrich and Martens have also reported that tricyclic antidepressants can be extracted and detected from human plasma by DI-SPME and GC–NPD.[11]

Phenothiazines

Chlorpromazine was introduced many years ago, but is still one of the most widely used drugs for treatment of schizophrenia. Until recently, many phenothiazine derivatives have been synthesized and used as antipsychotics (major tranquillizers), anti-parkinsonism drugs and also antihistaminics. These drugs are frequently encountered in clinical and forensic toxicology because of their relatively narrow safe therapeutic dose ranges.

Seno *et al.*[12] reported headspace SPME of promazine, chlorpromazine, triflupromazine, trimeprazine and methotrimeprazine from human whole blood and urine using a PDMS fiber (10 μm film thickness), followed by GC–FID. The calibration curves for the drugs extracted by headspace SPME from whole blood were linear in the range of 0.5–5 μg mL^{-1}. The detection limits of five drugs were about 100–200 ng mL^{-1} and 10–20 ng mL^{-1} for whole blood and urine, respectively.

Benzodiazepines

Benzodiazepines generate a significant proportion of sales in pharmaceutical markets in the world and are widely used as sedatives and hypnotics. The drugs cause intoxication due to accidental overdosage or intentional abuse, and are therefore very frequently encountered in forensic science practice.[13]

The drugs we tested were medazepam, fludiazepam, diazepam, midazolam, flunitrazepam, prazepam, nimetazepam, flurazepam, estazolam, alprazolam, etizolam, triazolam and brotizolam.[14] Spiked samples were analyzed by DI-SPME with a PDMS/DVB fiber (65 μm film thickness), with addition of 0.5 g of NaCl to a 1 mL urine sample. Although estazolam, alprazolam, etizolam, triazolam and brotizolam could not be extracted, other drugs were detected and were well resolved by GC-FID with an Rtx-5 Amine capillary column (30 m × 0.32 mm i.d., film thickness 1.0 μm, Restek Corp.). The calibration curves for medazepam, fludiazepam, diazepam, midazolam, fluni-trazepam and prazepam showed linearity in the range of 40–1000 ng mL^{-1}; those for nimetazepam and flurazepam were linear in the range of 200–1000 ng mL^{-1} and 100–1000 ng mL^{-1}, respectively. The detection limits were 10–20 ng mL^{-1} for the eight compounds other than nimetazepam (150 ng mL^{-1}).

Local Anaesthetics

Local anaesthetics are occasionally encountered in forensic science practice, especially in medical accidents. We reported that lidocaine, prilocaine, procaine, mepivacaine, tetracaine, bupivacaine, dibucaine, benoxinate, ethyl amino-benzoate and *p*-(butylamino)benzoic acid-2-(dithylamino)ethyl ester could be extracted from human whole blood by headspace SPME[6] and DI-SPME.[15]

To a 1 mL sample of human whole blood containing ten local anaesthetics, 1 mL of 1 M perchloric acid solution was added for deproteinization. After centrifugation, the clear supernatant was decanted into a 7.5 mL vial containing a magnetic stirring bar. The supernatant was mixed with 100 μL of 10 M NaOH solution and 1g of $(NH_4)_2SO_4$. We had tested addition of various salts, such as NaCl, K_2CO_3 and $(NH_4)_2SO_4$, to the alkaline solution; $(NH_4)_2SO_4$ gave the best results in partitioning rates for all drugs. After heating the vial at 100 °C for 15 min, a PDMS fiber (100 μm film thickness) was exposed in the headspace of the vial at 100 °C for 40 min. The drugs were determined by GC–FID with a DB-1 fused-silica capillary column (30 m × 0.32 mm i.d., film thickness 0.25 μm, J & W Scientific). They showed satisfactory resolution from each other and from impurities on the gas chromatograms (Figure 1C). The introduction rate (absolute recovery) of the drugs into GC instrument from samples was 0.37–11% with the lowest for procaine and the highest for lidocaine. Such low recovery percentages are common for SPME which is an equilibrium rather than exhaustive extraction method. These recovery percentages are not due to loss of the drugs during deproteinization with perchloric acid; more than 90% of the drugs could be recovered in the supernatant fraction after the protein precipitation. The calibration curves for ten drugs, with the exception of procaine, showed linearity in the range of 0.5–12 μg mL^{-1}, with *r*-values of 0.993–0.999. The detection limits of tetracaine, lidocaine, bupivacaine, *p*-(butylamino)benzoic acid-2-(dithylamino)ethyl ester and prilocaine were 58–250 ng mL^{-1}; those of dibucaine, mepivacaine, benoxinate and ethyl amino-benzoate 255–830 ng mL^{-1}; and that of procaine >2500 ng mL^{-1}.

For DI-SPME, the above supernatant was mixed with 0.5 g of NaCl and adjusted to about pH 7 with NaOH solution. The PDMS fiber was immersed directly into the sample solution of the vial at room temperature with stirring. All drugs were extracted and detected by SPME–GC (Figure 1E), and their absolute recoveries were 0.74–19.7% with the lowest for procaine and the highest for tetracaine. The *p*-(butylamino)benzoic acid-2-(dithylamino)ethyl ester, dibucaine, tetracaine and benoxinate extracted with DI-SPME were 3–6 times higher than those with the headspace SPME.

Meperidine

Meperidine is a narcotic analgesic drug, which is widely used in therapeutic practice. The therapeutic concentration range of meperidine in human plasma is reported to be 0.2–0.8 μg mL^{-1}, and toxic effects are usually associated with blood concentrations greater than 2 μg mL^{-1}.[10]

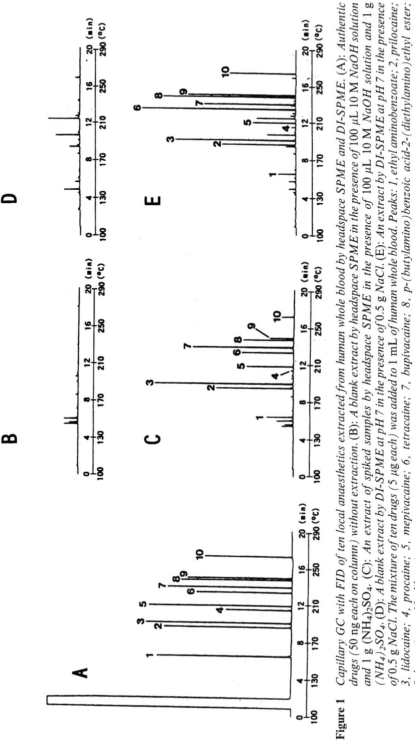

Figure 1 *Capillary GC with FID of ten local anaesthetics extracted from human whole blood by headspace SPME and DI-SPME. (A): Authentic drugs (50 ng each on column) without extraction. (B): A blank extract by headspace SPME in the presence of 100 μL 10 M NaOH solution and 1 g (NH₄)₂SO₄. (C): An extract of spiked samples by headspace SPME in the presence of 100 μL 10 M NaOH solution and 1 g (NH₄)₂SO₄. (D): A blank extract by DI-SPME at pH 7 in the presence of 0.5 g NaCl. (E): An extract by DI-SPME at pH 7 in the presence of 0.5 g NaCl. The mixture of ten drugs (5 μg each) was added to 1 mL of human whole blood. Peaks: 1, ethyl aminobenzoate; 2, prilocaine; 3, lidocaine; 4, procaine; 5, mepivacaine; 6, tetracaine; 7, bupivacaine; 8, p-(butylamino)benzoic acid-2-(diethylamino)ethyl ester; 9, benoxinate; 10, dibucaine*

Meperidine, together with diphenylpyraline as internal standard (IS), was added to whole blood and urine and extracted by headspace SPME.[16] Clear supernatant of whole blood after deproteinization with perchloric acid, and untreated urine, containing meperidine and IS, were heated at 100 °C in the presence of NaOH and NaCl in a vial. A PDMS fiber (100 μm film thickness) was then exposed in the headspace of the vial at the same temperature prior to capillary GC–FID. NaCl was effective for increasing efficiency of extraction for meperidine and IS in both samples; the amounts of meperidine and IS extracted were approximately 4.5 and 8 times better, respectively, than those without NaCl. The method gave linear calibration in the range of 0.125–1 μg mL^{-1} for whole blood, with an *r*-value of 0.993.

Lord and Pawliszyn have also mentioned recently that meperidine can be extracted from human urine by headspace SPME and DI-SPME.[17]

Cocaine

Cocaine is a naturally occurring stimulant alkaloid derived from the leaves of *Erythroxylon coca*, a tree or shrub indigenous to western South America. Cocaine may be taken intranasally, injected intravenously, or smoked as cocaine 'crack', and is one of the most frequently abused drugs.

Cocaine and cocapropylene as IS were added to human urine samples and extracted by DI-SPME with a PDMS fiber (100 μm film thickness) prior to capillary GC–NPD.[18] Both compounds were detected with high sensitivity and very low background noise levels by gas chromatography. The calibration curve for cocaine showed excellent linearity in the range of 30–250 ng 0.5 mL^{-1}, with an *r*-value of 0.9993. The detection limit was 6 ng 0.5 mL^{-1} urine.

Phencyclidine

Phencyclidine (PCP) was developed in the 1950s for use as an anaesthetic or analgesic agent. Its use has since been discontinued because of its hallucinogenic effects. It has been widely abused since the early 1970s and is known by street names, such as angel dust, elephant tranquilizer, killer weed and rocket fuel.

PCP has been determined in human whole blood and urine with relatively high sensitivity by headspace SPME with a PDMS fiber (100 μm film thickness) and GC–SID;[19] the drug contains a tertiary amino group and is very suitable for SID. Excellent linearity was observed in the range of 2.5–100 ng mL^{-1} for whole blood and 0.5–100 ng mL^{-1} for urine. The detection limits for whole blood and urine were 1.0 and 0.25 ng mL^{-1}, respectively.

Methamphetamine and Amphetamine

Methamphetamine and amphetamine are sympathomimetic amines producing central nervous system stimulation and mood elevation, and are significant drugs of abuse worldwide. Yashiki and colleagues have applied SPME for extraction of methamphetamine and amphetamine from human urine[20] and whole blood.[21] They have exclusively used headspace SPME with a PDMS fiber in spite of low absolute recoveries of these stimulants.

Ishii *et al.*[22] have recently obtained good results by DI-SPME with a PDMS/DVB fiber (65 μm film thickness) and detection by GC–NPD, for the presence of the stimulants in human urine samples. The extent of sorption to the fiber of methamphetamine and amphetamine from urine was several times higher than those by the headspace SPME method. The calibration curves for methamphetamine and amphetamine showed good linearity in the range of 10–161 and 9.18–147 ng mL^{-1} for urine, respectively. The detection limits were about 8 ng mL^{-1} for both stimulants.

Lord and Pawliszyn have also reported on method optimization of methamphetamine and amphetamine extraction from human urine by headspace SPME.[17]

1-Phenylethylamine

1-Phenylethylamine (1-PEA, α-phenylethylamine or α-methylbenzylamine) is a two-carbon homologue of amphetamine. 1-PEA is a CNS stimulant similar to the amphetamines, and is a possible drug of abuse.[23]

1-PEA, together with β-methylphenylethylamine as IS, were added to human urine samples and extracted by headspace SPME with a PDMS/DVB fiber (65 μm film thickness) prior to capillary GC–NPD.[24] The method produced excellent linearity of calibration in the range of 25–500 ng mL^{-1} for urine, with a detection limit of 20 ng mL^{-1}.

Diphenylmethane Antihistaminics and Their Analogues

Diphenylmethane antihistaminics are one of the most commonly used drug groups for the treatment of colds, asthma and other allergic diseases. They are freely available as over-the-counter medications at drug stores. Fatal cases involving their ingestion have been reported.[25]

Nishikawa *et al.*[26] reported headspace SPME of thirteen antihistaminic drugs and their analogues, *viz.* diphenhydramine, doxylamine, orphenadrine, terodiline, chlorpheniramine, carbinoxamine, diphenylpyraline, triprolidine, benactyzine, homochlorcylizine, cloperastine, clemastine and piperilate, from human whole blood and urine samples. The drugs were extracted with a PDMS fiber (100 μm film thickness) after alkalinization of both samples with NaOH, prior to capillary GC–FID. The drugs could be determined from whole blood samples with the exception of carbinoxamine, triprolidine, benactyzine and piperilate. The detection limits for diphenhydramine, orphenadrine and terodiline were 76–136 ng mL^{-1} whole blood, and that for diphenylpyraline was 473 ng mL^{-1}. Benactyzine and piperilate were not extracted from urine samples by headspace SPME. The other drugs were determined with detection limits of 13–186 ng mL^{-1} with the lowest for terodiline and the highest for doxylamine.

Pesticides and Disinfectants

Organophosphate Pesticides

Organophosphates are potent cholinesterase enzyme inhibitors and are widely used as insecticides and herbicides. Accidental or suicidal cases due to organophosphate exposure or intake are sometimes encountered in clinical toxicology and forensic science practice.[27–29]

We reported that organophosphate pesticides could be extracted and detected from human whole blood and urine samples by headspace SPME with a PDMS fiber (100 μm film thickness) and capillary GC–NPD.[30] The pesticides tested were fenitrothion, fenthion, methyl parathion, isoxathion, phosalone, ethion, malathion, ethyl p-nitrophenyl benzenethiophosphonate (EPN) and S-benzyl O,O-di-isopropyl phosphorothioate (IBP). The addition of 0.5 mL of distilled water, 100 μL of 6 M HCl and 0.4 g of NaCl to whole blood and urine samples (0.5 mL each) produced the highest efficiency of extraction for IBP, methylparathion, malathion, fenitrothion and fenthion while 0.5 mL of distilled water, 100 μL of 6 M HCl and 0.4 g of $(NH_4)_2SO_4$ gave the highest efficiency of extraction for isoxathion, ethion, EPN and phosalone. For simultaneous determination of the nine pesticides in urine, therefore, both NaCl and $(NH_4)_2SO_4$ were added to the acidified samples. The efficiency of extraction for the pesticides from whole blood in the presence of distilled water, HCl solution and both salts were, however, lower than those after acidification only, with the exception of malathion. The calibration curves for the pesticides showed excellent linearity in the range of 50–400 ng 0.5 mL^{-1} for whole blood other than malathion (100–400 ng 0.5 mL^{-1}), and 7.5–120 ng 0.5 mL^{-1} for urine, with the exception of phosalone (15–120 ng 0.5 mL^{-1}). The detection limits under conditions optimal for each of the pesticides extracted individually from whole blood were 2.2–40 ng 0.5 mL^{-1} for ethion, fention, isoxathion, fenitrothion, IBP, malathion, methyl parathion and EPN, and 200 ng 0.5 mL^{-1} for phosalone. For urine, the detection limits were 0.8–4.4 ng 0.5 mL^{-1}, with the exception of phosalone (12 ng 0.5 mL^{-1}).

Carbamate Pesticides

Carbamates are reversible inhibitors of cholinesterase and are widely used as pesticides. The drugs are occasionally encountered in forensic toxicology.[31]

Seno *et al.*[32] reported headspace SPME with a PDMS fiber (100 μm film thickness) for xylylcarb, 3,5-xylyl methylcarbamate (XMC), isoprocarb, fenobucarb, propoxur and carbofuran from human whole blood and urine samples. The headspace SPME–GC–FID gave intense peaks for each compound with low background noise. The calibration curves for the six pesticides extracted from whole blood were linear in the range of 0.4–10 μg mL^{-1} for isoprocarb and fenobucarb, and 2–10 μg mL^{-1} for XMC, xylylcarb, propoxur and carbofuran. The detection limits were 100–500 ng mL^{-1} for whole blood, and 10–50 ng mL^{-1} for urine.

Phenol and Cresols

Phenol and cresol are used commonly as disinfectants. A mixture of *o*-cresol, *m*-cresol, *p*-cresol and phenol as a minor constituent is commercially available. Suicidal cases due to their poisoning are occasionally encountered.[33]

We tested headspace SPME of *o*-cresol, *m*-cresol, *p*-cresol and phenol from human whole blood with a polyacrylate fiber (85 μm film thickness).[34] To a 7.5 mL vial containing a magnetic stirring bar were added 0.5 mL whole blood, the four compounds and 2,4-dimethylphenol as IS, 1.5 mL of distilled water and 0.4 g of NaCl. The gas chromatograms by GC–FID for the extracts of whole blood and urine, with use of an α-DEX 120 capillary column (30 m $\times$ 0.25 mm i.d., 0.25 μm film thickness, Supelco Inc.), showed satisfactory separation of the test peaks from each other and from impurities. The calibration curves for all compounds showed linearity in the range of 200–1000 ng 0.5 mL^{-1}, with *r*-values of 0.997–0.999. The detection limits of four compounds were 70–100 ng 0.5 mL^{-1} for whole blood.

In the field of environmental chemistry, Buchholz and Pawliszyn have reported DI-SPME for rapid screening of 11 phenolics from water and sewage samples.[35]

Solvents and Alcohols

Solvents

Thinner (solvent) abuse, especially by young people is now causing a serious social problem. Death due to asphyxia secondary to solvent sniffing is frequently encountered in forensic science practice.[36] Exposure to solvent vapors among paint workers is also a problem from an industrial hygiene point of view. Commonly used thinners contain 50–70% toluene, 10–20% ethyl acetate, 5–30% *n*-butyl acetate and minor components of other solvents. The composition differs according to different enduses or manufacturers.

We reported headspace SPME of five thinner components: toluene, benzene, *n*-butyl acetate, *n*-butanol and isoamyl acetate, from human whole blood and urine.[37] After heating a vial containing 0.5 mL of whole blood or urine together with five compounds and ethylbenzene as IS at 80 °C, a PDMS fiber (100 μm film thickness) was exposed to the headspace of the vial at 80 °C for 15 min prior to capillary GC–FID. The calibration curves for all compounds showed linearity in the range of 2–100 ng 0.5 mL^{-1} for both samples, with *r*-values 0.986–0.999. The detection limits of the compounds were 1.1–2.4 ng 0.5 mL^{-1} sample.

Other researchers have also published SPME methods for various solvents,[3,4,38,39] although these experiments were carried out only with water, groundwater or fruits.

Ethanol

Ethanol is one of the most important beverages for human beings. Ethanol determination in postmortem blood and urine is essential in forensic science practice, particularly following traffic accidents.

We reported that ethanol could be extracted from human whole blood and urine samples by headspace SPME.[40] Ethanol and isobutanol as IS were added to a 4 mL vial containing 0.5 mL of human whole blood or urine, 0.5 mL of distilled water and 0.8 g of $(NH_4)_2SO_4$. After heating the vial at 70 °C for 5 min, a Carbowax/DVB fiber (65 μm film thickness) was exposed to the headspace of the vial at 70 °C for 15 min prior to capillary GC–FID. The headspace SPME–GC gave intense peaks for ethanol and IS with very low background noises. The calibration curves for ethanol extracted from whole blood and urine showed satisfactory linearity in the range of 80–5000 μg mL^{-1} ($r = 0.9998$) and 40–5000 μg mL^{-1} ($r = 0.9995$), respectively. The detection limits of ethanol were 20 μg mL^{-1} for whole blood, and 10 μg mL^{-1} for urine.

Figure 2 shows gas chromatograms for non-extracted authentic ethanol and IS without extraction and for the headspace SPME extract from human whole blood obtained 30 min after oral administration of 37.5 g of ethanol; the ethanol concentration was 514 μg mL^{-1} whole blood.

Recently, we have found that ethanol can be determined for human whole blood and urine samples with much higher sensitivity by use of a Carboxen/

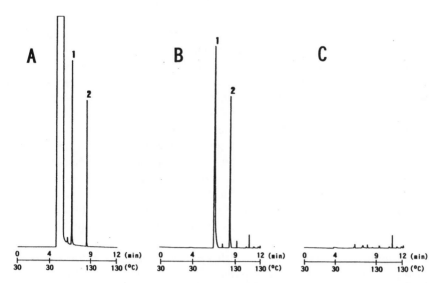

Figure 2. *Capillary GC with FID for headspace SPME extracts of whole blood (0.5 mL) obtained 30 min after drinking beer (5% ethanol, v/v) corresponding to 37.5 g ethanol. (A): Authentic ethanol and isobutanol dissolved in acetone (32 ng of ethanol and 12.5 ng of isobutanol on column) without extraction. (B): An extract of human whole blood after oral administration of ethanol. (C): A blank extract for control human whole blood. Peaks: 1, ethanol; 2, isobutanol. The amount of isobutanol as IS spiked to whole blood was 150 μg. The vertical scale of panels B and C are expanded 2× relative to panel A*

PDMS fiber (75 μm film thickness); the detection limits for whole blood and urine were 40 and 50 times lower, respectively, than those with the Carbowax/ DVB fiber.[41]

Methanol

Methanol is one of the most popular solvents and is available as a constituent of some antifreeze solutions, various paints, varnishes, gasoline additives and ethanol denaturants. Methanol is also present as a contaminant in many commercial wines and distilled liquors.

Methanol, together with acetonitrile as IS, was added to human whole blood samples and extracted by headspace SPME with a Carboxen/PDMS fiber (75 μm film thickness) prior to a capillary GC–FID.[42] The method produced linear calibration in the range of 12.5–400 μg 0.5 mL^{-1}, with an *r*-value of 0.999. The detection limit was 6 μg 0.5 mL^{-1} in whole blood.

4 Conclusions

A summary of clinical and forensic applications of SPME conducted by our groups is given in Table 1. The advantages of SPME are that the procedure is simpler and faster than those using liquid/liquid and solid-phase extractions, and much cleaner extracts can be obtained in blood and urine samples. It is also easier for automation. A disadvantage is that partitioning rates to the stationary phases coated on the fibers are quite different in different compounds. In addition, there are various experimental parameters affecting precision in SPME (see Chapter 1). To ensure good reproducibility, some amount of skillfulness in handling the SPME device is required.

The reduction in solubility of organic compounds in an aqueous phase produced by salts is commonly known as a 'salting-out' phenomenon. The addition of salting-out agents, such as $(NH_4)_2SO_4$ and/or NaCl, to the sample solutions contributes to improved efficiency of extraction in some drugs and poisons from biological samples. The absolute recoveries by SPME (an equilibrium method) are generally much lower than those by liquid/liquid and solid-phase extraction, which depend on exhaustive extraction. However, SPME methods generally produce higher chromatographic response and have much lower detection limits, compared with conventional extractions. This is because the entire amount of an analyte extracted by SPME is introduced to capillary GC column, versus a small fraction of the sample resulting from liquid/liquid or solid-phase extraction.

In view of its simplicity, low background noise, sensitivity and excellent quantitativeness, the SPME method is valuable for the determination of drugs and poisons from biological fluids in clinical toxicology, clinical pharmacology and forensic toxicology.

Table 1 *Summary of SPME methods in our laboratories for determination of drugs and poisons*

Chemical(s) analyzed	Sample[a]	HS or DI[b]	Additive[c]	Fiber	Vial temp. (°C)	Pre-heat time (min)	Extraction time (min)	GC detector	Detection limit	Ref.
Tricyclic antidepressants										
amitriptyline, imipramine, trimipramine and chlorimipramine	WB	HS	DW, NaOH	PDMS	100	30	60	FID / SID	16–25 ng 0.5 mL^{-1} / 90–69 pg 0.5 mL^{-1}	9
	U	HS	NaOH	PDMS	100	30	15	FID	24–38 ng mL^{-1}	5
Phenothiazines										
chlorpromazine, promazine, trimeprazine, triflupromazine and methotrimeprazine	WB	HS	DW, NaOH	PDMS	140	10	40	FID	100–200 ng mL^{-1}	12
	U	HS	NaOH	PDMS	140	10	40	FID	10–20 ng mL^{-1}	
Benzodiazepines										
medazepam, fludiazepam, diazepam, midazolam, flunitrazepam, prazepam, nimetazepam and flurazepam	U	DI	DW, NaCl	PDMS/DVB	R.T.[d]	–[e]	30	FID	10–20 ng mL^{-1} (nimetazepam, 150 ng mL^{-1})	14
Local anaesthetics										
lidocaine, bupivacaine, prilocaine, p-(butyl-amino)benzoic acid-2-	WB	HS	DW, NaOH, $(NH_4)_2SO_4$	PDMS	100	15	40	FID	58–830 ng mL^{-1} (procaine >2500 ng mL^{-1})	6
(diethylamino)ethyl ester, tetracaine, mepivacaine, procaine, benoxinate, ethyl aminobenzoate and dibucaine	WB	DI	NaCl (adjust pH 7)	PDMS	R.T.	–	40	FID	54–695 ng mL^{-1} (procaine >2000 ng mL^{-1})	15
Meperidine	WB	HS	NaOH, NaCl	PDMS	100	10	30	FID	100 ng mL^{-1}	16
	U	HS	NaOH, NaCl	PDMS	100	10	30	FID	20 ng mL^{-1}	
Cocaine	U	DI	NaF	PDMS	R.T.	–	30	NPD	6 ng 0.5 mL^{-1}	18
Phencyclidine	WB	HS	NaOH, K_2CO_3	PDMS	90	10	30	SID	1.0 ng mL^{-1}	19
	U	HS	NaOH, K_2CO_3	PDMS	90	10	30	SID	0.25 ng mL^{-1}	
Methamphetamine and amphetamine	U	DI	DW, Na_2CO_3	PDMS/DVB	65	5	30	FID	8 ng mL^{-1}	22

(continued)

Table 1 Continued

Chemical(s) analyzed	Sample[a]	HS or DI[b]	Additive[c]	Fiber	Vial temp. (°C)	Pre-heat time (min)	Extraction time (min)	GC detector	Detection limit	Ref.
1-Phenylethylamine	U	HS	NaOH, K_2CO_3	PDMS/DVB	90	5	30	NPD	20 ng mL^{-1}	24
Diphenylmethane antihistaminics										26
terodiline, diphenhydramine, orphenadrine and diphenylpyraline	WB	HS	DW, NaOH	PDMS	98	10	10	FID	76–473 ng mL^{-1}	
terodiline, clemastine, cloperastine, orphenadrine, homochlorcyclizine, diphenhydramine, diphenylpyraline, triprolidine, chlorpheniramine, carbinoxamine and doxylamine	U	HS	NaOH	PDMS	98	10	10	FID	13–186 ng mL^{-1}	
Organophosphate pesticides										30
malathion	WB	HS	DW, HCl, NaCl, $(NH_4)_2SO_4$	PDMS	100	15	20	NPD	25 ng 0.5 mL^{-1}	
ethion, fenthion, isoxathion, fenitrothion, IBP, methyl parathion, EPN and phosalone	WB	HS	DW, HCl	PDMS	100	15	20	NPD	2.2–40 ng 0.5 mL^{-1} (phosalone, 200 ng 0.5 mL^{-1})	
ethion, fenthion, isoxathion, IBP, fenitrothion, EPN, malathion, methyl parathion and phosalone	U	HS	DW, HCl, NaCl	PDMS	100	15	20	NPD	0.8–4.4 ng 0.5 mL^{-1} (phosalone, 12 ng 0.5 mL^{-1})	

Compound										Ref.
Carbamate pesticides										
fenobucarb, carbofuran, isoprocarb, XMC, xylylcarb and propoxur	WB	HS	DW, NaCl	PDMS	70	10	30	FID	100–500 ng mL^{-1}	32
	U	HS	NaCl	PDMS	70	10	30	FID	10–50 ng mL^{-1}	
Phenol and cresols										
phenol, o-cresol, m-cresol and p-cresol	WB	HS	DW, NaCl	Polyacrylate	100	20	30	FID	70–100 ng 0.5 mL^{-1}	34
Solvents										
toluene, benzene, isoamyl acetate, n-butanol and n-butyl acetate	WB, U	HS	DW	PDMS	80	15	5	FID	1.1–2.4 ng mL^{-1}	37
Ethanol	WB, U	HS	DW, (NH$_4$)$_2$SO$_4$	Carbowax/DVB	70	5	15	FID	10–20 μg mL^{-1}	40
	WB, U	HS	DW, (NH$_4$)$_2$SO$_4$	Carboxen/PDMS	60	5	15	FID	0.2–0.5 μg mL^{-1}	41
Methanol	WB	HS	DW, (NH$_4$)$_2$SO$_4$	Carboxen/PDMS	60	5	10	FID	6 μg 0.5 mL^{-1}	42

[a] WB = whole blood; U = urine.
[b] HS = headspace; DI = direct immersion.
[c] DW = distilled water.
[d] R.T. = room temperature.
[e] No mention made.

Acknowledgement

We gratefully acknowledge collaboration of Drs. Hiroshi Seno and Akira Ishii, Department of Legal Medicine, Hamamatsu University School of Medicine, on the SPME studies.

References

1. C.L. Arthur and J. Pawliszyn, *Anal. Chem.*, 1990, **62**, 2145.
2. M. Chai, C.L Authur and J. Pawliszyn, *Analyst*, 1993, **118**, 1501.
3. D.W. Potter and J. Pawliszyn, *J. Chromatogr.*, 1992, **625**, 247.
4. D. Louch, S. Motlagh and J. Pawliszyn, *Anal. Chem.*, 1992, **64**, 1187.
5. T. Kumazawa, X.-P. Lee, M.-C. Tsai, H. Seno, A. Ishii and K. Sato, *Jpn. J. Forensic Toxicol.*, 1995, **13**, 25.
6. T. Kumazawa, X.-P. Lee, K. Sato, H. Seno, A. Ishii and O. Suzuki, *Jpn. J. Forensic Toxicol.*, 1995, **13**, 182.
7. T. Fujii and H. Arimoto, *Anal. Chem.*, 1985, **57**, 2625.
8. H. Hattori, T. Yamada and O. Suzuki, *J. Chromatogr. A*, 1994, **674**, 15.
9 X.-P. Lee, T. Kumazawa, K. Sato and O. Suzuki, *J. Chromatogr. Sci.*, 1997, **35**, 302.
10. A.C. Moffat, J.V. Jackson, M.S. Moss and B. Widdop, *Clark's Isolation and Identification of Drugs*, Pharmaceutical Press, London, England, 1986.
11. S. Ulrich and J. Martens, *J. Chromatogr. B*, 1997, **696**, 217.
12. H. Seno, T. Kumazawa, A. Ishii, M. Nishikawa, K. Watanabe, H. Hattori and O. Suzuki, *Jpn. J. Forensic Toxicol.*, 1996, **14**, 30.
13. A. Wodak and C. Pedersen, *Med. J. Aust.*, 1992, **156**, 814.
14. H. Seno, T. Kumazawa, A. Ishii, K. Watanabe, H. Hattori and O. Suzuki, *Jpn. J. Forensic Toxicol.*, 1997, **15**, 16.
15. T. Kumazawa, K. Sato, H. Seno, A. Ishii and O. Suzuki, *Chromatographia*, 1996, **43**, 59.
16. H. Seno, T. Kumazawa, A. Ishii, M. Nishikawa, H. Hattori and O. Suzuki, *Jpn. J. Forensic Toxicol.*, 1995, **13**, 211.
17. H.L. Load and J. Pawliszyn, *Anal. Chem.*, 1997, **69**, 3899.
18. T. Kumazawa, K. Watanabe, K. Sato, H. Seno, A. Ishii and O. Suzuki, *Jpn. J. Forensic Toxicol.*, 1995, **13**, 207.
19. A. Ishii, H. Seno, T. Kumazawa, K. Watanabe, H. Hattori and O. Suzuki, *Chromatographia*, 1996, **43**, 331.
20. M. Yashiki, T. Kojima, T. Miyazaki, N. Nagasawa, Y. Iwasaki and K. Hara, *Forensic Sci. Int.*, 1995, **76**, 169.
21. N. Nagasawa, M. Yashiki, Y. Iwasaki, K. Hara and T. Kojima, *Forensic Sci. Int.*, 1996, **78**, 95.
22. A. Ishii, H. Seno, T. Kumazawa, M. Nishikawa, K. Watanabe, H. Hattori and O. Suzuki, *Jpn. J. Forensic Toxicol.*, 1996, **14**, 228.
23. L.A. King, A.J. Poortman-van der Meer and H. Huizer, *Forensic Sci. Int.*, 1996, **77**, 141.
24. A. Ishii, H. Seno, F. Guan, K. Watanabe, T. Kumazawa, H. Hattori and O. Suzuki, *Jpn. J. Forensic Toxicol.*, 1997, **15**, 189.
25. P. Kintz, A. Tracqui and P. Mangin, *Bull. Int. Assoc. Forensic Toxicol.*, 1991, **21** (3), 38.

26. M. Nishikawa, H. Seno, A. Ishii, O. Suzuki, T. Kumazawa, K. Watanabe and H. Hattori, *J. Chromatogr. Sci.*, 1997, **35**, 275.
27. K. Futagami, N. Tanaka, M. Nishimura, H. Tateishi, T. Aoyama and R. Oishi, *Int. J. Clin. Pharm. Ther.*, 1996, **34**, 453.
28. A.M. Tsatsakis, P. Aguridakis, M.N. Michalodimitrakis, A.K. Tsakalov, A.K. Alegakis, E. Koumantakis and G. Troulakis, *Vet. Hum. Toxicol.*, 1996, **38**, 101.
29. C.T. Hsiao, C.C. Yang, J.F. Deng, M.J. Bullard and S.J. Liaw, *J. Toxicol. Clin. Toxicol.*, 1996, **34**, 343.
30. X.-P. Lee, T. Kumazawa, K. Sato and O. Suzuki, *Chromatographia*, 1996, **42**, 135.
31. P. Picotte and M. Perreault, *Bull. Int. Assoc. Forensic Toxicol.*, 1991, **21** (2), 38.
32. H. Seno, T. Kumazawa, A. Ishii, M. Nishikawa, K. Watanabe, H. Hattori and O. Suzuki, *Jpn. J. Forensic Toxicol.*, 1996, **14**, 199.
33. C. Lo Dico, Y.H. Caplan, B. Levine, D.F. Smyth and J.E. Smialek, *J. Forensic Sci.*, 1989, **34**, 1013.
34. X.-P. Lee, T. Kumazawa, S. Furuta, T. Kurosawa, K. Akiya, I. Akiya and K. Sato, *Jpn. J. Forensic Toxicol.*, 1997, **15**, 21.
35. K.D. Buchholz and J. Pawliszyn, *Anal. Chem.*, 1994, **66**, 160.
36. R.W. Mayes, *Bull. Int. Assoc. Forensic Toxicol.*, 1987, **19** (2), 4.
37. X.-P. Lee, T. Kumazawa and K. Sato, *Int. J. Legal Med.*, 1995, **107**, 310.
38. A.J. Matich, D.D. Rowan and N.H. Banks, *Anal. Chem.*, 1996, **68**, 4114.
39. P. Popp and A. Paschke, *Chromatographia*, 1997, **46**, 419.
40. T. Kumazawa, H. Seno, X.-P. Lee, A. Ishii, O. Suzuki and K. Sato, *Chromatographia*, 1996, **43**, 393.
41. X.-P. Lee, T. Kumazawa, K. Sato, H. Seno, A. Ishii and O. Suzuki, *Chromatographia*, 1998, **47**, 593.
42. X.-P. Lee, T. Kumazawa, T. Kurosawa, K. Akiya, Y. Akiya, S. Furuta and K. Sato, *Jpn. J. Forensic Toxicol.*, 1998, **16**, 64.

CHAPTER 36

On-fiber Derivatization for Analysis of Steroids by SPME and GC–MS

NICHOLAS H. SNOW

1 Introduction

This chapter describes the development of methods for clinical and pharmaceutical analysis by SPME, which has been an on-going interest in the author's laboratory. The specific problems involved in employing SPME for the analysis of low levels of drugs in biological fluids, such as urine or serum are described. They will provide some contrast with those addressed in the bulk of SPME applications, which, as evidenced by the distribution of topics in this text, are environmental or food and fragrance in nature. The bulk of the work described here was performed during 1994–1997 and is also described in three papers[1–3] and one dissertation.[4]

A typical SPME extraction procedure, as described in detail elsewhere in this text, involves the equilibration of the SPME fiber coating with either a liquid solution or its headspace followed by the thermal desorption of the extracted analytes into the inlet of a gas chromatograph. Owing to the small fiber volume, this assumes that analyte partition coefficients between the sample matrix and fiber coating are large and that the analytes must be volatile enough to be thermally desorbed under typical GC inlet conditions. Under these restrictions, the application of SPME to clinical and pharmaceutical analysis is limited, as many drugs are polar, making extraction potentially difficult and most are non-volatile, making desorption and GC analysis of the native extracts impossible. Also, the many interferences present in urine and blood samples may complicate extraction and fiber maintenance.

Classically, drugs have been extracted from urine and blood using solid phase extraction, and there are a wide variety of SPE chemistries and well developed applications.[5] SPE or liquid/liquid extraction analysis methods for compounds such as steroids are often time-consuming and labor intensive. Both types of extraction often require the use of large quantities of organic solvents and the exposure of workers to potentially infectious samples. Following extraction, the dried extracts must be derivatized in solution with an appropriate reagent in order to convert the non-volatile analyte into a volatile derivative. These reactions, while well described in many textbooks,[6,7] have suffered owing to the possibilities of multiple reactions of the analyte, reactions with matrix interferences and high sensitivity to reaction conditions. Thus, while for many applications, GC–MS remains the instrument of choice for drug analysis, the necessary derivatization reactions have fallen out of favor, forcing analysts to alternate methods. In our laboratories, we have been examining the extraction of steroids from a variety of matrices.

SPME provides an alternative to the classical solid phase or liquid/liquid extraction approach. Essentially, the steps are the same:

1. extract the analyte from the matrix
2. derivatize using an appropriate derivatizing reagent
3. inject the derivatives into a GC/MS for analysis and quantification.

Initially, it was believed that the extraction step could be performed satisfactorily if reasonable care was taken to adjust partition coefficients using pH, ionic strength and the fiber coating, as is described later in this chapter. It was also known that trimethylsilyl derivatives of steroids can be easily injected onto GC–MS and that they behave well, chromatographically. The difficult portion of this analysis was the development of the derivatizing scheme, as these reactions had typically been carried out in organic solutions, not in coated fibers. Using the procedure outlined above, the analytes would have to be extracted from the matrix into the coated fiber; the fiber would then be removed from the sample and would be exposed to the derivatizing reagent. Following this exposure, the fiber would then be deosrbed into the GC inlet.

2 Method Development and Application to Estrogens

Our initial work[1] focused on the analysis of estrogens from water. Figure 1 shows structures for several of these clinically important compounds and for the anabolic steroids described later in this chapter. They were chosen because both the parent compounds and the TMS derivatives are easily chromatographed, so the extent of any successful reactions could be judged. The initial choice for derivatizing reagent was bis(trimethylsilyl)trifluoroacetamide, BSTFA. BSTFA is a widely used silylating reagent for general applications and is especially reactive to alcohols.[6] It was hoped that BSTFA would be a strong enough derivatizing reagent that all hydroxyl groups on the structures shown in Figure 1 would be silylated.

The first attempts to extract 17-β-estradiol from water by SPME were successful, as shown in Figure 2. Initial samples were dissolved to a concentration of 1 μg mL^{-1} in deionized water. The extraction and chromatographic conditions are shown in the figure caption. It was seen that extraction was possible with a minimum of optimization, although this is addressed later. The

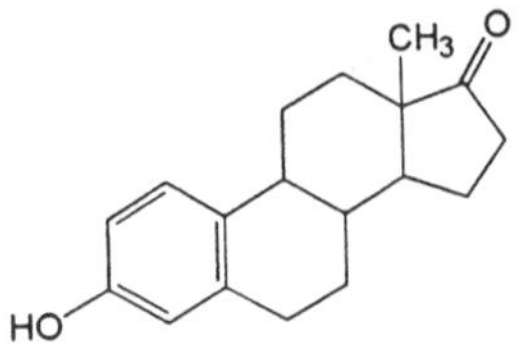

Estrone

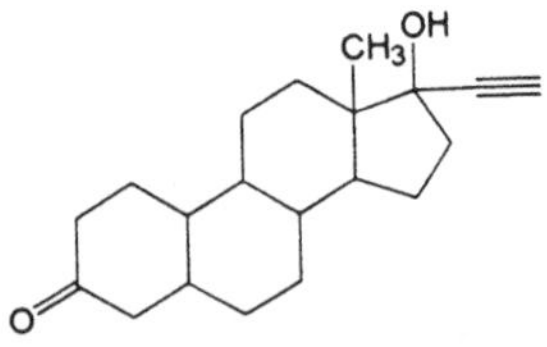

19-norethisterone

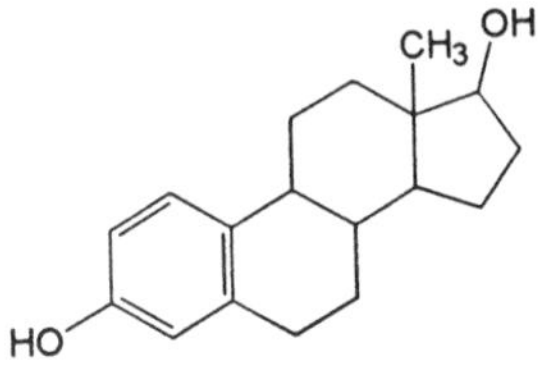

17-β-estradiol

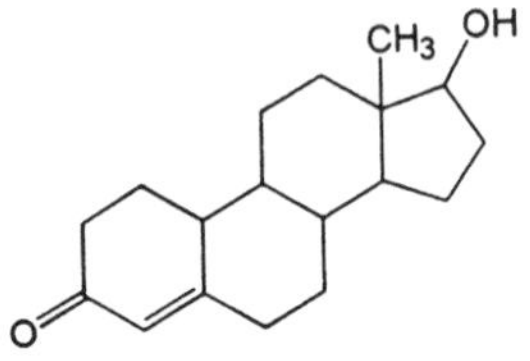

19-nortestosterone

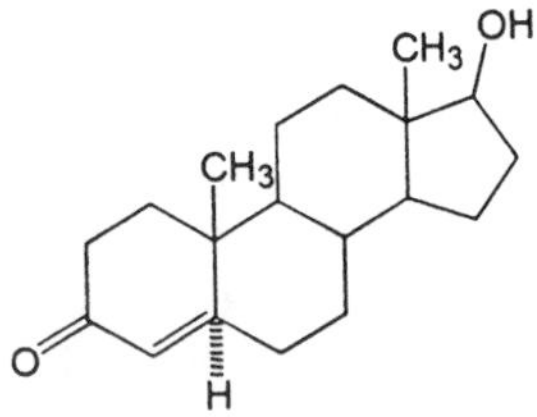

5α-androstan-17β-ol-one

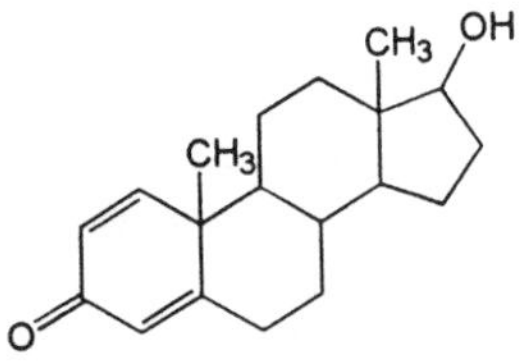

1-dehydrotestosterone

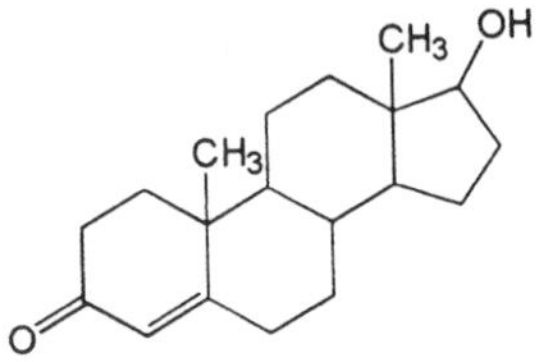

testosterone

Figure 1 *Structures for the estrogens and anabolic steroids used in this work*

derivatization step initially proved problematic. Repeated attempts to expose the extract-containing fibers directly to BSTFA led to destruction and, in some cases, dissolving of the fiber coating. We realized that it was not likely that direct exposure of SPME fibers to pure silylating reagents, or to organic solutions of them, would produce a successful derivatization.

Next, the fiber, following extraction as described above, was placed into the headspace of 10 μL of BSTFA in a small (1 mL total volume) conical reaction vial at elevated temperature (60 °C) and allowed to stand for 1 hour. The result of that analysis, with the same conditions as in Figure 2 (except for the headspace derivatization step and extraction from spiked human serum), is

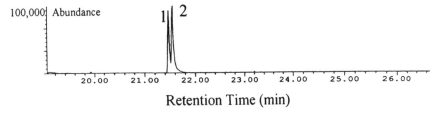

Figure 2 *Total ion chromatogram for underivatized estrogens following SPME extraction from water. Peak identification: 1: estrone, 2: 17-β-estradiol. Fiber: 85 μm polyacrylate; GC: Inlet: 250 °C, 2 mm liner, desorption time: 5 min; Column: 30 m × 0.25 mm × 0.5 μm DB-5MS; Temperature program: 50 °C/5 min, ramp 15 °C min^{-1} to 300 °C. Detector: Hewlett-Packard 5972A MSD*

shown in Figure 3. The peaks are easily identified by their mass spectra as 17-β-estradiol, the analyte, two other endogenous steroids and a TMS-derivatized phthalate. The small peak at about 26 minutes is a TMS derivative of cholesterol, which is endogenous in human serum.

This SPME derivatization technique, using the classifications described by Pan and Pawliszyn,[8] can be best described as post-extraction, on-fiber head-space derivatization. Having shown that it is possible to perform such reactions on compounds extracted into SPME fibers, it was necessary to examine the important analytical parameters involved with the new technique. The remainder of this chapter focuses on the optimization of factors such as extraction conditions: extraction time, analyte solution pH, analyte solution ionic strength; derivatization conditions: time and temperature; and GC injection conditions (liner diameter and temperature). The application of on-fiber head-space derivatization to the quantification of anabolic steroids, which are of concern in international athletic competition as drugs of abuse, is also described.

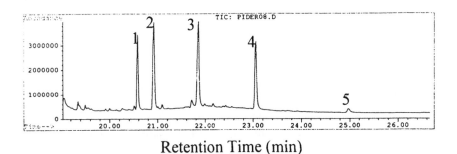

Figure 3 *Total ion chromatogram for derivatized 17-β-estradiol following SPME from serum with on-fiber headspace derivatization for 1 hour at 60 °C with BSTFA. GC and extraction conditions were the same as Figure 2. Peak identifications: 1 and 4: endogenous steroids, 2: di-iso-octyl phthalate (contaminant), 3: bis-(trimethylsilyl)-17-β-estradiol, 5: trimethylsilyl-cholesterol*

Extraction Conditions

Most of the SPME applications research has had the optimization of extraction conditions as a primary focus. The important variables are: analyte solution pH, analyte solution ionic strength, extraction time and choice of SPME fiber. Since many steroids are easily ionizable, it was hypothesized that pH and ionic strength would be especially important variables in the extraction process. We have observed that extraction equilibrium, with magnetic stirring, is achieved in about 30 minutes for extractions from water, urine and serum. The experimental details of the pH and ionic strength optimization for estrogens are shown in reference 2.

The pH effect on the extraction of estrogens is easily predictable and follows rules of thumb described by many workers in SPME. For partitioning from an aqueous phase to an organic phase to be effective, the pH must be adjusted to completely suppress analyte ionization. Since estrogens are weak organic acids, ionization is suppressed at low pH. Since the fibers are silica-based, they will not be affected by weakly acidic pH greater then 2. In our studies, a dramatic drop in analyte recovery was seen when the analyte solution pH was adjusted from weakly acidic to weakly basic conditions. Thus, for steroids, pH adjustment to suppress ionization of the compounds of interest is critical. It should be noted that the exact pH value is not critical, as long as suppression of ionization is assured.

The effect of ionic strength proved less predictable. Many classical extraction procedures[9] and SPME procedures[10] have described 'salting out' as a means for improving analyte transfer from the aqueous phase to the organic. Since the ionic strength of physiological samples can also vary, this effect was explored. A plot of peak area versus ionic strength for the extraction of estrogens is shown in Figure 4. It is seen that for addition of small quantities of salt to a water solution, the classical salting out is seen and recovery improves. However, at high salt concentration, recovery decreases sharply. In fact, it appears that a complementary process, 'salting in' is occurring. Salting in occurs when the salt contributes to increased analyte solubility in the liquid phase and is observed commonly in HPLC.[11,12] In this case, we believe that the salt content of the solution affects hydration of the ionizable steroids, causing their increased solubility in the aqueous phase.

Derivatization

Following extraction, the extracted steroids are derivatized by exposing the extract-containing fiber to the headspace of a small quantity of a volatile derivatizing reagent. A diagram showing the procedure is shown in Figure 5. In our initial estrogen work, the chromatogram shown in Figure 3 was obtained by exposing the fiber to the vapor of bis(trimethylsilyl)trifluoroacetamide (BSTFA) for 1 hour at 60 °C. The two factors that can most strongly affect the derivatization reaction, once an appropriate reagent is chosen,[6,7,13] are the incubation time and temperature.

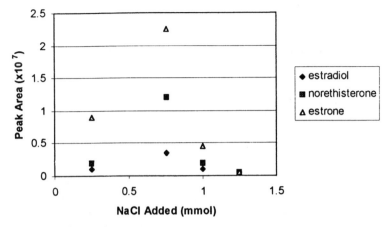

Figure 4   *Effect of analyte solution ionic strength on response for estrogens. Extraction and GC conditions as described for Figure 2, except for addition of salt. Sample volume:* 1.5 mL

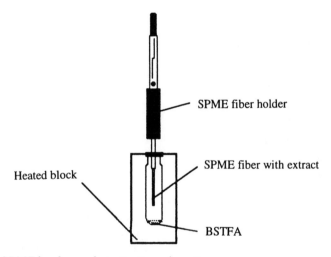

Figure 5   *SPME headspace derivatization schematic*

The incubation temperature for the derivatization reaction must be carefully controlled. Figure 6 shows the detector response *versus* derivatization temperature for several of the estrogens. Minima are seen at both low and high temperatures. Since BSTFA is relatively low boiling (boiling point 60 °C), it is volatile at room temperature. Therefore, even at low temperatures, the derivatizing reagent can contact the SPME fiber. It is believed that the slow kinetics of the reaction at low temperatures are causing the low recovery. At high temperature, the responses are also lower. Since the reaction products are nonpolar, they are more volatile than the parent compounds, thus, at elevated temperature, they may evaporate from the fiber as they are produced, resulting in losses. It is noted also that volatile compounds generally have lower

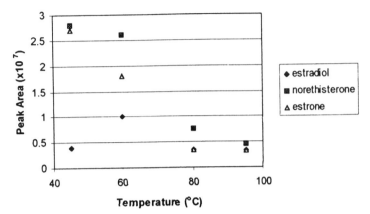

Figure 6   *Effect of incubation temperature on BSTFA derivatization of estrogens. Derivatization time: 1 h. Extraction and GC conditions as described in Figure 2*

vapor/fiber partition coefficients and higher diffusion coefficients at elevated temperatures, both favoring desorption. As can be seen from Figure 6, there is an optimum incubation temperature for each compound and these optimum temperatures may not coincide, resulting in compromises.

Derivatization time must be determined as a part of the usual method development. Recovery and regioselectivity of the reactions should be monitored as the incubation time is varied. For our analyses, 30 min–1 h incubations have been more than adequate to provide complete reaction of all active sites on the analytes with BSTFA. It is expected that shorter incubations will result in incomplete reactions, leaving some polar functional groups unreacted.

GC Inlet Conditions

The GC injection conditions employed for these studies were very straightforward. Since the analytes are semi-volatile, little optimization of the GC inlet was needed. Desorption time and temperature were 5 minutes and 250 °C for all of the analyses. A standard 2 mm inlet liner used for splitless injections was used in place of the usual narrow SPME liner with no deleterious effect. The inlet pressure was held to 15 psig with helium carrier gas. Since typical clinical analyte derivatives have low volatility, it is expected that by using a low analytical column initial temperature, the analytes can be re-focused on the column, if the injection bands are broad. Thus, the specific inlet conditions are not critical to these analyses. The general optimization of GC inlet conditions for SPME is described in Chapter 1 of this text and in papers by Okeyo and Snow[3] and by Langenfeld.[14]

3   Application to Anabolic Steroids

Quantification of low levels of anabolic steroids from biological fluids such as blood and urine is of major importance in clinical, pharmaceutical and forensic (drug testing) studies.[5] Classical methods for steroids often involve liquid/liquid or solid phase extraction, followed by extensive treatment and work-up prior to GC analysis. For forensic analysis, identifications generally must be confirmed by GC–MS. The extensive work-up, besides being time consuming, can be hazardous, as workers are exposed to potentially infectious samples and they must handle relatively large quantities of solvents. We have shown[2] that SPME can provide an alternative to solvent or solid phase extraction and, when combined with the headspace on-fiber derivatization, that it may eliminate most of the difficult work-up involved for clinical samples. Previous drug of abuse methods for SPME have concentrated on the volatile drugs that are amenable to GC analysis without derivatization.[15–17]

In the analysis of anabolic steroids from urine for drug of abuse testing, detection limits in the low part per billion range are often required. The procedure for analyzing anabolic steroids from urine is as follows. Steroid free urine, prepared as described in reference 18, was spiked with anabolic steroids at the concentrations indicated in the Figures. 1.5 mL of urine is placed into a 2 mL auto-sampler vial and crimp sealed. Internal standard ($[^2H_3]$testosterone) is added to give a concentration of 20 mg mL^{-1}. Extractions using a 60 μm carbowax-divinylbenzene fiber (Supelco, Inc.) are carried out at ambient temperature for 30 minutes with magnetic stirring. Following extraction, the fiber is exposed to the headspace of 10 μL of BSTFA in a conical vial and incubated at 60 °C for 30 minutes. The fiber is then desorbed into the inlet of a GC–MS (Hewlett-Packard 5890E Series II Plus GC with 5972 MSD) under splitless conditions for 5 minutes at 250 °C using a 2 mm inlet liner. The GC column is a 30 m by 0.25 mm by 0.50 μm DB-5MS (J&W Scientific, Folsom, CA) and was temperature programmed starting at 100 °C for the desorption period (5 minutes), ramped at 30 °C min^{-1} to 250 °C, ramped at 3 °C min^{-1} to 300 °C and held for 5 minutes. Using these conditions a total ion chromatogram for a steroid mixture, extracted from urine is shown in Figure 7. Several endogenous peaks are seen in this chromatogram, while water extracts are 'clean'. It is seen that the relative response for each compound is similar in both analyses. This is unexpected, as matrix differences between water and urine should be expected to have a larger effect.

In Figure 8, a calibration curve in the 1–1000 ng mL^{-1} range is shown for testosterone extracted from water, indicating that SPME can be used to quantify steroids at the low levels required for drug of abuse testing and metabolic studies. Testosterone-d_3 is an ideal internal standard for steroid analysis, as its base peak at $m/z = 363$ is unique. It can also be expected to exhibit nearly identical extraction recovery to testosterone. In this method, $[^2H_3]$testosterone was also used as internal standard for quantification of other steroids.[2] The use of a single internal standard is not viewed as a major limitation in this work, although multiple internal standards may be necessary for more complex analyses. The detection limit, defined as a signal three times the average noise, was determined to be approximately 70 pg mL^{-1} for testosterone, extracted from steroid-free urine.[2] We have shown calibrations curves for the other steroids[2] and they also exhibit linear behavior.

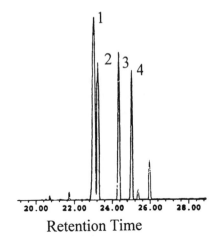

Retention Time

Figure 7   *TIC of anabolic steroids. Conditions as described in the text. Peak identification:
1. 5-α-androstan-17β-ol-one-TMS; 2. 19-nortestosterone-TMS; 3. Testoster-
one-TMS; 4. 1-dehydrotestosterone-TMS*

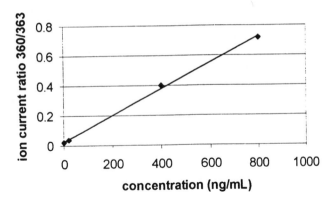

Figure 8   *Internal standard curve for testosterone from water. Experimental conditions as
described in the text*

As was observed for the estrogens, these derivatizations were highly regioselective, producing only one product per parent compound. This is potentially
important, as one of the classical complaints about solution phase derivatization reactions is that they are potentially subject to great variability in product
composition. We have observed only single derivatization products for the
SPME procedure. It is also noted that SPME fibers were re-used over 50 times in
developing the data presented here and in references 1 and 2. Thus, a sensitive,
straightforward and selective method for the analysis of non-volatile compounds by SPME and GC–MS has been developed and applied to the analysis
of steroids from water, urine and serum. This technique can be applied to any
compound that can be extracted into an SPME fiber, and has already been
applied to amphetamines by Brettell and co-workers.[19]

4 Conclusions

Using on-fiber derivatization, the applicability of SPME is greatly extended to include many classes of non- and semi-volatile analytes. The extraction, derivatization, qualitative and quantitative analysis of estrogens and anabolic steroids have been shown. The procedure for headspace derivatization of steroids extracted into a SPME fiber is straightforward and lends itself well to quantitative analysis. This technique should be applicable to a wide variety of sample types and analytes. Analytical detection limits of less than 50 pg mL^{-1} and linear range of several orders of magnitude are similar to other methods, with the advantages of this technique being minimal solvent usage, straightforward procedure and little contact with potentially infectious clinical samples. The technique is amenable to automation, so analytical throughput should be greater than for classical methods.

Acknowledgments

The author gratefully acknowledges the many contributions of Dr. Pius D. Okeyo, his first Ph.D. graduate, currently with DuPont Agricultural Products, Wilmington, DE, to the work described here. Supelco, Inc. is acknowledged for providing the SPME fibers. Financial contributions from the Seton Hall University Research Council and Merck and Company that allowed this work to begin and to flourish are also acknowledged.

References

1. P. Okeyo, S.M. Rentz and N.H. Snow, *J. High Resolut. Chromatogr.*, 1997, **20**, 171–175.
2. P. Okeyo and N.H. Snow, *J. Microcolumn Separations*, 1998, in press.
3. P. Okeyo and N.H. Snow, *LC-GC*, 1997, **15**, 1130.
4. P. Okeyo, *Solid Phase Microextraction: Optimization and Interface with GC/MS for Trace Analysis of Steroids in Biological Fluids*, Ph.D. Dissertation, Seton Hall University, 1997.
5. S.-H. Hsu, R.H. Eckerlein and J.D. Henion, *J. Chromatogr.*, 1992, **573**, 183.
6. D.R. Knapp, *Handbook of Analytical Derivatization Reactions*, John Wiley and Sons, New York, 1979, pp. 449–514.
7. K. Blau and J. Hallett (eds.), *Handbook of Derivatives for Chromatography*, 2nd Edn., John Wiley and Sons, New York, 1993.
8. L. Pan and J. Pawliszyn, *Anal. Chem.*, 1997, **69**, 196.
9. D.L. Pavia, G.M. Lampman and G.S. Kriz, *Introduction to Organic Laboratory Techniques*, Saunders College Publishing, Philaldelphia, 1988, pp. 541–550.
10. J. Pawliszyn, *Solid Phase Micro-extraction: Theory and Practice*, John Wiley and Sons, New York, 1997, pp. 130–131.
11. L. Snyder and J. Kirkland, *Introduction to Modern Liquid Chromatography*, John Wiley and Sons, New York, 1979, p. 288.
12. C. Horvath and W. Melander, *J. Chromatogr. Sci.*, 1977, **15**, 393.

13. A. Groppi, A. Polettini, C. Stramesi and M. Montagna, *Proc. 18th Intl. Symp. Capillary Chromatogr.,* Heidelberg: A. Huethig, 1996, pp. 922–930.

14. J. Langenfeld, S. Hawthorne and D. Miller, *J. Chromatogr.*, 1996, **740**, 139–145.

15. X.P. Lee, T. Kumazawa, K. Sato and O. Suzuki, *Chromatographia*, 1996, **42**, 135.

16. M. Yashiki, T. Kojima, T. Miyazaki, N. Nagasawa, Y. Iwzsaki and K. Hara, *Forensic Sci. Int.*, 1995, **76**, 169.

17. T. Kumazawa, X.P. Lee, M.C. Tsai, H. Seno, A. Ishii and K. Sato, *Jpn. J. Forensic Toxicol.*, 1995, **13**, 207.

18. S.K. Yap, G.A.R. Johnston and R. Kazlauskas, *J. Chromatogr.*, 1992, **573**, 183.

19. T. Brettell, M. Sinabaldi and E. McLaughlin, presented at the 36th Eastern Analytical Symposium, Somerset, NJ, November, 1997, paper 133.

CHAPTER 37

SPME–Quadrupole Ion Trap Mass Spectrometry for the Determination of Drugs of Abuse in Biological Matrices

BRAD J. HALL AND JENNIFER S. BRODBELT

1 Introduction

Quantitative analysis of drugs in biological fluids is an important extension of the solid-phase microextraction (SPME) technique with potentially numerous applications in the field of forensic and clinical toxicology. The validation of new SPME methodologies that are compatible with biological fluids is an area of ongoing interest. Successful SPME procedures for the determination of drugs in biological fluids involves using the techniques outlined in Chapter 1 with specific emphasis placed on the development of methods to enhance the partition of the drug(s) into the SPME fiber amidst high levels of endogenous components in the matrix. Accurate quantitative analysis is dependent on obtaining resolved chromatographic peaks that can be unequivocally assigned to the analyte of interest. Since SPME is a relatively non-selective extraction technique, often methods are employed to enhance the specific extraction and/ or the detection of the analyte. These methods include: performing SPME in the headspace of the biological fluid for analytes with sufficient vapor pressures at room temperature or some elevated temperature, pre-treatment of the sample such as protein precipitation or dialysis for whole blood or plasma, derivatization techniques either prior to performing SPME or in-fiber derivatization after the extraction, and use of mass spectrometry for sensitive identification of components. The use of mass spectrometric detectors in drug analysis is the standard for confident identification and quantitation, especially for forensic purposes. Gas chromatography–mass spectrometry (GC–MS) proves to be the ideal detector after SPME extraction/injection for both qualitative and quantitative analysis. Not only do the mass spectra provide key signatures of specific compounds and allow monitoring of selected ions to resolve overlapping peaks, but deuterated internal standards which specifically reflect the behavior of the analyte may be used throughout the SPME procedure to provide accurate quantitative results. If derivatization procedures are utilized to enhance SPME extraction, electrophilic derivatization reagents may be selected in combination with negative ion chemical ionization to attain extremely low levels of detectability. In addition, traditional positive ion chemical ionization with or without collisionally activated dissociation methods may be developed to reach a further degree of selectivity in analyte identification.

The first section of this chapter will discuss the two techniques used by SPME to sample drugs from biological systems, headspace and direct immersion extraction, with references to applications appearing in this field. The chapter will conclude with a discussion of quadrupole ion trap mass spectrometry as a detector for SPME–GC separations, with example applications from our laboratory including the determination of barbiturates in human urine, cannabinoids in human saliva, and a chemical ionization (CI)/collisionally activated dissociation (CAD) method in conjunction with SPME for confirmation of nordiazepam.

2   SPME Techniques in Biological Fluids

Headspace SPME

Performing SPME in the headspace of a biological fluid is ideal for those analytes which have a sufficient partition into the headspace of the sample vial. Volatile drugs are selectively enriched in the headspace and extracted by the SPME fiber, while the majority of the matrix components are left behind in the aqueous phase. This process results in clean chromatograms relatively free of interfering peaks. In order to obtain good quantitative results with headspace SPME, careful control of the extraction conditions such as the agitation and temperature is paramount. Typically an internal standard is present which compensates for minor variations in the sampling conditions. Several applications of headspace SPME for drugs in biological fluids have been reported.[1-10]

The detection of amphetamines and related compounds in urine was one of the earliest reports of headspace SPME for drugs of abuse in a biological fluid.[1-4] In addition, headspace SPME methods for phencyclidine,[5] tricyclic antidepressants,[6] antihistaminic drugs,[7] ethanol,[8] methylene chloride[9] and cyanide[10] have recently appeared in the literature.

Direct Immersion SPME

Performing SPME directly in a biological fluid is inherently more difficult owing to the nature of the fluid and the relative non-selectivity of currently available commercial SPME fibers. Attempting to extract trace quantities of a drug with direct immersion SPME amidst high concentrations of salt and organics in urine, high protein and lipid levels in plasma, and the viscous nature of saliva and plasma may be a difficult task. In addition, many drugs and especially metabolites are very polar and thus have a high affinity for the aqueous matrix and/or may be incompatible with gas chromatographic analysis. In many cases, sample pre-treatment such as protein precipitation is employed to free the drug and/or decrease the viscosity of the sample solution such that there is a higher partition into the SPME fiber. Derivatization methods accomplished in the aqueous matrix or SPME fiber might be used to enhance the extraction and chromatographic properties of a polar drug.

In order to exploit the benefits of SPME, such as the relative ease of the method and the elimination of organic solvent usage in the extraction, many reports have been published applying direct immersion SPME for the analysis of drugs in biological matrices. These include the determination of valprolic acid,[11] local anaesthetics,[12] benzodiazepines,[13,14] amphetamines[15,16] and antidepressant drugs.[17]

SPME and Gas Chromatography–Quadrupole Ion Trap Mass Spectrometry

Whether performing headspace or direct immersion SPME, mass spectrometry as a detector for SPME after chromatographic separations is ideal for those situations where unequivocal identification and accurate quantitation are desired for complex mixtures. The quadrupole mass filter has been used extensively in toxicology laboratories over the past two decades and has become the standard for many analytical methods for drugs of abuse. Quadrupole ion trap devices, first introduced commercially in the 1980s, have gained universal popularity as gas chromatographic detectors because of their ruggedness, simplicity and unparalleled sensitivity (see Figure 1). Because they are 'trapping' devices, quadrupole ion traps can be used to accumulate ions to maximize an ion signal that might otherwise be undetectable. Technological advances in quadrupole ion trap (QIT) hardware design and software control of the scan method are making the GC–QIT systems more attractive for toxicological laboratories with diverse needs and applications. These advances include the development of new ionization sources, including electrospray ionization and external chemical ionization sources that allow larger molecules to be ionized, and operation in the negative ion mode. Moreover, implementation of fast, efficient collision activated dissociation (CAD) methods allows acquisition of diagnostic fragmentation patterns for multi-component mixtures. An excellent discussion of practical aspects and considerations of GC–QIT systems has been published.[18] In addition, a good overview of the use of the QIT in toxicology work has recently been published.[19]

Barbiturates in Urine

Initial application of SPME in our lab addressed the analysis of a series of barbiturates, namely barbital, butabarbital, butalbital, amo- and pento-barbital, secobarbital, hexobarbital and phenobarbital, in human urine.[20] Method developmental studies were first conducted in pure water in which a 65 μm

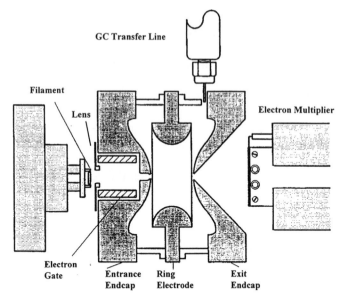

GC Transfer Line

Filament

Lens

Electron Multiplier

Electron Gate

Entrance Endcap

Ring Electrode

Exit Endcap

Figure 1 *Schematic of a standard GC-quadrupole ion trap using internal ionization. Ions are formed and stored within the ring electrode by gating electrons through the entrance endcap which subsequently interact with the neutral gas entering ring volume from the GC transfer line. The stored ions are ejected sequentially by mass through the exit endcap by ramping up the radio frequency voltage on the ring electrode. The ejected ions impinge on the electron multiplier upon which an ion current signal is recorded*
(Reprinted from ref. 18, © CRC Press, Boca Raton, Florida)

Carbowax/divinylbenzene fiber was found to most efficiently extract the barbiturates, and a Varian (Sugar Land, TX USA) Saturn GC–QIT was operated in the electron ionization (EI) mode. Utilizing twenty minute extraction conditions, a linear range was established from 10–1000 ng mL^{-1} with an overall method precision of 3% RSD. Further experimentation indicated that the method was easily extended to the analysis of barbiturates in urine. For all of the barbiturates studied except phenobarbital, the SPME recoveries from urine relative to the recoveries in pure water ranged from 83–99%. For phenobarbital, recovery was determined to be 50%.

A clinical sample which tested positive for barbiturates by an immunoassay method was subjected to the SPME procedure. Initial SPME tests with a 65 μm carbowax/divinylbenzene fiber indicated the abundant presence of butalbital. Figure 2 illustrates a total ion chromatogram along with a select ion chromatogram obtained using a DB-5MS (J&W Scientific, Folsom, CA USA) capillary column (i.d. 0.25 mm, d_f 0.25 μm) for butalbital after an eight minute extraction of the urine sample diluted 1:3 with deionized water. The amount of butalbital in this sample was quantified by internal standard calibration with pentadeuterated butalbital with the quadrupole ion trap operating in the electron ionization mode. The ratio of ion current for

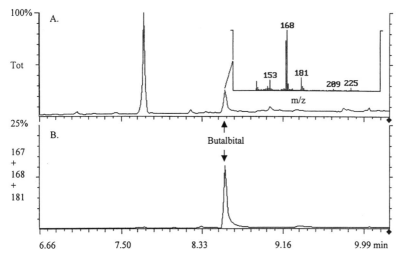

Figure 2 (A) *Total ion chromatogram after performing SPME on a clinical urine sample known positive for barbiturates. An eight minute extraction with a 65 μm Carbowax/divinylbenzene was conducted on 1 mL of urine diluted with 3 mL of deionized water. The electron ionization mass spectrum insert indicates the presence of butalbital.* (B) *Selected ion chromatogram for characteristic ions for butalbital at m/z 167, 168, and 181*

butalbital (monitoring 167^+, 168^+, and 181^+) and [$^2H_5$]butalbital (172^+, 173^+, and 186^+) was plotted against concentration for a series of butalbital spikes in deionized water ranging from 0.063 to 1.5 $\mu g\ mL^{-1}$. The resulting calibration curve is shown in Figure 3. The quantitation results by internal standard SPME along with external standard and standard addition SPME are summarized in Table 1. In addition, Table 1 lists the comparative results for a conventional solid-phase extraction method. The excellent agreement between the SPME internal standard, standard addition, and the SPE method indicate accurate quantitative results using the SPME methods with quadrupole ion trap mass spectrometry. The SPME method was found to be much less labor intensive and the use of solvent was eliminated in the extraction step, thus no concentration step was necessary.

Cannabinoids in Saliva

The determination of cannabinoids in human saliva by SPME–GC–QIT has recently been reported from our group.[21] Analysis for drugs in saliva is attractive to many researchers because sample collection is non-invasive and quantitative measurements may reflect the nonprotein-bound fraction of the drug in plasma. Saliva, although a complex mixture, is relatively free of interfering substances and has a much lower content of proteins than other physiological fluids. Furthermore, measurement of cannabinoids in saliva is reported to offer a more accurate value of concentration present during

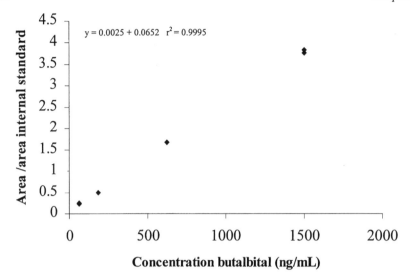

Figure 3 *Calibration curve for butalbital in 1:1 urine:deionized water. The ratio of the peak area for butalbital (ions 167^+, 168^+, and 181^+) and [$^2H_5$]butalbital (ions 172^+, 173^+, 186^+) is plotted on the y-axis. An eight minute extraction with a 65 μm Carbowax/divinylbenzene fiber was used to generate this plot*

Table 1 *Quantitation results for butalbital in a clinical human urine sample by SPME and solid-phase extraction*

	SPME (μg mL^{-1})[a,b]			Solid-phase extraction[c] (μg mL^{-1})
Compound	*External standard*	*Internal standard*	*Standard addition*[d]	
Butalbital	1.74 (4)	1.20 (3)	1.17	1.15 (3)

[a] Values in parentheses represent %RSD (*n* = 3).
[b] SPME conditions: 65 mm Carbowax/divinylbenzene fiber for eight minute extraction times.
[c] A 3M Empore™ extraction disk was used for the SPE procedure. [$^2H_5$]Butalbital was added as an internal standard to the urine sample prior to extraction.
[d] Based on a three point standard addition calibration with each point run in duplicate.

marijuana intoxication. From our initial experiments in pure water, the four cannabinoids, cannabinol, Δ^8-tetrahydrocannabinol, Δ^9-tetrahydrocannabinol (Δ^9-THC) and cannabidiol, exhibited efficient extraction with the series of PDMS SPME fibers, as predicted by the high degree of lipophilicity of the cannabinoids. Upon spiking drug-free saliva samples with the cannabinoids, recoveries relative to deionized water fell to 6–8%. This dramatic drop is attributed to both the viscid nature of the saliva matrix which hinders transport of the cannabinoids to the fiber surface, and to the fact that the cannabinoids may be binding to cellular or protein material, and thus not free to be absorbed by the SPME fiber. Saliva, although approximately 98% water, contains cellular debris, particulate matter and mucoprotein, which may interfere with the SPME process. In an effort find a simple procedure to enhance the extraction efficiency, it was discovered that the addition of glacial acetic acid effectively coagulated cellular debris and aided in the release of the cannabinoids for subsequent extraction. After the addition of acetic acid, the mixture was centrifuged and the clear liquid transferred to the SPME sample vial. Recoveries improved 3.5–5 times that of performing SPME in saliva with no pre-treatment. In addition, the liquid after centrifugation was less viscid and more free of interfering substances, therefore extending the lifetime of the fiber. Using this technique, good linearity was achieved from 5–500 ng mL^{-1} as illustrated in Figure 4 for Δ^9-THC, the main active constituent in marijuana. The precision averaged 15% RSD ($n = 6$) with no internal standard present.

In order to verify the accuracy of this technique, a saliva sample was collected from a volunteer after smoking a marijuana cigarette. Figure 5 illustrates the chromatograms after performing a ten minute extraction on the saliva sample collected prior to smoking marijuana and 30 minutes after smoking. The quadrupole ion trap was operated in the electron ionization mode with both the total ion chromatogram and reconstructed chromatogram for ions at m/z 231, 299, 314, which are characteristic Δ^9-THC ions. Δ^9-THC was readily detected in the saliva sample collected 30 minutes after smoking marijuana, and

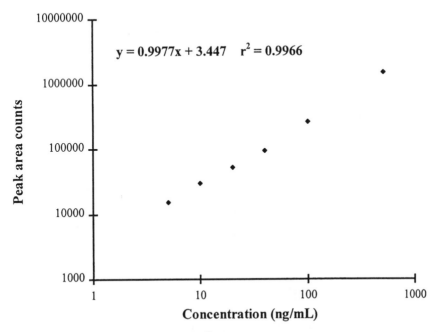

Figure 4 *Log-log calibration curve for Δ^9-THC in human saliva. A 100 μm polydimethyl-siloxane SPME fiber was used to extract Δ^9-THC after treatment of saliva with glacial acetic acid and centrifugation. The precision averaged 15% RSD (n = 6) with no internal standard present*

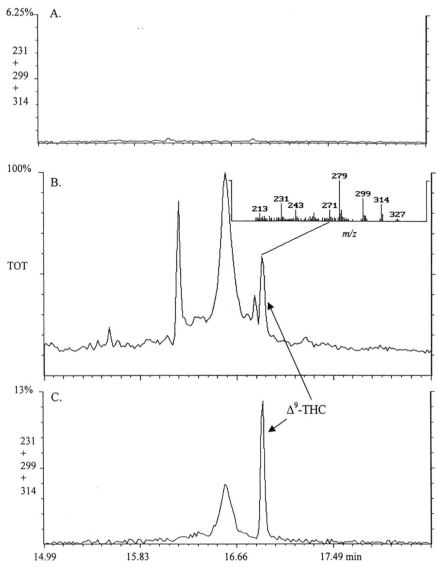

Figure 5 *Chromatograms after SPME of human saliva* (A) *prior to marijuana smoking, select ion plot for Δ⁹-THC.* (B) *30 minutes after smoking, total ion plot with mass spectrum insert for peak at 16.88 minutes indicating characteristic ions for Δ⁹-THC at 231, 243, 258, 271, 299, and 314 m/z.* (C) *Selected ion plot of chromatogram directly above*
(Reprinted from ref. 21, © 1998, American Chemical Society)

thereafter, quantitation was performed by SPME and a conventional liquid/ liquid extraction technique. The results are presented in Table 2. The internal standard SPME method is believed to be the most accurate value based on compensation for any matrix effects occurring during the sample preparation

Table 2 *Quantitation results for* Δ^9*-THC in a marijuana smoker's saliva by SPME and liquid/liquid extraction*

| | SPME (ng mL^{-1})^{a,b} | | | | Liquid/liquid extraction (ng mL^{-1}) | |
| | | | Standard addition | | | |
Compound	External standard	Internal standard	Trial 1	Trial 2	Trial 1	Trial 2
Δ^9-THC	7.70 (16)	9.54 (8.1)	12.2	7.12	7.50	9.52

[a] Values in parentheses represent %RSD ($n = 3$).
[b] SPME conditions: 100 μm PDMS fiber for 10 minute extraction times.
(Reprinted from ref. 21, © 1998 American Chemical Society)

and extraction. Standard addition methods are also accurate; however, in this case limited saliva quantity allowed only two single-point standard addition determinations. Trideuterated Δ^9-THC was utilized as the internal standard for both the SPME and liquid/liquid extraction method. In comparison with a previously reported liquid/liquid method of Δ^9-THC analysis, the SPME method retains excellent specificity and accuracy. The precisions listed in the previous liquid/liquid extraction method of 8.6–11.1% RSD utilizing an internal standard are directly comparable to the internal standard SPME precision of 8.1% RSD.

The results with SPME determination of cannabinoids in saliva obtained in our laboratory indicate the method be to applicable to a clinical setting. With optimized chromatographic settings, it is estimated that 3–4 analyses per hour may be accomplished. Automation of the SPME procedure would further simplify the analyses. Currently, autoinjectors for SPME are being developed and available from Varian (Sugar Land, TX USA) and Leap Technologies (Carrboro, NC USA).

Confirmation of Nordiazepam in Urine by SPME–Tandem Mass Spectrometry

QIT mass spectrometers have always been associated with the ability to perform multiple steps of mass spectrometric analysis (MS^n) sequentially as a function of time, and thus not requiring a multi-stage mass spectrometer. A typical sequence of events in a MS^n experiment involves the isolation of ions of a single mass-to-charge, then activation of the ions *via* energetic collisions to cause dissociation, a process known as collisionally activated dissociation (CAD). Particularly in combination with chemical ionization methods, which afford soft ionization and production of intact molecular ions, CAD offers an extra degree of selectivity in analyte identification by providing a fragmentation 'fingerprint' that is characteristic of a specific structure and may be complementary to or even more informative than an electron ionization mass spectrum. Positive chemical ionization (CI) methods are often used to verify the molecular weight of a substance by formation of protonated species.[22] Additionally, certain CI reagents undergo alternative structurally-specific reactions such as methylation, vinylation or acetylation, thus leading to other types of products that may provide key diagnostic information.[23] These characteristic adducts can be used for selected ion monitoring and for improved structural elucidation by CAD.

To illustrate the potential benefits of CI-CAD detection after SPME analyte introduction into the GC, a comparison was made against traditional EI detection for nordiazepam from a clinical urine specimen declared positive for benzodiazepines by immunoassay tests. Trimethyl borate (TMB) was selected as the chemical ionization reagent because of its ability to form both a protonated adduct and $[M + 73]^+$, which corresponds to the addition of $[B(OCH_3)_2]^+$ to the neutral analyte molecule.[24-27] The formation of the adduct at $[M + 73]^+$ from trimethyl borate CI is commonly observed with neutral molecules containing oxygen substituents, as in nordiazepam which possesses a carbonyl oxygen. Figures 6A and 6B illustrate the chromatographic results with the QIT operated in full scan EI detection and trimethyl borate CI–CAD detection mode, respectively. In both cases, direct immersion SPME was performed on the same urine specimen for a 10 minute extraction time with a 65 μm Carbowax/divinylbenzene SPME fiber. The EI full scan mode chromatogram consists of several distinct intense peaks, the majority of which are extractables from the urine sample. At a retention time of approximately 11.80 minutes, the mass spectrum indicates the presence of nordiazepam, tentatively identified based on the characteristic ion signals at m/z 269 and 242 and agreement of the retention time of standard nordiazepam solutions. However, the mass spectrum showed poor resolution under these conditions and contained additional ions which raised uncertainty in the assignment. Figure 6B illustrates the results of performing trimethyl borate CI and CAD. First, the ion trap was operated in the chemical ionization mode with TMB as the reagent. Secondly, m/z 343 (which corresponds to [nordiazepam + 73]$^+$) was isolated and subjected to CAD. Finally, the resulting CAD fragment ions were detected. The chromatogram in Figure 6B has fewer peaks and better signal-to-noise due to the selectivity of the tandem mass spectrometric sequence. A strong peak for nordiazepam is confirmed by the retention time and characteristic fragmentation pattern of [nordiazepam + 73]$^+$. This fragmentation pattern was independently confirmed by conducting TMB CI–CAD experiments on pure nordiazepam. The CI–CAD methodology benefits the SPME method in cases where background may be substantial and/or additional techniques are required for confirmation of a target analyte. Acceptance of CI–CAD in mass spectrometric strategies for detection of drugs of abuse is dependent on further exploration of novel combinations of CI with CAD and construction of databases for rapid identification of targeted analytes, as exist for traditional EI. Although these methods require an extra degree of sophistication, combining these techniques with SPME for cases in which high background may prohibit accurate quantitation provides a greater level of confidence that the results are based only on the target analyte response.

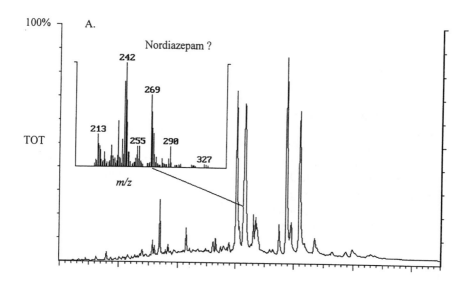

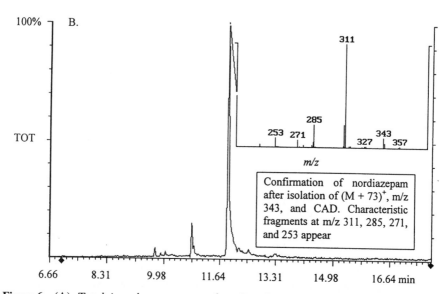

Figure 6 (A) *Total ion chromatogram of a clinical urine sample known positive for benzodiazepines after SPME with the mass spectrometer operated in electron ionization mode. (B) Total ion chromatogram for a separate run of the same urine sample with the mass spectrometer in CI–CAD mode. A ten minute extraction with a 65 μm Carbowax/divinylbenzene fiber was used for SPME. Details of the experimental methods are included in the text*

4 Conclusion

The number of applications of SPME for the analysis of drugs in biological matrices continues to grow each year. From the reports summarized herein, SPME–GC–QIT shows considerable promise as an alternative method of determination for certain classes of drugs. The advantages SPME offers in terms of reduced time and labor in analysis and elimination of solvent in the extraction are crucial to SPME methods being accepted in a clinical setting. When developing novel SPME methods for drug analysis in biological matrices, these advantages should be considered paramount while retaining the sensitivity and detection limits needed for clinical applications.

Through research conducted in our laboratory, quadrupole ion trap mass spectrometry was found to be an excellent method of detection after SPME introduction of several drugs of abuse into a gas chromatograph. The two methods complement each other well, particularly in cases when direct immersion SPME is applied and high levels of background complicate the interpretation of results. Employing chemical ionization and collisionally activated dissociation strategies in conjunction with SPME methods further refines this interpretation.

Acknowledgments

This work was supported by the National Science Foundation (CHE-9357422 and CHE-9421447), the Welch Foundation (F-1155), and the Texas Advanced Technology Program.

References

1. M. Yashiki, T. Kojima, T. Miyazaki, N. Nagasawa, Y. Iwasaki and K. Hara, *Forensic Sci. Int.*, 1995, **76**, 169.
2. N. Nagasawa, M. Yashiki, Y. Iwasaki, K. Hara and T. Kojima, *Forensic Sci. Int.*, 1996, **78** (2), 95.
3. H. Lord and J. Pawliszyn, *Anal. Chem.*, 1997, **69**, 3899.
4. C. Battu, P. Marquet, A.L. Fauconnet, E. Lacassie and G. Lachâtre, *J. Chromatogr. Sci.*, 1998, **36**, 1.
5. A. Ishii, H. Seno, T. Kumazawa, K. Watanabe, H. Hattori and O. Suzuki, *Chromatographia*, 1996, **43**, 331.
6. X. Lee, T. Kumazawa, K. Sato and O. Suzuki, *J. Chromatogr. Sci.*, 1997, **35**, 302.
7. M. Nishikawa, H. Seno, A. Ishii, O. Suzuki, T. Kumazawa, K. Watanabe and H. Hattori, *J. Chromatogr. Sci.*, 1997, **35**, 275.
8. Z. Penton, *Can. Soc. Forens. Sci. J.*, 1997, **30**, 7.
9. W.E. Brewer, R.C. Galipo, S.L. Morgan and K.H. Habben, *J. Anal. Toxicol.*, 1997, **21**, 286.
10. K. Takekawa, M. Oya, A. Kido and O. Suzuki, *Chromatographia*, 1998, **47**, 209.
11. M. Krogh, K. Johansen, F. Tønnesen and K.E. Rasmussen, *J. Chromatogr. B*, 1995, **673**, 299.
12. T. Kumazawa, K. Sato, H. Seno, A. Ishii and O. Suzuki, *Chromatographia*, 1996, **43**, 59.

13. M. Krogh, H. Grefslie and K.E. Rasmussen, *J. Chromatogr. B*, 1997, **689**, 357.
14. Y. Luo, L. Pan and J. Pawliszyn, *J. Microcolumn Sep.*, 1998, **10**, 193.
15. H.G. Ugland, M. Krogh and K.E. Rasmussen, *J. Chromatogr. B*, 1997, **701**, 29.
16. M. Chiarotti, S. Strano-Rossi and R. Marsili, *J. Microcolumn Sep.*, 1997, **9**, 249.
17. S. Ulrich and J. Martens, *J. Chromatogr. B*, 1997, **696**, 217.
18. N.A. Yates, M.M. Booth, J.L. Stephenson Jr. and R.A. Yost in *Practical Aspects of Ion Trap Mass Spectrometry*, Vol. 3, ed. R.E. March and J.F.J. Todd, CRC Press, Boca Raton, FL, 1995, 121.
19. L.D. Bowers in *Handbook of Analytical Therapeutic Drug Monitoring and Toxicology*, ed. S.H.Y. Wong and I. Sunshine, CRC Press, Boca Raton, FL, 1997, pp. 173–199.
20. B.J. Hall and J.S. Brodbelt, *J. Chromatogr. A*, 1997, **777**, 275.
21. B.J. Hall, M. Satterfield-Doerr, A.R. Parikh and J.S. Brodbelt, *Anal. Chem.*, 1998, **70**, 1788.
22. A.G. Harrison, *Chemical Ionization Mass Spectrometry*, 2nd Edn., CRC Press, Boca Raton, FL, 1992.
23. M. Vairamani, U.A. Mirza and R. Srinivas, *Mass Spectrom. Rev.*, 1990, **9**, 235.
24. H. Suming, C. Yaozu, J. Longfei and X. Shuman, *Org. Mass Spectrom.*, 1985, **20**, 719.
25. D.T. Leeck, T.D. Ranatunga, R.L. Smith, T. Partenen, P. Vainiotalo and H. Kenttamaa, *Int. J. Mass Spectrom. Ion Processes*, 1995, **141**, 229.
26. B.J. Hall and J.S. Brodbelt, *Int. J. Mass Spectrom. Ion Processes*, 1996, **155**, 123.
27. E.C. Kempen and J.S. Brodbelt, *J. Mass Spectrom.*, 1997, **32**, 846.

CHAPTER 38

Analysis of Drugs in Biological Fluids Using SPME

AKIRA NAMERA, MIKIO YASHIKI AND
TOHRU KOJIMA

1 Introduction

In clinical and forensic investigation,[1] identification and quantification of organic chemicals which may be the cause of poisoning are most important for diagnostics. The analysis of organic chemicals usually involves a separation from the biological materials, followed by isolation, purification and identification of any chemicals prior to chromatographic analysis. It is difficult, complex and laborious using conventional methods in which the target compounds are separated from biological samples. In addition, clinical and forensic analysis are restricted by a lack of biological samples as well as by the need to provide results quickly. Simple, rapid and sensitive extraction methods are required in clinical and forensic fields.

In our laboratory, to extract organic chemicals from biological samples without the influence of endogeneous matrices, the headspace-SPME technique is used and applied to intoxication cases.

This chapter describes the advantages, applications and problems of drug analysis in biological fluids using SPME.

2 Advantage of Solid Phase Microextraction for Drug Analysis in Biological Samples

Comparing SPME with Conventional Methods

In order to analyze drugs in biological samples qualitatively and quantitatively, the extraction of the target drugs from endogenous matrices is very important. A great number of methods for drug analysis have been published,[2] with liquid/liquid or solid phase extraction methods being the most widely used methods for extracting the target drugs from the biological samples. These methods are thought to have several disadvantages. Liquid/liquid extraction produces an emulsion, and requires a large amount of organic solvent to extract the target drugs. Solid phase extraction methods also usually require large amounts of organic solvents and an evaporation step. These sample preparation procedures are laborious, intensive and very costly. In addition, the organic solvents used are toxic in both the human body and the environment. As another disadvantage, sample contamination is sometimes observed either from the solvent or from the apparatus used for the extraction procedures. In forensic analysis, the potential for contamination must be reduced or eliminated. This could be achieved using the SPME method, because only a vial and an SPME assembly are required for extraction.

Using conventional methods, a large amount of sample is necessary if the concentration of the drug in the biological samples is low. However, when using a large sample, background influences are more likely to have an adverse effect on the analysis. For analysts, the SPME method is convenient because it can extract a usable amount from even a small sample.

Considerations for Extraction

The absolute recovery rate with SPME is less than for conventional methods because it is an equilibrium method and therefore the entire quantity of the target drug cannot be extracted. It has been suggested that interaction between drugs and proteins influences the recovery.[3-8] In order to separate chemicals from proteins, the sample may be deproteinized by gel filtration or by adding deproteinizing reagents such as trichloroacetic acid before SPME sampling.

When drugs in biological samples are analyzed, it is important to consider the decomposition of the target drugs during storage in the refrigerator or during repeated thawing and refreezing of the sample. It is well known that organo-phosphorus agricultural chemicals decompose during refrigerated storage.[9] In our laboratory, we have experienced the decomposition of malathion in a medico-legal urine sample, due to these problems. To avoid this, it is preferable that samples are divided and stored in separate aliquots, and analyzed as quickly as possible.

3   Approach for Biological Samples

The typical SPME extraction procedure is described in Chapter 1. When drugs are extracted from biological samples, it is possible to follow this procedure almost identically. However, the extraction of drugs from a biological sample is very complex compared with extraction from water. With biological samples, a headspace method is preferred.

In our laboratory, sample volumes from 0.2 g to 1.0 g and 12 mL vials made from strengthened glass are used. Where possible, deuterium labeled analogues of the target drugs are used as internal standards. Typically an extraction temperature above the boiling point of the target drug is selected. Several additional conditions should be optimized during method development.

Extraction Mode

If target drugs are extracted from the headspace of biological samples using SPME, non-volatiles cannot be extracted. In this case a direct immersion method is used, endogenous substances may be co-extracted and the analysis equipment may become contaminated. Extractions from headspace are cleaner than those from direct immersion. In headspace SPME, drugs may actually be extracted from steam in the vial, as the extraction temperature is occasionally 100 °C or more.

Extraction Time

In order to obtain the maximum extraction for each target drug, temperature and exposure times were adjusted.

We first applied SPME to biological samples[10] for determining amphetamines in urine, using the headspace method.[11,12] In many reports, samples were heated to equilibrium before extracting with SPME, without taking into consideration the character of the compound. For volatiles, a preheating is certainly effective. Semi-volatiles, however, were transferred to the headspace slowly, and therefore the preheating step was thought to be unnecessary for extracting them.

After reaching equilibrium extraction, the amount extracted theoretically becomes constant, when plotted against the extraction time.[13] In our experience however, the amount extracted from a headspace decreased after equilibrium had been reached (Figure 1).[10] This phenomenon was observed in other drugs such as herbicides, tetracyclic antidepressants and local anesthetics. Although

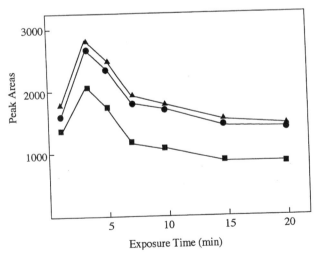

Figure 1   *Correlation of the extracted amount of amphetamines with the exposure time, for SPME analysis in the gas phase:* ■, *amphetamine;* ▲, *methamphetamine;* ●, *[²H₅]methamphetamine*

the reason was not explained clearly, it is suggested that the adsorbed drugs were released from the fiber as it became heated and the fiber–headspace partition coefficient was reduced.

Sample to Improve Extraction

When drugs are extracted from a headspace, the extraction may not be made optimal simply by diluting the sample in water. When basic drugs are extracted, the absolute extraction rate can be increased by increasing sample pH. In addition, the extraction rate can be raised even further through the addition of salts such as sodium chloride. In our laboratory, herbicides could not be extracted from biological samples as they precipitated in the vial when sodium chloride was added. In this case, ammonium citrate was used.

Derivatization

When free amphetamines were analyzed using SPME, difficulties were encountered in sensitivity and reproducibility, because of interaction with the GC column. In order to improve sensitivity and reproducibility, a new method was developed in which the extracted amphetamines were derivatized in the injection port of the GC (on-column derivatization).[14] There are many derivatizing methods available for target drugs: alkylation, acylation, silylation *etc.* Acylation was thought to be better in terms of reaction speed and low contamination of the GC. The best result was obtained using heptafluorobutyric anhydride. In this method,[14] sensitivity was improved remarkably and we were able to measure 0.01 μg g^{-1} of amphetamines in blood. The method reported by Ugland *et al.*[15] is also effective for derivatizing amphetamines. In this method the amphetamines were converted into butylchloroformate derivatives in the vial. The derivatives were then extracted using SPME.

In these reported methods, all the target drugs are amines. However, many metabolites of drugs have an aromatic and/or aliphatic hydroxyl group in the molecule. If these hydroxyl groups could be derivatized, it should be useful for drug screening.

4   Application for Clinical and Medico-Legal Cases

Application of this method has increased over the last few years in the clinical and forensic science fields. It is now possible to analyze many drugs and toxic agents in urine, blood and serum. A list of these is provided in Tables 1–3.[16–43]

The compounds which we analyzed in our laboratory are summarized in Table 4 and examples are discussed below.

Agricultural Chemicals[34,45,46]

In Japan, agricultural chemicals such as pesticides and herbicides are frequently used in suicide attempts. SPME detection methods for the organophosphorus

Table 1 *Published methods for SPME of biological fluids*

Compounds	Specimen[a]	Type[b]	Addition	Vial temp.[c] (°C)	Preheat time[c] (min)	Extraction time (min)	Extraction mode[d]	Detection	Fiber	Ref.
Alcohol										
Ethanol	S	S	NaCl	60	–	15	HS	GC–FID	85 μm PA	16
Ethanol	B, U	P	$(NH_4)_2SO_4$	70	5	15	HS	GC–FID	65 μm CW/DVB	17
Ethanol	U	P	NaCl	60	20	5, 10	HS	GC–MS	100 μm PDMS, 85 μm PA	18
Ethanol	B, U	P	NaF	–	–	3	HS	GC–FID	65 μm CW/DVB	19
Amphetamines[e]										
AM, MA	U	P	K_2CO_3	80	20	5	HS	GC–MS	100 μm PDMS	10
AM, MA	B	P	NaOH	80	20	5	HS	GC–MS	100 μm PDMS	14
AM, MA	U	S	Buffer (pH 9)	40	–	25	DI	GC–MS	100 μm PDMS	16
AM, MA	U	S	Na_2CO_3	65	5	30	DI	GC–NPD	65 μm PDMS/DVB	20
AM, MA	U	S	NaCl, NaOH	–	–	20	DI	GC–FID	100 μm PDMS	21
AM, MA, MDMA	U	S	NaCl	75	30	15	HS	GC–MS	100 μm PDMS	22
AM, MA	U	P	NaCl	60	–	15	HS	GC–FID	65 μm PDMS/DVB, 100 μm PDMS	23
AM, MA	U	S	Buffer (pH 10.8), NaCl			14	DI	GC–NPD	100 μm PDMS	15
Antidepressants										
Amitriptyline, *etc.*	U	S	NaOH	100	30	15	HS	GC–FID	100 μm PDMS	24
Amitriptyline	U	S	Buffer (pH 9), NH_4OH	55	–	25	DI	GC–MS	100 μm PDMS	16
Amitriptyline, *etc.*	B	S	NaOH	100	30	60	HS	GC–FID	100 μm PDMS	25
Imipramine, *etc.*	P	S	NaOH	22	–	10	DI	GC–NPD	100 μm PDMS	3

[a] B, Blood; S, serum; U, urine.
[b] Sample type: S, spiked sample; P, real poisoning or medico-legal sample.
[c] Time for preheating samples before SPME exposure.
[d] Extraction method: HS, headspace; DI, direct immersion.
[e] AM, amphetamine; MA, methamphetamine; MDMA, 3,4-methylenedioxymethamphetamine.

Table 2 *Published methods for SPME of biological fluids*

Compounds	Specimen	Type	Addition	Vial temp. (°C)	Preheat time (min)	Extraction time (min)	Extraction mode	Detection	Fiber	Ref.
Benzodiazepines										
Diazepam	U	S	Buffer (pH 9.1)	55	–	25	DI	GC–MS	100 μm PDMS	16
Diazepam	P[a]	S	Buffer (pH 7)	–	–	4	DI	GC–NPD	7, 100 μm PDMS, 85 μm PA	4
Diazepam, *etc.*	U	S	NaCl	r.t.[c]	–	30	DI	GC–FID	65 μm PDMS/DVB	26
Gasoline, Thinner										
Toluene, *etc.*	B, U	S	Water	80	15	5	HS	GC–FID	100 μm PDMS	27
Toluene, *etc.*	B	P	NaOH	90	10	5	HS	GC–MS	100 μm PDMS	28
Gasoline, kerosene	S		–	80		1–5	HS	GC–FID	100 μm PDMS	29
Toluene, *etc.*	B, U	P	Water	25	25	5	HS	GC–MS	carbon	30
Gasoline	S		–	40	30	20	HS	GC–FID	100 μm PDMS	31
Local anesthetics										
Lidocaine, *etc.*	B[b]	S	NaOH, (NH₄)₂SO₄	100	15	40	HS	GC–FID	100 μm PDMS	5
Lidocaine, *etc.*	B[b]	S	NaCl, NaOH	r.t.[c]	–	40	DI	GC–FID	100 μm PDMS	6
Pesticides										
Organophosphates	B, U	S	(NH₄)₂SO₄, NaCl, HCl or HCl	100	15	20	HS	GC–NPD	100 μm PDMS	32
Carbamates	B, U	S	NaCl	70	10	30	HS	GC–FID	100 μm PDMS	33
Malathion	B	P	(NH₄)₂SO₄, H₂SO₄	90	15	5	HS	GC–MS	100 μm PDMS	34

[a] Plasma and trichloroacetic acid mixture was centrifuged and the supernatant was analyzed.
[b] Blood and perchloric acid mixture was centrifuged and the supernatant was analyzed.
[c] Room temperature.

Table 3 *Published methods for SPME of biological fluids*

Compounds	Specimen	Type	Addition	Vial temp. (°C)	Preheat time (min)	Exposure time (min)	Extraction mode	Detection	Fiber	Ref.
Others										
Nicotine, Cotinine	U	P	K_2CO_3	80	20	5	HS	GC–MS	100 μm PDMS	35
Nicotine, Cotinine	U	S	Buffer (pH 9), NH_4OH	40	–	25	DI	GC–FID	100 μm PDMS	16
Valproic acid	P	P	H_3PO_4	r.t.[b]	–	3	DI	GC–FID	100 μm PDMS	36
Cocaine	U	S	NaF	–	–	30	DI	GC–NPD	100 μm PDMS	37
Meperidine	B[a], U	S	NaOH, NaCl	100	10	30	HS	GC–FID	100 μm PDMS	7
Promethazine, *etc.*	B, U	S	NaOH	140	10	40	HS	GC–FID	100 μm PDMS	38
Phencyclidine	B[a], U	S	NaOH, K_2CO_3	90	10	30	HS	GC–SID	100 μm PDMS	8
n-Butyl nitrite	B	S	–	20–23	60–180	5–20 [60 °C]	HS	GC–FID	100 μm PDMS, 85 μm PA	39
Methaqualone	U	S	Buffer (pH 9.1), NH_4OH	55	–	25	DI	GC–MS	100 μm PDMS	16
Fenfluramine	U	P	Na_2CO_3	–	–	20	DI	GC–MS	30 μm PDMS	40
Cresol(o-, m-, p-)	B	S	NaCl	100	20	30	HS	GC–FID	85 μm PA	41
Antihistaminics	B, U	P	NaOH	98	10	10	HS	GC–FID	100 μm PDMS	42
Opiates	U	S	Buffer (pH 12) or KOH	22, 60	–	15	HS	GC–FID	65 μm PDMS/DVB	23
							DI		100 μm PDMS	
Barbiturates	U	P	NaCl			20	DI	GC–MS	65 μm CW/DVB	43

[a] Blood and perchloric acid mixture was centrifuged and the supernatant was analyzed.
[b] Room temperature.

Table 4 *Summary of methods employed in our laboratory for headspace-SPME in clinical and medico-legal cases*

Compounds	Specimen[a]	Additives	Vial temp.[b] (°C)	Preheat time (min)	Extraction time (min)	Detection	Linear range[d] (µg mL^{-1})	Detection limit[d] (µg mL^{-1})	Fiber	Ref.
Agricultural chemicals										
Malathion	B (0.2 g)	0.1N H$_2$SO$_4$ (2 mL) (NH$_4$)$_2$SO$_4$ (0.2 g)	90	15	5	MS (EI)	2.5–50	1	100 µm PDMS	34
Propanil	S (0.2 mL)	1N NaOH (0.5 mL)	90	–	45	MS (EI)	0.25–10	0.25	100 µm PDMS	45
Cartap	S (0.5 mL) U (0.5 mL)	1N NaOH (1.5 mL)	70	–	30	MS (EI)	0.05–5.0	0.01	65 µm PDMS/DVB	46
Alkaloids										
Nicotine	U (1 mL)	K$_2$CO$_3$ (0.7 g)	80	20	5	MS (EI)	0.01–0.2	0.005	100 µm PDMS	35
Cotinine							0.5–10	0.3		
Amphetamines[c]										
AM, MA	U (1 mL)	K$_2$CO$_3$ (0.7 g)	80	20	5	MS (EI)	0.2–10	0.1	100 µm PDMS	10
AM, MA	B (0.5 g)	1N NaOH (0.5 mL)	80	20	5	MS (EI)	0.01–2	0.01	100 µm PDMS	14
Antidepressants										
Setiptiline	B (0.5 g)	1N NaOH (0.5 mL)	120	–	45	MS (EI)	0.005–5.0	0.002	100 µm PDMS	44
Inflammables										
Toluene, Xylene	B (0.5 g)	10% NaOH (0.5 g)	90	10	10	MS (CI)		0.1	100 µm PDMS	28
Hydrocarbons C$_9$–C$_{15}$								0.1		
Hydrocarbons C$_{16}$–C$_{30}$								1.0		
Local anesthetics										
Lidocaine	B (0.2 g)	5N NaOH (0.8 mL)	120	–	45	MS (EI)	0.1–20	0.01	100 µm PDMS	47
Mepivacaine							0.1–20	0.05		

[a] B, blood; S, serum; U, urine.
[b] See Table 1.
[c] AM, amphetamine; MA, methamphetamine.
[d] When blood was used as the specimen, the mass concentration was µg g^{-1}.

and carbamate pesticides have been published.[32,33] The headspace-SPME method was applied to a suicide case where the cause of death was suspected to be acute malathion poisoning. Figure 2 shows the SIM chromatograms of the case. The concentrations of the left and right heart blood respectively were 41.0 and 5.4 μg g^{-1} for malathion.

Suicide cases involving the ingestion of other agricultural chemicals has also been observed. Figure 3 shows the SIM chromatograms of an attempted suicide case in which the patient ingested a herbicide containing propanil and carbaryl. Propanil was detected in serum samples collected from the patient during the hospitalization in a range from 1.15 to 17.1 μg g^{-1}, respectively. Carbaryl was detected at only trace levels. Many agricultural chemicals are decomposed in high temperature and unstable in alkali or acidic conditions. Under these conditions, part of the extracted carbaryl may decompose in the vial and the injection port of the GC. Because of this, we were unable to distinguish whether 1-naphthol was present due to metabolization or decomposition.

Figure 4 shows the SIM chromatograms of an attempted suicide case in which the subject ingested a herbicide containing cartap.

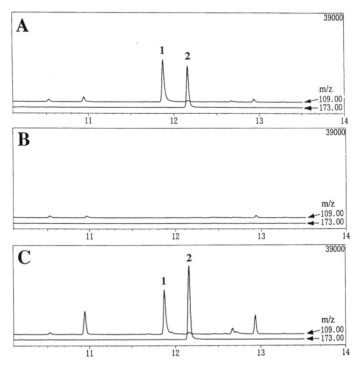

Figure 2 *The SIM chromatograms of spiked blood (A), blood blank (B) and blood sample (C) of a suicide case in which the subject was suspected to have acute malathion poisoning. Peaks: 1 fenitrothion (IS), 2 malathion*

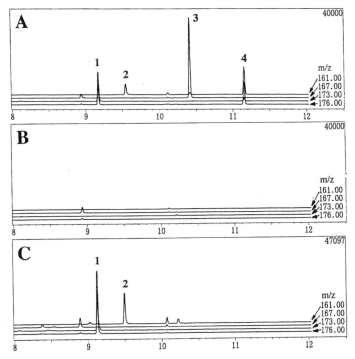

Figure 3 *The SIM chromatograms of spiked serum (A), serum blank (B) and serum sample (C) of a suicide case in which the subject ingested a herbicide containing propanil. Peaks: 1 propyzamide (IS), 2 propanil, 3 diphenamide, 4 butachlor*

Alkaloids[35]

Isolation and purification of nicotine and its principal metabolite cotinine in biological fluids are usually accomplished by conventional liquid/liquid and solid phase extraction methods. No reports on a headspace method for these compounds has been reported to date. We have applied headspace-SPME for the detection of nicotine and cotinine in urine.

A 47 year old male was found dead in the ruins of a fire in his house. At the medico-legal autopsy, soot was detected by naked eye in the trachea and bronchi. The carbon monoxide hemoglobin concentration in the heart blood sample was 42.5%. The victim was thought to be a smoker, because nicotine and cotinine levels in the urine were 97 and 3230 μg g^{-1} creatinine, respectively, measured by the headspace-SPME method. Therefore, careless smoking could be suspected as a cause of this fire.

Amphetamines[10,14]

Amphetamine and its derivatives are common drugs of abuse in many countries. In Japan, methamphetamine abuse is a large social problem. An accurate, simple and rapid method for analysis of methamphetamine and its

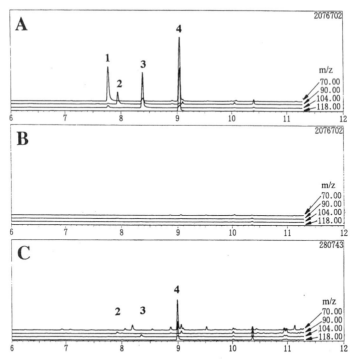

Figure 4   *The SIM chromatograms of spiked serum (A), serum blank (B) and serum
sample (C) of a suicide case in which the subject ingested a herbicide containing
cartap. Peaks: 1 nereistoxin, 2 2-methylthio-1-methylthiomethyl-ethylamine,
3 N-methyl-N-(2-methylthio-1-methylthiomethyl)ethylamine, 4 S,S'-dimethyl-
dihydronereistoxin*

principal metabolite is required for forensic, judicial and clinical purposes. The
detection of amphetamines in urine using a headspace method from an alkali
medium has been reported by Brandenberger,[11] and Tsuchihashi *et al.*[12] In this
method, a large sample volume was injected into the GC. Therefore, this
methodology could not be applied to a capillary gas chromatograph. The
headspace-SPME method was applied to detect amphetamines in the urine of
amphetamine abusers (Figure 5). The detection limit of this method was
0.2 μg mL^{-1} in urine.

Antidepressants[44]

Antidepressants are used for treating depression, and are commonly abused.
Methods for analyzing tricyclic antidepressants by SPME have previously been
reported.[16,24,25] We optimized the conditions for the detection of setiptiline, a
tetracyclic antidepressant, and were able to detect it at therapeutic levels.

We undertook a toxicological analysis of setiptiline in human whole blood
obtained from a suspected victim of setiptiline poisoning. The analysis was

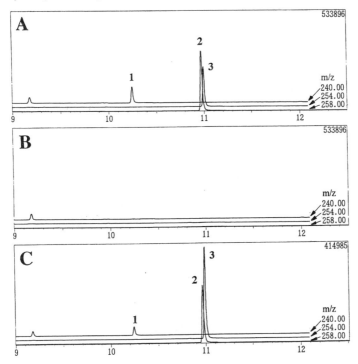

Figure 5 *The SIM chromatograms of spiked urine (A), urine blank (B) and urine sample collected from a methamphetamine abuser (C). Peaks: 1 amphetamine, 2 methamphetamine, 3 [²H₅]methamphetamine (IS)*

performed using the headspace-SPME method. Figure 6 shows SIM chromatograms of setiptiline from left heart blood samples of the victim. The concentrations of setiptiline of the left and right heart blood were 1.77 and 0.78 μg g^{-1}, respectively.

Inflammables[28]

In legal medicine, it is important to specify the inflammable substances present at a fire site. The presence of volatile substances such as benzene, toluene and xylenes in biological fluids has been detected using the headspace-SPME method. The presence of paraffin hydrocarbons, especially C_{13}–C_{20}, in biological fluids has only been detected using liquid/liquid extraction. There have been no reports for simultaneously detecting both volatile substances and paraffin hydrocarbons by the conventional headspace method. The headspace-SPME method was applied to detect inflammable substances in blood.

A 77 year old woman was found in the ruins of a house fire. At the medico-legal autopsy, soot was visible in the airway. The carbon monoxide hemoglobin concentration in the heart blood sample was 100%. Benzene, toluene, *m*- and/or *p*-xylene, *o*-xylene and paraffin hydrocarbons (C_9–C_{16}) were detected in the

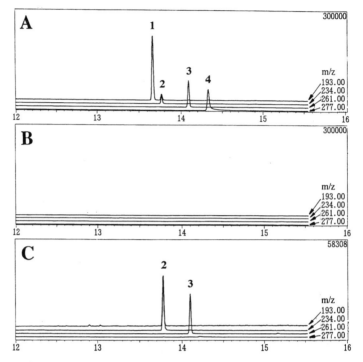

Figure 6   *The SIM chromatograms of spiked blood (A), blood blank (B) and blood sample (C) of a suicide case in which the subject ingested a herbicide containing setiptiline. Peaks: 1 mianserin, 2 imipramine (IS), 3 setiptiline, 4 maprotiline*

blood sample of the victim using the SPME method (Figure 7). It was thus suspected that an inflammable substance such as kerosene was used to set the fire.

Local anesthetics[47]

A toxicological analysis of local anesthetics in human whole blood, obtained from a suspected victim of local anesthetics poisoning, was performed using the headspace-SPME method. EI-SIM chromatograms of local anesthetics from the left heart blood samples of the victim are shown in Figure 8. Mepivacaine and lidocaine were detected in the left and right heart blood samples of the victim. The concentrations in the left and right heart blood were 18.6 and 15.8 μg g^{-1} for mepivacaine, 0.14 and 0.17 μg g^{-1} for lidocaine, respectively. The concentration of mepivacaine in the victim's blood was higher than the recommended therapeutic level (2.0–4.0 μg g^{-1}).[48]

5   Conclusion

It is clear that the SPME technique is sensitive and specific for detecting organic chemicals, not only in urine but also in various other biological materials.

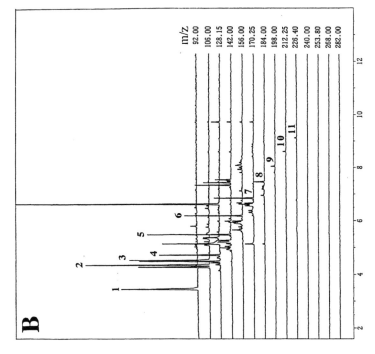

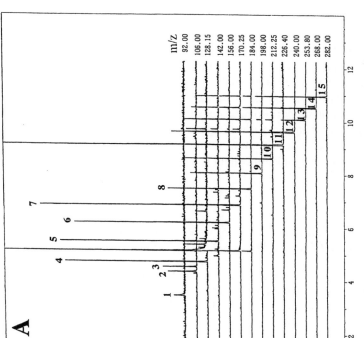

Figure 7 *The SIM chromatograms of spiked blood (A) and blood sample (B) of the victim. Peaks: 1 toluene, 2 xylene(m-, p-), 3 o-xylene, 4 n-nonane, 5 n-decane, 6 n-undecane, 7 n-dodecane, 8 n-tridecane, 9 n-tetradecane, 10 n-pentadecane, 11 n-hexadecane, 12 n-heptadecane, 13 n-octadecane, 14 n-nonadecane, 15 n-eicosane*

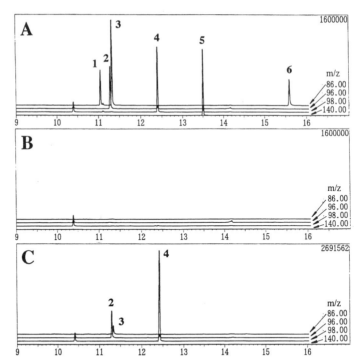

Figure 8   *The SIM chromatograms of spiked blood* (A), *blood blank* (B) *and blood sample* (C) *of the victim. Peaks: 1 prilocaine, 2 [²H₁₀]lidocaine (IS), 3 lidocaine, 4 mepivacaine, 5 bupivacaine, 6 dibucaine*

Although SPME was very useful for monitoring the identified compounds in biological samples, it is slightly difficult to use for identifying compounds that may be the cause of poisoning because the recovery is low. If the SPME method can be used for screening, chemists will be freed from the complex and time consuming sample preparation methods that are currently used. We hope to see an increase in the number of chemicals that can be detected in biological fluids using this method. The time saving and potential improvement in analytical quality could be of significant benefit in forensic and clinical fields.

References

1. H. Brandenberger and R.A.A. Maes, *Analytical Toxicology for Clinical, Forensic and Pharmaceutical Chemists*, Walter de Gruyter, Berlin, 1997.
2. R.P. Muller, *Toxicological Analysis*, Verl. Gesundheit, Berlin, 1991.
3. S. Ulrich and J. Martens, *J. Chromatogr. B*, 1997, **696**, 217.
4. M. Krogh, H. Grefslie and K.E. Rasmussen, *J. Chromatogr. B*, 1997, **689**, 357.
5. T. Kumazawa, X.-P. Lee, K. Sato, H. Seno, A. Ishii and O. Suzuki, *Jpn. J. Forensic Toxicol.*, 1995, **13**, 182.
6. T. Kumazawa, K. Sato, H. Seno, A. Ishii and O. Suzuki, *Chromatographia*, 1996, **43**, 59.

7. H. Seno, T. Kumazawa, A. Ishii, M. Nishikawa, H. Hattori and O. Suzuki, *Jpn. J. Forensic Toxicol.*, 1995, **13**, 211.
8. A. Ishii, H. Seno, T. Kumazawa, K. Watanabe, H. Hattori and O. Suzuki, *Chromatographia*, 1996, **43**, 331.
9. T. Kojima and M. Yashiki, *Jpn. J. Forensic Toxicol.*, 1989, **7**, 7.
10. M. Yashiki, T. Kojima, T. Miyazaki, N. Nagasawa, Y. Iwasaki and K. Hara, *Forensic Sci. Int.*, 1995, **76**, 169.
11. H. Brandenberger, *Mass Spectrometry in Drug Metabolism*, New York, 1977.
12. H. Tsuchihashi, K. Nakajima, M. Nishikawa, S. Suzuki, K. Shiomi and S. Takahashi, *Anal. Sci.*, 1991, **7**, 19.
13. Z. Zhang and J. Pawliszyn, *Anal. Chem.*, 1993, **65**, 1843.
14. N. Nagasawa, M. Yashiki, Y. Iwasaki, K. Hara and T. Kojima, *Forensic Sci. Int.*, 1996, **78**, 95.
15. H.G. Ugland, M. Krogh and K.E. Rasmussen, *J. Chromatogr. B*, 1997, **701**, 29.
16. F. Degel, *Clin. Biochem.*, 1996, **29**, 529.
17. T. Kumazawa, H. Seno, X.-P. Lee, A. Ishii, O. Suzuki and K. Sato, *Chromatographia*, 1996, **43**, 393.
18. W.E. Brewer, R.C. Galipo, S.L. Morgan and K.H. Habben, *J. Anal. Toxicol.*, 1997, **21**, 286.
19. Z. Penton, *Can. Soc. Forensic Sci. J.*, 1997, **30**, 7.
20. A. Ishii, H. Seno, T. Kumazawa, M. Nishikawa, K. Watanabe, H. Hattori and O. Suzuki, *Jpn. J. Forensic Toxicol.*, 1996, **14**, 228.
21. K. Ameno, C. Fuke, S. Ameno, H. Kinoshita and I. Ijiri, *Can. Soc. Forensic Sci. J.*, 1996, **29**, 43.
22. F. Centini, A. Masti and I.B. Comparini, *Forensic Sci. Int.*, 1996, **83**, 161.
23. H.L. Lord and J. Pawliszyn, *Anal. Chem.*, 1997, **69**, 3899.
24. T. Kumazawa, X.-P. Lee, M.-C. Tsai, H. Seno, A. Ishii and K. Sato, *Jpn. J. Forensic Toxicol.*, 1995, **13**, 25.
25. X.-P. Lee, T. Kumazawa, K. Sato and O. Suzuki, *J. Chromatogr. Sci.*, 1997, **35**, 302.
26. H. Seno, T. Kumazawa, A. Ishii, K. Watanabe, H. Hattori and O. Suzuki, *Jpn. J. Forensic Toxicol.*, 1997, **15**, 16.
27. X.-P. Lee, T. Kumazawa and K. Sato, *Int. J. Legal Med.*, 1995, **107**, 310.
28. Y. Iwasaki, M. Yashiki, N. Nagasawa, T. Miyazaki and T. Kojima, *Jpn. J. Forensic Toxicol.*, 1995, **13**, 189.
29. T. Kaneko and M. Nakada, Reports of the National Research Institute of Police Science, 1995, **48**, 107.
30. F. Mangani and R. Cenciarini, *Chromatographia*, 1995, **41**, 678.
31. K.G. Furton, J.R. Almirall and J.C. Bruna, *J. Forensic Sci.*, 1996, **41**, 12.
32. X.-P. Lee, T. Kumazawa, K. Sato and O. Suzuki, *Chromatographia*, 1996, **42**, 135.
33. H. Seno, T. Kumazawa, A. Ishii, M. Nishikawa, K. Watanabe, H. Hattori and O. Suzuki, *Jpn. J. Forensic Toxicol.*, 1996, **14**, 199.
34. A. Namera, M. Yashiki, N. Nagasawa, Y. Iwasaki and T. Kojima, *Forensic Sci. Int.*, 1997, **88**, 125.
35. M. Yashiki, N. Nagasawa, T. Kojima, T. Miyazaki and Y. Iwasaki, *Jpn. J. Forensic Toxicol.*, 1995, **13**, 17.
36. M. Krogh, K. Johansen, F. Tønnesen and K.E. Rasmussen, *J. Chromatogr. B*, 1995, **673**, 299.
37. T. Kumazawa, K. Watanabe, K. Sato, H. Seno, A. Ishii and O. Suzuki, *Jpn. J. Forensic Toxicol.*, 1995, **13**, 207.

38. H. Seno, T. Kumazawa, A. Ishii, M. Nishikawa, K. Watanabe, H. Hattori and O. Suzuki, *Jpn. J. Forensic Toxicol.*, 1996, **14**, 30.

39. J. Tytgat and P. Daenens, *Int. J. Legal Med.*, 1996, **109**, 150.

40. M. Chiarotti, S. Strano-Rossi and R. Marsili, *J. Micro. Sep.*, 1997, **9**, 249.

41. X.-P. Lee, T. Kumazawa, S. Furuta, T. Kurosawa, K. Akiya, I. Akiya and K. Sato, *Jpn. J. Forensic Toxicol.*, 1997, **15**, 21.

42. M. Nishikawa, H. Seno, A. Ishii, O. Suzuki, T. Kumazawa, K. Watanabe and H. Hattori, *J. Chromatogr. Sci.*, 1997, **35**, 275.

43. B.J. Hall and J.S. Brodbelt, *J. Chromatogr. A*, 1997, **777**, 275.

44. A. Namera, T. Watanabe, M. Yashiki, Y. Iwasaki and T. Kojima, *J. Anal. Toxicol.*, in press.

45. A. Namera, T. Watanabe, M. Yashiki, Y. Iwasaki and T. Kojima, *Forensic Sci. Int.*, in press.

46. A. Namera, T. Watanabe, M. Yashiki, T. Urabe and T. Kojima, unpublished data.

47. T. Watanabe, A. Namera, M. Yashiki, Y. Iwasaki and T. Kojima, *J. Chromatogr. B*, 1998, **709**, 225.

48. F.P. Meyer, *Int. J. Clin. Therap.*, 1994, **32**, 71.

CHAPTER 39

SPME–Microcolumn LC: Application to Toxicological Drug Analysis

KIYOKATSU JINNO, MASAHIRO TANIGUCHI,
HIROKAZU SAWADA AND MAKIKO HAYASHIDA

The technique of interfacing SPME (solid phase micro extraction) and microcolumn LC (micro-LC) has been evaluated for benzodiazepines analysis in human urine. Five SPME fibers were evaluated: polyacrylate (PA), Carbowax/template (CW/TEP), sol-gel C_{11} polydimethylsiloxane (sol-gel PDMS), poly(dimethylsiloxane) (PDMS) and methyl-octyl poly(dimethyl-siloxane) (C_8 PDMS). Method parameters considered were extraction efficiency, extraction time and total analysis time. Using the CW/TEP fiber and a 1.0 mm i.d. LC column for separation, the analysis of benzodiazepines in patient's urine can be performed in 2 hours, with very low solvent consumption for the whole analytical procedures (less than 1.5 mL solvent).

1 Introduction

The screening and confirmation of drug presence in clinical samples is important to clarify their therapeutic and toxic effects. Benzodiazepines are used clinically as a tranquilizer of the central nervous system. They will also produce toxic side effects in the case of overdose. Thus, toxicological drug confirmation is needed for these drugs to provide valuable information to physicians who are faced with drug poisoning patients.[1]

In human fluids analysis, a sample preparation technique is often necessary to extract and concentrate organic compounds of interest from the biological matrix. One technique for this purpose is solid phase microextraction (SPME), a solvent free sample preparation technique, which has been recently developed.[2] In SPME, the outer surface of a solid fused-silica fiber is coated with a selective polymeric material. Extraction is carried out by simply dipping the coated fiber into the sample matrix and allowing time for the partition equilibrium to be established. The amount of an analyte extracted by the coating can be described by Nernst's partition law. The partition coefficient of an analyte between the coating and the sample matrix also determines the sensitivity of the method. Selective extraction can be achieved using appropriate polymeric materials that exhibit high affinity toward the target analytes. Therefore the choice of coating is important in SPME method development. SPME is normally followed by gas chromatography (GC), in which the extracted analytes are thermally desorbed in the GC injector port, for introduction onto a GC column.[3–7] However, GC analysis is not suitable for non-volatile or thermally unstable compounds including some benzodiazepines. In order to apply SPME to these compounds, it must be coupled with other separation methods such as liquid chromatography (LC) or capillary electrophoresis (CE). Recently SPME coupled with LC and CE has been reported.[8–13] In our previous study SPME–semi-microcolumn LC (1.5 mm i.d.) was evaluated for drug analysis and the work showed that the method was useful for benzodiazepines in human urine without tedious and complex analytical procedures.[14] The decrease in LC column diameter brings several advantages. The most important feature is the combination of micro-column LC and mass spectrometry (MS) which enables the use of less solvent as mobile phase because MS can provide useful infomation about unknown compounds. In addition, mass sensitivity in LC analysis is improved and trace analysis can be realized by microcolumn LC separations (micro-LC).

In this chapter, small diameter LC columns (1.0 mm i.d.) were used in conjunction with SPME (SPME–micro-LC) in order to reduce the consumption of organic solvents. This SPME–micro-LC system can be a very important analytical method for this kind of toxic drug analysis in terms of speed, labor required and ecological impact.

2 Experimental

Materials and Reagents

The holder and the SPME assembly for manual sampling were purchased from Supelco (Bellefonte, PA, USA). The fibers used were 1.0 cm long polyacrylate, 85 μm thickness (PA), Carbowax/template resin, 100 μm thickness (CW/TEP) and poly(dimethylsiloxane), 100 μm thickness (PDMS). A new type sol-gel C_{11}PDMS fiber with 50 μm thickness and methyl-octylpoly(dimethylsiloxane) (C_8 PDMS) fiber with 50 μm thickness were provided by Supelco as a gift. Each new fiber was conditioned by immersing it in acetonitrile until interfering peaks were disappeared in LC chromatograms.

All solvents were reagent grade, purchased from Kishida Chemical (Osaka, Japan) and deionized water was obtained from a Milli-Q water system (Millipore, Tokyo, Japan).

Apparatus

Micro-LC was performed with an Nanospace SI-1 (Shiseido, Tokyo, Japan), which consists of a pump, a UV–Vis detector, a column oven and a degasser.

Capcell pak ODS columns (250 mm × 1.0 mm i.d., and 150 mm × 1.0 mm i.d.) and a Rheodyne 7125 injector (Cotati, CA, USA) with a 1 µL loop were used. The flow rate of the mobile phase was 50 µL min^{-1} and the column temperature was controlled at 35 °C. The mobile phase was acetonitrile/5 mM phosphate buffer (40/60) and the detection wavelength was 220 nm. BORWIN chromatography software (Jasco, Tokyo, Japan) was used for data acquisition and handling.

SPME–Micro-LC Procedure

The two steps in an SPME extraction are equilibrium between analytes and the fiber coating, and desorption to the mobile phase. In SPME–GC the fiber is transferred to GC injector as soon as the extraction is finished, and the analytes are thermally desorbed, separated on the column and quantified by the detector. Although the LC extraction procedure is similar to that used for GC analysis, the desorption procedure is different. For SPME–LC a module to desorb analytes from the fiber coating and introduce them on to the column has previously been described.[13] We have constructed a similar interface to couple SPME with micro-LC. Our group previously used a similar interfacing device to analyze pesticides in environmental water samples by an SPME–semimicro-LC system.[8] The big difference between the device above[13] and ours[8] is the volume of the desorption chamber. In this investigation the interface was constructed to have a smaller volume for increasing extraction efficiency and sensitivity, when micro-LC is used. The interface used has a volume of less than 30 µL. The interface consists of stainless-steel tee, connecting fittings, stainless-steel tubing, PTFE tubing and ferrules, which are parts normally used for GC, LC and SFE. The schematic diagram of this interface is shown in Figure 1. This interface is connected to a regular six-port injection valve in the position of the injection loop. Fibers are loaded when the injection valve is in the load position.

After the extraction step, the SPME fiber is withdrawn and inserted into the desorption device filled with mobile phase as the desorption solvent. After desorption for a prescribed amount of time, the injection valve is changed to the load position and a certain amount of the solvent is flushed through the interface. Immediately the desorption solvent containing sample analytes is introduced on to the column and the injection valve is changed to the injection position.

In this work, extraction was performed in a sealed 20 mL vial containing 15 mL of Milli-Q water spiked with three standard drugs (each 100 ppb), a cylindrical-shaped stirring bar (4 × 6 mm) and 6 g sodium chloride. The fiber was introduced carefully into the aqueous phase and the sample was stirred at 1200 rpm at 60 °C for 1.0 h. The fiber was then withdrawn and analytes were desorbed for a period of 30 min in the specially designed interface described above, filled with the mobile phase as the desorption solvent, for micro-LC analysis. The details of this procedure were described previously.[13]

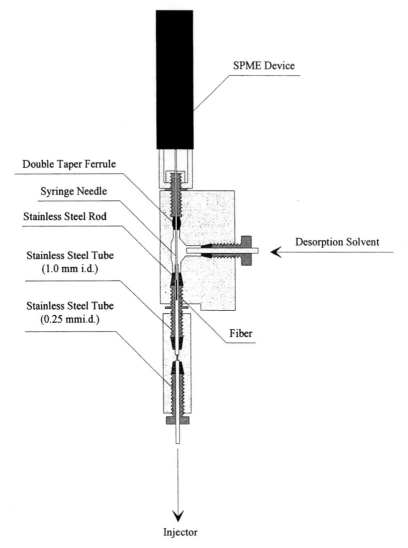

Figure 1 *Schematic of the SPME–micro-LC interface device*

3   Results and Discussion

The choice of an appropriate fiber coating is essential in the SPME method development. Depending on the molecular weight and polarity of the analytes to be extracted, the sensitivity of each fiber coating is different. Five types of fiber coatings were evaluated: sol-gel C_{11} poly(dimethylsiloxane) (sol-gel PDMS), polyacrylate (PA), Carbowax/template (CW/TEP), methyl-octyl poly(dimethylsiloxane) (C_8 PDMS) and poly(dimethylsiloxane) (PDMS). The chemical structures of these are shown in Figure 2.

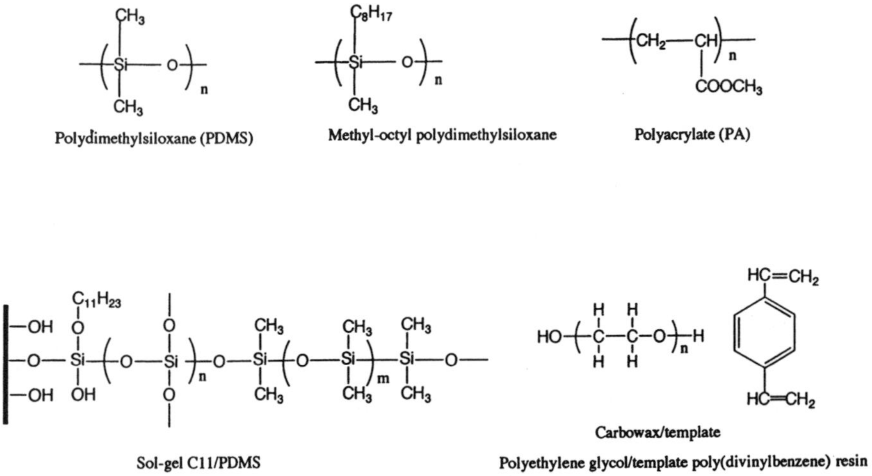

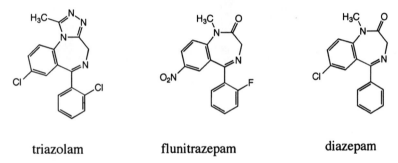

Figure 2   *Structures of polymeric materials for SPME fiber coatings*

The conditions established to investigate the performance of the five fiber coatings were as follows: extraction temperature was 60 °C, the extraction time was 60 min, the matrix was modified with phosphate buffer (pH 6.8) and saturated NaCl, the desorption time was 30 min and the desorption solvent was the mobile phase with a flush volume of 6 μL. Most of these parameters were optimized in our previous work.[13] Saturation with salt can be used to not only improve the extraction efficiency, but also to normalize random salt concentration in human fluids. The 1,4-benzodiazepines such as triazolam, flunitrazepam and diazepam were selected as target compounds. Their structures are shown in Figure 3. These compounds are considered to be weak bases.

Although these compounds have a certain degree of hydrophobicity, they can form ionic species easily, and thus can be classified as ionogenes.[15] Therefore the adjustment of matrix pH to promote the neutral forms of the drugs is important to effectively extract this group of compounds. The extraction performance of five fibers for three compounds was evaluated under the optimum SPME conditions and the results are summarized in Figure 4. As can be seen in

triazolam	flunitrazepam	diazepam

Figure 3   *Structures of 1,4-benzodiazepines used in this study*

Figure 4, the sol-gel C_{11} PDMS fiber gives the highest extraction efficiency of the five coatings for benzodiazepines. The sol-gel type coating enhances the surface area and has very strong thermal stability relative to the commercial PDMS fiber coating. The presence of the hydroxyl group in the sol-gel structure makes the coating more polar relative to PDMS coating (Figure 2). Therefore the sol-gel type coating can extract both polar and non-polar compounds.[16] Although the more polar CW/TEP fiber coating was successful in extracting benzodiazepines, the amounts of the analytes extracted by the CW/TEP were lower than for the sol-gel C_{11} PDMS fiber coating in this evaluation. The CW/TEP fiber coating, however, showed a higher extraction efficiency than the PA fiber, which was used to extract benzodiazepines previously.[13] CW/TEP is the more polar than PA and has been used to extract alkylphenol ethoxylate surfactant.[17] The PDMS and the C_8 PDMS did not effectively extract these compounds relative to other coatings. These coatings are more suited for extracting non-polar analytes. It is expected that the C_8 PDMS may extract more effectively than PDMS for non-polar analytes owing to a higher hydrophobicity with the C_8

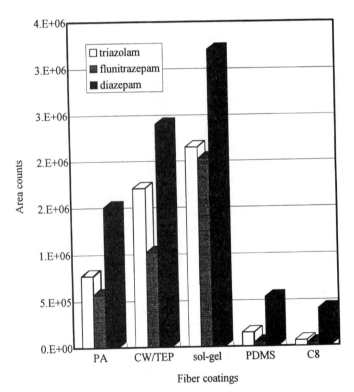

Figure 4   *Extracted amounts estimated by the peak area counts in LC chromatogram with various fiber coatings. Concentration of each analyte, 0.1 μg mL^{-1}; extraction time, 60 min; extraction temperature, 60 °C; desorption time, 30 min; desorption solvent, mobile phase (acetonitrile: 5 mM phosphate buffer, 40:60); desorption temperature, r.t.*

PDMS. It has been concluded by these results that the sol-gel C_{11} PDMS and the CW/TEP fibers are much better coatings than PA for benzodiazepine analysis.

The effect of matrix pH for sol-gel C_{11} and CW/TEP was next investigated in this study. As can be seen in Figures 5a and 5b, the matrix pH considerably affects the extraction efficiency. These results are consistent with those in the previous paper.[13] As pH increases, more of the basic compounds are present in the neutral form, which partitions into the coating, resulting in higher extraction efficiency. For full conversion of basic species to neutral form, the pH should be at least two units above the pK_a of an analyte.

The absorption–time profiles of the sol-gel C_{11} PDMS and the CW/TEP fibers were then studied by monitoring the LC peak area counts as a function of extraction time, as shown in Figure 6. It takes 120 min to reach equilibrium when the sol-gel C_{11} PDMS fiber was used (Figure 6a), while the CW/TEP requires 60 min to reach the equilibrium except the case of diazepam (Figure 6b). The extraction speed in SPME is controlled by mass transfer from the sample matrix to the fiber coating. The fiber coating thickness influences the time required to reach equilibrium, and the amount of analyte extracted.[17] As the thicknesses of these two fibers are the same, 50 μm, the difference in equilibration time is due to differing partition coefficients. Details of these behaviors will be found in pp. 61–68 in Pawliszyn's book.[18]

Carryover, described as the ratio of the amount of analytes remaining on the fiber after the first desorption to the amount of the total analytes absorbed, was also investigated. After the first 30 min desorption, the carryover of triazolam, flunitrazepam and diazepam with the CW/TEP fiber was 9.1, 14.4 and 18.6%, respectively. For the sol-gel C_{11} PDMS fiber the carryover was 20.3, 9.9, and 13.3%, respectively. Although the carryover is reduced by a longer desorption time, desorption times longer than 30 min are too long for practical use. Recently Pawliszyn *et al.*[19] reported that heating during the desorption process in SPME–LC was very effective to reduce the carryover on the fiber coating. This fact means that increasing the temperature in the desorption process can reduce the distribution constant between the fiber coating and the desorption solvent and increase the diffusion rate in the process.

Figure 7 illustrates the effect of the desorption solvent temperature on the carryover. In this work, heated mobile phase buffer solution was used as desorption solvent, and the carryover using two sol-gel C_{11} and the CW/TEP coatings was measured. The results are summarized as follows: the carryover of triazolam, flunitrazepam and diazepam at 60 °C desorption was 10.6, 6.5, and 10.1% with the sol-gel C_{11} (Figure 7a) and 6.3, 8.3 and 11.9% with the CW/TEP (Figure 7b), respectively. The results clearly show that preheated desorption solvent reduces carryover and enhances the total efficiency in SPME. Therefore, a mobile phase preheated at 60 °C was used with the desorption solvent in further evaluation.

In order to see the actual performance of this SPME–micro-LC system for the analysis of benzodiazepines in urine, the CW/TEP coating was used because of its better extraction speed and reasonable extraction efficiency, even though the

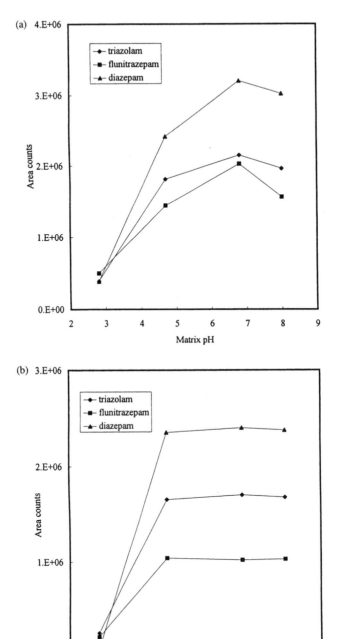

Figure 5   *The effect of matrix pH on the extraction efficiency with sol-gel C_{11} PDMS fiber coating (a) and CW/TEP fiber coating (b). Other conditions are the same as in Figure 4*

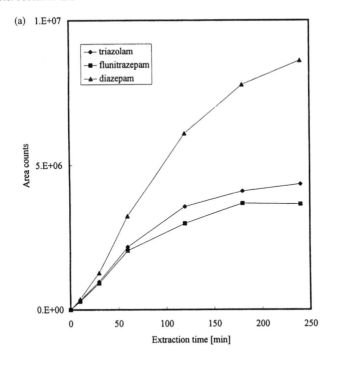

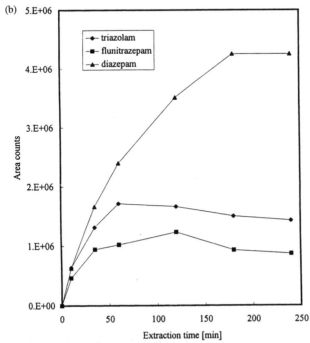

Figure 6 *Extraction time profiles with* (a) *sol-gel C_{11} PDMS fiber coating and* (b) *CW/ TEP fiber coating. Other conditions are the same as in Figure 4*

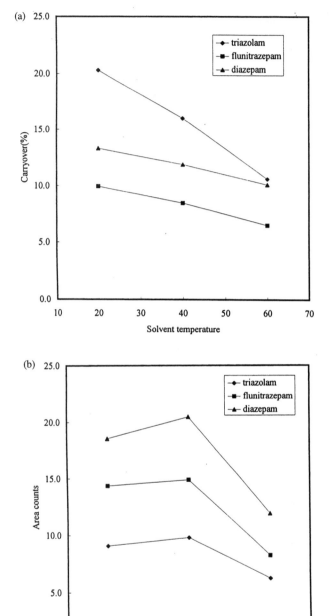

Figure 7   *Effect of desorption solvent temperature on analyte carryover with sol-gel C_{11} PDMS fiber coating (a) and CW/TEP fiber coating (b). Other conditions are the same as in Figure 4*

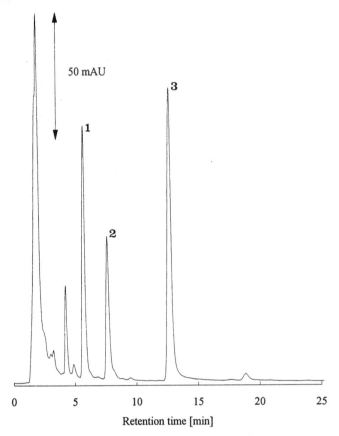

50 mAU

Figure 8 *Chromatogram of controlled-urine sample spiked by standard drugs (0.1 μg mL^{-1}). Column; CAPCELL PAK ODS 1.0 mm i.d. × 150 mm long. Peaks: 1, triazolam; 2, flunitrazepam; 3, diazepam*

sol-gel C$_{11}$ coating gave better extraction efficiency. As for the first experiments, three benzodiazepines (triazolam, flunitrazepam, and diazepam) were spiked in a controlled-urine at 0.1 μg mL^{-1} and the sample was then diluted 10-fold by the phosphate buffer (pH = 6.8) for the actual SPME–micro-LC analysis. Figure 8 shows the chromatogram by SPME–micro-LC analysis for the urine sample. The results clearly indicate that SPME–micro-LC is very effective for this type of analysis.

The usefulness of the method for analysis of actual patient urine was next investigated. The sample was taken from the patient who was suspected as positive for benzodiazepines by a previous screening. A chromatogram for analysis of patient urine by SPME–micro-LC analysis is shown in Figure 9. By comparison to the chromatogram for direct injection of a standard drugs sample, the peak appearing at 9.5 min retention time can be identified as triazolam. Although the identification in this case is only based on a retention time match to the standard sample, the SPME–micro-LC has been shown to

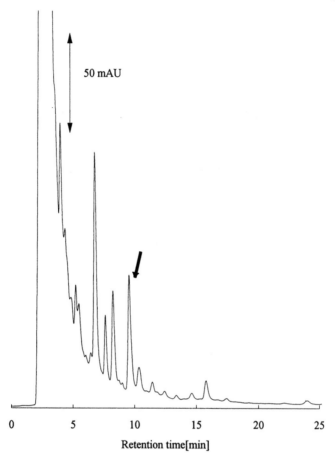

Figure 9 *Chromatogram of actual patient urine sample by SPME–micro-LC analysis.*
Column; CAPCELL PAK ODS 1.0 mm i.d. × 250 mm long

offer much better feasibility to couple to other spectroscopic identification
methods such as mass spectrometry (MS), because the low mobile phase flow
rate can provide easier interfacing between LC and MS.[14]

4 Conclusion

In this chapter it has been clearly suggested that a small internal diameter LC
column (1.0 mm i.d.) can be useful for the analysis of benzodiazepines in a
human urine sample by coupling to the solventless sample preparation technique, SPME, with a reduction in the total consumption of organic solvent in
analytical procedure. Five SPME fiber coatings (PA, CW/TEP, sol-gel C_{11}
PDMS, PDMS and C_8 PDMS) were compared for their extraction performance, based on extraction efficiency and extraction time required. Although the
sol-gel C_{11} PDMS was experimentally found to offer the highest extraction

efficiency for benzodiazepines in aqueous media, the CW/TEP fiber coating can give more rapid extraction equilibrium. Therefore the latter is more suitable for the actual sample analysis because analytical speed is the first priority in clinical emergency room analysis and many toxicological analytical situations. By using the CW/TEP coating as the SPME fiber and 1.0 mm i.d. column for the separation a rapid and nearly solventless analytical procedure is realized for toxicological and forensic drug analysis. The method proposed requires an analysis time of less than 2 hours with consumption of organic solvents of less than 1.5 mL.

References

1. R.B. Taylor, R.G. Reid, R.H. Behrens and I. Kanfer, *J. Pharm. Biomed. Anal.*, 1992, **10**, 867.
2. C.L. Arthur and J. Pawliszyn, *Anal. Chem.*, 1990, **62**, 2145.
3. M. Chai and J. Pawliszyn, *Environ. Sci. Technol.*, 1995, **29**, 693.
4. T. Górecki, R. Mindrup and J. Pawliszyn, *Analyst*, 1996, **121**, 1381.
5. W.H.J. Vaes, E.U. Ramos, H.J.M. Verhaar, W. Seinen and J.L.M. Hermens, *Anal. Chem.*, 1996, **68**, 4463.
6. I. Valor, J.C. Molto, D. Apraiz and G. Font, *J. Chromatogr. A*, 1997, **767**, 195.
7. P.A. Martos and J. Pawliszyn, *Anal. Chem.*, 1997, **69**, 206.
8. K. Jinno, T. Muramatu, Y. Saito, Y. Kiso, S. Magdic and J. Pawliszyn, *J. Chromatogr. A*, 1996, **754**, 137.
9. J.L. Liao, C.M. Zeng, S. Hjerten and J. Pawliszyn, *J. Microcolumn Sep.*, 1996, **8**, 1.
10. R. Eisert and J. Pawliszyn, *Anal. Chem.*, 1997, **69**, 3140.
11. A.L. Nguyen and J.H.T. Luong, *Anal. Chem.*, 1997, **69**, 1726.
12. K. Jinno, Y. Han, H. Sawada and M. Taniguchi, *Chromatographia*, 1997, **46**, 309.
13. J. Chen and J. Pawliszyn, *Anal. Chem.*, 1995, **67**, 2530.
14. K. Jinno, M. Taniguchi and M. Hayashida, *J. Pharm. Biomed. Anal.*, 1998, **17**, 1081.
15. A. Cataby, M. Taniguchi, K. Jinno, J. J. Pesek and E. Williamsen, *J. Chromatgr. Sci.*, 1998, **36**, 111.
16. S.L. Chong, D. Wang, J.D. Hayes, B.W. Wilhite and A. Malik, *Anal. Chem.*, 1997, **69**, 3889.
17. A.A. Boyd-Boland and J. Pawliszyn, *Anal. Chem.*, 1996, **68**, 1521.
18. J. Pawliszyn, *Solid Phase Microextraction: Theory and Practice*, Wiley-VCH, New York, 1997.
19. H. Daimon and J. Pawliszyn, *Anal. Commun.*, 1997, **34**, 365.

CHAPTER 40

Optimization of Drug Analysis by SPME

HEATHER LORD

1 Introduction

To date, most efforts in method optimization for SPME drug analysis have focused on improving sensitivity and limits of detection for forensic applications. Sensitivity, while often important, is not always the overriding goal to address. Endpoints for optimisation may be method accuracy and precision, stabilization of the active ingredient, speed and simplicity of the method to maximize throughput, or in automating the overall method as extensively as possible. In some pharmaceutical analyses, adjustment of the extraction mixture to account for formulation components may be the overriding goal. In other cases, the quantification of degradation products, or other contaminants such as solvent residues, may be of greater significance than the analysis of the active ingredient itself. In the analysis of drug mixtures, for instance combinatorial product mixtures, the emphasis may be on achieving some minimal level of extraction for a broad range of products, rather than maximizing extraction of any one component. In other cases, the emphasis may be on selectively extracting products exhibiting an affinity for an extracting phase which has been selected to mimic biological endpoints such as a specific drug receptor or degree of hydrophobicity. Thus the goal for optimization, in relation to the application at hand, should be clear before decisions are made in the selection of method parameters.

A general understanding of the parameters which impact extraction is important for method development, regardless of the specific goals. In the following pages, we discuss the use of SPME for the analysis of drugs by both GC and HPLC methods, with regard to the most commonly studied extraction parameters and impacts they may have for different method applications.

Extraction Temperature

The temperature of the extraction mixture has two important effects on extraction. First, increased temperature decreases the distribution of drug between the fibre coating and the extraction mixture. Therefore a lower equilibrium amount of drug is extracted at elevated temperature. With an increase in extraction temperature, diffusion is also enhanced, which shortens the equilibrium extraction time. Thus, it is often reported that an increase in extraction temperature results in an increased amount of drug extracted. This occurs because, at lower temperature, extraction is further from equilibrium, and therefore a low level of analyte is extracted. At higher temperatures under the same extraction time, however, the absorption–time profile will be closer to equilibrium, and therefore the amount extracted is generally greater. This effect also explains why several researchers have found that extraction reaches a maximum at a certain temperature, with additional increases in extraction temperature resulting in a lowering of amount of drug extracted.[1,2] Figure 1 presents experimental data, from the headspace extraction of methamphetamine, and demonstrates this relationship.[3] If one were to study the effect of temperature on amount extracted with a 5 minute extraction, the result would be a fairly low level of extraction at room temperature, a higher level at 40 °C, an additional increase in extraction at 60 °C, and then a lowering of extraction level at 73 °C. For a 60 minute extraction, 40 °C would produce maximal extraction. The room temperature curve would eventually cross the 40 °C curve, so that when all extractions are at equilibrium, room temperature would produce the highest level of extraction.

As is seen here, equilibrium extraction may require many hours, which is impractical for most applications. This is often encountered for compounds having relatively high partition coefficients. Employing either elevated extrac-

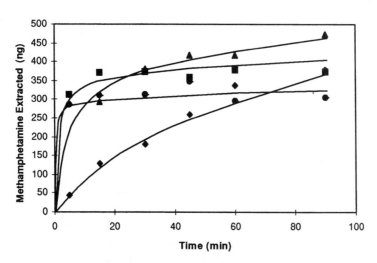

Figure 1 *Effect of extraction temperature on equilibrium time and amount extracted, for headspace methamphetamine analysis. Extraction conditions: PDMS 100 μm fibre, 0.5M KOH, saturated NaCl, 2 mL sample in a 4 mL vial, extraction temperature as shown, extraction time 15 min, analysis by GC–FID, desorption time 15 min. Key: (◆): 22 °C; (▲): 40 °C; (■): 60 °C; (●): 73 °C*

tion temperature or non-equilibrium extraction normally circumvents this problem. Occasionally elevated extraction temperature is not practical. It may speed compound degradation, or may not sufficiently reduce equilibrium extraction time. Non-equilibrium extraction can produce poor method precision; however, if extraction time and agitation conditions are carefully controlled or automated, this is not normally a factor. Extractions with very long equilibrium extraction times generally have small initial slopes of extraction *versus* time. This can be seen by comparing the initial slopes of the curves in Figure 1. The 20 °C extraction profile, which has a very long equilibrium extraction time, also has the lowest initial increase in amount extracted per unit time. Thus minor variances in extraction time at cooler extraction temperatures would not produce the same variance in amount extracted as would be seen at higher temperatures. Thus, under non-equilibrium extraction where the initial rate of increase is slow, method precision is better than for compounds with steeper initial extraction slopes.

In some cases, a thermally labile compound may require a lower extraction temperature, at the expense of method speed or sensitivity. Conversely, a drug contained in a hydrophobic formulation, such as a cream or ointment, may require an elevated extraction temperature to ensure a homogeneous extraction phase, as extraction from a heterogeneous extraction matrix can present several complications.

If the extraction matrix is composed of several immiscible phases, such as may be the case for a cream or ointment, the ability of the fibre to extract analyte will be related to the capacities of other phases present. While the distribution constant of an analyte between the fibre coating and the sample (K_{fs}) is independent of the number of phases existing during extraction, an immiscible organic liquid phase may have a very high capacity for the analyte of interest. If the diffusion coefficient of the analyte in this phase is low, then the mass transfer may be slow and the extraction process may be kinetically limited.[4] Additionally, if one of the phases is an immiscible solvent or oil, it may interact with the coating itself, swelling it or altering its physical properties, such that the distribution of the analyte into the coating is altered. In our experience, an adjustment of the matrix temperature and solvent concentration can improve the homogeneity of the extraction matrix.

Extraction pH

The pH of the extraction mixture is particularly important for drugs possessing a pH dependent dissociable group. It is only the undissociated form of the drug that will be extracted by an absorptive-type of fibre coating (PDMS or PA). This is important for extraction as drug that has partitioned into the fibre coating does not participate in the Henderson–Hasselbalch equilibrium between acid and base forms of a drug in an aqueous extraction mixture. In the case where the extraction mixture pH is controlled with a buffer, as the undissociated form of the drug is extracted by the fibre, more dissociated drug undissociates and is therefore available for extraction. Thus, in a buffered extraction mixture, more

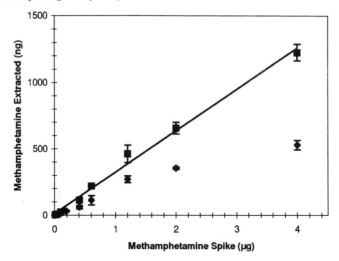

Figure 2 *Effect of pH adjustment method on methamphetamine calibration using head-space SPME. Conditions: PDMS 100 μm fibre, extraction for 15 minutes at 60 °C, saturated NaCl, 2 mL sample in a 4 mL vial, base adjustment by (◆): 0.5M KOH (pH > 12) or (■): 0.25M phosphate buffer pH 12. Analysis by GC–FID, desorption time 15 min*

drug can be extracted by an absorptive fibre coating than in an extraction mixture where the pH is not buffered. In an unbuffered extraction mixture, the ratio of undissociated to dissociated forms of the drug can vary. Therefore, one does not achieve the continual transfer of drug from the dissociated to the undissociated form, and then to the fibre coating. This effect is seen in Figure 2, which compares the amount of methamphetamine extracted from a solution where the pH is base adjusted with KOH, *versus* one where the pH is controlled at 12 with phosphate buffer.[3] Note that there is a significant deviation from linearity in amount extracted as drug concentration increases, in the non-buffered system.

Many drug formulations designed to be taken orally utilize the salt of the active ingredient, to aid dissolution in the stomach. Extraction pH will be quite significant in such a case, to ensure the undissociated form of the drug is present for extraction. Additionally, the formulation matrix of a pharmaceutical preparation may itself act to buffer the extraction matrix, and final extraction pH should be checked before extraction.

Salt Concentration in the Extraction Matrix

The addition of inorganic salt to the extraction mixture will initially increase the amount of undissociated drug extracted by the fibre coating. Salt ions in solution have two opposing effects on analytes in solution. As salt is added to solution, water molecules are tied up in hydration spheres around the salt ions. This reduces the availability of free water molecules for dissolving the analyte.

Thus the activity, or effective concentration of analyte in solution, increases and more analyte will distribute into the fibre. Conversely, as salt concentration continues to increase, the salt may itself interact with the analyte in solution, possibly either electrostatically, covalently as in the case of organic salts or through ion-pairing.[5,6] This will reduce the ability of the drug to move into the fibre coating, reducing the amount extracted. As dissociated drug will readily interact with salt ions in solution, this interaction is significant when the dissociated form is present. Thus salt normally decreases the affinity of the dissociated form of a drug for the fibre coating, whereas the affinity of the undissociated form of a drug is normally enhanced initially, and may then be decreased.

Recently the effect of salt on the SPME extraction of a series of barbiturates has been demonstrated.[7] In the series of drugs tested, most had pK_a values close to 8.0. Phenobarbital had the lowest pK_a value of the series, at 7.2.[8,9] Salt concentrations of 0, 25, 50, 75 and 100% saturated were tested. At all salt concentrations, phenobarbital showed the lowest enhancement of extraction relative to 0% salt, and for phenobarbital only salt concentrations above 50% produced a reduction in amount of drug extracted. All others in the series showed enhancement of extraction at all salt concentrations. For barbiturates, the acid form of the drug is undissociated. In these experiments, pH was not controlled, and extractions were carried out in aqueous solution. All drugs were added in the acid form. From the data, it would appear that phenobarbital is subject to both the enhancement of extraction by the hydration sphere effect and the inhibition of extraction due to the direct interaction of the salt with the analyte. Extraction of the other drugs by comparison appears predominantly effected by the hydration sphere effect. Thus it is possible that the overall effect of salt on drug extraction is dependent on the pK_a of the acid–base dissociation of the drug.

Organic Solvent Concentration during Extraction

The presence of solvent in an extraction mixture often causes an adverse effect on amount of drug extracted. As solvent miscible with an aqueous extraction mixture is added, the hydrophobicity of the aqueous solution increases. Thus the difference in hydrophobicity between the fibre coating and the solution is reduced, causing a reduction in the partition coefficient between the fibre coating and the solution, and a reduced amount of drug extracted by the fibre. The effect of this phenomenon was observed in our amphetamines work.[3] Figure 3a shows the amounts of amphetamines extracted with increasing amounts of methanol in the extraction mixture. This experiment was repeated at several different drug levels with the same trend seen in each case. As all of these extractions were carried out at equilibrium, an effective partition coefficient for PDMS may be calculated and plotted against percentage of methanol added. Here the effective partition coefficient is calculated as the ratio of the amount of the drug spike extracted to the amount of the drug spike not extracted. A true partition coefficient cannot be calculated in this case, as the drug concentration in the headspace, while significant, is unknown. In addition, the headspace concentration cannot be calculated according to Henry's law as the extraction matrix cannot be defined according to the assumptions of Henry's law. However, because the volumes of the fibre coating, the headspace and the sample, and the ratio of the drug concentration in the headspace to the sample would all be constant, this effective partition coefficient can be used. Figure 3b shows the relationship between the effective partition coefficient and the percentage of methanol added, for a range of drug spikes from 20 and 2000 ng mL^{-1}. Of interest, in both Figures 3a and 3b, the proportion of drug extracted becomes relatively constant above 0.2% v/v. Other researchers have also reported a reduction in amount of drug extracted with increasing percent solvent in the extraction mixture,[10] although not at levels below 1% v/v. It is possible that the extreme extraction conditions used in the amphetamine work amplified the solvent effect, and that in other cases the solvent effect would not be so significant at such low levels.

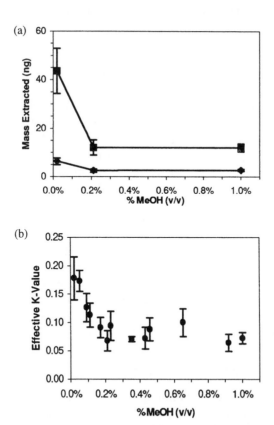

Figure 3 *Effect of methanol concentration on amount of drug extracted. (a) Single drug concentration at 0.1 μg mL^{-1}. (b) Various drug concentrations (0.02 to 2 μg mL^{-1}), equilibrium extractions with effective partition coefficients calculated and plotted versus methanol concentration*

In the opposite case, Rasmussen *et al.* have shown an enhancement of amount of diazepam extracted by polyacrylate fibres, by exposing the fibres to hydrophobic solvents prior to extraction.[11] This effect may be due either to a simple swelling of the fibre coating, with no effect on partition coefficient, or an altering of the hydrophobicity of the coating, and thus of the partition coefficient also. 1.5 and 3.0 μL of 1-octanol and 2-octanone respectively (as determined by GC–FID) were loaded onto the fibres. This represented the maximum amount of solvent that could be loaded in each case. No loss of solvent was observed during the extraction. As the 85 μm coating has a volume of just 0.5 μL, this doping may also have caused a significant swelling of the fibre coating, which on its own would produce an increase in amount of drug extracted. The authors reported a near doubling of the amount of diazepam extracted for both solvents, and there was no significant difference in amount extracted between the two solvents. For the effect to be due to fibre swelling only, the fibre in each case would have to be swollen to exactly the same extent, in order to produce the same enhancement of extraction. If the degree of swelling was not the same, the effect would have to be due in part to an altering of the properties of the fibre/solvent combination, relative to fibre only, thus changing the partition coefficient for diazepam in this system, enhancing its extraction into the fibre coating.

In many pharmaceutical or forensic analyses where solvent does have a negative effect on extraction, the presence of some amount of solvent must be accepted. For example, drug calibration standards and internal standards are often supplied as methanol solutions, and so some amount of solvent must be included in all extraction mixtures they are used with. In addition, some pharmaceutical preparations include a significant proportion of ethanol. In the case of a hydrophobic drug, for instance one designed to be absorbed across a mucosal membrane; aqueous solubility may be so low that solvent must be added to dissolve the compound so that it is available for extraction by the fibre. Also, if the drug is present in a hydrophobic matrix such as an ointment, an amount of solvent may be required to ensure a homogeneous extraction matrix. In these cases, the fibre coating may be swollen during extraction, so it is important to control the conditions such that extraction matrix components and hence degree of fibre coating swelling is the same in each extraction. It is also important to remember that the 100 and 30 μm PDMS coatings are non-bonded phases, whereas the 7 μm coating is bonded. Thus the 7 μm coating will swell to a relatively lesser degree than the other two.

Desorption of Analyte from the Fibre

In both GC and LC, the goal of optimizing the desorption process is typically to eliminate carryover and improve peak shape. A desorption time and temperature are selected such that any analyte remaining on the fibre after desorption will not cause variance in results outside of normal method precision. Chemicals with a low partition coefficient do not require a rigorous desorption process to achieve this goal. For applications where the analyte has a high partition coefficient, however, affinity to the fibre coating may be so high that removing all traces prior to a subsequent injection requires an inconvenient desorption process. Also, the required desorption temperature may be close to the temperature tolerance of the fibre coating, which may shorten the life of the fibre. If desorption requires both very high temperatures and a long time, fibre bleed may interfere with the analysis. To address this problem, Hall and Brodbelt[7] have recently introduced a cleaning step after each injection, for their barbiturates analysis. After desorption in the GC injector, the fibre is first cooled, then soaked in a solution of methanol–water (20:80), and then re-introduced to the hot injector for a short time. This process reduced the carryover of phenobarbital, the analyte with the highest partition coefficient, to less than 2%.

2 Introduction to SPME–HPLC

While the majority of the work to date on drug analysis by SPME has focused on GC or GC–MS analysis, the use of SPME coupled to HPLC and LC–MS is also practical. While most of the considerations for optimization of extraction conditions are the same, regardless of analysis method, there are some considerations dependent on the type of analysis being performed. Because the reports of drug analysis by SPME–LC are somewhat more limited at the time of writing, the fundamental methods of performing SPME–LC will be reviewed here, prior to the discussion of optimization.

Many drugs have low volatility or are thermally unstable, and therefore unsuitable for GC analysis, and so coupling to LC is logical. Additionally, sample salts or other contaminants are conveniently separated from the analytes of interest prior to injection. This has the potential to extend column life, and reduce matrix suppression of analyte spectra in LC–MS analysis. This is particularly important in flow injection analysis, or in cases where the analyte elutes early.

There are two modes by which SPME may be coupled to LC, and there are some minor variations in the way interfaces are incorporated, depending on the LC manufacturer. Either conventional fibre coupling, or the newer in-tube SPME may be incorporated, by placing the interface in the position of the sample loop. With conventional fibre coupling, analysts are currently limited to performing manual extractions and desorptions. For automated extraction and analysis, in-tube SPME is relatively simple to implement.

For conventional fibre coupling, a three-way tee is used, with two of the ports connected in the position of the sample loop. In the third position, a finger-tight fitting which compresses a standard 0.4 mm i.d. GC ferrule around the inner stainless steel tube of the fibre assembly gives a convenient port through which to introduce a fibre to fluid flow, for analyte desorption. A schematic of this interface, as incorporated into the Hewlett-Packard 1100 LC system, is shown in Figure 4. Depending on the analytes, desorption may be accomplished using only flow of mobile phase, if desorption is fast enough to provide a sharp peak. If analytes do not desorb quickly enough in mobile phase, another desorption

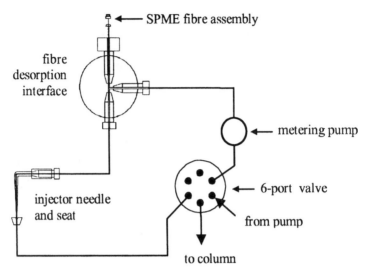

Figure 4 *Schematic of fibre SPME interface for SPME–LC and SPME–LC–MS using Hewlett-Packard HP1100 LC–MS*

solvent may be introduced to the interface to aid desorption, and then transferred with the analytes to the column. To achieve acceptable results, it is important to select a tee with sufficiently small internal diameter, so as to allow a high linear flow rate of mobile phase past the fibre, and desorption into as small a volume of liquid as possible. We find a 0.75 mm i.d. stainless steel Valco tee to be suitable. We have not had problems with coating damage using this i.d., even when solvent swells the coating.

For in-tube SPME, a section of fused silica GC column, coated on the inside with an appropriate material, is placed between the sample loop and the injector needle, again in the Hewlett-Packard 1100 LC system. A schematic of this arrangement is shown in Figure 5. To date, sections of standard GC capillaries, primarily poly(dimethylsiloxane) and poly(ethylene glycol) based phases have been employed. The existing sample loop is normally left in place, with the capillary connected between it and the injection needle. This prevents contamination of the metering pump during the extraction phase. The capillary is connected by adding a sleeve of appropriate i.d. PEEK tubing over the ends of the capillary, and then adding standard stainless steel fittings and ferrules.

During extraction, sample is aspirated from the sample to the capillary, and then dispensed back into the sample. This process is repeated until either an equilibrium extraction has been accomplished, or sufficient analyte is extracted to allow the desired method sensitivity. While this method is limited to relatively clean matrices, owing to the potential for plugging the system if dirty samples are extracted, the method has a clear benefit in the automation of analysis.

We have observed that the $[M + Na]^+$ ion is often much more abundant in both in-tube and fibre SPME–LC–MS electrospray spectra, relative to direct

In-Tube SPME capillary

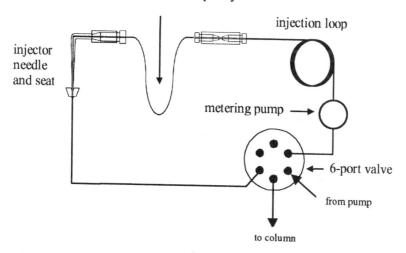

Figure 5 *Schematic of in-tube SPME interface for SPME–LC and SPME–LC–MS using Hewlett-Packard HP1100 LC–MS*

injection of the drug in solvent. This is likely due to the high levels of sodium ions present in the extraction matrix, either from added NaCl or from buffer salts.

Extraction Time Profile/Aspirate–Dispense Step Profile

Whether for fibre or in-tube SPME, it is beneficial to understand the extraction profile over time. This is because if analyses are performed under equilibrium conditions, extraction conditions such as extraction time and agitation can be ignored as they no longer have an effect on the amount of analyte extracted. In the case of fibre SPME, an absorption–time profile is constructed by performing several extractions at different extraction times. Figure 6 shows extraction time profiles for methamphetamine and amphetamine, for headspace fibre SPME with GC–FID detection. The amount of analyte extracted normally increases rapidly at first, and then slows until, at equilibrium extraction, no further increase in amount extracted occurs. Compounds with a low partition coefficient are characterized by shorter equilibrium extraction times, and lower equilibrium amounts extracted. Compounds with a high partition coefficient by contrast take longer to reach equilibrium, and have higher equilibrium amounts extracted. In Figure 6 methamphetamine has the higher partition coefficient. In this figure we see that methamphetamine does not reach equilibrium extraction within the 180 minutes tested. Amphetamine by contrast, has reached equilibrium extraction after about 1 hour.

In the case of in-tube SPME, the number of aspirate/dispense steps is varied, instead of the extraction time, in order to construct an extraction profile. Figure 7 shows an example of this for the extraction of dexamethasone with Hewlett-Packard FFAP capillary incorporated into the Hewlett-Packard

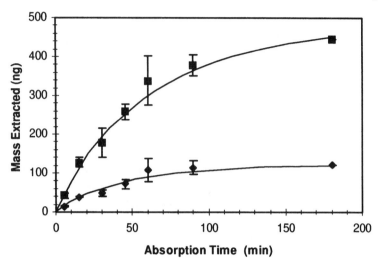

Figure 6 *Absorption–time profiles, amphetamines extraction by headspace SPME–GC, to compare equilibrium extraction times and amounts. Extraction at room temperature, 0.5M KOH, and saturated NaCl, 2 mL sample in a 4 mL vial. Key: (■): methamphetamine; (◆): amphetamine*

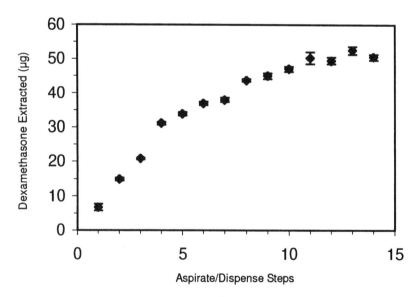

Figure 7 *Aspirate/dispense profile for the extraction of dexamethasone $(0.01\,mg\,mL^{-1})$ from aqueous (1% v/v MeOH) solution, using Hewlett-Packard FFAP capillary (0.2 mm i.d. × 0.3 μm film × 60 cm length), flow injection, mobile phase isocratic 50% methanol and 0.05% formic acid. Quantitation based on sum of ions 373 $[M-HF]^{+}$ and 415 $[M+Na]^{+}$*

1100 autoinjector, and MS detection. The very small error values seen in this figure are representative of method precision for automated extraction and injection. Because of this, non-equilibrium extraction is quite practical to improve throughput, where the added sensitivity possible with equilibrium extraction is not important.

In some cases, it has proven difficult to establish equilibrium extraction for in-tube SPME. It is assumed that this is due to the fact that in every aspirate/dispense step, a plug of analyte-free mobile phase follows the plug of sample during its travel through the capillary. Thus, during the dispense step, analyte previously extracted by the capillary coating may desorb into the mobile phase plug. While this can complicate the determination of equilibrium extraction levels, extraction at equilibrium levels is not necessary, as discussed above.

Encouraging results have been obtained using this technique for pesticides,[12] dexamethasone (Figure 8a) and ranitidine (Figure 8b). The results presented in Figure 8 were generated by flow injection analysis, *i.e.* no column was used for separation. The goal of the work was to maximize throughput for analysis of single component samples. By using this technique, analyte is conveniently separated from matrix components prior to analysis. For fibre SPME, the extraction is a manual process requiring of the order of 30 min. For in-tube SPME, the extraction process is automated and requires 10–15 min. While the flow injection technique does not produce the sharp symmetrical peaks that would be obtained by passing the flow through a column prior to detection, flow injection analysis for single component samples allows for much higher throughput.

Extraction Coating Selection—in-tube and Fibre

With both fibre and in-tube SPME the decision of which extraction phase to select must be balanced between attaining equilibrium extraction conditions, *versus* maximal sensitivity, *versus* ruggedness of the coating. Figure 9 shows the results of headspace fibre SPME extraction of amphetamines, with analysis by GC–FID. In this case, the PDMS coating would appear to exhibit the lowest partition coefficient for these drugs, based on the fact that both drugs reach equilibrium extraction quickly (amphetamine, < 5 min; methamphetamine, 15 min). It is expected that both the poly(dimethylsiloxane)/divinylbenzene (PDMS/DVB) and PA coatings would eventually reach equilibrium extraction at levels higher than those achieved by PDMS. While it is difficult to predict from these data whether PDMS/DVB or PA would eventually show the higher distribution coefficient for these compounds, that information would only be relevant if maximal sensitivity without regard to extraction time was the overriding goal. If one wanted to maximize sensitivity with a shorter extraction time, and was able to accurately control extraction time and agitation conditions, the PDMS/DVB fibre would provide the highest levels of drug extracted in the times tested, and thus would produce the best limits of detection. If method precision was the overriding determinant, the PDMS coating would be

(a)

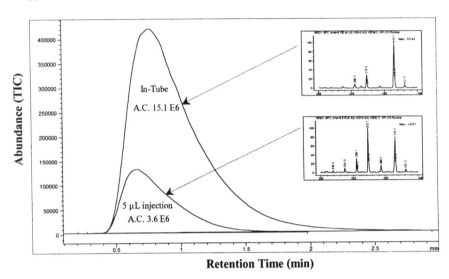

(b)

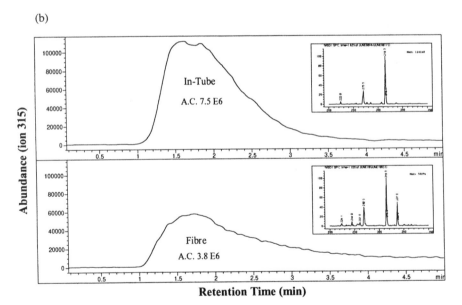

Figure 8 *Drug analysis by SPME–flow injection analysis with the HP1100 LC–MS.*
(a) In-tube SPME versus 5 μL injection of dexamethasone in water
(0.01 mg mL^{-1}). In-tube extraction: 15 aspirate/dispense steps at
100 μL min^{-1}. Mobile phase: 0.05% formic acid, 50% MeOH, 0.2 mL min^{-1}.
(b) In-tube SPME versus fibre SPME of ranitidine. In-tube extraction: 10
aspirate/dispense steps at 100 μL min^{-1}. Sample: ranitidine (10 μg mL^{-1}) in
water. Fibre extraction: PDMS/DVB, 50°C, 45 min. Sample: ranitidine
(50 μg mL^{-1}) in 1 mL Tris-HCl buffer, 0.1M, pH 8.5, saturated NaCl. Mobile
phase: MeOH/2-ProOH/5M AcONH₄ (50:50:1), 0.1 mL min^{-1}

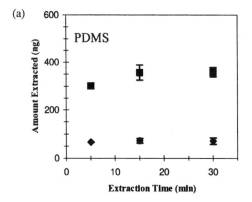

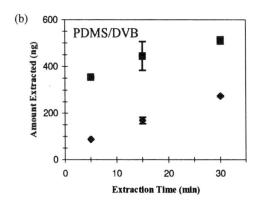

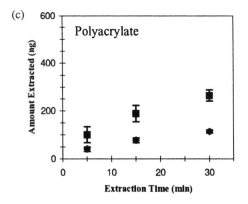

Figure 9 *Comparison of various fibres based on their extraction–time profiles for the headspace extraction of amphetamines. (a) Polydimethylsiloxane fibre, 100 μm; (b) PDMS/DVB fibre, 65 μm; (c) Polyacrylate fibre, 85 μm. Key: (■): methamphetamine; (◆): amphetamine. Conditions: extraction for 15 minutes at 60°C, saturated NaCl, 2 mL sample in a 4 mL vial, base adjustment by 0.5M KOH (pH > 12). Analysis by GC–FID, desorption time 15 min*

preferable, as equilibrium extractions will normally provide the best precision. It is convenient to select an extraction time equal to or less than the separation/detection time, to maximize throughput. For GC analyses, extractions in excess of 30–45 minutes are generally impractical as this is often in excess of the separation/detection time required by the instrument. Separation/detection times are often even shorter for LC analyses, particularly when flow injection or very short columns are used.

If fibre ruggedness was of prime concern (maximizing the number of extractions possible per fibre), the porous polymer coatings generally prove to be less rugged than the others. Polyacrylate probably has an advantage over PDMS in terms of ruggedness of the coating. Sections of the PDMS coating are sometimes removed from the fibres if they not handled carefully. This is not generally a problem with polyacrylate fibres. PA fibres though are not as flexible as PDMS fibres, and so are more susceptible to snapping away from the stainless steel tubing to which they are fastened.

Another consideration in fibre selection is the ability of the fibre coating to withstand the extraction medium. For instance, the porous polymer coatings are entirely incompatible with basic conditions, as the coating will be stripped from the fused silica core under basic conditions (pH ∼ 10). The PDMS coating by comparison is somewhat more tolerant of mildly basic conditions, as it shows visual degradation after repeated exposure to these conditions, as opposed to being immediately stripped from the fused silica core. For the analysis of some acidic compounds (phenols), acidic conditions are required. Acidic conditions below about pH 2 are generally not compatible with SPME fibres. In all cases, acidic or basic strength also affects fibre performance.

In the case of in-tube SPME, capillary selection is primarily determined by the shape of the extraction profile (profile of extraction versus number of aspirate/dispense steps). In addition to monitoring the extent to which various coatings will extract the analyte of interest, it is also important to monitor the coating durability under exposure to the necessary mobile phases, and any desorption solvents required. This is more difficult than monitoring a fibre coating, as one cannot directly visualize the coating on the inside of the tube. Thus it is necessary to monitor for loss in response to known analytes.

Agitation during Extraction

The degree of agitation in an extraction affects the dimensions of the boundary layer around the extraction phase. As the extraction phase removes analyte from the sample, the sample in the immediate vicinity of the extraction phase becomes depleted of analyte, and the size of the resulting boundary layer becomes limiting in terms of equilibrium extraction time. Efficient agitation reduces the size of the boundary layer and thus enhances the speed of extraction. With conventional fibre SPME, agitation is typically achieved in manual extractions by magnetic stirring. A vortex in the sample vial, which extends three-quarters or more of the distance down the sample is generally a good indication of efficient agitation. In automated fibre SPME, the Varian 3200 autosampler uses fibre vibration, which has also been shown to be efficient at enhancing extraction. For in-tube SPME, the rate of the aspirate/dispense fluid flow affects the speed of extraction. There is an upper limit to the speed at which the fluid can be drawn along the capillary. At some point, the draw rate is high enough that bubbles start to form inside the capillary, which causes a drop in extraction rate. At lower rates of fluid flow during aspirate/dispense steps, extraction time becomes very long. For 0.2–0.25 mm i.d. capillary, we typically use flow rates of 60 to $100 \, \mu L \, min^{-1}$. We have investigated the possibility of a physical mixing of the sample with the mobile phase at their interface in the extraction capillary, and have determined that no significant mixing occurs at a flow rate of $100 \, \mu L \, min^{-1}$ with a 0.2 mm i.d. capillary.

In some cases agitation is either not necessary, or dispensed of for other reasons. In the case of headspace amphetamines extraction, stirring rate was seen to have no effect on amount extracted. It is postulated that the extraction in this case was almost entirely from headspace, with very little requirement for mass transfer from the liquid. Stirring of the liquid will only affect the rate of mass transfer from the liquid to an adjacent phase (headspace in this case), and so it is reasonable that stirring would have no effect in a system independent of mass transfer from liquid to headspace. In other situations, such as on-site or process monitoring, agitation is too difficult to implement or control, and so it is beneficial to select extraction conditions which allow for equilibrium extraction so that agitation efficiency no longer effects amount extracted.

3   Conclusion

The application of SPME methods to drug analysis is a relatively new field, with significant numbers of publications first appearing in about 1995. As was the case in other fields, the first successful applications have been with the more volatile compounds, amenable to analysis by GC. We expect that as expertise and extraction phases advance, the range of compounds accessible to analysis with SPME techniques will expand significantly. While its initial applications are likely to be primarily in laboratory settings, we expect that it may also prove valuable where analyses are not currently practical owing to the constraints of the current extraction methods. Potential in these areas would include on-site analysis for instance in physician office laboratories or point-of-care situations in health services, or in process control and on-line quality assurance for pharmaceutical production.

Acknowledgements

We gratefully acknowledge financial support for this work from the Hewlett-Packard Company, Supelco Inc., Varian Associates Inc., and the National Science and Engineering Research Council of Canada.

References

1. Koide, *J. Chromatogr. B*, 1998, **707**, 99.
2. N. Nagasawa, M. Yashiki, Y. Iwasaki, K. Hara and T. Kohima, *Forensic Sci. Int.*, 1996, **18**, 95.
3. H. Lord and J. Pawliszyn, *Anal. Chem.*, 1997, **69**, 3899.
4. J. Pawliszyn, *Solid Phase Microextraction: Theory and Practice*, Wiley-VCH, New York, 1997, p. 47.
5. J.E. Gordon and R.L. Thorne, *J. Phys. Chem.*, 1967, **71**, 4390.
6. Y. Marcus, *Ion Solvation*, Wiley, New York, 1985, p. 306.
7. B.J. Hall and J.S. Brodbelt, *J. Chromatogr. A*, 1997, **777**, 275.
8. *Disposition of Toxic Drugs and Chemicals in Man*, 3rd Edn., Baselt and Cravey eds., Year Book Medical Publishers Inc., Chicago.
9. D.D. Breimer, *Pharmacokinetics of Hypnotic Drugs*, Drukkerij Brakkenstein, Nijmegen, The Netherlands, 1974, 314 pp.
10. B.J. Hall, A.R. Parikh and J.S. Brodbelt, 'Determination of cocaine, cocaethylene and metabolites by solid-phase microextraction (SPME) and ion-trap GC–MS', manuscript in preparation.
11. M. Krogh, H. Grefslie and K. Rasmussen, *J. Chromatogr. B*, 1997, **689**, 357.
12. R. Eisert and J. Pawliszyn, *Anal. Chem.*, 1997, **69**, 3140.

CHAPTER 41

Applications of SPME for the Biomonitoring of Human Exposure to Toxic Substances

MAURIZIO GUIDOTTI AND MATTEO VITALI

1 Introduction

Humans are continuously exposed to toxic substances, from the pollutants present in urban air, to specific compounds diffused in workplaces, to food or drinking water contaminants. The evaluation of human exposure to chemicals is necessary both for professional exposures and for the general population, but it assumes great importance in the field of industrial hygiene for the higher levels of absorption. Because of the sanitary risks connected with the exposure/absorption of many compounds frequently used in industrial activities, both environmental and biological monitoring are necessary for assessing health hazards of exposed workers.

Environmental assessments are usually performed by the evaluation of the concentration of contaminants in the workplace air (mainly the breathing zone air) or dust. Data are compared with acceptable levels, expressed as TLVs (Threshold Limit Values), obtained from epidemiological and clinical studies, and applied to eight hours exposure for five days a week.

Biological monitoring concerns an assessment of overall exposure to chemicals through measurement of appropriate determinants, such as the chemical itself or its metabolites, in biological specimens. Frequently the selected determinants are 'more or less specific' urinary metabolites, excreted after a known time from the exposure. As an example we can consider total urinary phenol for exposure to benzene, methylhippuric acid for xylenes, mandelic acid for styrene, 4-chlorophenol for 4-chlorobenzene.

The Biological Exposure Index (BEI), which serves as a reference value, represents the level of determinant which is likely to be found for an exposure to the chemical of interest at the TLV level in the workplace air. Biological Exposure Indices are calculated based on eight hours of exposure for five days a week.

The analytical procedures proposed in the literature for the determination of urinary metabolites are complex and not completely reliable. In effect, particularly for exposure to low levels of chemicals, the evaluation of exposure by means of urinary metabolites is subject to mistakes and misunderstandings due to several factors:

- the intraindividual and interindividual variability of metabolic activity makes it impossible to establish 'normal' excretion for many metabolites, and consequently the delineation between non-hazardous and hazardous levels is not sharply traceable;
- the numerous biological substances present in urine, in which the determinants must be analyzed, can affect the identification/quantitation of the compound considered;
- individual lifestyle such as smoking, alcohol consumption, drug intake, *etc.* can cause increased or decreased levels of 'normal excretion' of metabolites or other determinants (*e.g.* benzene intake for tobacco smokers).

For several substances, the appropriate determinant can be the chemical itself, measured in venous blood or exhaled air. It is evident that, for this kind of measurement, sampling appears to be the main problem. Firstly, it is difficult to find enough exposed workers willing to allow blood drawing; secondly, as the chemical concentration in exhaled air changes rapidly during the expiration, it is necessary to refer to mixed- or end-exhaled air.

Owing to these problems, until recently, biological monitoring has been considered as supplementary to air monitoring in assessing exposure. According to the American Conference of Governmental Industrial Hygienists (ACGIH) guidelines, biological monitoring should be conducted only when it offers advantages or can substantiate data on air concentrations of chemicals.

Nevertheless, only biological monitoring can give information about the real assumption of a specific compound through all the routes (*e.g.* skin), and not only through inhaled air. Besides, the exposure to air contaminants can frequently change owing to different industrial processes, movements of the worker in the workplace area, *etc.* These changes can occur during the same day, or on different days, offsetting the extrapolation of data from air monitoring to the real assumption.

In the opinion of many researchers (the authors included) the two types of monitoring are complementary in giving information for the exposure/assumption levels. If both are conducted simultaneously, health risks for the exposed persons could be better assessed.

Recently, several papers proposed the urinary concentration of the unmetabolized form of the chemical as its biological index, with the aim to simplify the procedures of sampling and analysis, and consequently to allow more useful measurements. These works experimentally demonstrated the direct relationships between the urinary levels and the breath air average levels for the compounds under consideration (expecially solvents).

The analytical procedures for these determinations apply GC after headspace sampling, liquid/liquid or SP extraction. In this field of analysis SPME applications appear advantageous.

2 Experimental

General Considerations

Standard Preparation

Analytical standards preparation requires a matrix (urine, blood) obtained from donors not exposed to the specific target analyte(s). Physiological and health status, both with lifestyle characteristics, may differ substantially from each donor; it is recommended to collect together at least five samples from five different donors, mix them, and evaluate background levels of the analytes before spiking.

Solvent content of the spiking volume must be minimized, to avoid interferences with the SPME fiber. After spiking, determinants must be allowed to diffuse and/or bind to coordinating groups present in the matrix for at least 10 min.

Method Parameters

Calibration ranges have to include BEI values; sample amount and treatment are planned accordingly. Consequently, LODs obtained with the procedures may be higher than in other applications. Lower LODs may be obtained by reducing headspace volumes, increasing sample volumes or extraction temperatures *etc.*

Carryover phenomena must be always considered, through checking the possibility of an incomplete desorption of analytes after GC injection.

Particular care must be paid to fiber cleaning after its use with 'dirty samples' (*e.g.* urine or blood), to remove organic compounds adsorbed on the fiber itself. Usually 2 min of immersion in distilled, organics-free water is sufficient. For phenols, distilled water is adjusted to pH = 2 with HCl.

A Hewlett-Packard (HP) gas-chromatograph 5890 series II, and a mass spectrometry detector (MS) HP 5971 were used for all determinations.

Theoretical LODs were calculated as the amount giving a peak three times taller then the noise level.

BTEX Determination in Urine

Introduction

Volatile aromatic compounds are used in different industrial activities: mainly as general solvents, but also in the composition of paints, enamels and lacquers, resins, adhesives, and as intermediate products for the chemical industry.[1]

Commonly, 'technical grade' mixtures are used, containing ethylbenzene, xylenes and/or toluene. The use of benzene is strictly limited, owing to its evidenced carcinogenic properties.[2] Toluene, ethylbenzene and xylene produce toxic effects in the central nervous system, liver, and kidneys, and are irritants for the skin.[3,4]

Urinary levels of toluene are proposed by ACGIH as a new determinant for toluene exposure,[5] while urinary benzene results correlate better then all the other determinants with air concentrations.[6-8] Perbellini et al.[9] and Imbriani et al.[10] report a significant relationship between xylene exposure and xylene urinary excretion.

Materials, Method, Procedure, Calibration, Blank Tests

HP 5MS capillary column, 25 m × 0.20 mm i.d., 0.25 μm film thickness was set at 40 °C for 5 min, then ramped at 8 °C min^{-1} to 240 °C, and held for 10 min, with He carrier gas at 0.8 mL min^{-1} flow. Source temperature was 173 °C, and the splitless injector temperature 240 °C.

5 mL of urine was placed into a 10 mL vial, NaCl (1 g) and a Teflon magnetic stirring bar were added, and the vial sealed with a pierceable Teflon-lined cap. The vial was heated at 60 °C under magnetic stirring for 1 hour, and then the SPME fiber was inserted into the headspace and allowed to equilibrate for 15 min.

Cumulative working standards were prepared by spiking a homogeneous urine sample collected from non-smoking donors, not exposed to workplace solvents, to obtain a range of concentrations from 0.05 to 500 μg L^{-1} for each benzene, toluene, ethylbenzene or xylene component. Blank tests were performed on the urine sample, showing the presence of benzene at 0.147 μg L^{-1} ± 3.9%, while toluene, ethylbenzene and xylene were not found (levels under the LOD).

Results and Conclusions

Data obtained from the application of the method are reported in Table 1. Considering these data, the proposed method is suitable for applications in determining environmental exposure to a single aromatic solvent or to a mixture, by means of biological monitoring. The sensitivity and precision of the overall procedure allow for worker exposure to be monitored at low air levels of aromatic solvents.

Styrene Determination in Urine and Venous Blood

Introduction

Styrene is a hydrocarbon widely used in numerous industrial processes for the production of polymers and reinforced plastics.[11] The major portion of absorbed styrene is metabolized in the liver by enzymatic systems, and a smaller amount is excreted unmodified in urine.[12,13] Styrene shows toxic effects

Table 1 Data obtained from the application of the described SPME methods to biological monitoring

Compound	SPME fiber	Desorption time (min)	Monitored Ions (m/z)	Correlation coefficient as r² (concentrations range)	Precision as %RSD on 10 consecutive determinations (concentration)	LOD (μg L⁻¹) Theoretical	LOD (μg L⁻¹) Experimental
BTEX in urine	PDMS 100 μm	1			10 μg L⁻¹ / 100 μg L⁻¹		
Benzene			78, 52, 77	0.981 (0.05–500 μg L⁻¹)	3.5 / 3.3	0.093	0.128
Toluene			91, 92, 65	0.975 (0.05–500 μg L⁻¹)	7.4 / 6.9	0.049	0.061
Ethylbenzene			91, 106, 105	0.946 (0.05–500 μg L⁻¹)	11.5 / 9.4	0.036	0.044
m+p-Xylenes			91, 106, 105	0.954 (0.05–500 μg L⁻¹)	11.2 / 9.9	0.033	0.039
o-Xylene			91, 106, 105	0.962 (0.05–500 μg L⁻¹)	9.8 / 7.1	0.027	0.035
Styrene in urine	PDMS 100 μm	1	140, 103, 78	0.998 (0.05–300 μg L⁻¹)	12.8 (0.1 μg L⁻¹)	0.028	0.046
Styrene in venous blood	PDMS 100 μm	1	140, 103, 78	0.995 (0.05–5000 μg L⁻¹)	11.5 (0.1 μg L⁻¹)	0.150	0.217
Chlorinated-VOC in urine	PDMS 100 μm	1			1.0 μg L⁻¹		
Methylene chloride			49, 84, 86	0.998 (0.5–500 μg L⁻¹)	11.6	0.200	
Trichloroethylene			132, 130, 95	0.998 (0.5–500 μg L⁻¹)	7.2	0.033	
Tetrachloroethane			83, 85, 95	0.997 (0.5–500 μg L⁻¹)	11.5	0.021	
Chlorinated-VOC in blood	PDMS 100 μm	1			10 μg L⁻¹		
Chloroform			83, 85, 47	1.000 (5.0–7500 μg L⁻¹)	10.2	0.714	
Trichloroethylene			132, 130, 95	0.999 (5.0–7500 μg L⁻¹)	13.6	0.625	
Tetrachloroethylene			166, 129, 94	0.999 (5.0–7500 μg L⁻¹)	7.6	0.400	
Tetrachloroethane			83, 85, 95	1.000 (5.0–7500 μg L⁻¹)	6.6	0.429	
Methylethylketone in urine	PDMS/DVB 65 μm	1	43, 72, 57	0.999 (0.10–10.0 mg L⁻¹)	5.6 (1.0 mg L⁻¹)	16.0	33.5
Methanol in urine	Carboxen/PDMS 75 μm	0.5	31, 29, 30	1.000 (0.5–30.0 mg L⁻¹)	9.8 (2.0 mg L⁻¹)	320	422
Pentachlorophenol in urine	Polyacrylate 85 μm	3	266, 264, 268, 270	0.997 (0.01–10.0 mg L⁻¹)	13.8 (10 μg L⁻¹) / 7.4 (2.0 mg L⁻¹)	0.43	0.76
Phenols in urine	Polyacrylate 85 μm	3					
p-Chlorophenol			128, 99, 130	0.999 (0.5–100.0 mg L⁻¹)	11.9 (0.5 mg L⁻¹) / 8.7 (25 mg L⁻¹)	210	8.0
p-Nitrophenol			129, 65, 81	1.000 (0.1–20.0 mg L⁻¹)	10.4 (0.1 mg L⁻¹) / 9.6 (5.0 mg L⁻¹)	15	18.2
Mercury in urine	PDMS 100 μm	3		(0.2–500 μg L⁻¹)	10 μg L⁻¹ / 0.2		
Mercury (diethylHg)			231, 260, 202	0.998	10.8	0.093	
Methylmercury (methylethylHg)			217, 246, 202	0.997	12.5	0.303	

on the central nervous system, irritation for pulmonary and digestive tissues, and can produce anaemia. Styrene is also a suspected carcinogen.[2,14]

A BEI indicated by ACGIH for styrene is 0.55 mg L^{-1} of unmodified compound in venous blood (sampled at the end of work shift), or 0.02 mg L^{-1} (prior to the next shift).[5] It is however demonstrated that environmental concentration and urinary excretion of styrene are significantly related.[15,16]

Materials, Method, Procedure, Calibration, Blank Tests

Supelco SPB-1 capillary column, 30 m $\times$ 0.25 mm i.d., 1.0 μm film thickness was set at 50 °C for 3 min, then ramped at 15 °C min^{-1} to 250 °C, and held for 10 min, with He carrier gas at 1.2 mL min^{-1} flow. The source temperature was 173 °C, and the splitless injector temperature 230 °C.

5 mL of urine was placed in a 10 mL vial and NaCl (1.5 g) and a Teflon magnetic stirring were added. The vial was sealed with a pierceable Teflon-lined cap and heated at 50 °C under magnetic stirring for 30 min, and then the SPME fiber was inserted in the headspace and allowed to equilibrate for 10 min.

0.200 mL of heparinated venous blood was placed in a 10 mL vial and a Teflon magnetic stirring bar was added. The vial was sealed with a pierceable Teflon-lined cap, and heated at 50 °C under magnetic stirring for 30 min, and then the SPME fiber was inserted in the headspace and allowed to equilibrate for 10 min.

Urine working standards were prepared by spiking a homogeneous urine sample collected from non-smoking donors, not exposed to solvents in the workplace, at a range of concentrations from 0.05 to 300 μg L^{-1}. Venous blood working standards were prepared by spiking a homogeneous heparinated blood sample collected from non-smoking donors, not exposed to solvents in the workplace, at a range of concentrations from 0.05 μg L^{-1} to 5.0 mg L^{-1}. The wide range is necessary to include both BEI values and low exposures. Blank tests were performed on urine and blood samples, showing the presence of styrene at 0.121 μg L^{-1} and 0.065 μg L^{-1} respectively.

Results and Conclusions

Data obtained are reported in Table 1 and shown in Figure 1. The proposed method is suitable for the evaluation of environmental exposure to styrene. The large linear range of the method allows for the surveying of subjects exposed to TLV and lower air levels of styrene.

Chlorinated-VOC Determination in Urine and Blood

Introduction

Chlorinated-VOCs (C-VOC) are used as industrial solvents and in the production of other organochlorine chemicals. They are also used as dry-cleaning agents, in degreasing processes and in the manufacturing of plastics and

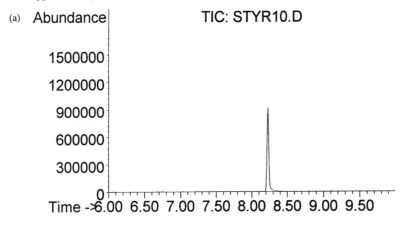

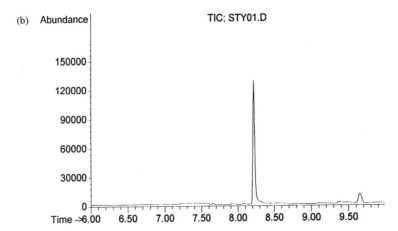

Figure 1 (a) *Total ion chromatogram of a standard urine sample spiked at* 10.0 μg L^{-1} *with styrene;* (b) *total ion chromatogram of a standard blood sample spiked at* 100 μg L^{-1} *with styrene*

textiles.[17] Because of their lipophilicity, C-VOC accumulate mostly in tissues with high lipid content, such as adipose tissue, brain and kidney. C-VOC show toxic effects for the central nervous system, liver, and kidney, and are irritants for the skin.[3] Some C-VOC are suspected carcinogens.[18]

ACGIH reports a BEI of 1 mg L^{-1} for tetrachloroethylene in venous blood, sampled prior to the last shift of the work week, and proposes to reduce this limit to 0.5 mg L^{-1}.[5] Several researchers and organizations recommend the monitoring of urinary and blood levels of unmodified chlorinated compounds for a better evaluation of multiple exposures. We developed analytical procedures for the determination of methylene chloride, trichloroethylene and tetrachloroethane in urine, and chloroform, trichloroethylene, tetrachloroethylene and tetrachloroethane in venous blood.

Materials, Method, Procedure, Calibration, Blank Tests

Supelco SPB-1 capillary column, 30 m × 0.25 mm i.d., 1.0 μm film thickness was set at 50 °C for 3 min, then ramped at 15 °C min^{-1} to 250 °C, and held for 10 min, with He carrier gas at 1.2 mL min^{-1} flow. The source temperature was 173 °C, and the splitless injector temperature 230 °C.

5 mL of urine was placed in a 10 mL vial and NaCl (1.5 g) and a Teflon magnetic stirring bar were added. The vial was sealed with a pierceable Teflon-lined cap and heated at 50 °C under magnetic stirring for 30 min, and then the SPME fiber was inserted in the headspace and allowed to equilibrate for 10 min.

0.400 mL of heparinated venous blood was placed into a 2 mL vial and NaCl (0.100 g) and a Teflon magnetic stirring bar were added. The vial was sealed with a pierceable Teflon-lined cap and heated at 50 °C under magnetic stirring for 30 min, and then the SPME fiber was inserted in the headspace and allowed to equilibrate for 10 min.

Cumulative working standards in urine were prepared by spiking a homogeneous urine sample collected from non-smoking donors, not exposed to solvents in the workplace, at a range of concentrations from 0.500 to 500 μg L^{-1} for each C-VOC. Venous blood working standards were prepared by spiking a homogeneous heparinated blood sample collected from non-smoking donors, not exposed to solvents in the workplace, and obtaining a range of concentrations from 5.0 μg L^{-1} to 7.5 mg L^{-1} for each C-VOC. Blank tests were performed on urine and blood samples, and all the target C-VOC were present at levels below LODs.

Results and conclusions

The results obtained from calibration data on analytical standards are shown in Table 1 and Figure 2; the proposed method is suitable for applications in determining environmental exposure to a single C-VOC or to a mixture, by means of biological monitoring. The sensitivity and precision of the overall procedure allows for the monitoring of subjects exposed to low air levels.

Methyl Ethyl Ketone Determination in Urine

Introduction

Methyl ethyl ketone (MEK) is widely used in numerous industrial activities for its chemical-physical characteristics and its solvent properties.[3] The ACGIH BEI for MEK is 2.0 mg L^{-1} of unmodified compound in urine sampled at the end of the work shift.[5]

Materials, Method, Procedure, Calibration, Blank Tests

HP Innowax capillary column, 30 m × 0.25 mm i.d., 0.25 μm film thickness was set at 40 °C for 2 min, then ramped at 15 °C min^{-1} to 250 °C, and held for

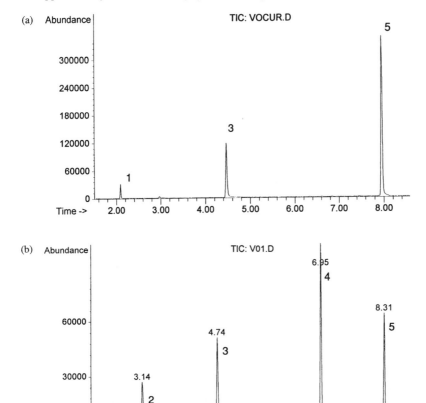

Figure 2 (a) *Total ion chromatogram of a standard urine sample spiked with methylene chloride (1), trichloroethylene (3), tetrachloroethane (5) at 50.0 µg L^{-1} each; (b) total ion chromatogram of a standard blood sample spiked with chloroform (2), trichloroethylene (3), tetrachloroethylene (4), tetrachloroethane (5) at 100.0 µg L^{-1} each*

10 min, with He carrier gas at 1.0 mL min^{-1} flow. The source temperature was 173 °C, and the splitless injector temperature 240 °C.

2 mL of urine sample was placed into a 5 mL vial and NaCl (0.6 g) and a Teflon magnetic stirring bar were added. The vial was sealed with a pierceable Teflon-lined cap and heated at 50 °C under magnetic stirring for 30 min, and then the SPME fiber was inserted in the headspace and allowed to for 10 min.

Urine working standards were prepared by spiking a homogeneous urine sample collected from non-smoking donors, not exposed to solvents in the workplace, at a range of concentrations from 0.100 to 10.0 mg L^{-1}. Blank tests were performed using urine samples, and MEK was shown not to be present (under LOD of the method).

Results and Conclusions

Data are reported in Table 1, from which it is concluded that the proposed
method, in the studied linearity range, is suitable for the evaluation of workplace exposure to MEK at air levels around its TLV.

Methanol Determination in Urine

Introduction

Methanol (M) is mainly used as a reaction intermediate in specific industrial
processes.[17,19] It shows toxic effects on the central nervous system, eyes and
respiratory system, and acts as an irritant to eyes, mucous and skin.[3,20]

The ACGIH BEI for methanol is 15.0 mg L^{-1} of unmodified compound in
urine sampled at the end of the work shift.[5]

Materials, Method, Procedure, Calibration, Blank Tests

HP Innowax capillary column, 30 m × 0.25 mm i.d., 0.25 μm film thickness was
set at 40 °C for 2 min, then ramped at 10 °C min^{-1} to 250 °C, and held for
10 min, with He carrier gas at 1.0 mL min^{-1} flow. The source temperature was
173 °C, and splitless injector temperature 240 °C.

2 mL of urine was placed into a 5 mL vial and NaCl (0.6 g) and a Teflon
magnetic stirring bar were added. The vial was sealed with a pierceable Teflon-
lined cap, and heated at 50 °C under magnetic stirring for 30 min, and then the
SPME fiber was inserted in the headspace and allowed to equilibrate for 10 min.

Urine working standards were prepared by spiking a homogeneous urine
sample collected from non-smokers who were not alcohol consumers, and were
not exposed to alcohols and solvents in the workplace. The tested range of
concentrations was from 0.500 to 30.0 mg L^{-1}. Blank tests were performed on
urine samples, and methanol was never found (below LOD of the described
method).

Results and Conclusions

Data obtained are reported in Table 1. As shown in the table, the proposed
method is suitable for the evaluation of workplace exposure to methanol at air
levels around its TLV.

Pentachlorophenol Determination in Urine

Introduction

Chlorinated phenols are used as biocides, preservatives, pesticides, and industrial organic chemicals. Pentachlorophenol (PCP) is the most widely used
chlorophenol in industry, mainly for wood and tanning processes.[17] This
compound is under investigation as a possible carcinogenic agent for man.[21]

In humans, PCP may cause liver damage[22] and changes in the immune system.[23] PCP is metabolized only to a limited extent, and consequently the target for biomonitoring of exposure is PCP itself in serum or urine. The ratio of metabolized-PCP/unchanged-PCP is pratically constant and PCP is excreted largely as free PCP (about 74% of the total intake) and protein-conjugated PCP (about 12%) through urine.[24] Therefore urinary levels can conveniently be monitored to assess the total intake of the compound for exposed workers.

ACGIH fixes for PCP a BEI at 2 mg total PCP g^{-1} of creatinine in urine, or 5 mg L^{-1} in blood.[5]

Materials, Method, Procedure, Calibration, Blank Tests

Supelco PTE-5 capillary column, 30 m × 0.25 mm i.d., 0.25 μm film thickness was set at 70 °C for 3 min, then ramped at 25 °C min^{-1} to 280 °C, and held for 10 min, with He carrier gas at 1.0 mL min^{-1} flow. The source temperature was 173 °C, and the splitless injector temperature 280 °C.

1 mL conc. HCl and 100 mg NaHSO$_3$ were added to a 4 mL urine sample in a culture tube. The tube was then placed in a boiling waterbath for 1 h, to hydrolyse conjugated PCP to free PCP. After this treatment 0.5 mL of the hydrolysed sample was transferred to a 10 mL vial and NaCl (1.5 g) and organics-free water (4.5 mL) were added. pH was adjusted to 2 with conc. H$_2$SO$_4$ and the vial was closed with a Teflon-lined septum cap. The SPME fiber was inserted into the liquid and extracted for 20 min under magnetic stirring before thermal desorption.

Working standard solutions were prepared by spiking a homogeneous urine sample to obtain a range of concentrations from 0.010 to 10.0 mg L^{-1}. The urine sample was analysed by the described method, and PCP was never found, always being below LOD of the described method.

Results and Conclusions

Data obtained from the application of the method described are reported in Table 1 and Figure 3. The proposed method is suitable for the evaluation of workplace exposure to PCP at air levels around its TLV.

p-Chlorophenol and *p*-Nitrophenol Determination in Urine for *p*-Chlorobenzene and *p*-Nitrobenzene Exposure

Introduction

Chlorobenzene (CB) is mainly used in the production of 2- and 4-nitrochlorobenzene. It is also used as a solvent and, in small amounts, for the production of some pesticides.[17] Nitrobenzene (NB) is used extensively in a variety of industrial processes such as the manufacturing of dyes, ammunitions and explosives, and in the production of polymers.[25.]

CB after metabolic transformation is mainly excreted in urine as *p*-chloro-

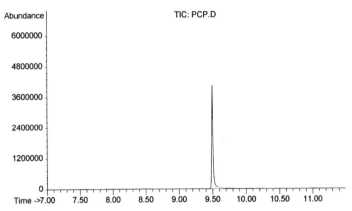

Figure 3 *Total ion chromatogram of a standard urine sample spiked at 500.0 μg L^{-1} with pentachlorophenol*

phenol (*p*-CP), while NB is excreted as *p*-nitrophenol (*p*-NP). ACGIH proposes to assess the exposure to CB and NB by measuring the urinary concentration of total *p*-CP and *p*-NP, fixing the relative BEI at 25 mg g^{-1} of creatinine and 5 mg g^{-1} of creatinine respectively.[5] Toxic effects of both CB and NB are on the central nervous system, blood, and digestive system. CB is also a suspected carcinogen for man.[3]

Materials, Method, Procedure, Calibration, Blank Tests

HP 50 + capillary column, 15 m × 0.25 mm i.d., 0.25 μm film thickness was set at 55 °C for 2 min, then ramped at 15 °C min^{-1} to 280 °C, and held for 10 min, with He carrier gas at 1.0 mL min^{-1} flow. The source temperature was 173 °C, and the splitless injector temperature 280 °C.

2 ml urine was placed into a 7 mL vial, with 1 ml of concentrated HCl. The vial was sealed and heated at 50 °C for 1 hour. The sample was cooled to room temperature, and then 100 μL was transferred to a 10 ml vial, with 9 mL distilled water, 3 g NaCl and a Teflon magnetic stirring bar. The vial was immediately sealed and put on a magnetic stirrer at 1000 rpm. The fiber was then lowered into the sample and exposed for 30 minutes.

Urine samples obtained from donors not exposed in the workplace to CB or NB were tested to confirm the absence of *p*-CP and *p*-NP. Standard working solutions of *p*-CP and *p*-NP were prepared by spiking urine samples in order to obtain a range of concentrations from 0.5 to 100.0 mg L^{-1} for *p*-CP and from 0.1 to 20.0 mg L^{-1} for *p*-NP.

Results and Conclusions

Data obtained are reported in Table 1 and Figure 4. They show that the proposed method is very selective and sensitive. Its sensitivity makes it suitable to screen subjects exposed to low concentrations of CB and NB.

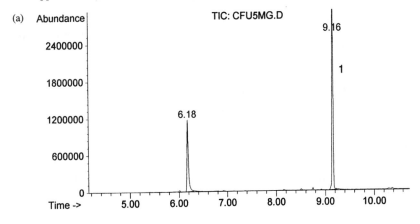

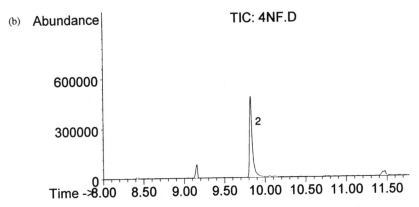

Figure 4 (a) *Total ion chromatogram of a standard urine sample spiked at 5.0 mg L^{-1} with p-nitrophenol (1); (b) total ion chromatogram of a standard urine sample spiked at 5.0 mg L^{-1} with p-chlorophenol (2)*

Mercury and Methylmercury Determination in Urine

Introduction

Mercury (Hg), used in many different industries and widely diffused in the environment, is very toxic and poses serious risks to workers exposed to its vapours.[26] Hg is easily derivatized to methylated forms (mono- and di-methylHg) in aquatic environments and bioconcentrated in fishes. Owing to these facts, Hg uptake in humans is mainly by air, food or beverage contamination, while methylmercury (MeHg) exposure is primarily from fish consumption.

Urinary Hg concentration is one of the most important indices of exposure, as it relates directly to Hg uptake in the organism. The ACGIH adopted 35 μg Hg g^{-1} of urinary creatinine, evaluated in urine sampled prior to the next shift, as the BEI for mercury.[5] Hg and MeHg show a strong affinity to organic compounds, and this results in a great tendency to be complexed as organic-bonded compounds in biological specimens. Hg and MeHg are decomplexed and then derivatized to their ethylated forms, to obtain compounds appropriate for gas chromatography.[27,28]

Materials, Method, Procedure, Calibration, Blank Tests

HP 5MS capillary column, 30 m × 0.25 mm i.d., 0.25 μm film thickness was set at 50 °C for 2 min, then ramped at 12 °C min^{-1} to 220 °C, and held for 10 min, with He carrier gas at 1.0 mL min^{-1} flow. The source temperature was 180 °C, and the splitless injector temperature 220 °C.

A fresh solution of sodium tetraethylborate (STEB) at 1% (w/v) was prepared daily with nano grade water and filtered through a 0.5 μm filter before use.

5 ml urine was put into a 20 mL vial, with 200 μL of concentrated HCl, sealed and heated at 50 °C for 1 hour. This treatment de-conjugates Hg and MeHg from organic compounds, and at the same time guarantees the stability of MeHg. When the sample had cooled to room temperature, the following were added: 200 μL of concentrated NaOH, 10 mL of buffer solution at pH 4, 200 μL of STEB solution at 1% and a magnetic stirring bar. The vial was then immediatly closed and put on a magnetic stirrer at 1000 rpm for 10 min to ensure the proper mixing of the solution. The SPME fiber was lowered into the headspace of the sample and exposed for 20 min.

The urine sample, obtained from donors not exposed to Hg in the workplace, was tested with the described procedure to assess the background content of Hg and MeHg.

Working standard solutions were prepared by spiking the urine sample with both Hg^{2+} and $MeHg^{+}$, obtaining the following concentrations: 0.2, 2.0, 20, 50, 100, 500 μg L^{-1} for each compound. Both the buffer and the STEB solutions were tested for the presence of Hg^{2+}, with negative results.

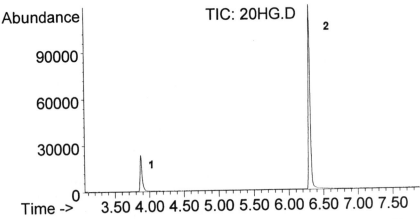

Figure 5 *Total ion chromatogram of a standard urine sample spiked with Hg^{2+} and $MeHg^{+}$ at 20.0 μg L^{-1} each; peak (1) is EtMeHg, peak (2) is Et$_2$Hg*

Results and Conclusions

Diethylmercury and methylethylmercury present in the urine after derivatization show great affinity for the PDMS phase of the fiber.

Data obtained are reported in Table 1 and Figure 5. They show that this method is adequate to meet the requirements for monitoring Hg in urine of exposed workers and Hg and MeHg in the unexposed population. It can also be useful for analyses in environmental medicine.

References

1. S. Ghirotti, M. Imbriani, G. Pezzagno and E. Capodaglio, *Am. Ind. Hyg. Assoc. J.*, 1987, **48**, 786.
2. A.J. McMichael in *Environmental Carcinogens Methods of Analysis and Exposure Measurement—Vol. 10: Benzene and Alkylated Benzenes*, IARC Sci. Publ. No 85, 1988.
3. W.J. Mann, *Academic Laboratory Chemical Hazards Guidebook*, Van Nostrand Reinhold, New York, NJ, 1990.
4. R. Lauwerys, in *Human Biological Monitoring of Industrial Chemical Series*, Commission of the European Communities, Luxembourg, 1983.
5. American Conference of Governmental Industrial Hygienists, *Threshold Limit Values and Biological Exposure Indices*, Cincinnati, OH, 1996.
6. O. Inoue, K. Seiji, M. Kasahara, H. Nakatsuka, T. Watanabe, S.N. Yin, G.L. Li, C. Jin, S.X. Cai, X.B. Wang and M. Ikeda, *Br. J. Ind. Med.*, 1986, **43**, 692.
7. C.N. Ong, P.W. Kok, H.Y. Ong, C.Y. Shi, B.L. Lee, W.H. Phoon and K.T. Tan, *Occup. Environ. Med.*, 1996, **53**, 328.
8. A.M. Medeiros, M.G. Bird and G. Witz, *J. Toxicol. Environ. Health*, 1997, **51**, 519.
9. L. Perbellini, F. Pasini, G.B. Faccini, B. Danzi, M. Gobbi, A. Zedde, P. Cirillo and F. Brugnone, *Med. Lav.*, 1988, **79**, 460.
10. M. Imbriani, S. Ghittori, G. Pezzagno and E. Capodaglio, *Med. Lav.*, 1987, **78**, 239.
11. R. Volpi, F. Bauleo, R. Pasquini, S. Monarca, C. Cicioni, G. Angeli and F. Blasi, *Med. Lav.*, 1988, **79**, 136.
12. E. Wigaeus, A. Lof, R. Bjurstrom and M. Byfalt Nordqvist, *Scand. J. Work Environ. Health*, 1983, **9**, 479.
13. J.C. Ramsey and M.E. Andersen, *Toxicol. Appl. Pharmacol.*, 1984, **73**, 159.
14. P. Hotz, M.P. Guillemin and M. Lob, *Scan. J. Work Environ. Health*, 1980, **6**, 241.
15. L. Perbellini, L. Romeo, G. Maranelli, G. Zardini, C. Alexopoulos and F. Brugnone, *Med. Lav.*, 1990, **81**, 382.
16. G. Pezzagno, M. Imbriani, S. Ghittori and E. Capodaglio, *Scand. J. Work Environ. Health*, 1985, **11**, 371.
17. Task Force on Water Quality Guidelines of the Canadian Council of Resource and Environmental Ministers, *Canadian Water Quality Guidelines*, Chapter 6: Parameter-specific background information, 1989.
18. K.P. Cantor, R. Hoover, T.J. Mason and L.J. McCabe, *J. Natl. Cancer Inst.*, 1978, **61**, 979.
19. D. Ferry, W.A. Temple and E.G. McQueen, *Int. Arch. Occup. Environ. Health*, 1980, **47**, 155.
20. V. Sedivec, M. Mraz and J. Flek, *Int. Arch. Occup. Environ. Health*, 1981, **48**, 1257.
21. L. Hardell, *Scand. J. Work Environ. Health*, 1981, **7**, 119.

22. E.E. McConnel, J.A. Moor, B.N. Gupta, A.H. Raken, M.I. Luster, J.A. Goldstein, J.K. Haseman and C.E. Paiker, *Toxicol. Appl. Pharmacol.*, 1980, **52**, 468.
23. N.I. Kerkvliet, L. Baecher-Steppan and J.A. Schmidt, *Toxicol. Appl. Pharmacol.*, 1982, **62**, 55.
24. W.B. Deichman, *Toxicology Occupational Medicine. Development in Toxicology and Environmental Science*, Elsevier, Amsterdam, 1979.
25. R. Lauwerys in *Biological Indicators for the Assessment of Human Exposure to Industrial Chemicals*, ed. L. Alessio, A. Berlin, M. Boni and R. Roi, Commission of the European Community, Luxembourg, 1988.
26. R. Lauwerys and J.P. Buchet, *Arch. Environ. Health*, 1973, **27**, 65.
27. S. Rapsomanikis, O.F.X. Donard and J.H. Weber, *Anal. Chem.*, 1986, **58**, 35.
28. Y. Cai and J.M. Boyana, *J. Chromatogr.*, 1995, **696**, 113.

CHAPTER 42

Applications of SPME in Criminal Investigations

TSUYOSHI KANEKO

This chapter introduces the applications of SPME in criminal investigations. In the author's laboratory, SPME techniques for obtaining information from physical evidence collected from crime scenes have been studied. SPME with gas chromatography is suitable to detect volatile compounds such as residual accelerants in fire debris at arson scenes.

1 Introduction

Crime scenes hold physical evidence that often provides detectives with information about criminals, or aids in solving crimes. In criminal investigations, detectives require immediate analysis of the physical evidence to obtain information from it. However, physical evidence is in many cases limited in quantity, and it must be analyzed efficiently. The smell perceived from physical evidence sometimes gives us a great deal of important information and may constitute part of the supplementary reference data in the analysis and discrimination of physical evidence. In this context, the author's laboratory has been conducting studies for the immediate determination of volatile compounds of such substances as aromatic components extracted from physical evidence and providing results for criminal investigations.

2 Forensic Discrimination of Household Cleaning Products

Common household cleaning products familiar to us, such as commercial dishwasher detergent, may sometimes be used as a tool in crimes such as theft or rape, and may be left at the crime scene. Although a great many reports of component analysis of household cleaning products are available and systematic analysis methods[1,2] have been established, they generally require a large sample size, complicated pretreatment, and a long time for analysis. However, in criminal investigations, immediate and simple discrimination of physical evidence is very often preferred to that of a precise component analysis. In this context, the author now reports the results of a method developed for the analysis of trace amounts of aromatic compounds spiked into household cleaning products, using the SPME technique. The possibility of simply discriminating the products by using the measurement results as an index is also discussed.

Experimental

For analysis of aromatic compounds in household cleaning products, a 65 μm-polydimethylsiloxane/divinylbenzene (PDMS/DVB) coated fibre was used. Twenty-two household cleaning products which could be obtained in Chiba City were purchased from local commercial outlets and were used without modification. The standards of aromatic compounds used in this study were all purchased from Tokyo Kasei Co. Ltd. Measurement of the aromatic compounds in household cleaning products by the SPME technique was conducted by adding 10 μL of a product to a 7 mL vial, sealing the vial hermetically and maintaining it at room temperature. Extraction of the aromatic compounds was performed in the headspace of the vial at room temperature for 20 minutes. Extracted compounds were separated and identified by GC and GC–MS. The GC analysis was performed on a HP-6890 gas chromatograph (Hewlett Packard) equipped with a flame ionization detector. The injector temperature was set at 250 °C. For separation of aromatic compounds, a 30 m × 0.25 mm i.d., OV-17 capillary column (0.25 μm film thickness, Quadrex) was used. The oven temperature was programmed to rise from 50 °C (held for 3 minutes) to 250 °C at 5 °C min^{-1}. Helium was used as carrier gas at 1.3 mL min^{-1} flow rate in a constant flow mode. The SPME fibre was subjected to thermal desorption in the injector for 1 minute, which corresponded to the splitless time. GC–MS analysis was performed on a GCMS-QP5000 gas chromatograph–mass spectrometer system (Shimadzu). Compound identification was based on comparison of GC retention indices and mass spectra with those of authentic compounds.

Results and Discussion

For fibre selection experiments, measurement by the SPME technique was conducted using an aqueous solution spiked with 30 standard aromatic compounds. Individual aromatic compounds in the water solution (1% ethanol solution) were mixed and adjusted so each had a concentration of about 1 ppm. The 65 μm PDMS/DVB fibre was more efficient in the extraction of the aromatic compounds in comparison with other fibres (85 μm polyacrylate, 100 μm polydimethylsiloxane, and 65 μm Carbowax/divinylbenzene). Figure 1 shows chromatograms of the aromatic compounds obtained by the SPME technique, using the 65 μm PDMS/DVB fibre for extraction of the standards mixture. As indicated in the figure, almost all of the 30 components in

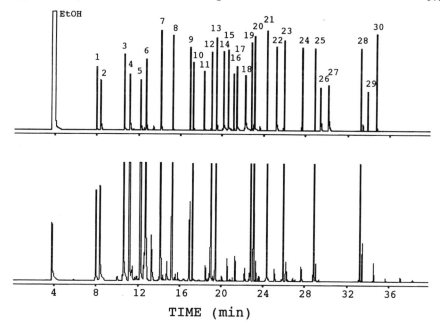

Figure 1 *Gas chromatograms of a mixture of standard aromatic compounds by direct injection (top), and by SPME technique (bottom). 1, ethyl butyrate; 2, n-hexyl aldehyde; 3, α-pinene; 4, allyl isothiocyanate; 5, n-heptyl aldehyde; 6, myrcene; 7, (−)-limonene; 8, benzaldehyde; 9, linalool; 10, n-nonyl aldehyde; 11, ethyl levulinate; 12, citronellal; 13, l-menthol; 14, 2-phenylethyl alcohol; 15, α-terpineol; 16, diethyl succinate; 17, nerol; 18, geraniol; 19, (+)-pulegone; 20, cuminaldehyde; 21, trans-anethole; 22, trans-cinnamaldehyde; 23, benzyl n-butyrate; 24, cis-jasmone; 25, β-ionone; 26, vanillin; 27, ethyl vanillin; 28, α-n-amylcinnamaldehyde; 29, triethyl citrate; 30, benzophenone*

the mixture were detected in the headspace of the solution. However, four components, ethyl levulinate, vanillin, ethyl vanillin, and triethyl citrate could not be detected by any fibre.

Next, in order to examine the effects of the existence of surfactants on the measurement of aromatic compounds, polyethylene glycol was added to the sample matrix of the above solution, and the variation in the quantity of the aromatic compounds detected was measured. Figure 2 indicates the amount of aromatic compounds detected when the concentration of polyethylene glycol in the medium was varied from 10% to 20% and 30%. As shown in the figure, an increase the concentration of polyethylene glycol in the sample caused an increase in extracted quantities of ethyl butylate and *n*-hexyl aldehyde. However, *n*-heptyl aldehyde, (+)-pulegone, *trans*-anethole, β-ionone and 20 components which are not indicated in the figure showed decreased extraction. This phenomenon can be explained by the variation of solubility of individual aromatic compounds with the variation of the polyethylene glycol concentration in the sample matrix. As reported previously,[3] for the headspace analysis of

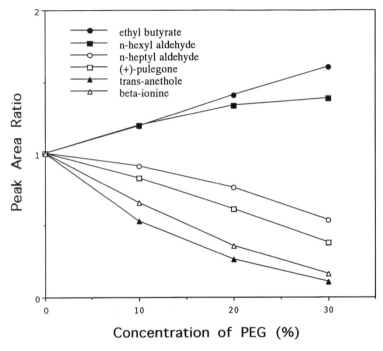

Figure 2 *Sample matrix effects on SPME extraction of standard aromatic compounds*

cosmetics, the amount of aromatic compounds emanating from the cosmetics varies according to matrix differences. It can be assumed that even a detergent to which the sample perfumes are added may show different chromatogram patterns if its matrix varies.

Figure 3 shows gas chromatograms obtained by the SPME analysis of volatile compounds in household cleaning products, using the 65 μm PDMS/ DVB fibre. Aromatic compounds consisting mainly of limonene were detected from all 22 products measured, and it was found that citrus fragrances were being added to all of the products. Volatile compounds derived from raw materials other than fragrances were also detected in the chromatograms. It was possible to specify the manufacturer of a product by identifying these components.

3 Analysis of Accelerants

Detection of accelerants from fire debris at fire scenes or the clothing of a suspect in an arson case is important physical evidence. In an arson case, petroleum products such as gasoline or kerosene are very often used as accelerants. If an accelerant is detected in the clothing of a suspect, this is an important fact which connects the suspect to the crime. However, the accelerant residue in many cases exists in trace amounts, and it is necessary to avoid

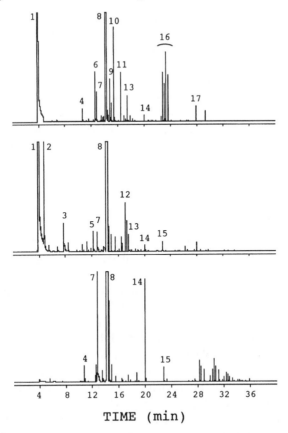

Figure 3 *Gas chromatograms of extracted volatile compounds from household cleaning products by SPME, using the 65 μm PDMS/DVB fibre. All products were dishwasher detergents of 'A Company'(top), of 'B Company'(middle) or of 'C Company'(bottom). 1, ethyl alcohol; 2, methylethylketone; 3, propylene glycol; 4, α-pinene; 5, octanone; 6, β-pinene; 7, myrcene; 8, limonene; 9, p-cymene; 10, terpinene; 11, terpinolene; 12, carene; 13, dodecane; 14, tridecane; 15, tetra-decane; 16, isomers of tetradecene; 17, methyl laurate*

damage to the physical evidence from heating or washing with an organic solvent in extraction of the accelerant, as other inspections are frequently required for that physical evidence. As a method to satisfy these requirements, the author reports here the detection of trace level accelerants (0.1 μL L^{-1}) by the SPME technique at room temperature.

Experimental

For analysis of accelerants, 100 μm polydimethylsiloxane (PDMS) and 75 μm Carboxen/polydimethylsiloxane (CX/PDMS) coated fibres were used. Gasoline, kerosene, and lighter fluid (for Zippo lighter) were purchased from local commercial outlets. Vapour samples were prepared as follows; 0.1 μL of gasoline, kerosene, or lighter fluid were placed separately in 1 L glass bottles and vaporized. Accelerants extracted by SPME were separated and identified by GC. The analysis was performed on a GC-14B gas chromatograph (Shimadzu) equipped with flame ionization detector. The injector temperature was set at 250 °C. For separation of accelerants, a 15 m × 0.53 mm i.d., DB-1 column (5 μm film thickness, J&W Scientific) was used, and the column was set in an inlet for packed column. The oven temperature was programmed to rise from 40 °C (held for 5 minutes) to 200 °C at 5 °C min^{-1}. Helium was used as carrier gas at 9 mL min^{-1} flow rate. The SPME fibre was subjected to thermal desorption in the injection for 1 minute.

Results and Discussion

Figure 4 shows the chromatogram obtained from the extraction of a trace level kerosene vapour by SPME. Kerosene was detected with high sensitivity using the 100 μm PDMS fibre.

Vapour samples of gasoline and lighter fluid could not be detected sensitively by the SPME technique using the 100 μm PDMS fibre. This indicates that the absorption ability of the 100 μm PDMS fibre is not sufficient for more volatile compounds. For that reason, the 75 μm CX/PDMS fibre, which is designed for the detection of very small molecules, was used for analysis of these compounds. Figures 5 and 6 show chromatograms obtained from extraction of trace level gasoline and lighter fluid vapours. Each accelerant could be detected with high sensitivity by the SPME technique, using the 75 μm CX/PDMS fibre. Although limits of detection were not determined, it was proved that the SPME technique could detect each accelerant without prior vapour concentration, at analyte concentrations which are difficult to analyse by the conventional headspace method.[4-8]

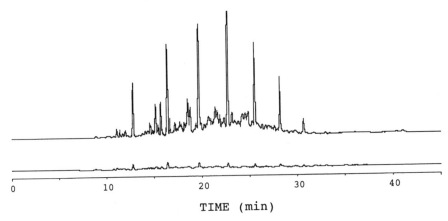

Figure 4 *Gas chromatograms of kerosene vapour by SPME, using the 100 μm PDMS fibre (top), and by the conventional static headspace method (bottom)*

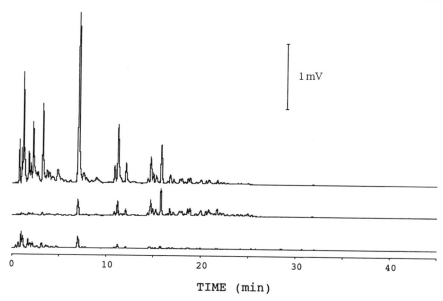

Figure 5 *Gas chromatograms of trace level gasoline vapour by SPME, using the 75 μm CX/PDMS fibre (top), 100 μm PDMS fibre (middle), and by the conventional static headspace method (bottom)*

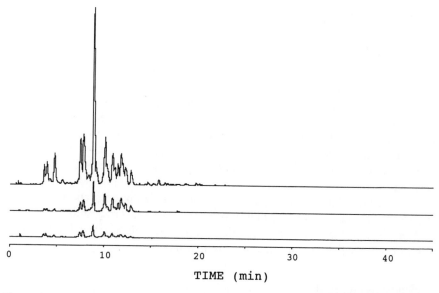

Figure 6 *Gas chromatograms of trace level lighter fluid vapour by SPME, using the 75 μm CX/PDMS fibre (top), 100 μm PDMS fibre (middle), and by the conventional static headspace method (bottom)*

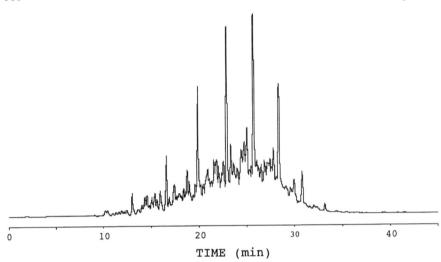

Figure 7 *Gas chromatogram of accelerant extracted from physical evidence by SPME*

Case Example

This technique was successfully applied to routine casework. Figure 7 shows a chromatogram obtained by the SPME technique for actual physical evidence. This physical evidence was the clothing of a suspect in an arson case. The physical evidence was put into a sealed container, and analysed at room temperature by the SPME technique, using the 100 μm PDMS fibre. Kerosene was positively identified.

4   Conclusions

For the measurement of aromatic components in household cleaning products, reproducible chromatograms could be obtained by maintaining constant extraction conditions with regard to sample size, container, and extraction time. By comparing the chromatograms obtained, immediate and simple discrimination of the products from a trace amount was possible.

In forensic inspection of fire debris for an accelerant, quantification is not important, as the major purpose is to identify the type of accelerant. The SPME technique can detect any accelerant without damaging the physical evidence and moreover, since it does not use any organic solvent for the extraction, it does not interfere with the detection of highly volatile compounds in gasoline.

Therefore, it is expected that the SPME technique will be applied widely in the forensic science field, including criminal investigation.

Acknowledgement

The author wishes to express his gratitude for guidance and encouragement received from Professor Masahiro Nakada of Chiba Institute of Technology.

References

1. M.J. Rosen and H.A. Goldsmith, *Systematic Analysis of Surface-Active Agents*, J. Wiley & Sons, NY, 1972.
2. H. Nishiya, *Bunseki*, 1993, **228**, 28.
3. I. Kimes and D. Lamparsky, *Perfümrie und Kosmetik*, 1978, **59**, 407.
4. R. Saferstein, *Criminalistics*, Prentice Hall, Englewood Cliffs, NJ, 1990.
5. J.D. Twibell and J.M. Home, *Nature*, 1977, **268**, 711.
6. K. Dynes and D.T. Burns, *J. Chromatogr.*, 1987, **396**, 183.
7. D.J. Tranthim-Fryer, *J. Forensic Sci.*, 1990, **35**, 271.
8. M. Kärkkäinen, I. Seppälä and K. Himberg, *J. Forensic Sci.*, 1994, **39**, 186.

Reaction Monitoring

CHAPTER 43

SPME–GC–MS Detection Analysis of Maillard Reaction Products

W.M. COLEMAN III

1 Introduction

The oldest and probably the most widely used method of food processing and preparation is heat treatment. The primary aim of this process is to create, produce, or impart the desired flavor, taste and textural nature of the food material. Embedded within all natural products, both plant and animal, are two abundant classes of compounds: (1) nitrogen containing molecules such as amino acids and amines and (2) monosaccharides such as fructose and glucose. While the major roles of these reagents are well established in living tissue their role in heat treated natural products takes on a unique perspective. That is, under the appropriate conditions, these reagents react to produce a complex series of volatile, semi-volatile and non-volatile compounds some of which have very powerful sensory impact at very low concentrations.[1] In particular, the reaction between amino acids and sugars was first investigated in detail by Maillard around the turn of the century and due to his pioneering work the reaction between amino acids and sugars is generally referred to as the Maillard reaction.[2-6]

Through the use of reaction variables such as pH, moisture content, temperature, concentration and time a wide array of volatile, semi-volatile and non-volatile compounds, including aldehydes, ketones, pyrazines, pyridines, furans and diketones, can easily be produced using model or natural systems.[7-27] Optimal reaction variables are easily attained during baking, frying and/or roasting of natural products such as meat, coffee, cocoa, licorice, peanuts, tobacco, soybeans *etc.* Thus, the presence of a complex array of volatile, semi-volatile and non-volatile compounds can be expected as a natural occurrence due to heat treatment of a wide variety of natural products.

Analytical approaches employed to provide qualitative and quantitative characteristics of both natural and model Maillard systems have for the most part involved extensive sample work-up *via* solvent extraction, liquid/liquid extraction, headspace, or cold trapping of the reaction products followed by analysis by either gas chromatography–mass spectrometry (GC–MS) or high performance liquid chromatography (HPLC).[28] In the early 1990s, an alternative approach to the use of the extensive sample work-up appeared, termed solid-phase microextraction (SPME).[29] Since its introduction, SPME has found application across a diverse array of analytical determinations including organic solvents in water,[30] explosives,[31] flavors[32–34] and pesticides.[35] These applications have shown SPME to be a convenient and efficient extraction method. Theory and applications have shown the fibers to provide for quantitative analysis in equilibrated and non-equilibrium situations.[36] The extraction phenomenon is based on using a thin polymer film coating of selected and variable polarities on a fine silica fiber to extract analytes of interest from a sample matrix. Mass diffusion from the matrix to the SPME fiber is considered the rate determining step. Thus, analytes of interest are extracted by the fiber and can, subsequently, be effectively thermally desorbed within the confines of a gas chromatographic injection port. Once desorbed, the analytes can be separated *via* capillary GC, followed by analysis using a variety of information rich detectors.

Owing to the demonstrated characteristics and performance of various SPME fibers of different polarities, for a wide array of analytes, investigations have been undertaken to demonstrate the potential of SPME for the rapid/convenient analysis of aqueous solutions/suspensions of volatile and semi-volatile products of Maillard reactions.[37,38] A discussion of very recent findings as well as new discoveries relating to mechanistic understandings and flavor preparation will be presented in this report. The findings based on results from SPME–GC–SIM–MSD analyses provide the first evidence for the unambiguous assignment of a mechanism for the formation of pyrazines in microwave heat treated natural product suspensions.

2 Experimental

Reagents

The pyrazines were obtained from Aldrich Chemical Company. Reagent grade NaCl and NaClO$_4$ were obtained from Fisher Chemical. Standard solutions were prepared gravimetrically using deionized water. Methylene chloride was obtained from Burdick and Jackson. The ^{15}N labeled compounds were obtained from Cambridge Isotope Laboratories. All reagents were used as received.

Cocoa powder was obtained from E. D. F. and Man Cocoa Products. Licorice powder was obtained from Mafco Worldwide Corp. Green Colombian coffee beans were obtained from local sources.

Sample Preparation

Suspensions

Suspensions of cocoa and licorice powders were prepared as 15% aqueous suspensions by adding 4.5 g of the powder to 25.5 g of deionized water contained within a vessel especially designed for microwave heat treatment (see below). After gently swirling, the suspension was placed in the microwave system for heat treatment.

Microwave System

The microwave reactions were performed in a CEM Corporation, Model MES-1000, Microwave Extraction System. The microwave power was set at maximum, 950 ± 50 watts, at a frequency of 2450 MHz. The reactions were performed in sealed vessels at 175 °C for a period of 30 minutes. The reaction temperature was attained slowly over a period of 10 minutes employing the software capabilities of the CEM instrument. The pressure of the sealed vessel was monitored, 100–150 psi, but not controlled. Specially designed microwave transparent lined vessels, supplied by CEM, were employed and assembled strictly following the manufacturer's instructions. The sealed vessels were placed on a turntable within the oven. The turntable was cycled back and forth during the course of the run to insure even distribution of the microwaves. After heat treatment, the samples were allowed to come to room temperature prior to opening. Once removed from the microwave, the samples were either analyzed or stored in a refrigerator prior to analysis.

Analysis Instrumentation/Methodology

Manual SPME–Gas Chromatography–Mass Selective Detection–Flame Ionization Detection

A Hewlett Packard (HP) 5880 gas chromatograph (GC) fitted with a HP 5970 mass selective detector (MSD) operating at 70 eV in the electron impact mode and a flame ionization detector (FID) were employed. The column effluent split ratio for the MSD–FID was approximately 80 : 20. The MSD was used for structure confirmation and quantitative analysis, while the FID area counts were used for qualitative purposes. The MSD interface temperature was set at 225 °C and the GC injection port was set at 250 °C. The GC was furnished with a DB-1701 fused-silica column (30 m × 0.32 mm i.d., 1.0 μm film thickness, J&W Scientific). The column linear velocity was set at approximately 40 mL min^{-1}. The GC oven was temperature-programmed from 10 to 220 °C at 20 °C min^{-1}. Splitless manual injections were used and the split valve was opened after 2 minutes.

A second GC–MSD system consisting of an HP 5890 Series II Plus GC equipped with a HP 5972 MSD operating in the electron impact mode at 70 eV was used to gain assistance in establishing the limits of detection for the analytes

of interest. This GC was fitted with a DB-Wax fused-silica column (30 m ×
0.25 mm i.d., 0.5 μm film thickness, J&W Scientific). The MSD interface and
GC injection port temperatures were 250 °C. The GC oven was temperature-
programmed from 40 to 240 °C at 5 °C min⁻¹. Splitless manual injections were
made and the split valve was opened after 1 minute.

Manual SPME fibers were obtained from Supelco and used strictly following
the manufacturer's instructions for use and conditioning prior to and after each
injection. After exposure to the sample of interest, thermal desorption of the
coated fiber in the injection port occurred without delay. The fiber remained in
the injection port for 10 minutes. For both liquid and headspace sampling,
10 mL of the aqueous solution containing the analyte(s) of interest was placed in
a 20 mL vial (2 cm diameter × 6.5 cm height) charged with a 7 × 2 mm magnetic
stir bar. The vial was sealed and allowed to stand overnight prior to sampling.
When liquid and headspace sampling were performed the vortex of the sample
was held constant at approximately 1 cm. Liquid sampling involved immersing
the entire fiber in the water while headspace sampling involved suspending the
fiber approximately 0.5 cm above the stirring water solution. Fresh samples
were used for every injection.

Automated SPME–Gas Chromatography–Mass Selective Detection–Flame Ionization Detection

A Varian 8200 CX AutoSampler with SPME III Sample Agitation was mounted
atop a HP 5890 Series II Plus GC equipped with a HP 5972 MSD operating
either in the electron impact mode at 70 eV or in the selected ion monitoring
(SIM) mode. This GC was fitted with a DB-Wax fused-silica column
(30 m × 0.25 mm i.d., 0.5 μm film thickness, J&W Scientific). The MSD interface and GC injection port temperatures were 250 °C. The GC oven was
temperature-programmed from 40 to 140 °C at 5 °C min⁻¹, then to 220 °C at
10 °C min⁻¹ and held there for 4 minutes. Splitless injections were made and the
split was opened after 1 minute. The fiber was automatically submerged in the
solution, vibrated for 0.75 minutes, removed, injected, and held in the injection
port for 30 minutes employing parameters set *via* the operating software.

SPME fibers for automated injections were obtained from Supelco and
employed strictly following the manufacturer's instructions for use and activation.

Prior to analysis by automated SPME the aqueous heat treated suspensions
were manually filtered through a Whatman Autovial equipped with a 0.45 μm
PVDF filter and designed for use with aqueous solutions. Then, to 1.8 mL vials
equipped with Teflon lined septa was added via a Rainin EDP Plus Motorized
Microliter Pipet, 1.7 mL of the filtered solution. Strict attention to consistency
in the addition of 1.7 mL was necessary to obtain reproducible results. The
charged vials were loaded on the sample carousel and automatically sampled
employing the instrumentation software provided by Varian and HP. In some
cases it was necessary to dilute the 15% heat treated suspensions to obtain
reproducible fiber performance.

Selected ion monitoring was used for the quantitative determination of selected pyrazines in the heat treated suspensions. The selected compounds and accompanying selected ions (*m/z*) are listed as follows: methylpyrazine, 94, 95, 96; C2 pyrazines, 107, 108, 109, 110; C3 pyrazines, 121, 122, 123, 124; and C4 pyrazines, 135, 136, 137, and 150. The C2, C3 and C4 notations used preceding pyrazines is used to denote a class of pyrazines. For example, C2 pyrazines would include substituted pyrazines such as all of the dimethyl-pyrazines as well as ethylpyrazine. In all of these cases the pyrazines have two carbons (C2) attached in some fashion to the fundamental pyrazine molecule. Identical arguments are used for the C3 and C4 terms.

Fresh samples were used for every injection.

3 Results and Discussion

Through a series of experiments employing both manual and automated approaches to solid-phase microextraction (SPME), the characteristics of a number of fibers will be described in terms of their ability to provide for both qualitative and quantitative analysis of selected products of the Maillard reaction chemistries. More specifically, selected polar and non-polar fibers will be studied for their performance toward such volatile and semi-volatile compounds of the Maillard reaction as pyrazines, thiazoles, furans and pyridines. These compounds were selected because they are known to possess very low odor threshold values and are powerful sensory compounds carrying aromatic notes associated with cooked products such as breads, meats, peanuts and other natural products. Two sets of experiments were executed, one involving use of the SPME fibers in the manual mode and the second involving use of the SPME fibers in the automated mode.

Manual SPME–Gas Chromatography–Mass Selective Detection–Flame Ionization Detection Experiments

Manual SPME headspace and liquid sampling were examined for aqueous solutions of a series of aroma and flavor compounds closely associated with the Maillard reaction chemistries and sugar thermal degradation. Two SPME fibers of differing polarities, 65 μm film thickness Carbowax/divinylbenzene (CW/DVB) and 100 μm film thickness poly(dimethylsiloxane) (PDMS), polar and nonpolar, respectively, were employed. PDMS fibers are absorptive and CW/DVB fibers are adsorptive. Preliminary trials had shown that an exposure time of 10 min was sufficient to establish a consistent amount of extracted material at a concentration of 50 ng μL^{-1}. Thus, using an exposure time of 10 minutes, accompanied by stirring, the number of nanograms extracted by these fibers was determined, Table 1. The number of nanograms was calculated from response factors obtained from splitless injections of methylene chloride solutions of the compounds of interest. It is important to note (see below) that the values contained in Table 1 were determined from aqueous solutions containing multiple compounds of the given class. For example, the values for the

Table 1 *Nanograms sorbed for selected compounds as a function of concentration and SPME fiber type*

Concentration	Thiazole (P)	Thiazole (C)**	Methylthiazole (P)	Methylthiazole (C)	Dimethylthiazole (P)	Dimethylthiazole (C)
50	16.1	46.5	60.0	125.0	175.4	259.0
5	1.5	7.1	7.6	28.3	23.2	66.5
0.5	0.1	0.8	0.7	4.6	2.1	11.4

Concentration	Trimethylthiazole (P)	Trimethylthiazole (C)	Ethoxythiazole (P)	Ethoxythiazole (C)
50	513.5	490.3	259.2	606.9
5	82.7	127.0	28.1	144.3
0.5	8.1	20.9	1.7	19.2

Concentration	Pyridine (P)	Pyridine (C)	4-Methylpyridine (P)	4-Methylpyridine (C)	Pyridine alone (P)	Pyridine alone (C)
50	64.1	81.7	141.3	152.2	74.8	152.7
5	9.8	11.4	22.7	29.6	9.1	24.1
0.5	0.9	2.3	2.7	7.2	Not detected	1.7

Concentration	4-Ethylpyridine (P)	4-Ethylpyridine (C)	5-Ethyl-2-methylpyridine (P)	5-Ethyl-2-methylpyridine (C)
50	452.8	426.8	1072.7	800.8
5	72.0	96.7	163.8	184.7
0.5	7.3	19.1	16.0	25.6

Concentration	Pyrazine (P)	Pyrazine (C)	Methylpyrazine (P)	Methylpyrazine (C)
50	24.9	61.2	37.4	96.5
5	2.9	7.6	4.5	17.7
0.5	Not detected	1.4	0.4	2.4

Concentration	2,3-Dimethylpyrazine (P)	2,3-Dimethylpyrazine (C)	Trimethylpyrazine (P)	Trimethylpyrazine (C)
50	75.3	149.0	146.3	262.5
5	10.2	33.0	19.9	59.8
0.5	0.1	5.1	0.2	9.0

Concentration	2-Methoxypyrazine (P)	2-Methoxypyrazine (C)	2-Methoxy-3-methylpyrazine (P)	2-Methoxy-3-methylpyrazine (C)	Acetylpyrazine (P)	Acetylpyrazine (C)
50	135.9	424.5	282.2	574.2	28.2	129.8
5	17.4	114.5	36.5	146.6	2.3	24.9
0.5	2.1	19.9	3.6	22.7	Not detected	3.1

Concentration	Furfural (P)	Furfural (C)	5-Methylfurfural (P)	5-Methylfurfural (C)
50	36.6	285.2	83.5	729.2
5	2.0	50.8	7.2	135.9
0.5	Not detected	6.7	Not detected	17.2

*(P) = Polydimethylsiloxane fiber, **(C) = Carbowax/divinylbenzene fiber.

pyrazines were determined from aqueous solutions containing all of the pyrazines listed in Table 1.

Fiber Selectivities, Capacities and Sample Matrix Influences

The polar and nonpolar fibers provided for some marked differences in a variety of areas. As the amount of alkyl substitution increased, both fibers extracted more material even though the concentration of each analyte was constant at $50\,ng\,\mu L^{-1}$. Specifically, for example, the $65\,\mu m$ CW/DVB fiber extracted 61.21 ng of pyrazine and 149.05 ng of 2,3-dimethylpyrazine from an aqueous solution containing nominally $50\,ng\,\mu L^{-1}$ of each reagent. For the same solution, the $100\,\mu m$ PMDS fiber extracted slightly less (Table 1, Figure 1). These results point directly to meaningful differences in the partition coefficients as a function of the structure of molecules. A full discussion of partition coefficients (distribution ratio of analyte between the sample matrix and the fiber coating material) in terms of SPME/analyte behavior can be found in Chapter 1 of this monograph.

The aroma of a selected formulation is attributed to the presence of compounds in the headspace above the given formulation.[39] In toasted and roasted products such a bread, coffee and peanuts these sensory responses are

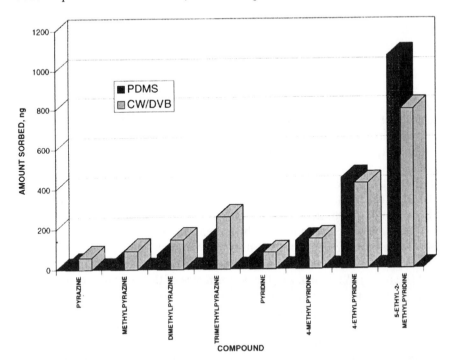

Figure 1 *Nanograms sorbed from aqueous solutions of selected compounds by selected SPME fibers*
(Reproduced from the *Journal of Chromatographic Science*, by permission of Preston Publications, a Division of Preston, Industries, Inc.)

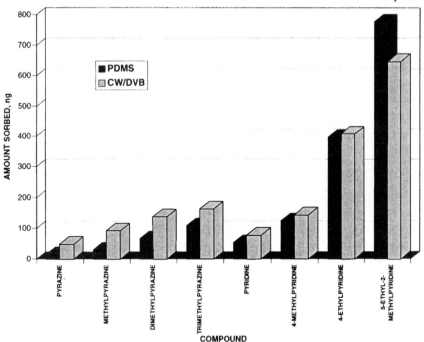

Figure 2 *Nanograms sorbed from the headspace above aqueous solutions of selected compounds by selected SPME fibers*
(Reproduced from the *Journal of Chromatographic Science*, by permission of Preston Publications, a Division of Preston, Industries, Inc.)

keyed by the presence of pyrazines, thiazoles and furans in the headspace above the products. An examination of the headspace above the aqueous solutions of selected compounds using the polar and nonpolar fibers revealed similar trends in their sorptive capacities and trends as found from the aqueous solutions. Figure 2 displays similar performances of the two fibers toward alkyl substituted derivatives when placed in the headspace above the aqueous solutions.

In addition to the presence of alkyl substituted compounds in the overall mix of compounds in heat treated products, compounds with functional groups such as methoxy and acetyl groups may be present. Figures 3 and 4 indicate that the polar and nonpolar fibers possess distinctive extraction characteristics toward these types of compounds. Notice the dramatic difference toward 5-methylfurfural (Figure 4). A summary of the performance of the two fibers to an array of compounds can be found in Figure 5. The extraction characteristics of the two fibers toward the more polar derivatives display the widest disparity.

If one or more water soluble organic species is contained within an aqueous based flavor formulation, it is almost intuitive that competition for adsorption into the fiber will occur between the species. In two relatively simple aqueous solutions: (1) of pyridine alone and (2) of pyridine in the presence of equal concentrations of 4-methylpyridine, 4-ethylpyridine and 5-ethyl-2-methylpyridine, the adsorption characteristics toward pyridine for the polar and nonpolar

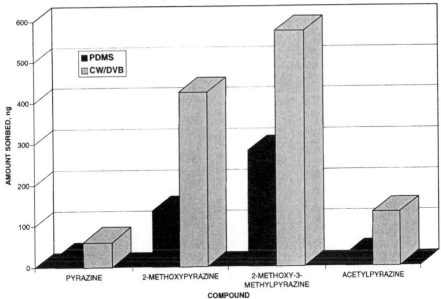

Figure 3 *Nanograms sorbed from aqueous solutions of selected pyrazines by selected SPME fibers*
(Reproduced from the *Journal of Chromatographic Science*, by permission of Preston Publications, a Division of Preston, Industries, Inc.)

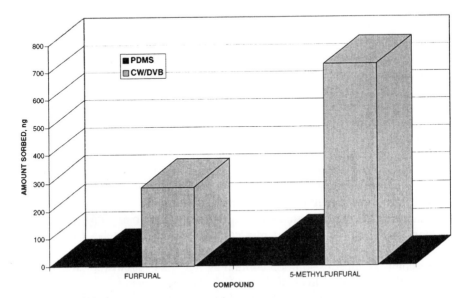

Figure 4 *Nanograms sorbed from an aqueous solution of furfurals by selected SPME fibers*
(Reproduced from the *Journal of Chromatographic Science*, by permission of Preston Publications, a Division of Preston, Industries, Inc.)

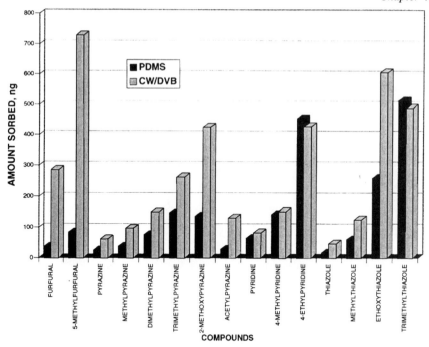

Figure 5 *Nanograms sorbed from aqueous solutions of selected compounds by selected SPME fibers*
(Reproduced from the *Journal of Chromatographic Science*, by permission of Preston Publications, a Division of Preston, Industries, Inc.)

fibers changed meaningfully (Figure 6). This was the first indication that the sample matrix could play a very significant role(s) in the overall performance of any selected fiber. This came to be especially important when quantitative analyses were conducted on the heat treated natural products.

Given the nature of natural products, the presence of notable amounts of alkali and alkaline earth metals as water soluble salts can be reasonably expected. Thus, in addition to the matrix impact due to other organic species, variations in ionic strength came to be very important in the quantitative analysis of the natural products. Table 2 and Figure 7 confirm that the presence of such components as multiple organic compounds and varying ionic strengths can influence the performance of both polar and nonpolar SPME fibers. For example, in the 20 component mix containing approximately $50 \, ng \, \mu L^{-1}$ of each component, the CW/DVB fiber sorbed 82.58 ng of 4-methylthiazole, but in a less complex aqueous solution containing only thiazoles at the same concentration the CW/DVB fiber sorbed 125.01 ng of 4-methylthiazole. With an average RSD of ± 5–10%, differences such as these were statistically significant. Likewise, the impact of ionic strengths of aqueous solutions on the adsorptivity of selected SPME fibers,[37,38,40,41] (Table 2) is clear. Notice the powerful impact of the use of a $0.2 \, g \, mL^{-1}$ aqueous solution of NaCl to prepare the samples

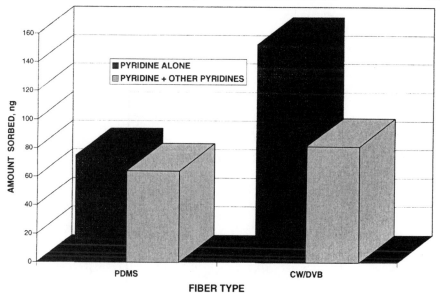

Figure 6 *Sorptive characteristics of pure and multicomponent mixtures by two SPME fibers*
(Reproduced from the *Journal of Chromatographic Science*, by permission of Preston Publications, a Division of Preston, Industries, Inc.)

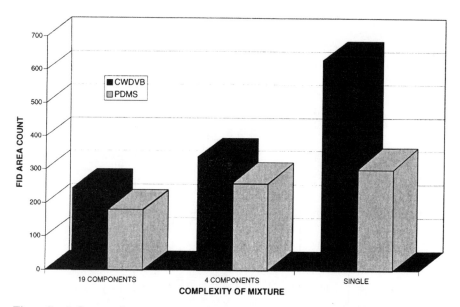

Figure 7 *Influence of component complexity on the amount of sorbed pyridine*
(Reproduced from the *Journal of Chromatographic Science*, by permission of Preston Publications, a Division of Preston, Industries, Inc.)

Table 2 *Behavior of selected SPME fibers to solutions of Maillard reaction products*

Component	20 Component mix ng CWDVB*	PDMS**	5 Component mix ng CWDVB*	PDMS**	20 Component mix ng CWDVB* 0.2 g mL⁻¹ NaClO4	PDMS** 0.2 g mL⁻¹ NaClO4	20 Component mix ng CWDVB* 0.2 g mL⁻¹ NaCl	PDMS** 0.2 g mL⁻¹ NaCl
Pyrazine	30	13	61	25	2	21	Not detected	22
Thiazole	37	19	46	16	63	10	66	46
Pyridine	59	44	82	64	86	13	84	80
4-Methylthiazole	83	44	125	60	110	51	130	98
Methylpyrazine	37	25	96	37	47	30	55	45
2,3-Dimethylpyrazine	56	40	149	75	63	42	90	104
4,5-Dimethylthiazole	172	129	259	175	225	141	316	326
4-Ethylpyridine	233	255	427	453	313	260	448	602
2,4,5-Trimethylthiazole	360	430	490	513	509	518	842	1305
2,3,5-Trimethylpyrazine	97	79	262	146	110	82	196	212
5-Methylfurfural	278	60	729	83	281	60	369	141
5-Ethyl-2-methylpyridine	362	493	801	1073	477	532	860	1551
2-Acetylpyrazine	75	27	130	28	53	27	82	42

*CWDVB = Carbowax/divinylbenzene fiber, PDMS** = Polydimethylsilsiloxane fiber.

versus deionized water. In some selected cases, the amount of material sorbed increased by a factor of 3 by employing the NaCl solution.

Thus, it appears that (1) both polar and nonpolar SPME fibers are well suited for the qualitative analysis of aqueous solutions of products closely associated with Maillard reaction chemistries and (2) influences such as the complexity of the matrix and matrix ionic strength can sway the performance of the SPME fiber.

Quantitative Aspects

Based on the impact of both the presence of multiple organic species and ionic strength (see above), any investigation of the potential for SPME fibers to perform quantitative analyses must include attention to these factors. In most all instances, use of a selected SPME fiber in concert with selected ion monitoring (SIM) mass selective detection (SIM-MSD) will precipitate an increase in both specificity and sensitivity for volatile and semi-volatile analytes. Thus, to gain an indication of the linear range and detection limits of the polar and nonpolar fibers toward the analytes of interest, a series of experiments using GC–SIM–MSD was performed. For the CW/DVB fiber a linear range of 0.005–0.5 ng μL^{-1} was discovered for an aqueous solution of the 20 components listed in Table 1. The linear range for the PDMS fiber was found to be 0.05–0.5 ng μL^{-1} for the same component mixture. The r^2 value for the linear plots for both fibers was consistently $\geqslant 0.99$ over these concentration ranges. Thus, the limits of detection for these fibers were different and appeared to be in the range of unit parts per billion for aqueous solutions of the Maillard reaction products examined.

Automated SPME–Gas Chromatography–Mass Selective Detection

Operating Parameters and Standards

With the advent of the agitated automated SPME injector, the collection of data has become much less labor intensive and as such has allowed for expansion of the studies on the behavior of the SPME fibers toward the Maillard reaction products.

Two experimental parameters were discovered in the initial studies with the automated system as being essential to obtaining reproducible results. First, and probably the most obvious, was the requirement to control the time of agitated exposure. This was easily accomplished through the use of the method development software available with the system. Optimization of this time was critical in so far as sensitivity and fiber lifetime issues were concerned. Extended adsorption/vibration times were found to produce premature undesired decomposition of the fibers mostly through cracking or breaking. An acceptable compromise between agitation time and required sensitivity for the analytes of interest here was established with an agitation time of 0.75 minutes. Under these

Table 3 *Percent RSD studies for selected SPME fibers using the automated approach*

Fiber type	Methyl	C2 pyrazines	C3 pyrazines	C4 pyrazines
100 μm PDMS	14.72	4.12	3.36	4.59
65 μm Carboxen/PDMS	5.30	7.85	8.07	8.73
65 μm CW/DVB	11.83	3.81	4.65	4.60

operating parameters no degradation in fiber performance was noted for at least 100 injections regardless of fiber type. Secondly, using the manufacturer recommended 1.8 mL sample vials dictated that the volume of solution within the vial must be held constant and the resulting solution volume must not completely fill the vial. A liquid volume of 1.7 mL was found to satisfy both criteria. The constant 1.7 mL liquid volume kept the headspace constant in the vial and allowed for the extensive movement of the solution within the vial during the agitation phase of the fiber exposure.

Three SPME fibers were evaluated using the automated agitation injection system: (1) 100 μm PDMS, (2) 65 μm CW/DVB and (3) 65 μm Carboxen/polydimethylsiloxane, Carboxen. Based on the performance of the PDMS and CW/DVB fiber in the manual studies, a range of aqueous solutions containing a group of selected substituted pyrazines in the 10–100 ng L^{-1} range was prepared. The performances of the three fibers toward these solutions were evaluated and linear responses could readily be attained for the three fibers over this concentration range. The r^2 values were consistently $\geq$0.99, 0.991–0.999, for all fibers in all solutions at 10–100 ng L^{-1}.

To obtain an appreciation for the measurement error of the automated agitated approach, the SIM area counts for six fresh injections of selected aqueous solutions containing approximately 50 ppb of selected pyrazines were collected. The data in Table 3 strongly indicated excellent analytical reproducibility with % RSD values consistently less than 10% in most all cases.

Embedded within Table 3 are indications that the three fibers perform significantly differently toward extraction of the pyrazines from deionized water solutions. For, example, notice that the average SIM area counts for the C2 pyrazines for the PDMS, Carboxen and CW/DVB fibers were 1395, 27 731 and 4217, respectively. For the PDMS and Carboxen fibers the range covered at least one order of magnitude. Similar behavior was found for the alkyl and functional substituted pyrazines. This trend in fiber performance in deionized solutions was very similar to that found with the manual approach (see above). The trend also speaks to significant differences in the partition coefficients of the analytes for the fibers investigated (see above).

Microwave Assisted Sample Preparation

The preparation of heat treated suspensions *via* microwave technology represents a relatively new approach for the synthesis of new flavor and aroma formulations.[42-48] The direct *versus* indirect heating via a microwave system offers advantages in the generation of new flavors. The use of microwave-transparent inert vessels such as PTFE versus metal pressure bombs benefits product stability and allows for enhanced uniform heating and cooling rates. The enhanced cooling rate may allow for the more successful isolation of thermally labile volatile compounds.

To date, the reports on Maillard chemistries in combination with microwave heating have focused on model systems. No reports exist describing the production of volatile and semi-volatile compounds in microwave assisted Maillard reactions on natural products. The following discussions will provide the first detailed descriptions of microwave assisted natural flavor formulation development from the Maillard reaction perspective. Specifically, SPME based experiments will be described which result in the determination of the amount of selected pyrazines in various microwave heat treated natural product suspensions. In addition, SPME based fundamental mechanistic studies will delineate the role(s) of selected nitrogen and sugar sources in the origin of the pyrazines and other Maillard reaction products.

Fiber Performance

Three fibers, 100 μm PDMS, 65 μm CW/DVB and 65 μm Carboxen were exposed to the filtered cocoa, licorice and coffee suspensions using the automated approach described earlier. Alkyl pyrazines and pyrazines with functional groups were monitored by GC–SIM–MSD. Table 4 contains the average SIM area counts for selected pyrazines. Based on the previous work with the manual fibers, the differences in SIM area counts for a specific analyte as a function of fiber type was not a surprise. The Carboxen fiber was found to produce the highest SIM area counts for the three fibers for virtually all of the analytes. Based on the performance of the three fibers the Carboxen and PDMS fibers were selected for use in quantitative analysis of the alkyl pyrazines in the filtered suspensions.

The method of standard addition was employed to determine the levels of methyl, C2, C3 and C4 pyrazines in the filtered suspensions. This approach was selected as a result of the influences discovered due to the presence of multiple organic species and the possible presence of alkali and alkaline earth metals in the aqueous suspensions.

The standard addition method was as follows: to 1.7 mL of the filtered heat treated suspension were added small known quantities, $\leq 10\,\mu$L, of aqueous solutions containing the following pyrazines: methylpyrazine; 2,5-dimethylpyrazine; 2-ethyl-3-methylpyrazine; and 2-ethyl-3,5-dimethylpyrazine. The addition of μL quantities of the standards was sufficiently small as to have no measurable impact on the vibration of the solution or on the concentration of the analytes. The 2,5-dimethylpyrazine, 2-ethyl-3-methylpyrazine and 2-ethyl-3,5-dimethylpyrazine served as surrogates for the total C2, C3 and C4 concentration in the suspension. Thus, using the method of standard addition, it was possible to determine the amount of pyrazines produced per gram of

Table 4 *SIM area counts for alkylpyrazines in 15% heat treated suspensions for selected SPME fibers*

Fiber type	Heat treated suspension	Methyl	C2 pyrazines	C3 pyrazines	C4 pyrazines	MeO	2-MeO-3-Me	2-MeO-6-Me	Acetyl
100 μm PDMS	Cocoa	1161138	102058	71938	68106	611	ND*	ND	7332
65 μm Carboxen/PDMS		918490	649042	229876	100642	1300	1513	1670	2354
65 μm Carbowax/DVB		351165	283412	140021	81180	286	ND	ND	2452
100 μm PDMS	Licorice	642678	333065	106047	17854	910	ND	ND	2167
65 μm Carboxen/PDMS		4039787	1273506	212787	21510	2003	ND	ND	16566
65 μm Carbowax/DVB		1164343	452342	70465	7564	ND	ND	ND	6972
100 μm PDMS	Coffee	432785	273083	49655	11526	836	ND	ND	ND
65 μm Carboxen/PDMS		4016642	2539062	381223	49031	6767	ND	ND	15905
65 μm Carbowax/DVB		938417	555487	76591	14721	1225	ND	ND	2568

*ND = Not detected.

Table 5 *Yield ($\mu g\ g^{-1}$) of selected pyra-*
zines from microwave assisted
heat treated suspensions of cocoa
and licorice powders

Selected pyrazine	Cocoa	Licorice
Methyl	34.40	203.00
C2 pyrazines*	26.70	56.42
C3 pyrazines**	3.23	7.17

*C2 pyrazines = pyrazines such as ethyl and dimethyl.
**C3 pyrazines = pyrazines such as ethylmethyl and trimethyl.

starting ground powder (Table 5). Under the microwave assisted heating conditions employed here, the licorice was found to yield more alkylpyrazines per unit mass than the cocoa. The pyrazines yielded for these natural products on a weight basis were similar to those obtained on model[15,17,18,22,23,49] and natural systems.[19,50] Thus, automated SPME/GC/SIM–MSD analysis has been shown to be a viable approach for the quantitation of alkylpyrazines produced in the heat treatment of natural products. Also, the production of an array of alkylpyrazines seems common for a diverse array of natural products. These findings are in concert with earlier results obtained from dynamic headspace–GC–MSD approaches.[9–12]

Mechanistic Studies, ^{15}N, Pyrazines Formation

Model studies have confirmed that nitrogen sources such as amino acids serve as nitrogen donors in the formation of pyrazines.[16] In addition, the carbon source forming the pyrazine arose from the sugar molecules. Selected volatile and semi-volatile aldehydes appearing in roasted natural products have been directly linked to the Strecker degradation of certain amino acids.[51] No reports have yet appeared describing the degree of incorporation of the nitrogen sources into pyrazines as a result of microwave assisted heat treatment of a natural system. By employing SPME–GC–SIM–MSD, the direct molecular link between selected nitrogen sources and pyrazine formation has been made. The approach is described below.

To gain a perspective on the issue of ^{15}N incorporation into the pyrazine molecules, a somewhat limited series of ^{15}N labeled amino acids as well as $^{15}NH_4OAc$, and [^{15}N]urea were employed as added reagents to microwave prepared licorice suspensions at 1% by weight relative to the mass of the licorice powder. ^{14}N is the most abundant atom in the nitrogen array, comprising approximately 99.6% of all nitrogen. Thus, use of ^{15}N isotopes would seem to be a very acceptable way to assess nitrogen incorporation into synthesized molecules due to the relatively simple increases in molecular weight.

For example, incorporation of one ^{15}N *versus* one ^{14}N into the pyrazine molecule would lead to an increase in molecular weight on one m/z unit to $m/z = 81$, while incorporation of two ^{15}N atoms into a pyrazine would result in an increase in molecular weight of two m/z units to $m/z = 82$ over the naturally abundant ^{14}N containing species, $m/z = 80$. For example, the molecular weight of pyrazine and the most abundant ion in the electron impact mass spectrum of pyrazine is $m/z = 80$ due in part to the natural abundance of ^{14}N being 99.6%. Therefore, any contribution from a ^{15}N atom in a naturally occurring arrangement would be very minimal resulting in very few species with m/z of 81 or 82. Such was observed in the mass spectrum of pyrazine. Identical arguments may be made for any other pyrazine. Therefore, since the molecular weights and resulting molecular ions of the pyrazines would be changing systematically upon incorporation of the ^{15}N, SIM was employed to speciate the array of pyrazine molecular structures produced in the presence of added ^{15}N reagents.

Before proceeding to the data containing the results of the SPME–GC–SIM–MSD experiments on ^{15}N incorporation into pyrazines, it is important to examine very closely the ion fragmentation patterns for the naturally occurring pyrazines. The fragmentation pattern presented by each pyrazine molecule can and will influence the ability to define exactly the amount of ^{15}N incorporated into the molecule. For methylpyrazine the fragmentation pattern is relatively simple and as such clear interpretations of minimal ^{15}N incorporation are possible. For example, the electron impact mass spectrum of methylpyrazine has only three ions, 94, 95, 96, in the close range of the molecular weight of methylpyrazine. Furthermore, the abundance of $m/z = 94$ is very large relative to $m/z = 95$ or 96 and thus any increase in the abundances at $m/z = 95$ and/or 96 would be easily detected. Such increases would directly confirm the incorporation of the ^{15}N label. However, for example, the fragmentation pattern for one of the C3 pyrazines is much more complex and contains ions at m/z of 121, 122, 123 and 124. Contributions to the abundance of these ions can come from an array of several combinations of ^{15}N and ^{14}N in the C3 pyrazine, thereby complicating the clear assignment of an ion as arising from a molecule containing only ^{15}N. To illustrate this complicating impact, the ion at $m/z = 121$ can only be due to the parent C3 pyrazine containing two ^{14}N atoms minus a proton. However, the ion at $m/z = 122$ can arise from the parent ^{14}N^{14}N C3 pyrazine as well as a C3 (^{14}N^{15}N) pyrazine minus a proton. Similar arguments are valid for the other pyrazines. Thus, minimal ^{15}N incorporation values can be obtained for the C2 and C3. The estimates of minimal incorporation were gathered from the data on C2 pyrazines using ion abundances at $m/z = 107, 109$ and 110, and from the C3 pyrazines using ions at $m/z = 121, 123$ and 124.

The following ^{15}N labeled compounds were employed in the investigation of ^{15}N incorporation into volatile pyrazines for cocoa and licorice suspensions: leucine, glycine, phenylalanine, alanine, ammonium acetate, asparagine, urea and lysine. Selected examples employing these reagents will be discussed to illustrate the mechanistic implications. Table 6 reveals clearly and unambiguously that ^{15}N was being incorporated in most of the volatile pyrazines studied and that each ^{15}N reagent contributes some of its ^{15}N atoms to the pyrazines

Table 6 *Minimal percent of pyrazine molecules containing at least one* ^{15}N

Licorice suspension

1% ^{15}N Ammonium acetate added

Methylpyrazine	C2 pyrazines			C3 pyrazines		C4 pyrazines	
14.86 min	16.56 min	16.75 min	16.87 min	18.38 min	18.58 min	20.0 min	20.49 min
10.22	6.61	8.21	6.93	6.37	8.10	ND	ND

1% ^{15}N Asparagine added

Methylpyrazine	C2 pyrazines			C3 pyrazines		C4 pyrazines	
14.86 min	16.56 min	16.75 min	16.87 min	18.38 min	18.58 min	20.0 min	20.49 min
38.57	24.38	30.43	19.19	16.45	15.20	3.97	2.97

1% ^{15}N Lysine added

Methylpyrazine	C2 pyrazines			C3 pyrazines		C4 pyrazines	
14.86 min	16.56 min	16.75 min	16.87 min	18.38 min	18.58 min	20.0 min	20.49 min
1.69	5.95	2.95	4.98	5.11	9.44	ND	ND

Cocoa suspension

1% ^{15}N Leucine added

Methylpyrazine	C2 pyrazines			C3 pyrazines		C4 pyrazines	
14.86 min	16.56 min	16.75 min	16.87 min	18.38 min	18.58 min	20.0 min	20.49 min
19.03	19.64	16.74	13.59	14.08	12.27	2.71	1.57

1% ^{15}N Urea added

Methylpyrazine	C2 pyrazines			C3 pyrazines		C4 pyrazines	
14.86 min	16.56 min	16.75 min	16.87 min	18.38 min	18.58 min	20.0 min	20.49 min
31.43	17.46	24.68	17.06	11.54	15.62	2.37	1.54

1% ^{15}N Lysine added

Methylpyrazine	C2 pyrazines			C3 pyrazines		C4 pyrazines	
14.86 min	16.56 min	16.75 min	16.87 min	18.38 min	18.58 min	20.0 min	20.49 min
13.33	15.33	13.99	12.01	11.94	11.63	2.81	2.10

*ND = No ^{15}N incorporation detected.

formed in the licorice experiment. For example, the percent of ions containing at least one ^{15}N atom in one of the C2 pyrazines (16.56 min) for the formulation containing ^{15}N leucine was 15.96%. Also contained within Table 6 is data clearly indicating that all N sources are not created equal in their ability to serve as reagents for pyrazines synthesis. Note that asparagine was very effective in providing ^{15}N for incorporation into the licorice derived pyrazines. Asparagine possesses two N atoms per molecule and this could possibly account for its increased performance. On the other hand, while still effective in its own right, lysine was much less effective. Little evidence was found for the incorporation of one ^{15}N into any of the licorice derived C4 pyrazines. Only the asparagine and glycine containing experiments indicated any evidence for one ^{15}N incorporation.

Table 6 also contains virtually the same information on the incorporation of one ^{15}N into cocoa derived pyrazines. Again all of the ^{15}N containing reagent produced pyrazines having at least one ^{15}N incorporated. Addition of ^{15}N urea to the cocoa yielded the most incorporation of one ^{15}N. For example, for methylpyrazine, at least 31% of the molecules contain at least one ^{15}N in the ^{15}N urea experiment and the next highest incorporation for methylpyrazine was much lower at approximately 20% for [^{15}N]leucine. In contrast to the information from the licorice experiments, ample evidence was found for incorporation of at least one ^{15}N into the cocoa derived C4 pyrazines. All of the ^{15}N sources contributed a small portion of their ^{15}N to form the C4 pyrazines.

These results with licorice and cocoa suspensions were consistent with previous findings[6,22–24,26] on model systems using selected nitrogen sources for the formation of pyrazines. However, this is the first demonstration, in two widely used natural products, of the incorporation of at least one ^{15}N into a variety of volatile pyrazines from a wide variety of ^{15}N sources including amino acids, ammonium salts and urea.

With the evidence obviously in place for the incorporation of at least one ^{15}N into an array of volatile pyrazines from both the licorice and cocoa experiments, the ion abundance data was examined for possible evidence for the incorporation of two ^{15}N atoms into the volatile pyrazines. The fragmentation patterns for the volatile pyrazines were very informative and reasonably straightforward to interpret when scanning for indications of two ^{15}N atom incorporation. For example, in the specific case of a C3 pyrazine, the parent ion m/z is 122 for the naturally abundant ^{14}N compound. Incorporation of two ^{15}N atoms into the molecule would yield a parent ion at $m/z = 124$ corresponding to an increase in mass of two m/z units. Thus, simply the difference in the ion abundance percentage at $m/z = 124$ between the natural and ^{15}N containing experiments will yield the percentage of C3 pyrazine molecules containing two ^{15}N atoms.

Table 7 contains data on the effectiveness of the ^{15}N containing reagents in their ability to produce pyrazines containing two ^{15}N atoms. Both data sets, licorice and cocoa, provide conspicuous information confirming that the ^{15}N reagents were very effective in producing volatile pyrazines containing two ^{15}N atoms. This was particularly true for the lower molecular weight pyrazines. For example, in the asparagine experiments with the licorice, the percentage of

Table 7 *Percent of pyrazine molecules containing two ^{15}N atoms*

Licorice suspension

1% ^{15}N Leucine added

Methylpyrazine 14.86 min	C2 pyrazines			C3 pyrazines		C4 pyrazines	
	16.56 min	16.75 min	16.87 min	18.38 min	18.58 min	20.0 min	20.49 min
7.19	8.77	6.31	5.91	5.41	8.91	ND*	ND

1% ^{15}N Asparagine added

Methylpyrazine 14.86 min	C2 pyrazines			C3 pyrazines		C4 pyrazines	
	16.56 min	16.75 min	16.87 min	18.38 min	18.58 min	20.0 min	20.49 min
16.65	10.43	10.80	6.37	4.73	7.38	ND	ND

1% ^{15}N Lysine added

Methylpyrazine 14.86 min	C2 pyrazines			C3 pyrazines		C4 pyrazines	
	16.56 min	16.75 min	16.87 min	18.38 min	18.58 min	20.0 min	20.49 min
1.95	8.68	4.30	6.12	5.85	10.51	ND	ND

Cocoa suspension

1% ^{15}N Glycine added

Methylpyrazine 14.86 min	C2 pyrazines			C3 pyrazines		C4 pyrazines	
	16.56 min	16.75 min	16.87 min	18.38 min	18.58 min	20.0 min	20.49 min
6.26	9.41	12.53	7.88	8.49	7.51	ND	ND

1% ^{15}N Ammonium acetate added

Methylpyrazine 14.86 min	C2 pyrazines			C3 pyrazines		C4 pyrazines	
	16.56 min	16.75 min	16.87 min	18.38 min	18.58 min	20.0 min	20.49 min
13.22	15.83	16.72	14.42	12.85	9.73	2.02	1.39

1% ^{15}N Urea added

Methylpyrazine 14.86 min	C2 pyrazines			C3 pyrazines		C4 pyrazines	
	16.56 min	16.75 min	16.87 min	18.38 min	18.58 min	20.0 min	20.49 min
11.45	10.66	12.46	7.98	8.99	8.52	0.14	0.05

*ND = not detected.

methylpyrazine molecules containing two ^{15}N atoms was 16.65. Ammonium ions from ammonium acetate were also very efficient in the production of pyrazines with two ^{15}N atoms in the cocoa suspension. For example, for the C3 pyrazine at 18.38 min, the percentage of molecules with two ^{15}N atoms was 12.85. The percent of C4 pyrazines containing two ^{15}N atoms was found to be much less than the other pyrazines for both the cocoa and licorice heat treated samples. These facts substantiate the effective incorporation of nitrogen into volatile pyrazines from both amino acids, covalently bound nitrogen such as urea and ammonium ions. Thus, there are at least two pathways: (1) the reaction of sugars with ammonium ions and (2) the reaction of sugars with covalently bound nitrogen in amino acids and urea.

In addition to the production of alkyl substituted pyrazines in both natural and model systems, pyrazines containing functional groups have been reported.[52,53] These functional groups have included methoxy and acetyl groups resulting in alkylmethoxy and alkylacetyl-pyrazines. These types of functional pyrazines were, somewhat surprisingly, not detected in the products of microwave assisted heat treatment cocoa and licorice studied here. With detection limits around 0.1 ppm, formation of these types of pyrazines must not represent a significant reaction pathway(s) in these systems.

4   Summary and Conclusion

Quantitative and mechanistic information on the formation of volatile components associated with the Maillard and sugar–amine chemistries in microwave assisted heat treated natural products has been gathered employing solid phase microextraction–gas chromatography–selected ion monitoring–mass selective detection. Manual and automated evaluation of a series of SPME fibers on such model molecules as low molecular weight pyrazines, furans, pyridines and thiazoles revealed the fibers to display unique adsorption behavior toward the molecules. Sample matrix was found to be very important, requiring the method of standard addition for quantitative analyses. The effectiveness of selected ^{15}N sources as sources for the formation of selected volatile pyrazines in microwave assisted heat treated natural products was convincingly demonstrated using automated SPME–GC–SIM–MSD. The ^{15}N was found to have been incorporated into at least 50% of certain pyrazine molecules. Ammonium ions, covalently bound nitrogen and amino acids were demonstrated to be capable of yielding N for incorporation into pyrazines. These results shed new light on the possibilities of designing and synthesizing new aroma/flavor formulations from natural products *via* microwave assisted heat treatment. Furthermore, these findings confirm that manual and automated SPME–GC–SIM–MSD approaches are very viable techniques for qualitative and quantitative determination of the mechanisms associated with the formation of volatile products from heated natural products.

References

1. T.F. Stewart, *A Survey of the Chemistry of Amino Acid-Reducing Sugar Reactions in Relation to Aroma Production*, British Food Manufacturing Industries Research Association, Scientific and Technical Surveys, No. 61, 1969, pp. 1–41.
2. J.E. Hodge, *J. Agric. Food Chem.*, 1953, **1**, 928.
3. T.M. Reynolds, *Adv. Food Res.*, 1963, **12**, 1.
4. T.M. Reynolds, *Adv. Food Res.*, 1965, **14**, 167.
5. T.M. Reynolds, *Non-enzymatic Browning Sugar-amine Interrelation*, in *Symposium on Foods: Carbohydrates and Their Role*, ed. Schultz, Cain and Wrolstad, Avi Publ. Co., Westport, CN.
6. G. Lu, *Generation of Flavor Compounds by the Reaction of 2-Deoxyglucose with the Selected Amino Acids*, Ph.D. Dissertation, Rutgers University, October 1996, UMI Dissertation Services.
7. F.B. Whitfield, *Crit. Rev. Food Sci. Nutr.*, 1992, **31**, 1.
8. R.J. Clarke, *Develop. Food Sci.*, **3B**, *Food Flavors, Part B*, 1988, 1.
9. W.M. Coleman, III, *J. Chrom. Sci.*, 1992, **30**, 159.
10. W.M. Coleman, III, J.L. White and T.A. Perfetti, *J. Chrom. Sci.*, 1994, **32**, 323.
11. W.M. Coleman, III, J.L. White and T.A. Perfetti, *J. Agric. Food Chem.*, 1994, **42**, 190.
12. W.M. Coleman, III, J.L. White and T.A. Perfetti, *J. Sci. Food Agric.*, 1996, **70**, 405.
13. M.M. Leahy, *The Effects of pH, Types of Sugars, Amino Acids and Water Activity on the Kinetics of the Formation of Alkyl Pyrazines*, Ph.D. Dissertation, 1985, University of Minnesota, MN.
14. T.A. Rohan and T. Steward, *J. Food Sci.*, 1966, **31**, 202.
15. T. Shibamoto and R.A. Bernhard, *Agric. Biol. Chem.*, 1977, **41**, 143.
16. T. Shibamoto and R.A. Bernhard, *J. Agric. Food Chem.*, 1977, **25**, 609.
17. T. Shibamoto and R.A. Bernhard, *J. Agric. Food Chem.*, 1976, **24**, 847.
18. P.E. Koehler and G.V. Odell, *J. Agric. Food Chem.*, 1970, **18**, 895.
19. G.A. Reineccius, P.G. Keeney and W. Weissberger, *J. Agric. Food Chem.*, 1972, **20**, 202.
20. R. Teranishi, R.A. Flath and H. Sugisawa, *Flavor Research, Recent Advances*, Marcel Dekker, New York, USA, 1981.
21. G.R. Waller and M.S. Feather, *The Maillard Reaction in Foods and Nutrition*, ACS Symposium Series No. 215, American Chemical Society, Washington, DC, USA, 1983.
22. H. Hwang, T.G. Hartman and C.-T. Ho, *J. Agric. Food Chem.*, 1995, **43**, 179.
23. H. Hwang, T.G. Hartman and C.-T. Ho, *J. Agric. Food Chem.*, 1995, **43**, 2917.
24. A. Arnoldi, C. Arnoldi, O. Baldi and A. Griffini, *J. Agric. Food Chem.*, 1988, **36**, 988.
25. W.W. Weeks, *Chemistry of Tobacco Constituents Influencing Flavor and Aroma*, in *Recent Advances in Tobacco Science, Highlights of Current Chemical Research on Tobacco Composition*, 39th Tobacco Chemist's research Conference, Montreal, Canada, October, 1985.
26. J.M. Wong and R.A. Bernhard, *J. Agric. Food Chem.*, 1988, **36**, 123.
27. J. Mauron, *Prog. Food Nutr. Sci.*, 1981, **5**, 5.
28. S. Porretta, *J. Chromatogr.*, 1992, **624**, 211.
29. Z. Zhang, M.J. Yang and J. Pawliszyn, *Anal. Chem.*, 1994, **66**, 844A.
30. B.L. Whittkamp and D.C. Tilotta, *Anal. Chem.*, 1995, **67**, 600.
31. J.Y. Horng and S.D. Huang, *J. Chromatogr.*, 1994, **678**, 313.
32. X. Xang and T. Peppard, *J. Agric. Food Chem.*, 1994, **42**, 1925.

33. A.D. Harmon, *Food Sci. Technol.*, 1997, **79**, 81.

34. W.M. Coleman, III and B.M. Lawrence, *Flavor Fragrance J.*, 1996, **44**, 1.

35. R. Eisert and K. Levsen, *J. Am. Soc. Mass Spectrom.*, 1995, **6**, 1119.

36. J. Ai, *Anal. Chem.*, 1997, **69**, 1230.

37. W.M. Coleman, III, *J. Chromatogr. Sci.*, 1996, **34**, 213.

38. W.M. Coleman, III, *J. Chromatogr. Sci.*, 1997, **35**, 245.

39. H. Maarse, Ed., *Volatile Compounds in Foods and Beverages*, Marcel Dekker, New York, NY, 1991.

40. Z. Zhang and J. Pawliszyn, *Anal. Chem.*, 1994, **65**, 1843.

41. L. Pan, M. Adams and J. Pawliszyn, *Anal. Chem.*, 1995, **67**, 4396.

42. H.C.H. Yeo and T. Shibamoto, *J. Agric. Food Chem.*, 1991, **39**, 370.

43. H.C.H. Yeo and T. Shibamoto, *J. Agric. Food Chem.*, 1991, **39**, 948.

44. B.I. Peterson, C.-H. Tong, C.-T. No and B.A. Welt, *J. Agric. Food Chem.*, 1994, **42**, 1884.

45. F.J. Hidalgo and R. Zamora, *J. Agric. Food Chem.*, 1995, **43**, 1023.

46. R. Zamora and F.J. Hidalgo, *J. Agric. Food Chem.*, 1995, **43**, 1029.

47. C.R. Strauss and R.W. Trainor, *Application of New Microwave Reactors for Food and Flavor Research*, in *Biotechnology for Improved Foods and Flavors*, ACS Symposium Series No. 637, G.R. Takeoka, R. Teranishi, P.J. Williams and A. Kobayashi, eds., American Chemical Society, Washington, DC, 1996, Ch. 26, p. 272.

48. T.H. Parliament, in *Food Flavors, Ingredients and Composition*, G. Charalambous, ed., Elsevier, Amsterdam, 1993, p. 657.

49. Y.-C. Oh, C.-K. Shu and C.-T. Ho, *J. Agric. Food Chem.*, **39**, 1991, 1553.

50. T.-H. Yu, Y.-N. Chen and L.-Y. Lin, *Flavor Formation in Fried Shallot via Thermal Reactions of Nonvolatile Flavor Precursors*, in *Contribution of Low- and Non-Volatile Materials to Flavor*, 1996, p. 227.

51. A. Arnoldi, C. Arnoldi, O. Baldi and A. Griffini, *J. Agric. Food Chem.*, 1987, **35**, 1035.

52. I. Flament, *Coffee, Cocoa and Tea*, in *Volatile Compounds in Foods and Beverages*, H. Maarse, ed., Marcel Dekker, Inc., New York, 1991, Ch. 17.

53. D.S. Mottram, *Flavor Compounds Formed during the Maillard Reaction*, in *Thermally Generated Flavors, Maillard, Microwave and Extrusion Processes*, ACS Symposium Series, No. 543, T.H. Parliament, M.J. Morello and R.J. McGorrin, eds., Americal Chemical Society, Washington, DC, 1994, Ch. 10.

CHAPTER 44

SPME Investigation of Intermediates Produced during Biodegradation of Contaminated Materials

JALAL A. HAWARI

1 Synopsis

The present study describes the applicability of SPME–GC–MS to analyze intermediates produced during the biodegradation of several environmental contaminants in aqueous media. Three representative compounds were selected: dibenzothiophene (DBT), a model bitumen compound, 2,4,6-trinitrotoluene (TNT), a widely used explosive, and hexadecane, a typical petroleum product. TNT was degraded by a consortium in an anaerobic sludge whereas in the case of DBT and hexadecane defined strains of bacteria were used, namely *Rhodococcus* sp. strain ECRD-1 and *Rhodococcus* sp. Q15, respectively. A time course study for the appearance and disappearance of the pollutant and its corresponding metabolites is carried out to establish degradation pathways.

2 Introduction

SPME is a solventless and rapid sample preparation technique that uses a polymer-coated fiber for the adsorption of organic compounds from an aqueous matrix followed by direct thermal desorption into a GC for subsequent detection and quantification. The technique is known for its sensitivity which enables detection in the $\mu g\,L^{-1}$ range. Further details on the theory and applications of the SPME technique can be found in refs. 1–5.

Although SPME has been widely used for the trace analysis of organic compounds in several aqueous based matrices, little is known about the applicability of the technique for monitoring organic biotransformations in biological matrices.[6] Until recently, lengthy sample preparation and separation techniques (*e.g.* liquid/liquid extraction followed by chromatographic clean-up procedures) were required to isolate and identify intermediates produced during biotransformation processes.[7,8] When such intermediates are formed in trace amounts, the previously mentioned traditional techniques are not rapid or sensitive enough for their detection, thus leading to the loss of valuable information on the transformation pathways.

The present study was undertaken to establish the practicality of SPME as a versatile technique suitable for monitoring a wide variety of pollutants and their corresponding metabolites under a variety of biological conditions. Three examples will be discussed: the biodesulfurization of a DBT (a model bitumen compound) by *Rhodococcus* sp. strain ECRD-1, the biotransformation of TNT (a widely used explosive) using anaerobic sludge, and finally the biodegradation of hexadecane (a typical petroleum product) using *Rhodococcus* sp. Q15.

3 Investigation of Biodesulfurization of DBT by SPME–GC–MS

In this example, we tested the suitability of a combined SPME–GC–MS to identify metabolites formed during desulfurization of a model thiophenic compound commonly found in fossil fuel, *i.e.* DBT by *Rhodococcus* sp. strain ECRD-1.[8] This compound was selected as an example since there exist large reserves of fossil fuels such as bitumen in the world, but their fuel value is low due in part to their high organic sulfur content. When such compounds are combusted, sulfur dioxide is released into the atmosphere causing acid rain. To increase the fuel value without causing harm to the environment the crude oil must be desulfurized without destroying the carbon skeleton, thus avoiding excessive reduction of its calorific value.[9–11] Biodesulfurization is environmentally safe but the extent of desulfurization and the identity of intermediate and end products has always been a challenge to researchers.[12] The present work describes the utility of SPME–GC–MS in the identification of key metabolites formed during the desulfurization of DBT using the *Rhodococcus* sp. ECRD-1 strain. A time profile of the appearance and disappearance of the detected metabolites was used to elucidate the desulfurization pathway of this model fossil fuel compound.

DBT ($10\,mg\,L^{-1}$) was incubated with *Rhodococcus* sp. strain ECRD-1 in a mineral salts medium (MSM) supplemented with glucose as a carbon source in 100 ml serum bottles on a shaker at 240 rpm at 27 °C for 4 days. A high concentration of DBT was used to insure that the substrate did not become limiting during the assay. Control flasks containing autoclaved cells were incubated under the same conditions to determine if any degradation of the substrate occurred abiotically. The cultures were sampled at intervals by removing aliquots (2 mL) for SPME–GC–MS analysis as described below.

SPME–GC–MS of DBT and Metabolites: Procedure

The thermodynamic equilibrium for the partitioning of DBT and its final desulfurized product 2-hydroxybiphenyl (2HBP) between the SPME sorbent and the aqueous phase was first established using fused silica fiber that was coated with an 85 μm polyacrylate polymer (Supelco, Bellefonte, PA) fitted to an autosampler assembly (Varian). A 20 min extraction followed by 10 min desorption inside a GC were found appropriate for reproducible analysis. The response was linear ($r^2 = 0.998$) over the concentration range 20–800 ppb. The detection limit (DL) was low (≤ 1 ppb) with an absolute recovery of $> 90\%$.

Aliquots of the cell suspension (2 mL) from the culture medium were acidified using nitric acid (pH 2) and filtered through a Millex-HV 0.45 μm filter for subsequent SPME extraction (20 min) and desorption (10 min) inside the injector port of a GC–MS. A time course study to monitor the formation and disappearance of metabolites during desulfurization, was carried out by SPME sampling at different time intervals ($t = 0$–72 h). A Varian GC–MS equipped with a Saturn II ion trap detector (70 eV) and a DB-5 capillary column (15 m $\times$ 0.25 mm i.d. $\times$ 0.25 μm film) was employed for the analysis using a mass range of 20–300 amu.

Metabolites from the Desulfurization of DBT by *Rhodococcus* sp. Strain ECRD-1

A typical SPME–GC–MS total ion chromatogram of DBT undergoing desulfurization, containing the starting material dibenzothiophene, DBT (I), and several other intermediate products, is shown in Figure 1. The metabolites were identified as DBT-sulfoxide (II), DBT-sulfone (III), dibenz[*c,e*][1,2]oxathiin 6-oxide (DBT sultine, IV) dibenz[*c,e*][1,2]oxathiin 6,6-dioxide (DBT sultone, V) and 2-hydroxybiphenyl (2-HBP, VI). These metabolites were identified by their retention times (r.t.), molecular mass ions (*m/z* amu) and by comparison with reference standards. DBT-sulfoxide (II) appeared after 13.40 min giving a molecular ion (*m/z*) of 200 amu and a base peak of $m/z = 184$ amu corresponding to the loss of an oxygen radical (16 amu). The second metabolite, DBT-sulfone (III), was observed at a r.t. of 13.50 min with a $m/z = 210$ amu which was also the base peak. The peak appearing at a r.t. of 12.23 min with molecular ion and base peak at 216 amu was attributed to DBT-sultine (IV). DBT-sultone (V) appeared at a r.t. of 13.24 min with both molecular ion (*m/z*) and base peak (bp) equal to 232 amu. The final product, 2-HBP (VI), was observed at a r.t. of 5.23 min with *m/z* 170 amu.

Metabolic Pathway of DBT Desulfurization

After establishing the suitability of SPME–GC–MS for the direct detection of intermediates formed during DBT biodesulfurization, a new culture medium with relatively high biomass content, *i.e.* high optical density (OD$_{600nm}$ 0.9), was prepared to produce sufficient amounts of DBT metabolites for time profiling. The SPME–GC–MS peak area counts of each detected metabolite were then used to determine the appearance and disappearance of these metabolites with time (Figure 2). DBT-sulfoxide first appeared at 20 minutes, then decreased

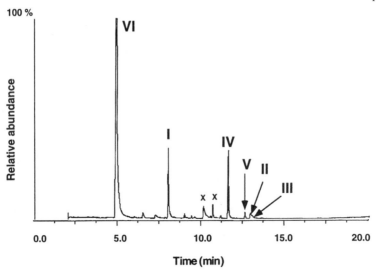

Figure 1 *A typical SPME–GC–MS total ion chromatogram of biodesulfurization of DBT by Rhodococcus sp. strain ECRD-1. The metabolites are identified as follows: DBT (I), DBT-sulfoxide (II), DBT-sulfone (III), DBT-sultine (IV), DBT-sultone (V) and 2-HBP (VI). x = unidentified*

rapidly to trace amounts after 24 h (Figure 2B). However, the raw data indicate that the sulfone first appeared in the sample after 1 h. DBT-sultine appeared at 1 h and continued to accumulate for the duration of the experiment. DBT-sultone first appeared at 3 h 40 min and did not decrease in quantity until 72 h (Figure 2B). The concentration of the end-product, 2-HBP, increased rapidly and reached a plateau after 24 h (Figure 2A). The SPME–GC–MS data shown in Figure 2 does not include DBT because the initial amount of DBT added to the culture medium (t = 0) was in excess of its water solubility by two orders of magnitude.

The data indicate a stepwise metabolism of DBT showing that DBT-sulfoxide (II) was the first metabolite formed. The second metabolite, probably DBT-sulfone (III) seemed to be converted rapidly into DBT-sultine (IV), the acid rearranged product of the corresponding sulfinic acid, and this was followed by the appearance of DBT-sultone (V), the acid rearranged product of the corresponding sulfinic acid. 2-HBP (VI) was the final product, and its final concentration was two orders of magnitude higher than any other metabolite.

Several studies have described the use of *Rhodococcus* sp. strain IGTS8 to desulfurize DBT to 2-HBP.[12–15] Piddington *et al.*[13] reported that DBT is first converted into the corresponding sulfone which in turn is desulfurized to produce 2-HBP as the end product. However, Olson *et al.*[12] and Denome *et al.*[14] reported that the desulfurization of DBT by the same *Rhodococcus* sp. IGTS8 produced several intermediates including DBT-sulfoxide, DBT sulfone, 2-hydroxybiphenyl-2′-sulfonic and 2-hydroxybiphenyl-2′-sulfinic acid before producing 2-HBP. All four intermediates detected by Olson *et al.*,[12] *i.e.* DBT-

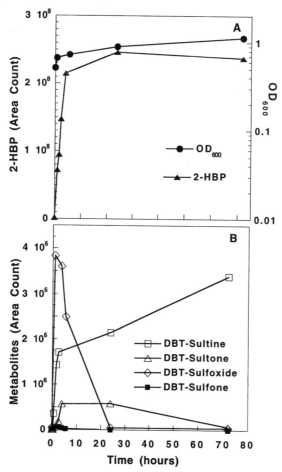

Figure 2 *A typical profile of DBT metabolites detected with Rhodococcus sp. strain ECRD-1. Culture density (OD₆₀₀), accumulation of the final product (2-HBP) (A), and the formation and disappearance of intermediates (B) were monitored for 72 hours*

sulfoxide, DBT-sulfone, the sultone and the sultine, were detected by the present SPME study. 2′-Hydroxybiphenyl-2-sulfonic acid and 2′-hydroxybiphenyl-2-sulfinic acid were not observed as acids but as the corresponding cyclized derivatives (*i.e.* the sultone and the sultine) because of the acidic conditions (pH 2) employed in preparing the sample.[12] The time profile showing the relationship among various detected intermediates is best described in Scheme 1, which closely resembles a hypothetically constructed pathway, known as the 4S desulfurization pathway. [12] For example, the SPME data indicated the following sequence: DBT, DBT-sulfoxide, DBT-sulfone, DBT-sultine, DBT-sultone, and 2-HBP. However, the 4S pathway indicates that desulfurization proceeds through the following sequence: DBT, DBT-sulfoxide, DBT-sulfone, DBT-sulfonate (a precursor to the corresponding sultone) and finally 2-HBP.

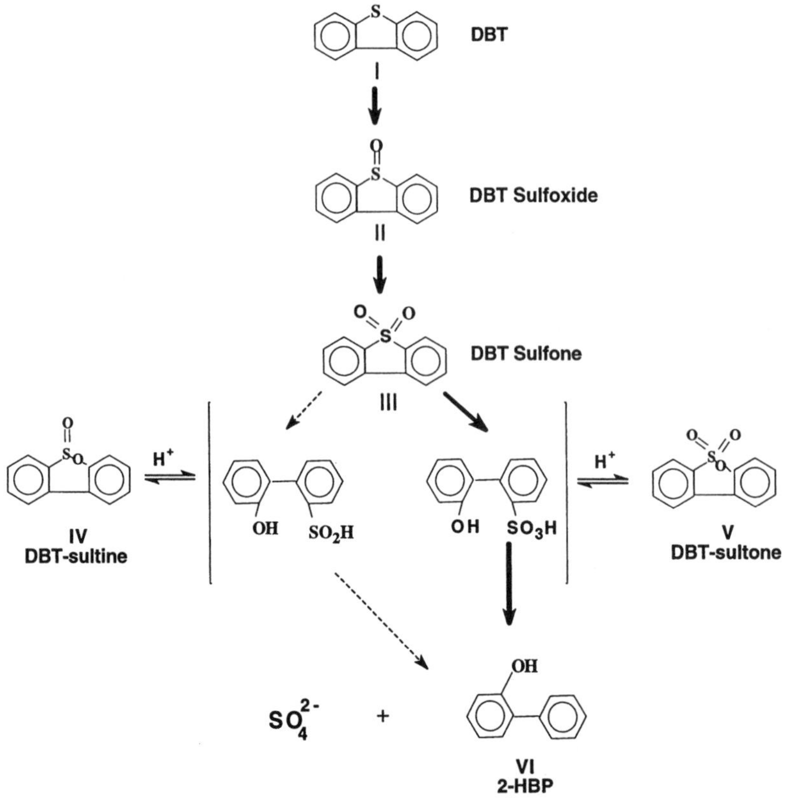

Scheme 1 *Proposed metabolic pathway of biodesulfurization of DBT by Rhodococcus sp. strain ECRD-1. The bold arrows represent the actual sequential conversion observed by the SPME–GC–MS time study. The dashed arrow represents the possible second route to 2-HBP via the sulfinic acid. The bracketed intermediates (2-hydroxyybiphenylsulfinic acid and 2-hydroxyybiphenylsulfonic acid) were not observed directly by SPME, but were inferred from the detection of their acid (pH 2) cyclic derivatives, DBT-sultine (IV) and DBT-sultone (V)*

4 Investigation of Biotransformation of TNT by SPME–GC–MS

In another example we attempted to verify the suitability of the SPME–GC–MS technique to monitor the fate of 2,4,6-trinitrotoluene (TNT) during biodegradation by a consortium contained in an anaerobic sludge. Presently contamination by highly energetic chemicals such as 2,4,6-trinitrotoluene (TNT), generated as wastes from the munitions and defence industries, is a significant worldwide environmental problem. TNT is mutagenic, toxic and has the tendency to persist in the environment.[16,17] It is estimated that TNT is produced in amounts close to 2 million tons a year.[18]

There have been several attempts to degrade TNT and to monitor its fate in the environment.[19–24] However, a thorough understanding of the biotransformation process depends on the availability of analytical techniques suitable for direct detection of transient metabolites during their formation. The HPLC based EPA SW-846 Method 8330 is the most widely used technique for the analysis of explosives; however, this method necessitates the use of organic solvents such as acetonitrile (AcN). [25] SPME–GC–MS is thus applied to analyze TNT and its related transformation products in water under anaerobic conditions. A time study for the formation and disappearance of these metabolites was also conducted to help understand the metabolic pathway for the transformation of TNT.

In a typical set-up, a serum bottle (100 mL) was charged with deionized water (44 mL), anaerobic sludge (5 mL) obtained from a baby food factory (Cornwall, ON, Canada), a mineral salt medium (MSM) (1 mL), molasses ($10 \, g \, L^{-1}$) as a carbon source and TNT ($50 \, mg \, L^{-1}$) as a nitrogen source for the degrading microorganisms. Some serum bottles (microcosms) were supplemented with a uniformly labeled $[U\text{-}^{14}C]$TNT (100,000 dpm) and then fitted with a small test tube containing 1.0 mL of 0.5 M KOH to trap liberated carbon dioxide ($^{14}CO_2$). [26] The headspace in each microcosm was flushed with nitrogen gas to maintain anaerobic conditions and then sealed with butyl rubber septa and aluminum crimp seals to prevent the loss of CO_2 and other volatile metabolites. Two control microcosms were prepared: one contained the sludge without TNT and the second contained TNT without the sludge to serve as an abiotic control. Each microcosm was wrapped with aluminum foil to protect the mixture against photolysis. Microcosms with $[U\text{-}^{14}C]$TNT were routinely sampled (daily or every two days) for the determination of $^{14}CO_2$ in the KOH trap using a Packard, Tri-Carb 4530 liquid scintillation counter (Model 2100 TR, Packard Instrument Company, Meriden, CT). Microcosms that did not receive ^{14}C-labeled TNT were reserved for SPME–GC–MS analysis of residual TNT and its metabolites (see below). Each microcosm preparation was carried out in triplicate.

SPME–GC–MS analysis of TNT Metabolites: Procedure

Fused silica capillary fibers (1 cm) coated with 85 μm polyacrylate or poly(dimethyl)siloxane fitted to an autosampler holder were used (Supelco, Bellefonte, PA, USA). Aliquots (2 mL) from the culture medium were centrifuged to remove the suspended material including biomass. The remaining clear supernatant was subjected to SPME analysis. Analytes were extracted (20 min) directly from the salty culture medium (about 1% w/v) onto the fiber and then thermally desorbed (10 min) inside the GC injector of a Varian Saturn GC–MS.

We first found that the thermodynamic equilibria for the partitioning of TNT and its most common amine products [2-amino-4,6-dinitrotoluene (2-ADNT), 4-amino-2,6-dinitrotoluene (4-ADNT), 2,4-diamino-6-nitrotoluene (2,4-DANT) and 2,6-diamino-4-nitrotoluene (2,6-DANT)] between the aqueous phase and the SPME coating were about 10 min. Using standard calibration procedures and applying the criteria of signal/noise ratio of 3/1 the DL for TNT, 2-ADNT, 4-ADNT, 2,4-DANT and 2,6-DANT were found to be 9, 20,

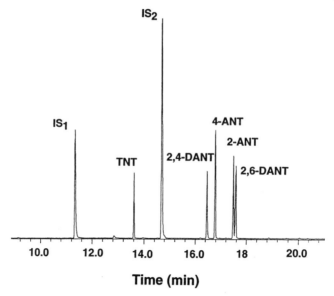

Figure 3   *SPME–GC–MS total ion chromatogram of TNT biotransformation after 16 hours of treatment with an anaerobic sludge*

10, 26 and 29 ppb, respectively. We also found that the linearity of the response for the analytes TNT, 2-ADNT, 4-ADNT, 2,4-DANT and 2,6-DANT to be high (r = 0.998, 0.997, 0.997, 0.966, 0.991, respectively) over the concentration range of 20–800 ppb.

Metabolites from TNT Biotransformation with Anaerobic Sludge

Figure 3 summarizes TNT biotransformation after 16 h of incubation with the anaerobic sludge. All intermediates in Figure 3 were identified as 2-ADNT, 4-ADNT, 2,4-DANT and 2,6-DANT by comparison with their corresponding standards using retention times (r.t), molecular mass ions (m/z) and base peak mass ions (bp). The parameters [r.t. (min), m/z (amu) and bp (amu)] for the four detected intermediates were 2-ADNT (17.70, 197, 180), 4-ADNT (16.98, 197, 180), 2,4-DANT (16.66, 167, 167) and 2,6-DANT (17.82, 167, 167), respectively. Another set of metabolites were detected which had much shorter lifetimes than the metabolites mentioned above. These included 2-hydroxylamino-4,6-dinitro-toluene (2-HADNT) and 4-hydroxylamino-2,6-dinitrotoluene (4-HDANT). Their presence was confirmed by their LC–MS spectra (both molecular mass ions appeared at 213 amu) and by comparison with authentic standards. Both 2-HADNT and 4-HADNT are considered prerequisites for the formation of the corresponding amine derivatives 2-ADNT and 4-ADNT, respectively.

Owing to its high polarity and solubility in water, triaminotoluene (TAT) was not detected by SPME–GC–MS and furthermore it eluted with the void volume in the case of HPLC (Method 8330).[25] The presence of TAT was finally confirmed by ion-pairing HPLC using octanesulfonic acid as ion pairing agent

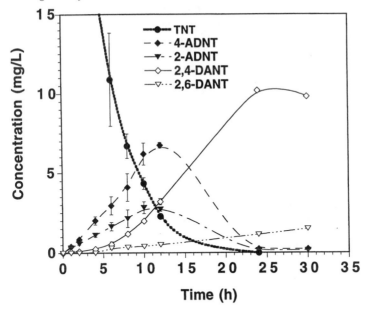

Figure 4 *A time study of TNT biotransformation as observed by SPME–GC–MS*

and a C_{18} column. The triamine metabolite, appearing after 12 h of TNT incubation, was identified by comparison with a known standard and by atmospheric positive chemical ionization (APcI+)–LC–MS (M^+H, 138 amu). TAT was detected in very high concentrations (160 μM), accounting for 73% of the initial concentration of TNT (220 μM).[26]

Metabolic Pathway of TNT Biotransformation

An SPME–GC–MS time study for the disappearance of TNT and the appearance and disappearance of its metabolites is shown in Figure 4. Whereas TNT disappeared rapidly in less than 15 h, its amine metabolites 4-ADNT, 2-ADNT, 2,4-DANT and 2,6-DANT formed in a stepwise and regioselective fashion favoring reduction of the NO_2 group at the *para*-position over that at the *ortho*-one. For example, during the first 15 h of incubation the ratios of 4-ADNT/2-ADNT and 2,4-DANT/2,6-DANT were 2.0 and 6.0, respectively. Stepwise reduction of polynitroaromatics under both biotic and abiotic conditions have been reported earlier but with varying degrees of regioselectivity.[27–30] Preuß *et al.* reported that the reduction of 2,4-DANT is the rate limiting step in the overall reduction process of TNT.[31] After 60 h, the di-aminotoluenes 2,4-DANT and 2,6-DANT disappeared leaving behind TAT (160 μM). Until this stage the total mineralization as measured by liberated $^{14}CO_2$ did not exceed 0.1% of the TNT that had disappeared, thereby indicating that TAT was implicated in other side reactions that had nothing to do with mineralization. TAT is known to be a very reactive compound that polymerizes or reacts with air, light and acids. This would explain the fact that, despite TNT's rapid

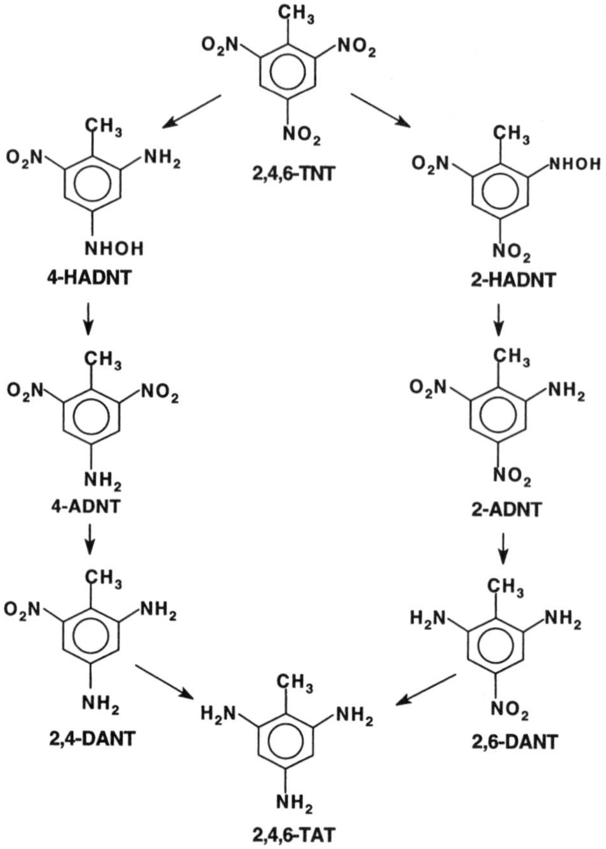

Scheme 2 *Constructed metabolic pathway for TNT biodegradation with anaerobic sludge*

conversion into TAT, negligible mineralization took place. Several other reports have also described the formation of TAT as a 'killer' metabolite that misrouted TNT from mineralization under anaerobic conditions.[17,32–34]

The sequence of events deduced from the above detailed time study was eventually translated into the transformation pathways depicted in Scheme 2. Most of the elements shown in Scheme 2 represent the biotic cycle of TNT biotransformation to TAT and have been frequently observed by other researchers.[31,35]

5 Investigation of Hexadecane Biotransformation with *Rhodococcus* sp. Q15

In this example, SPME–GC–MS was used to identify potential metabolic intermediates resulting from the treatment of hexadecane, an alkane model petroleum product, with *Rhodococcus* sp. Q15. The Q15 cells were incubated at 24 °C while shaking (150 rpm) in a mineral salt medium (MSM) (50 mL)

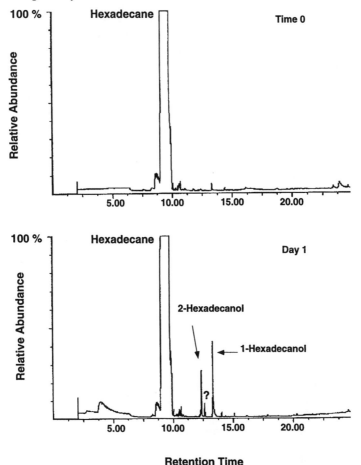

Figure 5 *A typical SPME–GC–MS total ion chromatogram of biodegradation of hexadecane with Rhodococcus sp. Q15 strain. Two metabolites were identified as 1-hexadecanol and 2-hexadecanol*

supplemented with 25 ppm yeast extract and hexadecane (50 ppm). The presence of metabolites in the culture medium was monitored daily by collecting 1 mL of the growth medium. The medium was centrifuged and the supernatant was subject to SPME–GC–MS using a polyacrylate coated fiber for analytes extraction (20 min) followed by desorption (10 min) inside the injector port (270 °C) of a Varian Saturn Ion Trap GC/MS. Separation was achieved in a DB 5 column (15 m × 0.25 mm × 0.25 μm film) that lead to a mass detector in the electron impact mode (70 eV). Sterile controls were used to confirm that the analytes detected were in fact metabolites. Figure 5 shows SPME–GC–MS total ion chromatogram of hexadecane after 24 h of incubation with *Rhodococcus* sp. showing two main metabolites. The two metabolites were identified as 1-hexadecanol and 2-hexadecanol by comparison with their corresponding standards using retention times and mass spectra. After 48 h of incubation, both hydroxy metabolites disappeared and traces of hexadecanoic acid were detected (data not shown).[35] These metabolites are the same as those previously reported for the degradation of *n*-alkanes.[36]

In conclusion, the present study demonstrates the practicality of SPME/GC/MS in the quantitative analysis of various contaminants and their corresponding metabolites during biodegradation. The solventless technique is versatile and universal and may be successfully used to analyze for multicontaminants and their corresponding biotransformation products with high selectivity and specificity. Our data presented signify that SPME is a very powerful tool that is suitable for field monitoring and can also provide an accurate assessment of the effectiveness of remediation technologies.

6 Acknowledgments

The author wishes to acknowledge the following: Dr. C. W. Greer and Dr. L. Whyte for their data on hexadecane, Dr. C. F. Shen, Dr. B. Spencer, Ms. L. Paquet, Mrs. A. Mihoc and Mrs. A Halasz for their work on TNT, Dr. A. M Jones, Mr. E. Zhou, and G. Wisse for their work on DBT and Dr. G. Sunahara for reviewing the manuscript.

7 References

1. C.L. Arthur and J. Pawliszyn, *Anal. Chem.*, 1990, **62**, 2145.
2. Z. Zhang, M.J. Yang and J. Pawliszyn, *Anal Chem.*, 1994, **66**, 844A.
3. T. Górecki and J. Pawliszyn, *Anal. Chem.*, 1995, **67**, 3265.
4. J. Pawliszyn, Chapter 1 of this book.
5. J. Pawliszyn, *Solid Phase Microextraction: Theory and Practice*, Wiley-VCH, New York, 1997, p. 229.
6. R.F. Dias and K.H. Freeman, *Anal. Chem.*, 1997, **69**, 944.
7. J. Pörschmann, Z. Zhang, F.D. Kopinke and J. Pawliszyn, *Anal. Chem.*, 1997, **69**, 597.
8. T. MacPherson, C.W. Greer, E. Zhou, A.M. Jones, G. Wisse, P.C.K. Lau, B. Sankey, M.J. Grossman and J. Hawari, *Environ. Sci. Tech.*, 1998, **32**, 421.
9. P.R. Dugan, *Resources, Conservation and Recycling*, 1991, **5**, 101.
10. W.R. Finnerty, *Curr. Opinion Biotechnol.*, 1992, **3**, 277.
11. J.J. Kilbane, *Trends Biotechnol.*, 1989, **7**, 97.
12. E.S. Olson, D.C. Stanley and J.R. Gallagher, *Energy & Fuels*, 1993, **7**, 159.
13. C.S. Piddington, B.R. Kovacevich and J. Rambosek, *Appl. Environ. Microbiol.*, 1995, **61**, 468.
14. S.A. Denome, E.S. Olson and K.D. Young, *Appl. Environ. Microbiol.*, 1993, **59**, 2837.
15. A.K. Gray, O.S. Pogrebinsky, G.T. Mrachko, L. Xi, D.J. Monticello and C.H. Squires, *Nature Biotechnol.*, 1996, **14**, 1705.
16. W.D. Won, L.H. Di Salvo and J. Ng, *Appl. Environ. Microbiol.*, 1976, **31**, 575.
17. P.G. Rieger and H.-J. Knackmuss, in *Biodegradation of Nitroaromatic Compounds*, ed. J.C. Spain, Plenum Press, New York, 1995, p. 1.
18. D.R. Harter, in *Toxicity of Nitroaromatic Chemicals*, Chemical Industry Institute of Toxicology Series, ed. D.E. Ricket, Hemisphere Publishing Corp., NY, 1985, p. 1.

19. D.F. Carpenter, N.G. McCormick, J.H. Cornell and A.M. Kaplan, *Appl. Environ. Microbiol.*, 1978, **35**, 949.

20. S.B. Funk, D.H. Roberts, D.L. Crawford and R.L. Crawford, *Appl. Environ. Microbiol.*, 1993, **59**, 2171.

21. E.A. Duque, F. Haidour, F. Godoy and J.L. Ramos, *J. Bacteriol.*, 1993, **175**, 2278.

22. J.F. Manning Jr., R. Boopathy and C.F. Kupla, *A Laboratory Study in Support of the Pilot Demonstration of a Biological Soil Slurry Reactor*, Report SFIM-AEC-TS-CR-94038, Bioremediation Group, Environmental Research Division, Argonne National Laboratory, Argonne, IL 60439.

23. R. Boopathy and C.F. Kupla, *Can. J. Microbiol.*, 1993, **34**, 430.

24. J. Hawari, C.W. Greer, A. Jones, C.F. Shen, S.R. Guiot, G. Sunahara, S. Thiboutot and G. Ampleman, in *Challenges in Propellants and Combustion 100 years after Nobel*, Proc. 4th Int. Sym. on Special Topics in Chem. Propulsion, Stockholm, Sweden, Bagell House Inc. Publishers, NY, 1996, pp. 135–144.

25. M.E. Walsh, T.F. Jenkins, P.S. Schnitker, J.W. Elwell and M.H. Stutz, *Evaluation of SW 846 Method 8330 for Characterization of Sites Contaminated with Residues of High Explosives*, CRREL Report 93–5; US Army Cold Regions Research and Engineering Laboratory, Hanover, NH, 1993.

26. J. Hawari, B. Spencer, A. Halasz, S. Thiboutot and G. Ampleman, *Appl. Environ. Microbiol.*, 1998, **64**, in press.

27. N.G. McCormick, E.F. Florence and L. Hillel, *J. Appl. Microbiol.*, 1976, **31**, 949.

28. R.P. Naumova, N.N. Amerkhanova and A. Shaikhutdinov, *Priklad. Biokhim. Mikrobiol.*, 1979, **15**, 45.

29. D.L. Kaplan, in *Advances in Applied Biotechnology*, ed. D. Kamely, A. Chakhrabarty and G.S. Omenn, Portfolio Publishing Co., Houston, TX, 1990, Vol. 4, p. 155.

30. E.J. Weber, M.S. Elcovitz, D.G. Truhlar, J.C. Cramer and S.E. Barrows, *Environ. Sci. Technol.*, 1996, **30**, 3029.

31. A. Preuß and P.-G. Rieger, in *Biodegradation of Nitroaromatic Compounds*, ed. J. Spain, Plenum Press, NY, 1995, p. 69.

32. A. Preuss, J. Fimpel and G. Diekert, *Arch. Microbiol.*, 1993, **159**, 345.

33. T.A. Lewis, S. Goszczynski, R.L. Crawford, R.A. Korus and W. Admassu, *Appl. Environ. Microbiol.*, 1996, **62**, 4669.

34. D.L. Kaplan and M.A. Kaplan, *Appl. Environ. Microbiol.*, 1982, **44**, 757.

35. L.G. Whyte, J. Hawari, E. Zhou, W.E. Inniss and C.W. Greer, *Appl. Environ. Microbiol.*, 1998, in press.

36. R.J. Watkinson and P. Mergan, *Biodegradation*, 1990, **1**, 79.

Related Techniques

CHAPTER 45

Infrared Spectroscopic Detection for SPME

DANESE C. STAHL AND DAVID C. TILOTTA

1 Introduction

Solid phase microextraction with infrared detection (SPME–IR) is a useful alternative to SPME–GC. As opposed to the GC application, which uses a cylinder of solid or liquid coated on a fiber, the IR application utilizes a small square of polymer film because it is more compatible with IR spectroscopic sample handling. Once the organics are extracted into the polymer film (using a similar procedure as with the syringe device), IR absorbance spectroscopy is used to quantitatively identify them directly in it.

Like SPME–GC, SPME–IR is also sensitive, selective, and fast. Additionally, it is environmentally-friendly since the films are reusable and the SPME–IR procedure does not require the use of any organic extraction solvents. As advantages, the instrumentation for SPME–IR is inexpensive and readily portable, and optical measurements are inherently simple to perform.

2 Infrared Spectroscopy

The Infrared Spectrum

Infrared spectroscopy is perhaps a technique that is unparalleled with respect to its ability to determine molecular structure. Unlike mass spectroscopy or other analytical methods, IR spectroscopy provides direct information on molecular connectivity, that is, on how the atoms in a molecule are arranged relative to one another. The absorption of infrared radiation in a molecule arises from changes in its dipole moment. Generally, any molecule that is not a homonuclear diatomic species will absorb radiation corresponding to its specific vibrational changes (*i.e.*, the movement of its atoms relative to one another).[1] Typically, these absorption 'bands' occur in the 2.5–40 μm (4000–400 cm^{-1}) region of the electromagnetic spectrum, and different vibrational changes absorb different energies (*i.e.*, wavelengths or wavenumbers). Of course, the pattern of these absorptions (*i.e.*, absorbance *versus* wavelength or wavenumber [cm^{-1}]) constitutes its infrared spectrum which functions as a 'fingerprint' for the molecule.

Infrared spectroscopy is widely used by chemists because it has two major advantages. First, it is relatively simple to implement requiring only minimal training in instrument operation and spectral interpretation (especially with modern advances in spectral libraries and search algorithms). Second, infrared spectroscopy employs relatively inexpensive instrumentation. For example, a basic Fourier transform infrared spectrometer (FTIR, discussed below) currently can be obtained for less than $20,000.

However, to be fair, IR spectroscopy also has two major disadvantages. First, and the more significant one, is the difficulty that IR spectroscopy has in speciating individual components in complex mixtures. This disadvantage arises from the band overlap from the various components in the mixture. Although advances in statistical data-handling methods (*e.g.*, partial least squares analysis and principal component regression) have improved this situation, it is still difficult to speciate more than a handful of components in a mixture. This feature has previously limited the application of IR spectroscopy in real world applications.

Another disadvantage of IR spectroscopy is its poor analyte sensitivity in aqueous solution. The sensitivity of infrared measurements can be in the ppm range as long as the analyte is in an infrared-clear matrix, for example air or CS_2 (which, although an IR active species, has an insignificant IR spectrum). However, the sensitivity of the measurements are severely degraded when the matrix absorbs infrared radiation, as in the case of organic analytes in water solutions. As shown in Figure 1, water has two very strong absorption bands that preclude the obtainment of spectral information unless the analyte is present in percent quantities. As will be shown below, the application of SPME in IR spectroscopy overcomes both major disadvantages.

Infrared Absorption Spectrometry

The quantitative aspects of infrared spectroscopy arise from the measurement of absorption of infrared energy of the sample at a given wavenumber. The relationship between the absorbance and the concentration is given by the Beer–Lambert law which states that:

$$A = \varepsilon b C \tag{1}$$

where A is the absorbance, ε is the molar absorption constant for the particular transition, b is the optical pathlength, and C is the concentration.[2] The Beer–Lambert law states that, for a given molecular vibration (in this case), the absorbance can be increased if either the pathlength is increased or the analyte concentration is increased.

Most infrared spectrometers are of the Fourier transform type. These instruments provide several advantages over older, grating based instruments that are useful for IR spectroscopy applied to SPME measurements.[3] Fourier

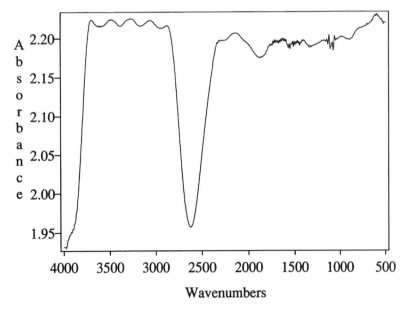

Figure 1 *Infrared absorbance spectrum of water acquired in a* 0.206 mm *pathlength ZnSe cell*

transform IR spectrometers have superior signal-to-noise ratios, typically in the sub milli-absorbance range (mA). Thus, analyte concentrations can be determined at fairly low levels providing the sample matrix is IR 'clear'. Second, FTIRs can acquire spectra quickly (*e.g.*, several per second). As discussed below, this advantage will have important ramifications with respect to sample volatility loss from the solid phase film. Other features of modern FTIR spectrometers include a small footprint and low power requirements (both of which are useful for field work).

3 Experimental Considerations

Apparatus and Procedures

Implementation of SPME with IR detection requires only a holder for the solid phase (shown in Figure 2), an extraction vessel, a stirrer plate, and an appropriate spectrometer.[4] The holder supports the polymer film (typically a 3.2 cm × 3.2 cm square) during the extraction procedure and also while acquiring the spectrum in the spectrometer. In order to accommodate the solid phase holder, the vessel used for extraction is a wide-mouth, glass jar with a Teflon-lined, screw lid. A hook on the lid is used to suspend the holder in the aqueous solution. Currently, 50–250 mL volumes of the aqueous solutions are extracted.[4-6]

The procedure for performing SPME using an IR spectrometer is straightforward. First, a single-beam infrared spectrum of the film is obtained prior to

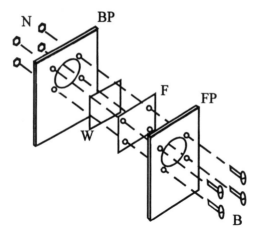

Figure 2 *Schematic diagram (exploded) of SPME–IR holder.* BP, *backplate;* N, *nuts;*
W, *wire spacer;* F, *polymer film;* FP, *front plate;* B, *bolts*

extraction. This spectrum serves as a reference background and is used to remove (through the ratioing process) those IR bands which are characteristic of the solid phase film. The film/holder assembly is then placed in an extraction vessel containing the stirred water sample. Analytes partition from the aqueous phase into the polymer solid phase. Following the partitioning step, the film/holder assembly is removed from the extraction vessel and another spectrum is acquired to serve as a sample single-beam spectrum. The ratio of sample spectrum to reference spectrum provides the resultant spectrum of the partitioned analytes.

Quantitative data are obtained from the SPME–IR experiment by measuring the absorbances of the characteristic infrared bands. The concentrations of the analytes are then determined from these absorbances with the use of calibration curves. These curves are prepared from standard solutions (*i.e.*, analytes in distilled water) because it has been shown that real water matrices do not appreciably affect the partitioning of the organics into the solid phase film as long as the samples are relatively clean (*e.g.*, dissolved solids of *ca.* $<2.5 \text{ g L}^{-1}$).[4–6] In order to reduce the analysis time, parallel extractions can be performed.

Solid Phases for SPME–IR

A solid phase suitable for determining organic compounds in water by IR transmission spectroscopy must possess several features. As for the application of SPME in gas chromatography, the solid phase must have an affinity for the analytes of interest and be insoluble in the water matrix. The solid phase must also have reasonable analyte equilibration times or be amenable to quantitation under nonequilibrium conditions. However, the use of IR transmission spectroscopy for detection additionally requires that the solid phase be structurally rigid or be bonded to a transparent support to hold its shape during the analysis, since the solid phase must be placed in the sample compartment of a conventional IR spectrometer. It must also be optically transparent in the region of the analyte infrared absorbance bands.

Of the many solid phases examined to date, three polymers have been found which fulfill the necessary requirements: Parafilm M^{TM}, poly(dimethylsiloxane), and Teflon PFA^{TM}. Figure 3 shows the infrared spectra for these three solid phases. All of these polymers are well suited for infrared transmission because they are manufactured as moderately uniform, thin films. Each polymer film has its own infrared absorbance spectrum. Those regions of the spectrum that strongly absorb infrared radiation are opaque and are not, therefore, useful for analytical determinations. A judicious match, however, of the solid phase to the analytes of interest provides a means of analyzing these compounds in aqueous solutions by IR transmission spectroscopy. Such determinations are otherwise impossible to achieve directly due to the strong absorbance of the water matrix.

Parafilm M^{TM}, the spectrum of which is shown at the bottom of Figure 3, is a proprietary wax-impregnated polymer/rubber composite that is present in nearly every laboratory setting. Optically clear regions in Parafilm are >3035 cm^{-1}, 2768–1500 cm^{-1}, 1335–1240 cm^{-1}, 1204–735 cm^{-1}, and 710–400 cm^{-1}. The spectral region of 1200–460 cm^{-1} is the most useful for the

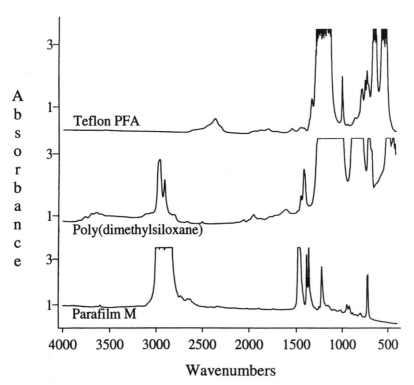

Figure 3   *Infrared absorbance spectra of three solid phase films. Baselines of the spectra have been offset from one another for clarity*

determination of organic compounds partitioned into Parafilm.[4] It encompasses the skeletal vibrations such as out-of-plane ring C–H bends (900–675 cm^{-1}), out-of-plane ring bends (710–675 cm^{-1}), methylene twisting and wagging (1350–1150 cm^{-1}) and the C–Cl stretching vibrations (850–550 cm^{-1}). This optical transparency provides Parafilm with the means to identify individual components of multicomponent mixtures.[4]

Poly(dimethylsiloxane), an important solid phase material in the SPME syringe technology, is also available as a translucent film that is commonly used as a cell growth membrane. Its infrared spectrum is shown in the middle of Figure 3. Optically clear regions in poly(dimethylsiloxane) are >3035 cm^{-1}, 2768–1470 cm^{-1}, 1408–1289 cm^{-1}, 958–906 cm^{-1}, 745–714 cm^{-1} and 658–523 cm^{-1}. This polymer absorbs infrared radiation more strongly in a wider area of the spectrum than does Parafilm. It is, therefore, very difficult to analytically determine individual components of complex mixtures. Poly(dimethylsiloxane) does, however, provide better detection limits and shorter equilibration times for perchloroethylene and is able to differentiate this compound in a mixture with trichloroethylene.[7] Detection limits are also quite good for trifluralin, an herbicide that could not be determined in Parafilm.

Teflon PFA^{TM}, the spectrum of which is shown at the top of Figure 3, is a perfluoroalkoxy Teflon polymer that is commercially available as a visibly transparent film. Optically clear IR regions in Teflon PFA^{TM} are 4000–2650 cm^{-1} and 2200–2000 cm^{-1}. The remaining regions, 2650–2200 cm^{-1} and 2000–400 cm^{-1}, absorb infrared radiation strongly and are therefore opaque. The spectral region of 4000–2650 cm^{-1} is the most useful for the determination of organic compounds partitioned into PFA.[8] It encompasses aromatic (3033 cm^{-1}), aliphatic methyl (2965 cm^{-1}), and aliphatic methylene (2932 cm^{-1}) C–H stretching absorbances. This optical transparency provides a means of determining aliphatic hydrocarbons in water.

Equilibrium Considerations

As discussed in other chapters, solid phase microextraction is an equilibrium method. The formal equilibration time in SPME–IR is defined as the minimum interval required for an absorbance signal to reach its maximum and become constant (to within the standard deviations of the measurements). The equilibrium time can be determined by extracting replicate solutions of one concentration for increasing periods of time. Equilibration times for these polymer films exceed the times of other SPME solid phases for the same analytes because they are thicker (*e.g.*, 100–200 μm) and have more cross-linking.

It is a significant disadvantage in performing analyses if the extraction times are greater than one hour. Indeed, a consideration when using the Teflon PFA^{TM} film is its long equilibration times, which are in excess of 3 hours.[8] It is, however, not necessary to make determinations at equilibrium provided the extractions are performed consistently.[4–6,9] Although nonequilibrium analyses are reproducible, the detection limits are generally higher and there is a consequent restriction of the linear dynamic range.

4 Selectivity and Sensitivity

The strength of standard chromatographic methods, such as GC and HPLC, has been their adeptness in separating components of complex mixtures. A primary role for SPME in the application of these methods has thus been to provide the means of directly sampling the system of interest and of preparing samples for analysis without extraction solvents. Chromatography is then used for analyte separation, identification, and quantitation.

The selectivity of SPME–IR for determining a specific analyte or class of analytes involves matching the properties of the analyte (or class) to the chromatographic properties of the solid phase. The ability of SPME–IR to determine individual compounds in mixtures is dependent upon both the properties of the solid phase and the IR spectroscopy, since several analytes in the mixture may readily partition into the solid phase. The three polymer films currrently in use for SPME–IR are all moderately nonpolar. Polar organic compounds do not, therefore, easily partition from the aqueous phase into the solid phase, and detection limits for these compounds are often higher.

The sensitivity of the SPME–IR measurements is governed principally by four factors:

1. the concentration of the analyte(s) in the solid phase film during the spectral measurement,
2. the absorptivity of the analytical infrared band (ε),
3. the film thickness (which is related to the pathlength, b),
4. and the noise of the IR spectrometer.

The first factor is influenced predominantly by the distribution constant (K_d) for the analyte/solid phase film system. The larger the K_d value, the better the detection limit. Table 1 shows some representative distribution constants for several analyte/solid phase combinations.[7] Table 1 also shows a comparison of the K_d values with the octanol/water partition constants (K_{ow}).[10] Generally, the K_ds follow the trends of the K_{ow}s, but they are often smaller in magnitude.

Table 1 *Molar absorptivity values and distribution constants for selected organic compounds*

			PDMS	*Parafilm M*[TM]	
Compound	ε^a (L mm^{-1} mol)	*Band* (cm^{-1})	K_d^b	K_d^c	K_{ow}^d
Benzene	7.64	3039	4.30	27.5	135
Chlorobenzene	36.8	741	40.6	72.6	832
Toluene	31.3	729	27.9	51.3	490
Trichloroethylene	31.6	933	8.80	54.9	263
o-Xylene	36.1	742	143	109	1319

[a] The ε values were obtained in CS$_2$ at the wavenumber shown in the following column.
[b] Distribution constant determined in 127 μm poly(dimethylsiloxane) film.
[c] Distribution constant determined in 130 μm Parafilm M[TM].
[d] The octanol/water partition coefficient at 25 °C [(mol L^{-1} octanol) (mol L^{-1} water)$^{-1}$].[10]

It should be noted that the sensitivity of the SPME–IR measurement can be degraded under some circumstances because the concentration of analyte in the film can decrease with respect to time. Specifically, evaporative loss of volatile organics into the air immediately occurs when the solid phase is removed from the water matrix.[4] Additionally, FTIR spectrometry often involves the co-addition of spectral scans in order to increase the signal-to-noise (S/N) ratio and improve detection limits. These two procedures, however, expose volatile compounds to air and to the IR beam resulting in analyte loss as a function of time. As an example of this loss, Figure 4 shows how the concentration of benzene in the film (a volatile analyte) decreases on exposure to both the air and the IR beam of the spectrometer. This decrease in analyte concentration has an important ramification with respect to acquiring spectra. For this experiment, noise from the co-added spectra initially decreases faster than the absorbance signal decreases. After a succession of coadded scans, the S/N ratio is then maximized because the absorbance signal decreases faster than the noise. Co-addition of more spectra beyond this maximum only serves to diminish the S/N ratio. Thus in contrast to conventional FTIR practice, shorter scan times (*i.e.*, fewer coadded scans) can result in better detection limits for volatile analytes.

The second factor affecting sensitivity is the molar absorptivity constant. Simply, the larger the value of ε, the greater the absorbance and sensitivity. However, Table 1 shows that the absorptivity constants for the various analytically-useful infrared spectral bands are moderately similar. That is, on the average they vary by < 10-fold.

The Beer–Lambert law dictates that the optical pathlength, the third factor affecting sensitivity, should be long in order to make the absorbance as high as possible. However, a long optical path would require a thick film which would, as a result, provide for very long analyte equilibration times. The current work employs films that have thicknesses of 100–150 μm. This range already results in

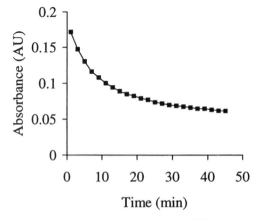

Figure 4   *Evaporative loss of benzene from Parafilm M^{TM} following a 90 minute extraction. The absorbance band at 673 cm^{-1} was monitored*

equilibration times greater than 30 minutes and, thus, should probably not be exceeded.

Finally, with regard to the noise of the infrared spectrometer (factor 4), most modern FTIRs have noise levels significantly below 1 mA. Of course, this number varies with respect to spectral region because of atmospheric transparency, film background, *etc.* Consequently, some analytes will have poorer detection limits than others simply because of the spectral location of their major analytical band, all other considerations being equal.

5 Applications

General

Solid phase microextraction coupled with IR spectroscopy is a useful alternative to SPME–GC. Indeed, many of the same analytes determined by SPME–GC can also be determined by SPME–IR. To date, SPME–IR has been applied to both the determination of individual components in complex mixtures and the determination of aggregate quantities in aqueous solutions (*i.e.*, gasoline range organics).

As a reference for the subsequent discussion, Table 2 shows analytical data for several classes of analytes determined in the three solid phase films. Multiple entries in this table for a given analyte imply that more than one film is useful. Conversely, the absence of an entry for a given analyte/film combination indicates that film is not suitable for the analysis.

The Determination of Aggregate Properties: Total Gasoline-Range Organics

Petroleum fuels are comprised of an assortment of many organic compounds. For example, gasoline fuels include the more volatile organics such as small-chain hydrocarbons and the BTEX compounds (benzene, toluene, ethylbenzene, xylenes). SPME–IR analysis using the C–H stretching region can be used as a method for determining total hydrocarbons. In fact, Teflon PFA^{TM} has been found to successfully extract gasoline-range organics from water.[8] This film provides a clear spectral region for identification and quantitation which the other two films do not. An absorbance spectrum of analytes extracted from a solution of unleaded gasoline in water is shown in Figure 5. Although PFA cannot be used to identify individual components of multicomponent mixtures, it is an effective technique for analysis of the mixture itself.

The Determination of Individual Components: Volatile Organic Compounds

Volatile organic compounds (VOCs) are common contaminants in ground and surface waters. These compounds include aromatics such as the BTEX compounds, and halocarbons such as carbon tetrachloride, chlorobenzene,

Table 2 *SPME–IR analytical data for selected compounds and compound classes using three solid phase films*

Compound/Class	Parafilm M™			Poly(dimethylsiloxane)			Teflon PFA™		
	Linear dynamic range, ppb[a]	Limit of quantitation,[b] ppb	Equilibration time, min	Linear dynamic range, ppb	Limit of quantitation, ppb	Equilibration time, min	Linear dynamic range, ppb	Limit of quantitation, ppb	Extraction time, min
VOCs[c]									
Benzene	750–100 000	182 (9)	90						
Carbon tetrachloride	640–80 000	200 (3)	60						
Chlorobenzene	600–50 000	187 (8)	70	20 000–160 000	5000 (8)	65			
Chloroform	4000–800 000	1290 (11)	50						
p-Chlorotoluene	225–37 500	66 (10)	70						
Ethylbenzene	480–30 000	182 (9)	60	10 000–80 000	3000 (2)	55			
Toluene	4000–25 000	752 (4)	30						
Trichloroethylene	5000–100 000	1830 (7)	300	4000–100 000	2680 (10)	60			
m-Xylene	140–7000	80 (4)	165	5000–100 000	3060 (7)	80			
o-Xylene	140–7000	102 (9)	165	5000–100 000	4450 (8)	85			
p-Xylene	140–7000	66 (3)	200	1000–40 000	780 (10)	55			
Perchloroethylene	2000–40 000	1560 (9)	275						
GRO[d] **mixtures**									
Aviation gasoline							2800–42 000	2800 (6)	30
Unleaded gasoline							2000–30 000	2000 (11)	30
Lighter fuel							1000–7200	700 (9)	30
Crop protection chemicals									
Lindane									
Atrazine									
Trifluralin				200–2000	188 (12)	75			

[a] Tested linear dynamic range.
[b] RSD in parentheses.
[c] Volatile organic compounds.
[d] Gasoline-range organics.

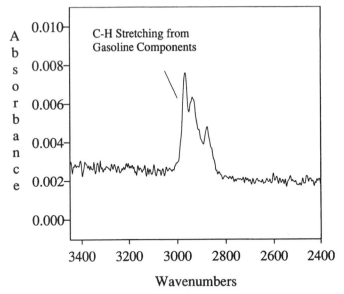

Figure 5 *Infrared absorbance spectrum of Teflon PFA^{TM} film following a 30 minute extraction of a water solution containing 20 ppm unleaded gasoline. The absorbance bands are due to C–H stretching from gasoline components partitioned into the film*

chloroform, and *p*-chlorotoluene. Figure 6 illustrates how ethylbenzene can be identified in an aqueous single-component solution using SPME–IR with a poly(dimethylsiloxane) film and a spectral library.

Parafilm M^{TM} has also been a particularly useful solid phase for the analytical determination of the VOCs. Strongly absorbing vibrational bands occur for these molecules in the range 804–673 cm^{-1}, which is a spectral region that is optically clear in Parafilm. Each compound has several unique vibrational bands in this region that are well resolved from one another, thus providing a means of distinguishing components of mixtures.

The BTEX compounds are ubiquitous environmental pollutants. They are also important constituents of gasoline, since they comprise 40% of the total sample by weight. SPME–IR analyses have demonstrated Parafilm's ability to distinguish four of the six alkylbenzenes (benzene, *o*-xylene, *m*-xylene, *p*-xylene) in a petroleum industry wastewater sample.[4] Quantitation by simple univariate calibration based on absorbances provided good agreement with purge and trap GC–MS standard methods. The ability to individually determine *m*- and *p*-xylene is noteworthy since these compounds are difficult to distinguish by GC–MS owing to chromatographic limitations. SPME–IR determinations of ethylbenzene and toluene are complicated by the spectral overlap of other components known to be in gasoline. Multivariate methods, such as partial least squares or principal component regression, may provide the means of quantifying all the components of such a complex mixture.

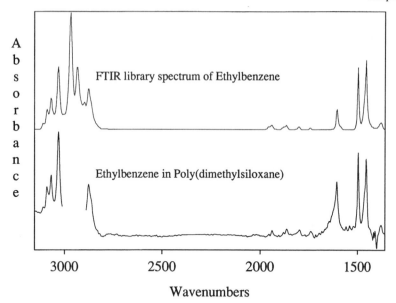

Figure 6 *Infrared spectra illustrating how SPME–IR can be used to determine an individual component in an aqueous solution. The lower spectrum is of a poly(dimethylsiloxane) film following a 55 minute extraction of a water solution containing 20 ppm ethylbenzene. The upper spectrum shows an FTIR library absorbance spectrum for ethylbenzene for comparison. It should be noted that the 'blanked-out' region centered at ca. 2960 cm^{-1} is due to the strong absorption of poly(dimethylsiloxane)*

6   Future Directions

The identification of new solid phase films is a continuing endeavor. Polymer films that are more polar in nature would expand the SPME–IR technique to a much wider range of analytes. Although poly(dimethylsiloxane) was successfully used to extract trifluralin from aqueous solutions, attempts to expand the procedure to other herbicides, using the current films, have not provided significant results.[7] The next phase of research will therefore emphasize the development of methods for analysis of pesticides and herbicides by SPME–IR.

Another goal of this project is to demonstrate that the laboratory methods are viable as field methods by conducting the analyses at actual sites. These field trials are scheduled to begin in late 1998.

Acknowledgments

The authors wish to thank the Environmental Protection Agency, grant number R825343-01-0, for providing financial support.

References

1. D.A. Skoog and J.J. Leary, *Principles of Instrumental Analysis*, Saunders College Harcourt Brace, Orlando, FL, 1992, Ch. 12.
2. D.A. Skoog and J.J. Leary, *Principles of Instrumental Analysis*, Saunders College Harcourt Brace, Orlando, FL, 1992, Ch. 7.
3. P.R. Griffiths and J.A. DeHaseth, *Fourier Transform Infrared Spectrometry*, Wiley & Sons, New York, 1986.
4. D.L. Heglund and D.C. Tilotta, *Environ. Sci. Technol.*, 1996, **30**, 1212.
5. B.L. Wittkamp, S.B. Hawthorne and D.C. Tilotta, *Anal. Chem.*, 1997, **69**, 1197.
6. B.L. Wittkamp, S.B. Hawthorne and D.C. Tilotta, *Anal. Chem.*, 1997, **69**, 1204.
7. S.H. Lubbad, S.A. Merschman and D.C. Tilotta, *J. Chromatogr. A*, 1999, accepted for publication.
8. D.C. Stahl and D.C. Tilotta, *Environ. Sci. Technol.*, 1999, accepted for publication.
9. J. Ai, *Anal. Chem.*, 1997, **69**, 1230.
10. R.P. Schwarzenbach, P.M. Gschwend and D.M. Imboden, *Environmental Organic Chemistry*, John Wiley & Sons, New York, 1993, pp. 618–625.

CHAPTER 46

SPME in Near-IR Fiber-optic Evanescent Field Absorption Spectroscopy: A Method for Rapid, Remote In situ Monitoring of Nonpolar Organic Compounds in Water

JOCHEN BÜRCK

1 Introduction

An associated hazard to utilizing, storing and transporting polluting agents such as chlorinated hydrocarbons is the environmental impact of leaks and spills that can contaminate soil and ground water at industrial sites. Due to this, there is a need for screening and monitoring of these compounds at contaminated areas.[1] Furthermore, the treatment of discharged process waters at production and storage facilities or at remediation sites, has to be verified to meet regulatory discharge limits for hydrocarbon (HC) contaminants. Traditionally, discontinuous batch sample collection followed by sample transport to the laboratory, sample preparation and GC laboratory analysis is used to demonstrate regulatory compliance.[2] This kind of analysis often is time consuming, expensive and can only provide a brief 'snapshot' of a treatment system's performance. Therefore, fast and simple HC analysis techniques are needed which allow the monitoring of pollutants continuously and *in situ*.

Recently new spectroscopic sensing techniques for HC pollutants have been developed which hold promise to fulfill these demands. These methods, which are all based on evanescent field absorption (EFA) measurements with optical fibers as sensing elements work either in the near-infrared (NIR)[3–5] or in the mid-infrared (MIR) spectral range.[6,7] Similar to SPME introduced by Pawliszyn, a polymer-clad fiber is used to obtain a solvent-free HC extraction and sample preparation. Instead of transferring the extracting fiber to a GC, the optical fiber is directly coupled to a spectrometer or photometer unit and the HC species extracted into the fiber cladding are monitored *in situ* by an absorptiometric measurement. This is possible due to the evanescent wave tail of the measuring light conducted in the fiber core, which extends into the fiber polymer cladding. In contrast to the very short fibers used in SPME (1 cm length), that are mounted in a syringe-like device, in EFA spectroscopy the interaction length of a silver halide sensing fiber with the sample has to be in the range of around 10 cm for MIR monitoring and very long quartz glass fibers between 10 and 30 m have to be used in the NIR region to get a suitable sensitivity. This is due to the fact that the intensity of C–H, N–H or O–H harmonics and combination oscillations of molecules occurring in the NIR is lower by a factor of 10–100 than the corresponding ground modes in the MIR range. In spite of this disadvantage and the higher amount of spectral information contained in MIR spectra, EFA measurements in the NIR are preferable, because fiber materials that are mainly used for MIR–EFA sensing of HCs are silver halides. Silver halide (AgClBr) fibers still suffer from high transmission losses and degradation under UV exposure and contact with metals,[6] so that they are hardly suitable for practical sensing applications or true remote analysis at their present stage of development. In the NIR spectral range, on the other hand, quartz glass optical fiber technology can be applied, which is fairly advanced due to numerous applications in the telecommunications industry. These fibers have rather good mechanical properties, are stable against chemical attack and are able to transmit the light over distances of up to a few hundred meters between spectrometer and sensing element, *i.e.*, *in situ* monitoring in deep wells of contaminated areas is possible.

In this chapter emphasis is put on describing the basic principles of NIR–EFA spectroscopy, the instrumental set-up used to do these kind of measurements, typical calibration data and the performance of the equipment for HC sensing obtained during field measurements.

2 Evanescent Field Absorption (EFA) Spectroscopy

The theory of internal reflection and EFA spectroscopy has been given in detail earlier[8,9] and will only be discussed briefly in this chapter. The measuring light transmitted in an optical fiber by total reflection at the core/cladding interface produces a standing wave whose evanescent field penetrates into the fiber cladding over some distance, which is in the range of the wavelength of light used for the measurement. The quartz glass fiber NIR–EFA sensor developed at the Institut für Instrumentelle Analytik (IFIA) has a hydrophobic silicone cladding that extracts nonpolar HCs from the aqueous phase.[4] For this extracting fiber all the principles and theoretical aspects of SPME described in Chapter 1 of this book are valid.[10]

The extracted HC species can absorb energy from the portion of light in the evanescent field. Such interactions, illustrated schematically in a longitudinal section view of the fiber in Figure 1, lead to a specific absorption of light intensity at the corresponding C–H overtone bands of HC compounds. Owing

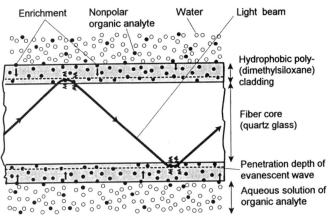

Figure 1   *Schematic view of the extraction of a nonpolar organic analyte from the aqueous phase into the cladding of a polymer-clad silica optical fiber and illustration of evanescent field absorption (EFA) sensing principle*

to the extraction step and the direct absorptiometric measurement in the cladding, interference from strong water O–H absorption bands or from turbidity in the solution is avoided. For low analyte concentrations ($\ll$ saturation concentration) the absorbance A obtained from an evanescent field measurement can be approximated by:[11]

$$A = \log \frac{I_0}{I} = \varepsilon_e \cdot L \cdot K_{fw} \cdot C_w^\infty$$

where I_0 is the transmitted light intensity with the sensor in pure water, I is the transmitted intensity with the sensor in analyte solution, ε_e is the effective absorptivity, L is the length of the sensor fiber, K_{fw} is the fiber/water distribution constant and C_w^∞ is the equilibrium concentration of analyte in water.

Because of the linear dependence of absorbance on fiber length and analyte concentration the equation has a formal similarity with Beer's law. However, the effective absorptivity $\varepsilon_e = \eta_m \cdot \varepsilon_m$ is dependent on η_m, which is the ratio of light intensity in the evanescent field to the total light intensity transported in the fiber core, and ε_m the molar absorptivity of the analyte species. Because η_m is influenced by refractive index changes in the cladding caused by analyte enrichment, ε_e cannot be considered to be constant at higher analyte concentrations and thus non-linearity of the absorbance/concentration values may occur.[4]

3   Instrumentation and Practical Aspects of EFA Sensing

Fiber-optic Sensor

A commercially available polymer-clad silica (PCS) fiber (Fiberguide Industries, NJ), which has a low-hydroxide quartz glass core of 210 μm diameter, a

10 μm poly(dimethylsiloxane) (PDMS) coating, and a 20 μm nylon protective jacket is used as the sensing element. The sensor is constructed by winding the jacketed fiber in a single or double coil on a support made of stainless steel and Teflon®. This support has four comb-like structures and the fiber of 12–30 m length is threaded through the teeth of these structures. Finally, it is fixed by four perforated Teflon® strips, which are pressed on the teeth of the comb structures. To turn the optical fiber into a sensor element for nonpolar HCs, the outer nylon protective jacket is chemically removed by dissolving it with 1,2-propanediol at 160 °C. Protection of the helical sensor fiber against mechanical damage is provided by a stainless steel perforated plate housing which has outer dimensions of 14 × 5 cm.

NIR (Spectro)photometer Units Used in EFA Sensor Systems

EFA absorption measurements of HC substances enriched in the fiber PDMS coating can be performed with any commercially available NIR spectrometer that has fiber-optic adaptation. Figure 2 shows a schematic view of the sensor coil, which can be connected over all-silica fibers (fibers with a silica core and a fluorine doped silica cladding) either to a NIR spectrophotometer (shown on the left side of Figure 2) or to a low-cost, portable filter photometer (right side) developed at the IFIA. For both sensor system configurations, light from a tungsten halogen lamp first is conducted over the input fiber into the sensor fiber. The light attenuated by the HC absorption in the sensing fiber is then

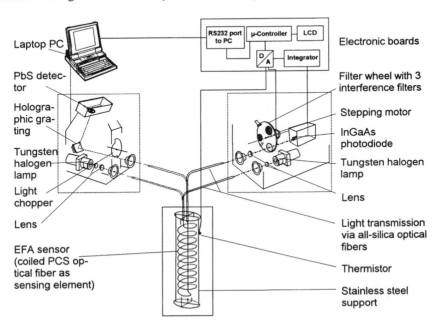

Figure 2 *Instrumental set-up of coiled fiber-optic sensor adapted either to a NIR spectro-photometer or a bandpass filter photometer unit*
(Reproduced with kind permission from Kluwer Academic Publishers[16])

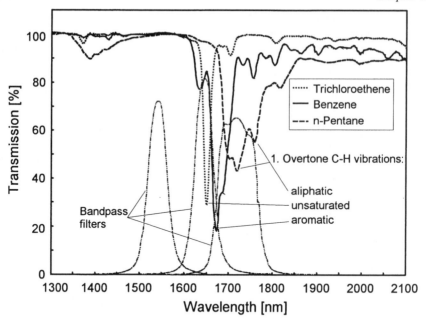

Figure 3   *NIR transmission spectra of hydrocarbon compounds and of the measuring and*
reference filters used in the photometer; measurements are done in the spectral
range of the different 1st overtone C–H absorption bands
(Reproduced with kind permission from *Field Anal. Chem. Technol.*, 1998,
J. Wiley & Sons[18])

transferred over the output fiber into the corresponding spectral evaluation
unit. Up to now sensor coils have been connected to a Guidedwave model 260
grating spectrometer and a Bruker Vector 22N FT-NIR instrument. Details of
the instrumental set-up used with the Guidedwave model 260 and the performance of the corresponding HC sensor system have been thoroughly described
in the literature.[3–5,12]

If the EFA sensor is connected to the filter photometer system, light coming
back from the sensor element is focused on a wheel with three bandpass filters.
Figure 3 depicts transmission spectra of typical HC pollutants and of the
measuring filters used in the instrument. While two filters select light in the
wavelength range of the unsaturated and saturated first overtone C–H absorption bands of HC compounds around 1646 and 1717 nm, the third filter is
transparent in the non-absorbing wavelength range around 1540 nm and
provides the reference light. The three filters are positioned sequentially in the
light beam by a stepping-motor driven, rotating wheel and the transmitted
intensities are measured by an InGaAs photodiode (*cf.* Figure 2). Automatic
control of the filter wheel stepping-motor, tungsten halogen source and InGaAs
photodiode is provided by a microcontroller. This controller also calculates
absorbance values from the intensities measured by the photodiode at the C–H
(I_M) and reference (I_R) wavelengths according to: $A = \log I_R/I_M$. Due to the fact
that the wavelength ranges of the measuring and reference filters are close

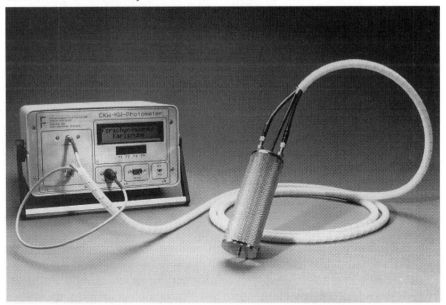

Figure 4 *Photograph of fiber-optic EFA in situ monitoring system for nonpolar hydro-carbons in water; coiled sensor fiber is protected by outer stainless steel perforated plate housing and adapted to NIR filter photometer via 3 m all-silica fibers; the temperature sensor can be seen at the bottom of the sensor housing* (Reproduced with kind permission from *Field Anal. Chem. Technol.*, 1998 J. Wiley & Sons[18])

together, drifts in the intensity of the light source, in the sensitivity of the photodiode and changes in the transmission of the optical components within the photometer are effectively compensated.

Up to 20 HC calibration functions can be stored on the microcontroller. With a stored calibration function the HC concentration is calculated directly from the corresponding absorbance value and indicated on the LCD display of the photometer. A photograph of the EFA sensor/filter photometer system is shown in Figure 4. The stand-alone instrument can be easily operated by simple function keys. Data are collected and stored automatically by the system. *Via* an RS232 interface, stored data are sent to a computer for evaluation and documentation purposes or new calibration data are transmitted from the computer to the photometer.

The strong dependence of the penetration depth of the evanescent wave light field in the sensor fiber on temperature leads to an offset in the absorbance signals given by the system, if the sensor temperatures are different during zero-setting and sample measurements. Studies on these temperature effects for EFA sensors adapted to a NIR spectrophotometer have been presented in the paper by Klunder *et al.*[13]

For the EFA sensor/filter photometer system these effects on the signal are automatically corrected by measuring the actual sample temperature by a thermistor attached to the sensor, whose signals are also transmitted to the microcontroller of the photometer. The absorbance/temperature curves relative to measured wavelengths, known from calibration measurements, can be fitted by polynomial functions, which are stored on the microcontroller. From the difference of the sample temperature and the temperature during zero-setting in pure water the controller calculates the absorbance offset and corrects it automatically using the stored polynomial functions.

Sensor Calibration and Field Measurements

For the described EFA sensor systems the absorbance zero-point is set by performing a measurement with the sensor in pure water. Afterwards, periodic measurements are started (minimum interval is 30 s), the baseline noise is checked and finally the probe is immersed directly into the aqueous HC solution. For analytes that have fast sorption rates, sensor signals are acquired until extraction equilibrium is reached after a few minutes. These equilibrium absorbance/concentration data are used for setting up calibration functions. For HC mixtures containing long-chain hydrocarbons, *e.g.* gasoline, the time to reach equilibrium would be too long.[12] In this case, a kinetic evaluation is made, by taking the absorbance values after a shorter contact time, *e.g.* 15 min, and plotting them *versus* concentration.

For *in situ* monitoring with the EFA sensor system the procedure is similar to laboratory calibration measurements. For continuous monitoring, the EFA sensor preferably should be installed in a flow cell with a volume of around 1 L. The inlet and outlet of this flow cell should be controlled by two stopcocks, so that the cell can be separated from the analyte stream and be filled from a third stopcock with pure water for absorbance zero-setting or with specified HC solutions to do recalibration. The flow cell is then installed in the bypass of a pipe with process water from a remediation facility or waste water of a production plant. Flow measurements should be always done with a turbulent water flow in the cell to minimize analyte sorption times (very turbulent flow does not impede sorption).[11]

In situ determination of the HC concentration in a deep monitoring well is performed by connecting the sensor by all-silica fiber cables with a longer length (50–100 m) to the spectral evaluation unit. After absorbance zero-setting in a beaker filled with pure water the sensor attached to the fiber cables is lowered into a pre-set water level and measurements are started.

4   EFA Sensor System Specifications

NIR–EFA Spectra and Sensitivity of EFA Measurements

HC species dissolved in a large excess of H_2O cannot be measured directly by NIR absorption spectroscopy using conventional transmission techniques due to the strong water O–H absorption bands at 1450 and 1940 nm, that overlap the much weaker C–H peaks in the 1600–1900 nm range.[3] For this reason NIR

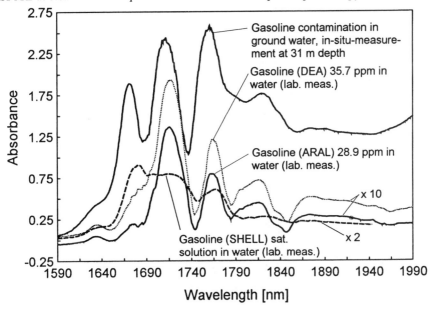

Figure 5 *Comparison of NIR–EFA spectra of gasoline contaminated aqueous solutions; laboratory samples and in situ measurement in a monitoring well using a 12 m fiber-optical sensor in combination with a NIR spectrophotometer*

(and also MIR) spectroscopy has not been applied directly to aqueous samples, *e.g.*, in environmental applications. EFA–NIR spectroscopy overcomes this problem by extraction of the nonpolar species into the hydrophobic coating, that is formed by the PDMS cladding of the fiber sensing element. If an EFA fiber-optic sensor is combined with a spectrometer one can exploit the full spectral information without any water interference.

In Figure 5 NIR–EFA spectra of aqueous solutions of gasoline from different companies measured in the laboratory are compared with an *in situ* spectrum collected in a monitoring well of a gasoline contaminated area. Generally, in all spectra the aliphatic CH_3 (1704 nm) and CH_2 bands (1724 and 1760 nm), as well as the aromatic CH band (1679 nm) can clearly be distinguished. From the intensities of the different bands one can see that branched and unbranched aliphatic HCs are the major components of two gasoline samples (DEA and Aral), while the aromatic CH vibrations are relatively weak and appear only as a shoulder in the flank of the aliphatic peaks. However, in the spectra of the other samples (Shell, gasoline contamination) the aromatic band is much more intense indicating a higher percentage of aromatics. From this it is obvious that EFA–NIR spectroscopy compared with a chromatographic method like GC has a much lower selectivity. However, this example also demonstrates that by using the EFA–NIR absorption spectrum it is at least possible to extract information about different HC groups contained in a mixture. Quantitative multi-component analysis of four or five compounds in a HC mixture can be

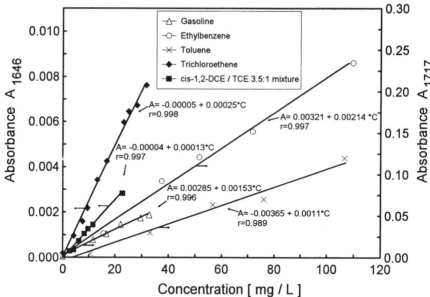

Figure 6 *Calibration functions for different nonpolar hydrocarbon compounds or mixtures using a 12 m EFA sensor in combination with the portable NIR filter photometer; A_{1646} = absorbance value obtained from the bandpass filter with central wavelength 1646 nm; A_{1717} = absorbance value obtained from the bandpass filter with central wavelength 1717 nm*
(Reproduced with kind permission from *Field Anal. Chem. Technol.*, 1998 J. Wiley & Sons[18])

obtained if chemometric calibration and data evaluation techniques are applied to NIR–EFA spectra.[14]

On the other hand, in many applications it is not the surveillance of single HC species, but rather the monitoring of a whole group of compounds which is required. Thus, for describing water quality often a sum parameter, *e.g.* the DOC value (concentration of dissolved organic carbon) or the AOX value (adsorbable organic halogen) is used.[15] The broadband EFA sensor/filter photometer system developed at the IFIA provides an absorbance sum signal (*i.e.* a kind of *in situ* sum parameter) for the total amount of nonpolar HCs extracted into the silicone cladding of the fiber.[16]

In Figure 6 calibration functions of different aqueous HC solutions are shown, which have been obtained using a 12 m EFA sensor adapted to the NIR filter photometer. From the distinct variations in the slope of the linear calibration functions of the pure compounds ethylbenzene, toluene and trichloroethene and of the hydrocarbon mixtures (gasoline and a 3.5:1 mixture of trichloroethene and *cis*-1,2-dichloroethene) it is obvious that the sensitivity strongly depends on the HC compound. From the fact that the molar extinction coefficients of these species at the C–H absorption bands differ only by a factor of 1.4 at the most, it can be established that the differences in sensitivity are mainly caused by the strongly varying concentration of the analytes in the

Table 1 *Fiber/water distribution constants K_{fw} and limit of detection (LOD) for nonpolar hydrocarbons in aqueous solution using different EFA sensor systems*

		Limit of detection		
HC compound	Fiber/water distribution constant K_{fw}	12 m EFA sensor/filter photometer (mg L^{-1})	30 m EFA sensor/filter photometer (mg L^{-1})	12 m EFA sensor/ spectrophotometer (mg L^{-1})
Dichloromethane	11	*	*	80
Trichloromethane	41	*	*	18
1,1-Dichloroethene	110	*	*	1.3
Trichloroethene	246	0.8	0.3	0.8
Toluene	260	0.9	*	0.9
Chlorobenzene	342	*	*	0.4
Ethylbenzene	490	0.5	*	*
p-Xylene	526	*	0.1	0.4
1,2-Dichlorobenzene	*	*	0.05	*
1.2.4-Trichlorobenzene	1421	*	*	0.1
Gasoline	*	0.7 **	*	2.2 **
Crude oil (emulsion in sea water)	*	*	*	1.3 **

* Value has not been determined.
** After 15 min contact time, kinetic measurement.

silicone cladding. This is due to the varying enrichment behavior of the substances reflected in the fiber/water distribution constant K_{fw}.[5,11] The K_{fw} value of ethylbenzene is approximately higher by a factor of two than the value of toluene (*cf.* Table 1); correspondingly the slope of the calibration function is increased by this factor. Generally, the EFA sensitivity for a HC compound will increase with lower polarity of the substance (which in a rough approximation can be correlated to the water solubility).

The limits of detection (LOD) for HC compounds in aqueous solution typically are in the range of a few mg L^{-1}. Data for nonpolar hydrocarbons obtained with different combinations of EFA sensor and spectrometer unit are given in Table 1. They have been determined from the threefold standard deviation of the spectral noise at the corresponding wavelength range and the slope of the calibration function. The LOD values have been improved recently to a few 100 μg L^{-1} by adapting a 30 m double coil sensing fiber to an EFA/NIR photometer unit.

Reponse Time and Reversibility of EFA Measurements

The sensor equilibration time for HC compounds in aqueous solution strongly depends on the HC compound and the hydrodynamic conditions. The $t_{95\%}$ values for volatile chlorinated hydrocarbons (CHC) like trichloroethene and the BTEX aromatic hydrocarbons typically vary between 2 and 30 minutes.[4,5] In

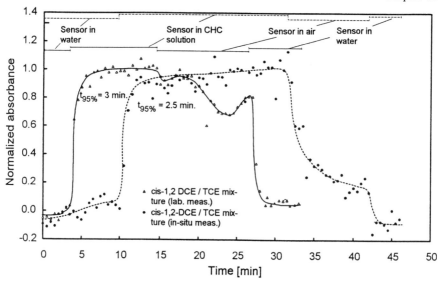

Figure 7 *Response signals versus time obtained with an EFA sensor/filter photometer system for measurements of aqueous CHC solutions with fast sorption rate (3.5:1 mixtures of cis-1,2-dichloroethene and trichloroethene); comparison of a calibration measurement in the laboratory with an in situ measurement in the feed of a ground water clean-up facility (Mühlacker, Germany)*

Figure 7 the response signals of EFA sensor/filter photometer systems to aqueous CHC solutions are compared for a laboratory calibration measurement and an *in situ* field experiment. Both *cis*-1,2-dichloroethene (*cis*-1,2-DCE)/ trichloroethene (TCE) mixtures were measured under turbulent flow conditions, *i.e.* the solution in the laboratory experiment was strongly stirred and a process water throughput $> 20 \, L \, min^{-1}$ was used in the flow cell during the field-trial. From the absorbance–time plots it can be seen that the equilibration times are similar for both situations and that the $t_{95\%}$ values are in the 2.5–3 min range.

A longer equilibration time ($t_{95\%} = 26$ min) is needed for a larger molecule like 1,2-dichlorobenzene and non-turbulent hydrodynamic conditions (see Figure 8). Here the sensor was placed in a flow cell with low dead volume (35 mL) and 1,2-dichlorobenzene solutions of varying concentration were pumped through the cell with a flow rate in the range of $0.125 \, L \, min^{-1}$. The increasing and decreasing concentrations and the corresponding equilibration steps are clearly traced by the sensor. This curve also demonstrates that the sorption of volatile hydrocarbon molecules in the silicone cladding is completely reversible, because the signal returns to the baseline if the sensor is contacted with pure water. Another possibility to regenerate the fiber sensor is to put it in air for evaporation of volatile compounds and then immerse it again in pure water (*cis*-1,2-DCE/TCE curves in Figure 7 and gasoline curve in Figure 8).

The need for a kinetic evaluation for long-chain molecules contained in gasoline, diesel fuels or mineral oils is obvious from the gasoline sorption curve

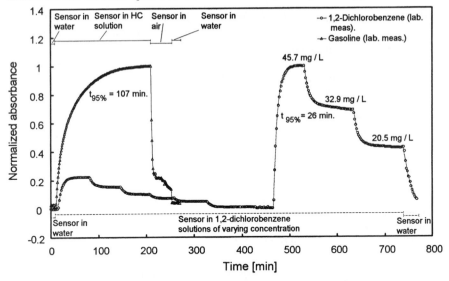

Figure 8 *Response signals versus time obtained with an EFA sensor/filter photometer system for measurements of aqueous hydrocarbon solutions with slower sorption rate (1,2-dichlorobenzene and gasoline); calibration measurements in the laboratory*

shown in Figure 8. A $t_{95\%}$ value of 107 min measured in a strongly stirred solution is far too long for practical applications. Therefore, the absorbance values after 15 min contact time are used to set up a calibration curve. If the sensor is used to measure diesel fuels or mineral oils it has to be regenerated by rinsing with petroleum ether or acetone owing to the high affinity between long-chain HC molecules and the PDMS cladding.[12]

5 Field Tests

In co-operation with the Lawrence Livermore National Laboratory (LLNL) field trials with an EFA sensor/Guidedwave NIR spectrometer system have been carried out in monitoring wells contaminated with TCE at LLNL and at the Savannah River Site (SRS), USA.[17] The TCE NIR–EFA spectra collected *in-situ* in wells with depths of up to 45 m were evaluated by means of a PLS (partial least-squares regression) calibration model using the full spectral information of the TCE first C–H overtone vibration band around 1650 nm, which had been set up previously with calibration samples in the 0.9–31 mg L^{-1} concentration range. A mean deviation of 23% (arithmetic mean of relative difference) was obtained for a comparison of EFA sensor *in situ* data and GC reference analyses.

The performance of an EFA sensor/filter photometer system for *in situ* monitoring of CHCs was first evaluated at the research facility for subsurface remediation VEGAS (Versuchseinrichtung zur Grundwasser und Altlastensanierung) of the University of Stuttgart, Germany.[18] In this case continuous *in*

Table 2 *Comparison of trichloroethene (TCE) EFA sensor/filter photometer in situ data with HPLC–UV reference analysis of off-line samples (experiment at VEGAS facility)*

Sample	EFAS in-situ TCE conc. (mg L^{-1})	HPLC/UV ref.* TCE conc. (mg L^{-1})	Deviation (mg L^{-1})	Rel. deviation ** (%)
1	606	587	19	3.2
2	499	470	29	6.2
3	575	838	−263	−31.4
4	674	673	1	0.1
5	670	743	−73	−9.8
6	599	622	−23	3.7

* Column: RP 18; eluent: acetonitrile/water 70:30, wavelength 195 nm.
** Concentration measured by HPLC–UV = 100%.

situ measurements were studied in an artificial aquifer with TCE as single pollutant and concentrations extending from the low mg L^{-1} range up to saturation. The experiments showed that the sensor system provides quantitative data over a measuring period of some weeks, without recalibration and a mean deviation of 9% compared with HPLC off-line analysis (see Table 2). The fiber sensor proved to be chemically stable even in the presence of 'free' TCE phase. The PDMS coating, which strongly swells in contact with pure nonpolar compounds, did not detach from the fiber core owing to covalent bonding.

The EFA sensor/filter photometer system also was tested at a former chemical landfill, where organic solvents were stored together with galvanic sludges (Mühlacker, Germany).[18] CHCs are found in the drainage and ground water below the site and the major pollutants appearing in the ground water are *cis*-1,2-DCE, TCE and dichloromethane. Other CHC compounds were only present at concentrations < 15 µg L^{-1}, as could be confirmed from GC reference analysis. A clean-up facility is installed at this site, where the CHCs are removed from the ground water by a stripping process and by adsorption on charcoal. Here, *in situ* measurements were performed in a bypass stream of the feed pipe of the remediation facility. For technical reasons within the facility no continuous measurements could be performed with the sensor system over longer periods. Therefore, at time intervals of two to three weeks *in situ* data were taken over a period of around one hour using measuring intervals of 30 s. Also during this time, process water was sampled from the bypass stream, which was analyzed by an independent laboratory according to the German DIN method (GC analysis) for the determination of CHCs.[2]

At the beginning of the field tests, the concentration ratio of *cis*-1,2-DCE and TCE in the feed water was approximately 3.5 : 1, and the dichloromethane concentration was at the 10–100 µg L^{-1} level. Owing to the fact that the sensor system has a much lower sensitivity for this compound[5] its contribution to the sensor signal could be neglected and calibration of the sensor sum signal was done according to the two major components' feed ratio. As can be seen from

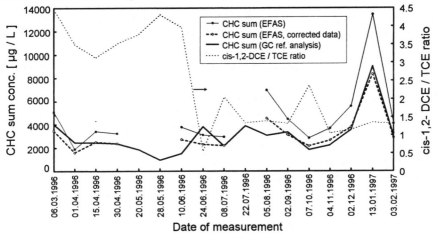

Figure 9 *CHC sum concentration measured in the feed of the ground water clean-up facility at a former chemical landfill (Mühlacker, Germany) during a number of field trials; comparison of EFA sensor in-situ data with GC reference analysis (Reproduced with kind permission from Field Anal. Chem. Technol., 1998 J. Wiley & Sons[18])*

the calibration functions of pure TCE solutions and of the *cis*-1,2-DCE/TCE mixture in Figure 6, the slope of the TCE calibrating plot is steeper by a factor of two compared with the plot of the mixture. This is due to the higher enrichment of TCE in the silicone sensor membrane compared with *cis*-1,2-DCE.[11] For the 3.5:1 mixture both components contribute to the sum signal measured by the sensor and because of the *cis*-1,2-DCE excess in the mixture the corresponding calibration function has a decreased slope.

Figure 9 shows the *in situ* CHC sum concentration data collected in a series of field measurements over a period of one year by a 12 m EFA sensor/filter photometer system. Reference data obtained from GC laboratory analysis of process water samples taken during the measurements are also given in Figure 9. From a comparison of the concentration plots shown in Figure 9 one generally can establish a good correspondence in the order of magnitude of the data given by the sensor system and the GC off-line data. In the first phase of the measurements the CHC sum concentrations indicated by the EFA sensor system show positive and negative deviations of typically 30% compared with the GC data. After June 1996, owing to ground water sampling in additional wells, the ratio of the two main contaminants in the feed changed drastically to a value of approximately 1.5:1 (the ratio values are also given in Figure 9). The calibration function of the EFA sensor system deliberately was not adjusted at this point to the new analyte ratio (*i.e.* a steeper slope of the calibration function), to see how 'wrong' the concentration data measured by the system would be. As expected, the CHC sum concentration recorded by the system after June 1996 is typically 50–100% higher compared with the corresponding GC data. However, despite being calibrated with the 'wrong' function the sensor signal follows the pattern in the HC sum concentration and a loss of function of a pump in one of the monitoring wells, leading to a distinct increase in the feed concentration, in January 1997, was clearly identified by the sensor system. From these results, it is obvious that the sum signal of the EFA sensor/filter photometer system provides quantitative information only for a relative constant analyte composition, but even in the event of moderate changes gives at least semi-quantitative information.

Besides the well-known problem of sampling highly volatile CHC compounds without losses, another source of error during the measurements in Mühlacker made it difficult to do an exact comparison between off-line analysis and *in situ* monitoring data: relatively fast fluctuations in the CHC compositon and concentration of the feed water were caused by pumping and mixing of ground water of eight different monitoring wells with varying CHC composition and concentration at the inlet pipe of the remediation facility. Therefore, the EFA sensor *in situ* concentration data presented in Figure 9 are half-hour or one-hour average values, while the off-line reference data reflect a distinct concentration value of a water sample taken during this measuring interval.

In spite of this shortcoming, the mean deviation between EFA sensor and GC data for all measurements shown in Figure 9 is 53.9%. If the absorbance values measured by the EFA sensor are evaluated with the correct slope of the calibration function, which depends on the actual *cis*-1,2-DCE/TCE ratio (known from the GC reference measurements at the corresponding date) the mean deviation decreases to 20.3%. Accordingly, the corrected EFA sensor concentration data are also presented in Figure 9 and match the GC data much better. In view of the experimental problems mentioned above the results obtained with the EFA sensor system are rather satisfying.

References

1. R.O.G. Franken, in *Contaminated Soil '93*, ed. F. Arendt, G.J. Annokée, R. Bosman and W.J. van den Brink, Kluwer Academic Publishers, Dordrecht, 1993, p. 845.
2. DIN 38409: Part 18 (February 1981), *Bestimmung von Kohlenwasserstoffen*, and DIN 38407: Part 4 (May 1988), *Bestimmung von leichtflüchtigen Halogenkohlenwasserstoffen (LHKW)*, Beuth Verlag GmbH, Berlin.
3. J. Bürck, J.-P. Conzen and H.J. Ache, *Fresenius' J. Anal. Chem.*, 1992, **342**, 394.
4. J.-P. Conzen, J. Bürck and H.J. Ache, *Appl. Spectrosc.*, 1993, **47**, 753.
5. J. Bürck, J.-P. Conzen, B. Beckhaus and H.J. Ache, *Sensors & Actuators B*, 1994, **18–19**, 291.
6. R. Krska, K. Taga and R. Kellner, *Appl. Spectrosc.*, 1993, **47**, 1484.
7. R. Kellner, R. Göbel, R. Götz, B. Lendl, B. Edl-Mizaikoff, M. Tacke and A. Katzir, *Proc. SPIE*, Vol. 2508, ed. A.V. Scheggi, SPIE, Bellingham, 1995, 212.
8. N.J. Harrick, *Internal Reflection Spectroscopy*, Harrick, New York, NY, 1979.
9. M.D. DeGrandpre and L.W. Burgess, *Appl. Spectrosc.*, **44**, 1990, 273.
10. J. Pawliszyn, Chapter 1 of this book.
11. J.-P. Conzen, Doctoral thesis, 1994, KfK Report No. 5302.
12. E. Sensfelder, J. Bürck and H.J. Ache, *Fresenius' J. Anal. Chem.*, 1996, **354**, 848.
13. G.L. Klunder, J. Bürck, H.J. Ache, R.J. Silva and R.E. Russo, *Appl. Spectrosc.*, 1994, **48**, 387.

14. J.-P. Conzen, J. Bürck and H.J. Ache, *Fresenius' J. Anal. Chem.*, 1994, **348**, 501.
15. W. Guhl and U. Werner, *Nachr. Chem. Tech. Lab.*, 1997, **45**(4), M15.
16. J. Bürck and M. Mensch, *Field Screening Europe – Proceedings of the First Int. Conf. on Strategies and Techniques for the Investigation and Monitoring of Contaminated Sites*, ed. J. Gottlieb, H. Hötzl, K. Huck and R. Nießner, Kluwer Academic Publishers, Dordrecht, 1997, pp. 243.
17. J. Bürck, J.-P. Conzen, G. Klunder, B. Zimmermann and H.J. Ache in *Contaminated Soil '93*, ed. F. Arendt, G.J. Annokée, R. Bosman and W.J. van den Brink, Kluwer Academic Publishers, Dordrecht, 1993, p. 917.
18. J. Bürck, M. Mensch and K. Krämer, *Field Anal. Chem. Technol.*, 1998, **2**, 205.

Subject Index